LATIN AMERICAN POLITICAL, ECONOMIC, AND SECURITY ISSUES

CUBA

POLITICS, ECONOMICS, U.S. RELATIONS AND DEMOCRACY ASSISTANCE EFFORTS

LATIN AMERICAN POLITICAL, ECONOMIC, AND SECURITY ISSUES

Additional books in this series can be found on Nova's website under the Series tab.

Additional E-books in this series can be found on Nova's website under the E-book tab.

GLOBAL POLITICAL STUDIES

Additional books in this series can be found on Nova's website under the Series tab.

Additional E-books in this series can be found on Nova's website under the E-book tab.

LATIN AMERICAN POLITICAL, ECONOMIC, AND SECURITY ISSUES

CUBA

POLITICS, ECONOMICS, U.S. RELATIONS AND DEMOCRACY ASSISTANCE EFFORTS

GUSTAVO NOGUEIRA
EDITOR

nova publishers
New York

For permission to use material from this book please contact us:
Telephone 631-231-7269; Fax 631-231-8175
Web Site: http://www.novapublishers.com

Library of Congress Cataloging-in-Publication Data

ISBN: 978-1-62618-970-6

Published by Nova Science Publishers, Inc. † New York

CONTENTS

PREFACE

Cuba remains a one-party communist state with a poor record on human rights. The country's political succession in 2006 from the long-ruling Fidel Castro to his brother Raul was characterized by a remarkable degree of stability. In February 2013, Castro was reappointed to a second five-year term as president, and selected a 52-year old former Education Minister Miguel Diaz-Canel as his First Vice President, making him the official successor in the event that Castro cannot serve out his term. Raul Castro has implemented a number of gradual economic policy changes over the past several years, including an expansion of self-employment. A party congress held in April 2011 laid out numerous economic goals that, if implemented, could significantly alter Cuba's state-dominated economic model. Few observers, however, expect the government to ease its tight control over the political system. This book provides an overview of Cuba's current political and economic situation; U.S. policy toward Cuba; and select issues in U.S.-Cuba relations.

Chapter 1 – Cuba remains a one-party communist state with a poor record on human rights. The country's political succession in 2006 from the long-ruling Fidel Castro to his brother Raúl was characterized by a remarkable degree of stability. In February 2013, Castro was reappointed to a second five-year term as president (until 2018, when he would be 86 years old), and selected a 52-year old former Education Minister Miguel Díaz-Canel as his First Vice President, making him the official successor in the event that Castro cannot serve out his term. Raúl Castro has implemented a number of gradual economic policy changes over the past several years, including an expansion of self-employment. A party congress held in April 2011 laid out numerous economic goals that, if implemented, could significantly alter Cuba's state-dominated economic model. Few observers, however, expect the government to ease its tight control over the political system. While the government reduced the number of political prisoners in 2010-2011, the number increased in 2012; moreover, short-term detentions and harassment have increased significantly.

Chapter 2 – Cuba is moving toward development of its offshore oil resources. While the country has proven oil reserves of just 0.1 billion barrels, the U.S. Geological Survey estimates that offshore reserves in the North Cuba Basin could contain an additional 4.6 billion barrels of undiscovered technically recoverable crude oil. The Spanish oil company Repsol, in a consortium with Norway's Statoil and India's Oil and Natural Gas Corporation, is expected to begin offshore exploratory drilling in early 2012, and several other companies are considering such drilling. At present, Cuba has six offshore projects with foreign oil companies. If oil is found, some experts estimate that it would take at least three to five years before production would begin.

In the aftermath of the Deepwater Horizon oil spill, some Members of Congress and others have expressed concern about Cuba's development of its deepwater petroleum reserves so close to the United States. They are concerned about oil spill risks and about the status of preparedness and coordination in the event of an oil spill. Dealing with these challenges is made more difficult because of the long-standing poor state of relations between Cuba and the United States. If an oil spill did occur in the waters northwest of Cuba, currents in the Florida Straits could carry the oil to U.S. waters and coastal areas in Florida, although a number of factors would determine the potential environmental impact. If significant amounts of oil did reach U.S. waters, marine and coastal resources in southern Florida could be at risk.

The Obama Administration has been making efforts to prepare for a potential oil spill in Cuban waters that could affect the United States. This has included: updating oil spill area contingency plans covering Florida and developing a broader offshore drilling response plan; engaging with Repsol over its oil spill response plans (including plans to inspect the oil rig that Repsol will use); and licensing U.S. companies to provide personnel and export equipment needed for oil spill preparedness and response. Some energy and policy analysts have called for the Administration to ease regulatory restrictions on the transfer of U.S. equipment and personnel to Cuba for oil spill preparedness and response. Some have also called for direct U.S.-Cuban government cooperation to minimize potential oil spill damage, looking at U.S. cooperation with Mexico as a potential model as well as information sharing and cooperation through multilateral channels under the auspices of the International Maritime Organization. In contrast, some policy groups call for the United States to focus on preventing Cuba from engaging in offshore oil exploration altogether.

In the 112th Congress, five legislative initiatives have been introduced taking varying approaches toward Cuba's offshore oil development, and there have been two oversight hearings. H.R. 372 would authorize the Secretary of the Interior to deny oil leases and permits to those companies that engage in activities with the government of any foreign country subject to any U.S. government sanction or embargo. S. 405, among its provisions, would require the development of oil spill response plans for nondomestic oil spills in the Gulf of Mexico, including recommendations for a joint contingency plan with Mexico, Cuba, and the Bahamas. H.R. 2047 would impose visa restrictions on foreign nationals and economic sanctions on companies that help facilitate the development of Cuba's offshore petroleum resources. S. 1836 and H.R. 3393 would provide that foreign offshore oil developers would be liable for damages from oil spills that enter U.S. waters.

Chapter 3 – Restrictions on travel to Cuba have been a key and often contentious component in U.S. efforts to isolate Cuba's communist government since the early 1960s. Under the George W. Bush Administration, restrictions on travel and on private remittances to Cuba were tightened. In March 2003, the Administration eliminated travel for people-to-people educational exchanges unrelated to academic coursework. In June 2004, the Administration further restricted family and educational travel, eliminated the category of fully-hosted travel, and restricted remittances so that they could only be sent to the remitter's immediate family. Initially there was mixed reaction to the Administration's June 2004 tightening of Cuba travel and remittance restrictions, but opposition to the policy grew, especially within the Cuban American community regarding the restrictions on family travel and remittances.

Chapter 4 – Report to the Chairman, Committee on Foreign Relations, U.S. Senate, on Cuba Democracy Assistance: USAID's Program Is Improved, but State Could Better Monitor Its Implementing Partners, dated January 2013.

In: Cuba
Editor: Gustavo Nogueira
ISBN: 978-1-62618-970-6

Chapter 1

CUBA: U.S. POLICY AND ISSUES FOR THE 113TH CONGRESS*

Mark P. Sullivan

SUMMARY

Cuba remains a one-party communist state with a poor record on human rights. The country's political succession in 2006 from the long-ruling Fidel Castro to his brother Raúl was characterized by a remarkable degree of stability. In February 2013, Castro was reappointed to a second five-year term as president (until 2018, when he would be 86 years old), and selected a 52-year old former Education Minister Miguel Díaz-Canel as his First Vice President, making him the official successor in the event that Castro cannot serve out his term. Raúl Castro has implemented a number of gradual economic policy changes over the past several years, including an expansion of self-employment. A party congress held in April 2011 laid out numerous economic goals that, if implemented, could significantly alter Cuba's state-dominated economic model. Few observers, however, expect the government to ease its tight control over the political system. While the government reduced the number of political prisoners in 2010-2011, the number increased in 2012; moreover, short-term detentions and harassment have increased significantly.

* This is an edited, reformatted and augmented version of Congressional Research Service, Publication No. R43024, dated March 29, 2013.

U.S. Policy

Over the years, Congress has played an active role in shaping policy toward Cuba, including the enactment of legislation strengthening and at times easing various U.S. economic sanctions. While U.S. policy has consisted largely of isolating Cuba through economic sanctions, a second policy component has consisted of support measures for the Cuban people, including U.S. government-sponsored broadcasting (Radio and TV Martí) and support for human rights and democracy projects. The Obama Administration has continued this similar dual-track approach. While the Administration has lifted all restrictions on family travel and remittances, eased restrictions on other types of purposeful travel, and moved to reengage Cuba on several bilateral issues, it has also maintained most U.S. economic sanctions in place. On human rights, the Administration welcomed the release of many political prisoners in 2010 and 2011, but it has also criticized Cuba's continued harsh repression of political dissidents through thousands of short-term detentions and targeted violence. The Administration has continued to call for the release of U.S. government subcontractor Alan Gross, detained in 2009 and sentenced to 15 years in prison in March 2011. Gross's continued detention has been a major impediment toward improved relations.

Legislative Activity

Strong interest on Cuba is expected to continue in the 113th Congress with attention focused on economic and political developments, especially the human rights situation, and U.S. policy toward the island nation. The continued imprisonment of Alan Gross remains a key concern for many Members. Now that Congress has completed action on FY2013 appropriations, it will soon be considering the Administration's FY2014 request for the State Department and Foreign Operations, which includes funding for Cuba democracy programs and Cuba broadcasting. For many years, U.S. sanctions, particularly restrictions on travel, remittances, and agricultural exports to Cuba, have been topics of congressional debate, and this could be possible again in the 113th Congress.

To date in the 113th Congress, seven initiatives on Cuba have been introduced. Several would lift or ease U.S. economic sanctions on Cuba: H.R. 214 and H.R. 872 (overall embargo); H.R. 871 (travel); and H.R. 873 (travel and agricultural exports). H.R. 215 would allow Cubans to play organized professional baseball in the United States. H.R. 778 would modify a 1998 trademark sanction. H.Res. 121 would honor the work of Cuban blogger Yoani Sánchez in challenging the oppression of the Castro regime.

RECENT DEVELOPMENTS

On March 28, 2013, the U.S. Department of State called for an "investigation with independent international observers into the circumstances leading to the death of [Cuban human rights activists] Oswaldo Payá and Harold Cepero" in July 2012. (See "Human Rights Conditions" below.)

On March 14, 2013, internationally known Cuban blogger Yoani Sánchez (on a multi-nation trip after receiving a new passport under Cuba's new travel policy) arrived in the United States, with stops in New York City and Washington, D.C. (including Capitol Hill) through March 21. She then traveled to Europe, but returned to the United States March 28, arriving in Miami. Sánchez is scheduled to speak at the Freedom Tower in Miami on April 1, 2013. (See "Human Rights Conditions" below; also see Sánchez's website, available at http://www.desdecuba.com/generaciony/)

On March 14, 2013, Amnesty International issued an urgent action appeal for prisoner of conscience Calixto Ramón Martínez Arias who began a hunger strike in early March. Calixto was imprisoned in September 2012 for reporting on a cholera outbreak. (See "Human Rights Conditions" below.)

On March 12, 2013, the State Department released its 2013 *International Narcotics Control Strategy Report*, which stated that Cuba maintained a significant level of anti-drug cooperation with the United States in 2012. The report also indicated that the U.S. government was still reviewing a draft bilateral counternarcotics cooperation accord that Cuba presented in 2011, and that such an accord, if structured appropriately, could advance counternarcotics efforts taken by both countries. (See "Anti-Drug Cooperation" below.)

On March 6, 2013, the *Washington Post* published an interview with Spanish politician Angel Carromero, who was convicted by a Cuban court in October 2012 for vehicular manslaughter in the death of two human rights activists, including internationally known dissident Oswaldo Payá. Carromero asserted in the interview that the car he was driving was struck from behind and that he had been heavily drugged when he admitted to driving recklessly. Many observers have called for an independent investigation into the accident. In July 2012, the U.S. Senate approved S.Res. 525 (Nelson) calling for an impartial third-party investigation of the crash. (See "Human Rights Conditions" below.)

On March 5, 2013, Venezuelan President Hugo Chávez died after battling several recurrences of cancer since mid-2011. Under President Chávez, Venezuela became a strong political and financial supporter of Cuba over the past decade, providing the island with some 100,000 barrels of oil per day. (See "Cuba's Foreign Relations" below.)

On February 24, 2013, Cuba's National Assembly, as expected, appointed Raúl Castro to a second five-year term as President. Most significantly, the Assembly also appointed 52-year old Miguel Díaz-Canel as First Vice President, making him Castro's constitutional successor. Díaz-Canel replaced outgoing 82-year old First Vice President José Ramón Machado Ventura. (See "Political Conditions" below.)

On January 14, 2013, Cuba's new travel policy went into effect whereby Cubans wanting to travel abroad no longer need an exit permit and letter of invitation. Under the new policy, travel requires only an updated passport and a visa issued by the country of destination, if required. Thousands of Cubans lined up at government migration offices and travel agencies on the first day. (See "Cuba Alters Its Policy Regarding Exit Permits" below.)

INTRODUCTION

Political and economic developments in Cuba and U.S. policy toward the island nation, located just 90 miles from the United States, have been significant congressional concerns for many years. Since the end of the Cold War, Congress has played an active role in shaping U.S. policy toward Cuba, first with the enactment of the Cuban Democracy Act of 1992 (P.L. 102-484, Title XVII) and then with the Cuban Liberty and Democratic Solidarity Act of 1996 (P.L. 104-114). Both of these measures strengthened U.S. economic sanctions on Cuba that had first been imposed in the early 1960s, but the measures also provided roadmaps for a normalization of relations dependent upon significant political and economic changes in Cuba. A decade ago, Congress partially modified its sanctions-based policy toward Cuba when it enacted the Trade Sanctions Reform and Export Enhancement Act of 2000 (P.L. 106-387, Title IX) allowing for U.S. agricultural exports to Cuba that led to the United States becoming a major source for Cuba's food imports.

Over the past decade, much of the debate over U.S. policy in Congress has focused on U.S. sanctions, especially over U.S. restrictions on travel to Cuba.

The George W. Bush Administration initially liberalized U.S. family travel to Cuba in 2003, but subsequently tightened restrictions on family and other categories of travel in 2004 because of Cuba's crackdown on political dissidents. In 2009, Congress took legislative action in an appropriations measure (P.L. 111-8) to ease restrictions on family travel and travel for the marketing of agricultural exports, marking the first congressional action easing Cuba sanctions in almost a decade. The Obama Administration took further action in April 2009 by lifting all restrictions on family travel and on cash remittances by family members to their relatives in Cuba and restarting semi-annual migration talks that had been curtailed in 2004. In January 2011, the Administration announced the further easing of restrictions on educational and religious travel to Cuba and on non-family remittances, and it also expanded eligible airports in the United States authorized to serve licensed charter flights to and from Cuba.

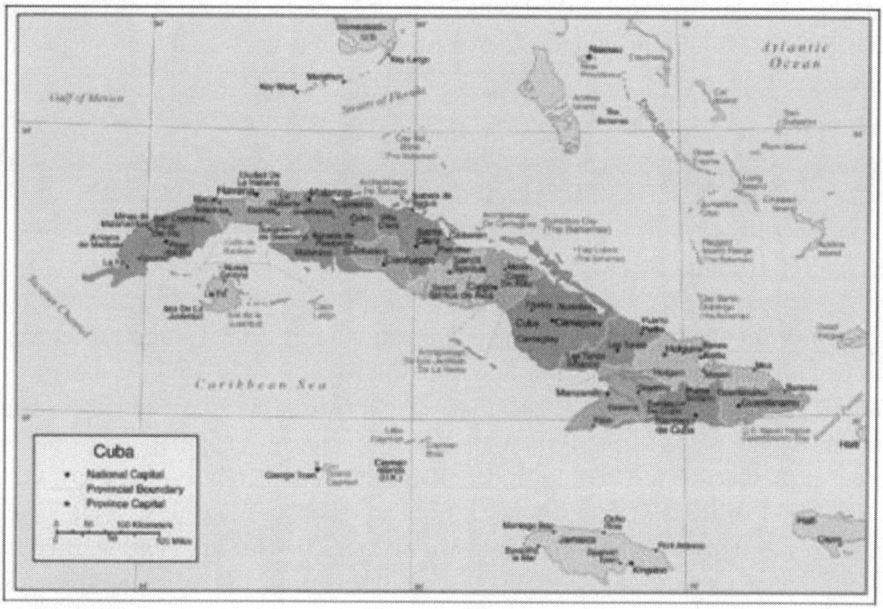

Source: CRS.

Note: The above map does not reflect 2010 changes, in which Cuba created two new provinces from the former province of Havana. See a current interactive provincial map of Cuba from *Juventud Rebelde* (Cuba), available at http://www.juventudrebelde.cu/multimedia/graficos/nueva-division-politico-administrativa/

Figure 1. Map of Cuba.

This report is divided into three major sections analyzing Cuba's political and economic situation, U.S. policy toward Cuba, and selected issues in U.S.-Cuban relations. The first section includes a brief historical political background on Cuba; a discussion on the current political situation under Raúl Castro, including human rights conditions; an examination of economic conditions and policy changes that have occurred to date under Raúl Castro; and Cuba's foreign relations. The second section on U.S. policy provides a broad overview of U.S. policy historically through the George W. Bush Administration and then a discussion of current policy under the Obama Administration. It then provides a brief discussion on the general policy debate regarding the direction of U.S. policy toward Cuba. The third section analyzes many of the key issues in U.S.- Cuban relations that have been at the forefront of the U.S. policy debate on Cuba and have often been the subject of legislative initiatives. While legislative initiatives are noted throughout the report where appropriate, a final section of the report provides a listing of current bills and resolutions introduced in the 113th Congress. An appendix also provides links to selected Executive Branch reports and web pages on Cuba.

CUBA'S POLITICAL AND ECONOMIC SITUATION

Brief Historical Background[1]

Cuba did not become an independent nation until 1902. From its discovery by Columbus in 1492 until the Spanish-American War in 1898, Cuba was a Spanish colony. In the 19th century, the country became a major sugar producer with slaves from Africa arriving in increasing numbers to work the sugar plantations. The drive for independence from Spain grew stronger in the second half of the 19th century, but it only came about after the United States entered the conflict when the USS *Maine* sank in Havana Harbor after an explosion of undetermined origin. In the aftermath of the Spanish-American War, the United States ruled Cuba for four years until Cuba was granted its independence in 1902. Nevertheless, the United States still retained the right to intervene in Cuba to preserve Cuban independence and maintain stability in accordance with the Platt Amendment[2] that became part of the Cuban Constitution of 1901. The United States subsequently intervened militarily three times between 1906 and 1921 to restore order, but in 1934, the Platt Amendment was repealed.

Cuba's political system as an independent nation was often dominated by authoritarian figures. Gerardo Machado (1925-1933), who served two terms as president, became increasingly dictatorial until he was ousted by the military. A short-lived reformist government gave way to a series of governments that were dominated behind the scenes by military leader Fulgencio Batista until he was elected president in 1940. Batista was voted out of office in 1944 and was followed by two successive presidents in a democratic era that ultimately became characterized by corruption and increasing political violence. Batista seized power in a bloodless coup in 1952 and his rule progressed into a brutal dictatorship. This fueled popular unrest and set the stage for Fidel Castro's rise to power.

Castro led an unsuccessful attack on military barracks in Santiago, Cuba, on July 26, 1953. He was jailed, but subsequently freed and went into exile in Mexico where he formed the 26th of July Movement. Castro returned to Cuba in 1956 with the goal of overthrowing the Batista dictatorship. His revolutionary movement was based in the Sierra Maestra and joined with other resistance groups seeking Batista's ouster. Batista ultimately fled the country on January 1, 1959, leading to more than 45 years of rule under Fidel Castro until he stepped down from power provisionally in July 2006 because of poor health.

While Castro had promised a return to democratic constitutional rule when he first took power, he instead moved to consolidate his rule, repress dissent, and imprison or execute thousands of opponents. Under the new revolutionary government, Castro's supporters gradually displaced members of less radical groups. Castro moved toward close relations with the Soviet Union while relations with the United States deteriorated rapidly as the Cuban government expropriated U.S. properties (see "Background on U.S.-Cuban Relations" below). In April 1961, Castro declared that the Cuban revolution was socialist, and in December 1961, he proclaimed himself to be a Marxist-Leninist. Over the next 30 years, Cuba was a close ally of the Soviet Union and depended on it for significant assistance until the dissolution of the Soviet Union in 1991.

From 1959 until 1976, Castro ruled by decree. In 1976, however, the Cuban government enacted a new Constitution setting forth the Cuban Communist Party (PCC) as the leading force in state and society, with power centered in a Political Bureau headed by Fidel Castro. Cuba's Constitution also outlined national, provincial, and local governmental structures. Since then, legislative authority has been vested in a National Assembly of People's Power that meets twice annually for brief periods. When the Assembly is not in session, a Council of State, elected by the Assembly, acts on its behalf.

According to Cuba's Constitution, the president of the Council of State is the country's head of state and government. Executive power in Cuba is vested in a Council of Ministers, also headed by the country's head of state and government, that is, the president of the Council of State.

Fidel Castro served as head of state and government through his position as president of the Council of State from 1976 until February 2008. While he had provisionally stepped down from power in July 2006 because of poor health, Fidel still officially retained his position as head of state and government. National Assembly elections were held on January 20, 2008, and Fidel Castro was once again among the candidates elected to the now 614-member legislative body. (As in the past, voters were only offered a single slate of candidates.) On February 24, 2008, the new Assembly was scheduled to select from among its ranks the members of the Council of State and its president. Many observers had speculated that because of his poor health, Fidel would choose not to be reelected as president of the Council of State, which would confirm his official departure from heading the Cuban government. Statements from Castro himself in December 2007 hinted at his potential retirement. That proved true on February 19, 2008, when Fidel announced that he would not accept the position as president of the Council of State, essentially confirming his departure as titular head of the Cuban government.

Political Conditions

After Fidel stepped down from power, Cuba's political succession from Fidel to Raúl Castro was characterized by considerable stability. After two and one half years of provisionally serving as president, Raúl Castro officially became Cuba's president on February 24, 2008, when Cuba's legislature selected him as president of the 31-member Council of State.[3]

While it was not a surprise to observers for Raúl to succeed his brother Fidel officially as head of government, the selection of José Ramón Machado Ventura as the Council of State's first vice president in February 2008 was a surprise. Born in 1930, Machado was part of the older generation of so-called *históricos* of the 1959 Cuban revolution along with the Castro brothers (Fidel Castro was born in August 1926, while Raúl Castro was born in June 1931). Described as a hard-line communist party ideologue, Machado reportedly was a close friend and confident of Raúl for many years.[4] The position of first vice president of the Council of State is significant because, according to the Cuban

Constitution, the person holding the office is the official successor to the president.

While Raúl Castro began implementing some economic reforms in 2008, there has been no change to his government's tight control over the political system and few observers expect there to be, with the government backed up by a strong security apparatus. Under Raúl, who served as defense minister from the beginning of the Cuban revolution until 2008, the Cuban military has played an increasing role in government with several key military officers and confidants of Raúl serving as ministers.

The Cuban Communist Party (PCC) held its sixth congress in April 2011. While the party concentrated on making changes to Cuba's economic model, some political changes also occurred. As expected, Fidel was officially replaced by Raúl as first secretary of the PCC, and First Vice President José Ramón Machado became the party's second secretary. The party's Political Bureau or Politburo was reduced from 23 to 15 members, with three new members, Marino Murrillo, Minister of Economy Adel Yzquierdo Rodriguez, and the first secretary of the party in Havana, Mercedes Lopez Acea. The party's Central Committee also was reduced from 125 to 115 members, with about 80 of those being new members of the committee.

At the April 2011 party congress, Raúl Castro proposed two five-year term limits for top positions in the party and in the government, calling for systematic rejuvenation, a change that was confirmed by a January 2012 national PCC conference. Cuba's revolutionary leadership has been criticized by many observers for remaining in party and government positions far too long, and for not passing leadership opportunities to a younger generation. Some observers had expected leadership changes and more significant reforms at the January 2012 PCC conference. While this did not occur, the PCC approved a resolution by which its Central Committee would be allowed to replace up to 20% of its 115 members within its five-year mandate.[5]

On February 3, 2013, Cuba held elections for over 600 members of the National Assembly of People's Power, the national legislature, as well as over 1,600 provincial government representatives, both for five-year terms. Under Cuba's one-party system, the overwhelming majority of those elected are PCC members. Critics maintain that elections in Cuba are a sham and entirely controlled by the PCC.

The new National Assembly met on February 24, 2013 to select the next president of the Council of State, Cuba's head of government. As expected, Raúl Castro was selected for a second five year-term as president (until February 2018, when Raúl will be 86 years old), but also indicated that this would be his last term in conformity with the new two-term limit for top officials. Most significantly, First Vice President José Ramón Machado, 82 years old, was replaced by 52-year old Miguel Díaz-Canel Bermúdez, who was serving as one of the Council of State's vice presidents. Díaz-Canel's appointment as the official constitutional successor to President Castro represents a move toward bringing about generational change in Cuba's political system. Díaz-Canel became a member of the Politburo in 2003 and also held top PCC positions in the provinces of Villa Clara and Holguín. He became education minister in 2009 until he was tapped to be a vice president of the Council of State. Díaz-Canel has been described in media reports as an experienced manager with good relations with the military and as someone that worked his way up through the party.[6]

In another significant move in February 2013, the National Assembly appointed Esteban Lazo Hernández as the new president of Cuba's National Assembly. Lazo, who is the Cuban government's highest ranking official of Afro-Cuban descent, replaced long-time National Assembly President Ricardo Alarcón, who was not a candidate in this year's National Assembly elections. Lazo has held top party positions in several provinces and has served as a vice president of the Council of State.

Human Rights Conditions

The Cuban government has a poor record on human rights, with the government sharply restricting freedoms of expression, association, assembly, movement, and other basic rights since the early years of the Cuban revolution. Two Cuban political prisoners conducting hunger strikes have died in recent years, Orlando Zapata Tamayo in February 2010 and Wilman Villar Mendoza in January 2012. Tamayo died after an 85-day hunger strike that he had initiated to protest inhumane conditions in Cuba's prisons. Villar Mendoza died following a 50-day hunger strike after he was convicted of "contempt" of authority and sentenced to four years in prison. The government has continued to harass members of the Ladies in White (*Damas de Blanco*) human rights group that was formed in 2003 by the female relatives of the so-called "group of 75" dissidents arrested that year in a massive crackdown.

While the Cuban government has released numerous political prisoners over the past several years, including more than 125 released in 2010-2011, the number of political prisoners has reportedly increased since 2012, according to the Havana-based Cuban Commission on Human Rights and National Reconciliation (CCDHRN). In January 2013, the group estimated that Cuba held at least 90 political prisoners, compared to an estimated 50 in

April 2012 and more than 200 estimated at the beginning of 2010.[7] Amnesty International currently lists two prisoners of conscience in Cuba: Marcos Máiquel Lima Cruz, detained in December 2010 and serving a three-year sentence for playing songs by a hip-hop group criticizing freedom of expression in Cuba; and Cuban journalist Calixto Ramón Martínez Arias arrested in September 2012 and imprisoned for reporting on a cholera outbreak.[8] In early March 2013, Martínez reportedly began a hunger strike to protest his detention and is being punished by being placed in solitary confinement.[9]

Short-term detentions for political reasons have increased significantly over the past three years, according to the CCDHRN, a reflection of the government's change of tactics in repressing dissent. The human rights group reports that there were at least 4,123 short-detentions for political reasons in 2011, compared to at least 2,074 in 2010, almost double. Short-detentions of dissidents for political reasons increased further in 2012, with at least 6,602 such detentions (a 60% increase over 2011), according to the CCDHRN. This included 1,158 detentions in March 2012 alone surrounding the visit of Pope Benedict XVI to Cuba.[10] In the first two months of 2013, CCDHRN reported at least 868 detentions for political reasons.[11]

In March 2012, Amnesty International (AI) published a report maintaining that "the Cuban government wages a permanent campaign of harassment and short-term detentions of political opponents to stop them from demanding respect for civil and political rights." The report maintained that the release of dozens of political prisoners in 2011 "did not herald a change in human rights policy." AI asserted that "the vast majority of those released were forced into exile, while in Cuba the authorities were determined to contain the dissidence and government critics with new tactics," including intimidation, harassment, surveillance, and "acts of repudiation," or demonstrations by government supporters targeting government critics.[12]

In July 2012, internationally known Cuban democracy and human rights activist Oswaldo Payá was killed in a car accident in the eastern province of Granma along with another Cuban human rights activist Harold Cepero. Payá, whose death prompted expressions of sympathy from around the world, is probably best known for his work founding the Varela Project in 1996. [13] Two Europeans accompanying Payá were also in the crash: Jens Aron Modig, president of the Swedish Christian Democrats youth wing, who was a passenger; and Angel Carromero Barrios, a leader of the Spanish Popular Party's youth organization, who was driving. Carromero was convicted in October 2012 of vehicular manslaughter for speeding and sentenced to four years in prison; after diplomatic efforts by Spain, he was released in late December to serve the rest of his term in Spain. In early March 2013, however, Carromero maintained in an interview with the *Washington Post* that the car he was driving was struck from behind just before the accident and that he was heavily drugged when he admitted to driving recklessly.[14] Payá's daughter has called for the United Nations to conduct an independent investigation into the crash. Modig maintains that he was asleep at the time of the accident and does not remember any details.[15] In July 2012, the U.S. Senate approved S.Res. 525 (Nelson), calling for an impartial third-party investigation of the crash. The U.S. Department of State issued a statement on March 28, 2013 calling for an independent international investigation into the circumstances leading to the death of Payá and Cepero.[16]

On November 7, 2012, Cuban security agents arrested government critic Antonio Rodiles, and charged him with "resisting authority." Rodiles was arrested along with some 20 dissidents who were going to the Department of State Security headquarters to inquire about the detainment of an independent lawyer and journalist, Yaremis Flores. Subsequently Flores was released on November 9, 2011, while the other dissidents, with the exception of Rodiles, were also subsequently freed. Amnesty International issued an urgent action appeal on November 15, 2012, calling for the Cuban government to immediately and unconditionally release Rodiles and to cease the harassment of other citizens peacefully exercising their rights to freedom of expression and association. Rodiles ultimately was released on November 26, 2012, and charges of "resisting authority" were dropped. According to Amnesty International, Rodiles was warned not to continue with his activism in Cuba.[17]

Rodiles is one of the coordinators of an initiative known as the "Citizens' Demand for Another Cuba," which was presented to Cuba's National Assembly in June 2012. The initiative calls on the Cuban government to implement the legal guarantees and polices in the Universal Declaration of Human Rights and to ratify two U.N. human rights treaties that the government signed in 2008—the International Covenant on Civil and Political Rights and the International Covenant on Economic, Social, and Cultural Rights. Rodiles also is the coordinator of a forum known as "Estado de Sats" (State of Sats)[18], which was begun in July 2010 in order to encourage open debate on cultural, social, and political issues.

On November 27, 2012, according to Amnesty International, Cuban authorities assaulted independent journalist Guillermo Fariñas on the street, and in a separate incident threatened Elizardo Sánchez, the head of the CCDHRN. Both had attended a meeting with other civil society groups.

Amnesty International issued an urgent action appeal to call on Cuban authorities to investigate these incidents and "to cease immediately the harassment of the two men and all other citizens who peacefully exercise their rights to freedom of expression and association."[19]

Over the past several years, numerous independent Cuban blogs have been established that are often critical of the Cuban government. The Cuban government has responded with its own team of official bloggers to counter and disparage the independent bloggers.[20] Cuban blogger Yoani Sánchez has received considerable international attention since late 2007 for her website, Generación Y, which includes commentary critical of the Cuban government. (Sánchez's website is available at http://www.desdecuba.com/generaciony/, and has links to numerous other independent Cuban blogs and websites). For several years, Sánchez was denied an exit permit to travel abroad to accept numerous prizes that she received for her work, highlighting what was a common practice of the Cuban government in forbidding citizens from leaving Cuba without official permission. With Cuba's elimination of the exit permit in January 2013, Sánchez applied and received a new passport enabling her to travel. In mid-February 2013 she began a multi-nation trip in Brazil and later traveled to the United States in mid-March 2013. In the 113th

Congress, H.Res. 121 (Hastings, FL), introduced March 15, 2013, would honor Sánchez "for her ongoing efforts to challenge political, economic, and social repression by the Castro regime."

Human Rights Reporting on Cuba

Amnesty International (AI), Human Rights in the Republic of Cuba, http://www.amnesty.org/en/region/cuba. Also see: *Routine Repression, Political Short-Term Detentions and Harassment in Cuba,"* available at http://www.amnesty.org/en/library/asset/AMR25/007/2012/en/647943e7-b4eb-4d39-a5e3-ea061edb651c/ amr250072012en.pdf.

The independent Havana-based Cuban Commission on Human Rights and National Reconciliation

(Comisión Cubana de Derechos Humanos y Reconciliación Nacional, CCDHRN) reports there were at least 4,123 short-term detentions for political reasons in 2011 (almost double the number in 2010), and 6,602 in 2012 (a 60% increase over 2011). In March 2012 alone, surrounding the visit of Pope Benedict XVI, there were 1,158 such detentions,

See: CCDHRN, "Cuba: Algunos Actos de Represion Politica en el Mes de Diciembre de 2012," January 3, 2013, available at http://www.procubalibre.org/informes/pdf /OVERVIEW_ DICIEMBRE_ 2012.pdf,

Human Rights Watch, http://www.hrw.org/en/americas/cuba. The human right group's 2013 World Report maintained that "Cuba remains the only country in Latin America that represses virtually all forms of political dissent," available at https://www.hrw.org/sites/default/files /wr2013_ web.pdf.

According to the **State Department's human rights report for 2011, issued in May 2012,** Cuba's "principal human rights abuses were: abridgement of the rights of citizens to change their government; government threats, intimidation, mobs, harassment, and detentions to prevent citizens from assembly peacefully; and a significant increase in the number of short-term detentions." Additional human rights abuses included: beatings, harsh prison conditions, selective prosecution and denial of fair trials, pervasive monitoring of private communications, and severe limitations on freedom of speech and press. See the report at http://www.state.gov/j/drl/rls/hrrpt/humanrightsreport/index.htm? dynamic_load_id=186505.

Economic Conditions and Reform Efforts

Cuba's economy is largely state-controlled, with the government owning most means of production and employing over 80% of the workforce. Key sectors of the economy that generate foreign exchange include the export of professional services (largely medical personnel to Venezuela); tourism, which has grown significantly since the mid-1990s, with 2.7 million tourists visiting Cuba in 2011; nickel mining, with the Canadian mining company Sherritt International involved in a joint investment project; and a biotechnology and pharmaceutical sector that supplies the domestic healthcare system and has fostered a significant export industry. Remittances from relatives living abroad, especially from the United States, have also become an important source of hard currency, with more than $1 billion sent to Cuba annually in remittances from families living abroad. The once-dominant sugar industry has declined significantly over the past 20 years; in 1990, Cuba produced 8.4 million tons of sugar while in 2012 it produced just 1.4 million tons and estimates for the 2013 harvest are 1.7 million tons.[21]

Cuba is highly dependent on Venezuela for its oil needs. In 2000, the two countries signed a preferential oil agreement that provides Cuba with some 100,000 barrels of oil per day. Cuba's goal of becoming a net oil exporter with the development of its offshore deepwater oil reserves was set back significantly in 2012 when three exploratory oil drills were unsuccessful.

Over the years, Cuba has expressed pride for the nation's accomplishments in health and education. According to the U.N. Development Program's 2011 Human Development Report, life expectancy in Cuba in 2011 was 79.1 years and adult literacy was estimated at almost 100%. The World Bank estimates that Cuba's per capita income level of $5,550 (2010, latest available) is in the upper-middle-income range, higher than a number of other countries in the Americas.[22]

In terms of economic growth, Cuba experienced severe economic deterioration from 1989 to 1993, with an estimated decline in gross domestic product ranging from 35% to 50% when the Soviet Union collapsed and Russian financial assistance to Cuba practically ended. Since then, however, there has been considerable improvement. From 1994 to 2000, as Cuba moved forward with some limited market-oriented economic reforms, economic growth averaged 3.7% annually. Economic growth was especially strong in the 2004-2007 period registering an impressive 11% and 12%, respectively, in 2005 and 2006 (see Figure 2). The economy benefitted from the growth of the tourism, nickel, and oil sectors, and support from Venezuela and China in terms of investment commitments and credit lines.

The economy was hard hit by several hurricanes and storms in 2008 and the global financial crisis caused further strains in 2009, with the government having to implement austerity measures. Economic growth, however, has improved since 2010, with 2.4% growth in 2010 and estimated growth rates of 2.8% in 2011 and 3.1% in 2012. Beyond that, some observers maintain that Cuba's economic reform efforts to expand the private sector and boost productivity could help increase average economic growth in the 2013-2017 period to over 4%.[23]

The government of Raúl Castro has implemented a number of economic policy changes, but there has been some disappointment that more far-reaching reforms have not been forthcoming. As noted above, the government employs a majority of the labor force, over 80%, but it has been allowing more private sector activities. In 2010, the government opened up a wide range of activities for self-employment and small businesses, identifying 178 categories of work allowed with 83 of those allowing small businesses to hire non-family members. By the end of 2012, the number of self-employed rose to some

400,000, compared to some 156,000 self-employed at the end of 2010.[24] In December 2012, the government authorized non-state cooperatives for some 200 small and medium enterprise businesses in such sectors as transportation, fishing, domestic services, and construction. While authorizing more private sector employment, the government also announced in 2010 that it would be laying off over a million state employees over a period of several years. Layoffs in the state sector, however, have lagged. Nevertheless, state payrolls were reportedly cut by 137,000 in 2011 and by 228,000 in 2012.[25]

When the Cuban Communist Party held its sixth congress in April 2011, it approved over 300 economic guidelines that, if implemented, include some potentially significant economic reforms. These include the liquidation of state enterprises with financial losses, the creation of special development zones for foreign investment, the gradual development of a tax system as a means to distribute income, and a gradual elimination of the ration system.[26] Some economic analysts, however, maintain that the proposed changes are too limited and late to deal with the severity of Cuba's difficult economic situation.[27]

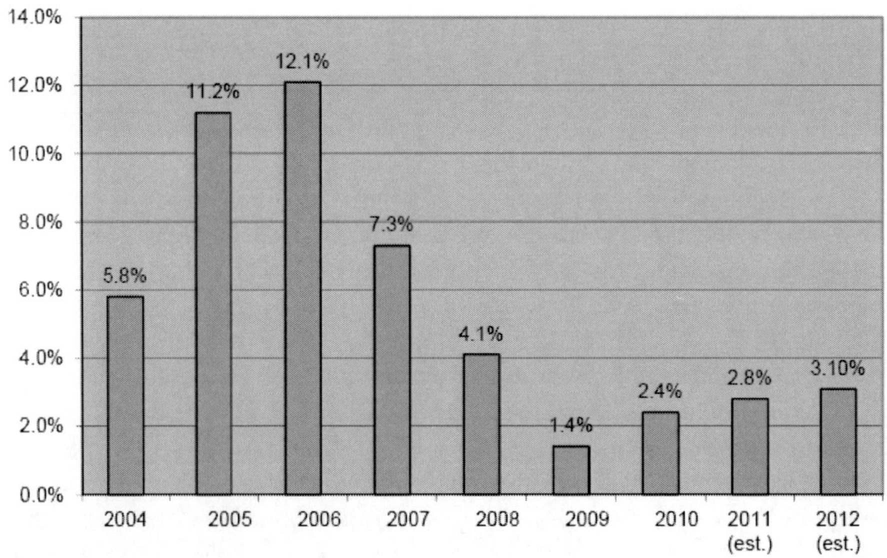

Source: Economist Intelligence Unit, Country Data Tool.

Figure 2. Cuba: Real GDP Growth (percentage), 2004-2012.

Among Cuba's significant economic challenges are low wages (whereby workers cannot satisfy basic human needs) and the related problem of how to unify Cuba's two official currencies circulating in the country. Most people are paid in Cuban pesos and the minimum monthly wage in Cuba is about 225 pesos (about $9 U.S. dollars[28]), but for increasing amounts of consumer goods, convertible pesos are used. Cubans with access to foreign remittances or who work in jobs that give them access to convertible pesos are far better off than those Cubans who do not have such access.

A significant reform effort under Raúl Castro has focused on the agricultural sector, a vital issue because Cuba reportedly imports some two-thirds of its food needs. In an effort to boost food production, the government has turned over idle land to farmers and given farmers more control over how to use their land and what supplies to buy. Despite these and other efforts, overall food production has been significantly below targets – overall agricultural production in 2011 was less than it was in 2007. In 2012, Cuba's coffee sector was hard hit by Hurricane Sandy in October, and overall agricultural production reportedly underperformed for the year.[29]

For Additional Reading on the Cuban Economy

Association for the Study of the Cuban Economy, Annual Proceedings, available at http://www.ascecuba.org/publications/proceedings/.

The Cuban Economy, La Economia Cubana, website maintained by Arch Ritter, from Carlton University, Ottawa, Canada, available at http://thecubaneconomy.com/profile/.

Philip Peters, "A Viewer's Guide to Cuba's Economic Reform," Lexington Institute, May 2012, available at http://www.lexingtoninstitute.org/library/resources/documents/Cuba/ResearchProducts/ViewersGuide.pdf; and **"Cuba's Entrepreneurs: Foundation of a New Private Sector,"** July 2012, available at http://www.lexingtoninstitute.org /library/resources /documents/Cuba/ResearchProducts/CubaEntrepreneurs.pdf **"Reforming Cuban Agriculture: Unfinished Business,"** October 2012, available at http://www.lexingtoninstitute.org/library/resources/documents/Cuba/ResearchProducts/CubanAgriculture.pdf

Rafael Romeu, **"Cuba: Reform Continues,"** in *Latin American Economic Perspectives, All Together Now: The Challenge of Regional Integration,* Latin American Initiative at Brookings, April 2012, available on pp. 60-68 in the following pdf: http://www.brookings.edu/~ /media /research/files/reports/2012/4/latin%20america%20perspectives/ 04_latin_america_economic_perspectives.pdf.

Revista Temas (Havana), links to the Cuban journal's articles on Economy and Politics, in Spanish available at http://www.temas.cult.cu /catalejo.php

Oficina Nacional de Estadísticas, República de Cuba, (Cuba's official economic statistics) available at http://www.one.cu/

Cuba's Foreign Relations

During the Cold War, Cuba had extensive relations with and support from the Soviet Union, with billions of dollars in annual subsidies to sustain the Cuban economy. This subsidy system helped fund an activist foreign policy and support for guerrilla movements and revolutionary governments abroad in Latin America and Africa. With an end to the Cold War, the dissolution of the Soviet Union, and the loss of Soviet financial support, Cuba was forced to abandon its revolutionary activities abroad.

As its economy reeled from the loss of Soviet support, Cuba was forced to open up its economy and economic relations with countries worldwide, and it developed significant trade and investment linkages with Canada, Spain, other European countries, and China. Over the past decade, Venezuela—under populist President Hugo Chávez—became a significant source of support for Cuba, providing subsidized oil (some 100,000 barrels per day) and investment. For its part, Cuba has sent thousands of medical personnel to Venezuela. In the aftermath of Chávez's death in early March 2013, Venezuela is facing new presidential elections scheduled for April 14. Acting Venezuelan President Nicolás Maduro, who is leading in presidential election polls, has indicated that Venezuela's strong support for Cuba would continue under his presidency while opposition candidate Henrique Capriles has vowed to end the shipment of subsidized oil to Cuba. (Also see "Cuba's Offshore Oil Development" below.)

In 2011, Cuba's leading trading partners in terms of Cuban exports were Venezuela, China, Canada, and the Netherlands (see Figure 3), while the leading sources of Cuba's imports were Venezuela, China, Spain, and Brazil (see Figure 4).

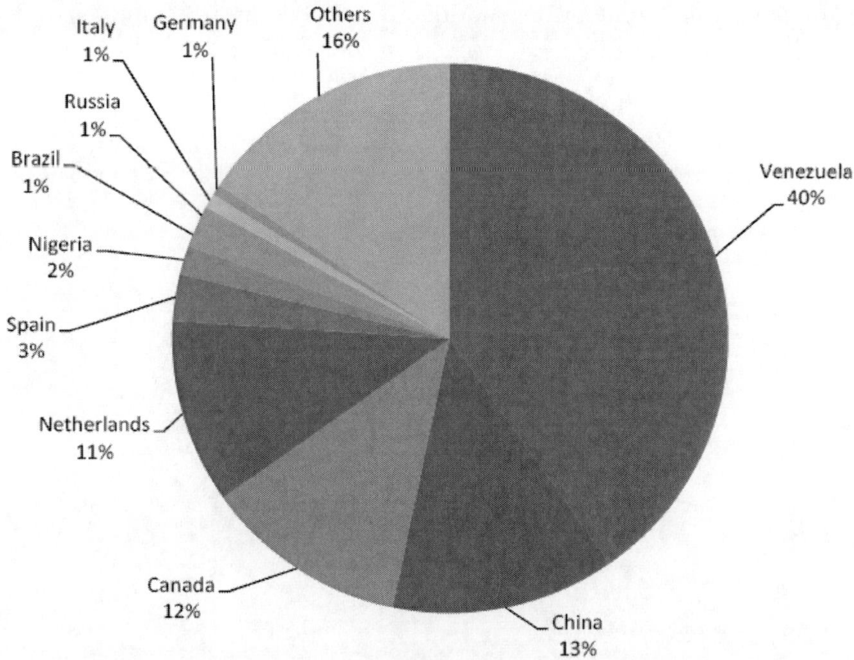

Source: Created by CRS based on information from República de Cuba, Oficina Nacional de Estadisticas, Anuario Estadístico de Cuba 2011, Table 8.5, http://www.one.cu/aec2011/esp/20080618_tabla_cuadro.htm.

Figure 3. Cuban Exports by Country of Destination, 2011.

Relations with Russia, which had diminished significantly in the aftermath of the Cold War, were strengthened with the November 2008 visit of then-Russian President Dmitry Medvedev to Havana, the visit of several Russian warships to Cuba in December 2008, and Raúl Castro's visit to Russia in January 2009. Castro visited Russia again in July 2012, with the goal reportedly to increase and diversify trade and investment. While trade relations between the two countries are not significant, two Russian energy companies have been involved in oil exploration in Cuba. Gazprom has been in a partnership with the Malaysian state oil company Petronas that conducted unsuccessful deepwater oil drilling off of Cuba's western coast in 2012, while the Russian oil company Zarubezhneft began drilling in Cuba's shallow coastal waters east of Havana in November 2012 (also see "Cuba's Offshore Oil Development" below). In February 2013, Russian Prime Minister Medvedev visited Cuba signing 10 bilateral accords, including a debt accord that reportedly will write off a majority of the $20 billion-$30 billion in Soviet era debt owed by Cuba.[30]

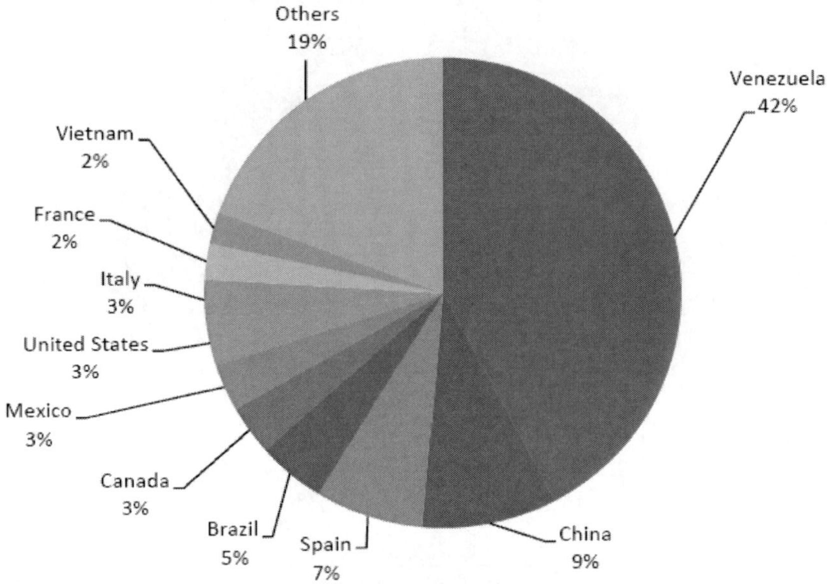

Source: Created by CRS based on information from República de Cuba, Oficina Nacional de Estadisticas, Anuario Estadístico de Cuba 2011, Table 8.6, http://www.one.cu/aec2011/esp/20080618_tabla_cuadro.htm.

Figure 4. Cuban Imports by Country of Origin, 2011.

Relations with China have also increased in recent years. Chinese President Hu Jintao visited Cuba in November 2008, signing a dozen agreements, while Chinese Vice President Xi Jinping visited Cuba in June 2011. During the Xi Jinping visit, China signed a letter of intent to invest in upgrading a Cuban oil refinery in Cienfuegos. In July 2012, President Castro visited China on a four-day visit; the two countries reportedly signed eight cooperation agreements, and talks reportedly focused on trade and investment issues. (After China, Castro visited Vietnam on a four-day trip. The two countries have long had good relations, and some observers speculate that Cuba looks to Vietnam for potential lessons in implementing economic reforms.)

With El Salvador's restoration of relations with Cuba in June 2009, all Latin American nations now have official diplomatic relations with Cuba. Cuba has increasingly become more engaged in Latin America beyond the already close relations with Venezuela. Brazilian President Luiz Inácio Lula da Silva visited Cuba twice in 2008 while President Dilma Rouseff visited in January 2012. Cuba is a member of the Bolivarian Alliance for the Americas, (ALBA), a Venezuelan-led integration and cooperation scheme founded in 2004. Cuba became a full member of the Rio Group of Latin American and Caribbean nations in November 2008, and a member of the succeeding Community of Latin American and Caribbean States (CELCAC) that was officially established in December 2011 to boost regional cooperation. In January 2013, CELAC held its first summit in Chile, and Raúl Castro assumed the presidency of the organization for one year.

Cuba had expressed interest in attending the sixth Summit of the Americas in April 2012 in Cartagena, Colombia, but ultimately was not invited to attend. The United States and Canada expressed opposition to Cuba's participation. Previous summits have been limited to the hemisphere's 34 democratically elected leaders, and the OAS (in which Cuba does not participate) has played a key role in summit implementation and follow-up activities. Several Latin American nations, however, have vowed not to attend the next Summit of the Americas to be held in Panama in 2015 unless Cuba is allowed to participate.

Cuba was excluded from participation in the OAS in 1962 because of its identification with Marxism-Leninism, but in early June 2009, the OAS overturned the 1962 resolution in a move that could eventually lead to Cuba's reentry into the regional organization in accordance with the practices, purposes, and principles of the OAS. While the Cuban government welcomed the OAS vote to overturn the 1962 resolution, it asserted that it would not return to the OAS.[31]

Cuba is an active participant in international forums, including the United Nations and the controversial United Nations Human Rights Council. Since 1991, the U.N. General Assembly has approved a resolution each year criticizing the U.S. economic embargo and urging the United States to lift it. Cuba also has received support over the years from the United Nations Development Programme (UNDP) and the United Nations Educational, Scientific, and Cultural Organization (UNESCO), both of which have offices in Havana. The U.N. has played a significant role in providing relief and recovery from Hurricane Sandy that struck in October 2012. In November 2012, the U.N. system launched a Plan of Action (developed in cooperation with the Cuban government) for $30.6 million to address the urgent needs of the population in the following sectors: shelter and recovery, water sanitation and hygiene, food security, health, and education.[32]

Among other international organizations, Cuba was a founding member of the World Trade Organization, but it is not a member of the International Monetary Fund, the World Bank, or the Inter-American Development Bank. Cuba hosted the 14th summit of the Non-aligned Movement (NAM) in 2006, and held the Secretary Generalship of the NAM until its July 2009 summit in Egypt.

U.S. POLICY TOWARD CUBA

Background on U.S.-Cuban Relations[33]

In the early 1960s, U.S.-Cuban relations deteriorated sharply when Fidel Castro began to build a repressive communist dictatorship and moved his country toward close relations with the Soviet Union. The often tense and hostile nature of the U.S.-Cuban relationship is illustrated by such events and actions as U.S. covert operations to overthrow the Castro government culminating in the ill-fated April 1961 Bay of Pigs invasion; the October 1962 missile crisis in which the United States confronted the Soviet Union over its attempt to place offensive nuclear missiles in Cuba; Cuban support for guerrilla insurgencies and military support for revolutionary governments in Africa and the Western Hemisphere; the 1980 exodus of around 125,000 Cubans to the United States in the so-called Mariel boatlift; the 1994 exodus of more than 30,000 Cubans who were interdicted and housed at U.S. facilities in Guantanamo and Panama; and the February 1996 shootdown by Cuban fighter jets of two U.S. civilian planes operated by the Cuban American group Brothers to the Rescue, which resulted in the death of four U.S. crew members.

Since the early 1960s, U.S. policy toward Cuba has consisted largely of isolating the island nation through comprehensive economic sanctions, including an embargo on trade and financial transactions. The Cuban Assets Control Regulations (CACR), first issued by the Treasury Department in July 1963, lay out a comprehensive set of economic sanctions against Cuba, including a prohibition on most financial transactions with Cuba and a freeze of Cuban government assets in the United States. The CACR have been amended many times over the years to reflect changes in policy, and remain in force today.

These sanctions were made stronger with the Cuban Democracy Act (CDA) of 1992 (P.L. 102- 484, Title XVII) and with the Cuban Liberty and

Democratic Solidarity Act of 1996 (P.L. 104- 114), the latter often referred to as the Helms/Burton legislation. The CDA prohibits U.S. subsidiaries from engaging in trade with Cuba and prohibits entry into the United States for any sea-borne vessel to load or unload freight if it has been involved in trade with Cuba within the previous 180 days. The Cuban Liberty and Democratic Solidarity Act, enacted in the aftermath of Cuba's shooting down of two U.S. civilian planes in February 1996, combines a variety of measures to increase pressure on Cuba and provides for a plan to assist Cuba once it begins the transition to democracy. Most significantly, the law codified the Cuban embargo, including all restrictions under the CACR. This provision is noteworthy because of its long-lasting effect on U.S. policy options toward Cuba. The executive branch is circumscribed in lifting the economic embargo without congressional concurrence until certain democratic conditions are met, although the CACR includes licensing authority that provides the executive branch with administrative flexibility (e.g., travel-related restrictions in the CACR have been eased and tightened on numerous occasions). Another significant sanction in the law is a provision in Title III that holds any person or government that traffics in U.S. property confiscated by the Cuban government liable for monetary damages in U.S. federal court. Acting under provisions of the law, however, Presidents Clinton, Bush, and now Obama have suspended the implementation of Title III at six-month intervals.

In addition to sanctions, another component of U.S. policy, a so-called second track, consists of support measures for the Cuban people. This includes U.S. private humanitarian donations, medical exports to Cuba under the terms of the Cuban Democracy Act of 1992, U.S. government support for democracy-building efforts, and U.S.-sponsored radio and television broadcasting to Cuba. In addition, the 106th Congress approved the Trade Sanctions Reform and Export Enhancement Act of 2000 (P.L. 106-387, Title IX) that allows for agricultural exports to Cuba, albeit with restrictions on financing such exports. This led to the United States becoming one of Cuba's largest suppliers of agricultural products.

Clinton Administration's Easing of Sanctions

The Clinton Administration made several changes to U.S. policy in the aftermath of Pope John Paul II's 1998 visit to Cuba, which were intended to bolster U.S. support for the Cuban people.

These included the resumption of direct flights to Cuba (which had been curtailed after the February 1996 shootdown of two U.S. civilian planes), the resumption of cash remittances by U.S. nationals and residents for the support of close relatives in Cuba (which had been curtailed in August 1994 in response to the migration crisis with Cuba), and the streamlining of procedures for the commercial sale of medicines and medical supplies and equipment to Cuba.

In January 1999, President Clinton announced several additional measures to support the Cuban people. These included a broadening of cash remittances to Cuba, so that all U.S. residents (not just those with close relatives in Cuba) could send remittances to Cuba; an expansion of direct passenger charter flights to Cuba from additional U.S. cities other than Miami (direct flights later in the year began from Los Angeles and New York); and an expansion of people-to-people contact by loosening restrictions on travel to Cuba for certain categories of travelers, such as professional researchers and those involved in a wide range of educational, religious, and sports activities.

Bush Administration's Tightening of Sanctions

The George W. Bush Administration essentially continued the two-track U.S. policy of isolating Cuba through economic sanctions while supporting the Cuban people through a variety of measures. However, within this policy framework, the Administration emphasized stronger enforcement of economic sanctions and further tightened restrictions on travel, remittances, and humanitarian gift parcels to Cuba. The Administration established an interagency Commission for Assistance to a Free Cuba in late 2003 tasked with identifying means "to help the Cuban people bring about an expeditious end of the dictatorship" and to consider "the requirements for United States assistance to a post-dictatorship Cuba."[34] In issuing its first report in May 2004, the Commission made recommendations to tighten restrictions on family visits and other categories of travel and on private humanitarian assistance in the form of remittances and gift parcels.[35] The Administration subsequently issued these tightened restrictions in June 2004, while in February 2005, it tightened restrictions on payment terms for U.S. agricultural exports to Cuba. The Commission issued a second and final report in July 2006 that made recommendations to hasten political change in Cuba toward a democratic transition and led to a substantial increase in U.S. funding to support democracy and human rights efforts in Cuba.

The Bush Administration continued to emphasize a continuation of the sanctions-based approach toward Cuba pending political change in Cuba. When Raúl Castro officially became head of state in February 2008, then-Secretary of State Condoleezza Rice issued a statement urging "the Cuban government to begin a process of peaceful, democratic change by releasing all political prisoners, respecting human rights, and creating a clear pathway towards free and fair elections."[36] In remarks on Cuba policy in March 2008, President Bush maintained that in order to improve U.S.- Cuban relations, "what needs to change is not the United States; what needs to change is Cuba." The President asserted that Cuba "must release all political prisoners ... have respect for human rights in word and deed, and pave the way for free and fair elections."[37]

Obama Administration Policy

In some respects, the Obama Administration has continued the dual-track policy approach toward Cuba that has been in place for many years. It has largely maintained U.S. economic sanctions and it has continued measures to support the Cuban people, such as U.S. government-sponsored radio and television broadcasting and funding for democracy and human rights projects.

At the same time, a significant shift in policy toward Cuba under the Obama Administration has occurred with its efforts to reach out to the Cuban people through the easing of restrictions on travel and remittances. In April 2009, the Obama Administration fulfilled a campaign pledge and announced it would lift all restrictions on family travel and remittances. The Administration went further in January 2011 when it announced new measures to ease travel restrictions and to allow all Americans to send remittances to Cuba. The measures increased purposeful travel to Cuba related to religious, educational, and journalistic activities, including people-to-people travel exchanges; authorized any U.S. person to send remittances to non-family members in Cuba; made it easier for religious institutions to send remittances for religious activities; and allowed all U.S. international airports to apply to provide flights to and from Cuba.

In 2012 congressional testimony, Assistant Secretary of State for Western Hemisphere Affairs Roberta Jacobson asserted that "the Obama Administration's priority is to empower Cubans to freely determine their own future." She maintained that "the most effective tool we have for doing that is building connections between the Cuban and American people, in order to give Cubans the support and tools they need to move forward independent of their government." The Assistant Secretary maintained that "the Administration's travel, remittance and people-to-people policies are helping Cubans by providing alternative sources of information, taking advantage of emerging opportunities for self-employment and private property, and strengthening civil society."[38]

When the Obama Administration took office in 2009, it initiated a policy to engage with the Cuban government in an effort to improve relations. At the April 2009 Summit of the Americas, President Obama announced that "the United States seeks a new beginning with Cuba." While recognizing that it would take time to "overcome decades of mistrust," the President said "there are critical steps we can take toward a new day." He stated that he was prepared to have his Administration "engage with the Cuban government on a wide range of issues—from drugs, migration, and economic issues, to human rights, free speech, and democratic reform." The President maintained that he was "not interested in talking just for the sake of talking," but said that he believed that U.S.-Cuban relations could move in a new direction.[39] In the aftermath of the Summit, there appeared to be some momentum toward improved relations. In July 2009, Cuba and the United States restarted the semi-annual migration talks that had been suspended by the United States five years earlier.

Relations took a turn for the worse in December 2009, however, when Alan Gross, an American subcontractor working on Cuba democracy projects funded by the U.S. Agency for International Development (USAID) was arrested in Havana state for providing Internet communications equipment to Cuba's Jewish community. Since that time, a key impediment to dialogue and improved relations has been Cuba's continued detention of Gross, who was convicted in March 2011, and sentenced to 15 years in prison. U.S. officials and Members of Congress have repeatedly raised the issue with the Cuban government and asked for his release. While the United States and Cuba continue to cooperate on such issues as antidrug efforts and oil spill prevention, preparedness, and response, improvement of relations in other areas will likely be stymied until Alan Gross is released from prison. While four rounds of migration talks have been held since 2009, and often included issues beyond migration, the last round was in January 2011 and it is unclear when additional talks may take place.

Human rights violations also have remained a fundamental concern regarding Cuba under the Obama Administration. President Obama and the State Department have continued to issue statements expressing concern about violations as they occur, including the death of hunger strikers Orlando Zapata Tamayo and Wilman Villar Mendoza in February 2010 and January 2012, respectively, and targeted repression against the Ladies in White human rights group. U.S. officials lauded the release of dozens of Cuban political prisoners in 2010 and 2011, but maintain that their release has not changed the government's poor human rights record as it continues to resort to repeated short-term detentions.[40] More recently, in response to Cuba's crackdown on the Ladies in White in December 2012, the State Department expressed deep concern about Cuba's "repeated use of arbitrary detention and violence to silence critics, disrupt peaceful assembly, and intimidate independent society." It called for Cuba to end such practices and asserted that "we look forward to the day when all Cubans can freely express their ideas, assemble freely, and express their opinions peacefully."[41]

Debate on the Direction of U.S. Policy

Over the years, although U.S. policymakers have agreed on the overall objectives of U.S. policy toward Cuba—to help bring democracy and respect for human rights to the island—there have been several schools of thought about how to achieve those objectives. Some have advocated a policy of keeping maximum pressure on the Cuban government until reforms are enacted, while continuing efforts to support the Cuban people. Others argue for an approach, sometimes referred to as constructive engagement, that would lift some U.S. sanctions that they believe are hurting the Cuban people, and move toward engaging Cuba in dialogue. Still others call for a swift normalization of U.S.-Cuban relations by lifting the U.S. embargo. Legislative initiatives introduced over the past decade have reflected these three policy approaches.

Over the past decade, there have been efforts in Congress to ease U.S. sanctions, with, one or both houses at times approving amendments to appropriations measures that would have eased U.S. sanctions on Cuba. Until 2009, these provisions were stripped out of final enacted measures, in part because of presidential veto threats. In March 2009, Congress took action to ease some restrictions on travel to Cuba, marking the first time that Congress has eased Cuba sanctions since the approval of the Trade Sanctions Reform and Export Enhancement Act of 2000.

In light of Fidel Castro's departure as head of government and the gradual economic changes being made by Raúl Castro, some observers called for a reexamination of U.S. policy toward Cuba. In this new context, two broad policy approaches have been advanced to contend with change in Cuba: a status-quo approach that maintains the U.S. dual-track policy of isolating the Cuban government while providing support to the Cuban people; and an approach aimed at influencing the attitudes of the Cuban government and Cuban society through increased contact and engagement.

In general, those who advocate easing U.S. sanctions on Cuba make several policy arguments. They assert that if the United States moderated its policy toward Cuba—through increased travel, trade, and dialogue—then the seeds of reform would be planted, which would stimulate forces for peaceful change on the island. They stress the importance to the United States of avoiding violent change in Cuba, with the prospect of a mass exodus to the United States. They argue that since the demise of Cuba's communist government does not appear imminent, even without Fidel Castro at the helm, the United States should espouse a more pragmatic approach in trying to bring about change in Cuba. Supporters of changing policy also point to broad international support for lifting the U.S. embargo, to the missed opportunities for U.S. businesses because of the unilateral nature of the embargo, and to the increased suffering of the Cuban people because of the embargo. Proponents of change also argue that the United States should be consistent in its policies with the world's few remaining communist governments, including China and Vietnam.

On the other side, opponents of changing U.S. policy maintain that the current two-track policy of isolating Cuba, but reaching out to the Cuban people through measures of support, is the best means for realizing political change in Cuba. They point out that the Cuban Liberty and Democratic Solidarity Act of 1996 sets forth the steps that Cuba needs to take in order for the United States to normalize relations. They argue that softening U.S. policy at this time without concrete Cuban reforms would boost the Castro government, politically and economically, and facilitate the survival of the communist regime. Opponents of softening U.S. policy argue that the United States should stay the course in its commitment to democracy and human rights in Cuba, and that sustained sanctions can work. Opponents of loosening U.S. sanctions further argue that Cuba's failed economic policies, not the U.S. embargo, are the causes of Cuba's difficult living conditions.

Issues in U.S.-Cuban Relations

For many years, Congress has played an active role in U.S. policy toward Cuba through the enactment of legislative initiatives and oversight on the many issues that comprise policy toward Cuba. These include: U.S. restrictions on travel and remittances to Cuba; U.S. agricultural exports to

Cuba with conditions; funding and oversight of U.S.-government sponsored democracy and human rights projects; the imprisonment of USAID subcontractor Alan Gross in Cuba since December 2009; funding and oversight for U.S.-government sponsored broadcasting to Cuba (Radio and TV Martí); terrorism issues; migration issues; bilateral anti-drug cooperation; and the status of Cuba's offshore oil development, including efforts to ensure adequate oil spill prevention, preparedness, and response efforts.

U.S. Restrictions on Travel and Remittances[42]

Restrictions on travel to Cuba have been a key and often contentious component of U.S. efforts to isolate the communist government of Fidel Castro for much of the past 40 years. Over time there have been numerous changes to the restrictions and for five years, from 1977 until 1982, there were no restrictions on travel. Restrictions on travel and remittances to Cuba are part of the CACR, the overall embargo regulations administered by the Treasury Department's Office of Foreign Assets Control (OFAC).

Under the George W. Bush Administration, enforcement of U.S. restrictions on Cuba travel increased, and restrictions on travel and on private remittances to Cuba were tightened. In 2003, the Administration eliminated travel for people-to-people educational exchanges that had begun under the Clinton Administration. In 2004, the Administration imposed further restrictions on travel, especially family travel and the provision of private humanitarian assistance to Cuba in the form of remittances and gift parcels.

Under the Obama Administration, Congress took legislative action in March 2009 easing restrictions on family travel (restoring the restrictions to as they were under the Clinton Administration) and on travel related to U.S. agricultural and medical sales to Cuba (P.L. 111-8, Sections 620 and 621 of Division D). In April 2009, the Obama Administration went further when the President announced that he was lifting *all* restrictions on family travel as well as restrictions on cash remittances to family members in Cuba.

In January 2011, the Obama Administration made a series of changes further easing restrictions on travel and remittances to Cuba. The measures: 1) increased purposeful travel to Cuba related to religious, educational, and journalistic activities, including people-to-people travel exchanges; 2) allowed any U.S. person to send remittances to non-family members in Cuba and make it easier for religious institutions to send remittances for religious activities;

and 3) allowed U.S. international airports to become eligible to provide services to licensed charter flights to and from Cuba.

In most respects, these new measures were similar to policies that were undertaken by the Clinton Administration in 1999, but were subsequently curtailed by the Bush Administration in 2003 and 2004. An exception is the expansion of airports to service licensed flights to and from Cuba. While the new travel regulations immediately went into effect for those categories of travel falling under a general license category, OFAC delayed processing applications for new travel categories requiring a specific license (such as people-to-people exchanges) until it updated and issued guidelines in April 2011.

The first people-to-people trips began in August 2011. In May 2012, the Treasury Department tightened its restrictions on people-to-people travel by making changes to its license guidelines. The revised guidelines require an organization applying for a people-to-people license to describe how the travel "would enhance contact with the Cuban people, and/or support civil society in Cuba, and/or promote the Cuban people's independence from Cuban authorities." The revised guidelines also require specification on how meetings with prohibited officials of the Cuban government would advance purposeful travel by enhancing contact with the Cuban people, supporting civil society, or promoting independence from Cuban authorities.[43] In September 2012, various press reports cited a slowdown in the Treasury Department's approval or reapproval of licenses for people-to-people travel since the agency had issued new guidelines in May. Companies conducting such programs complained that the delay in the licenses was forcing them to cancel trips and even to lay off staff.[44] By early October 2012, however, companies conducting the people-to-people travel maintained that they were once again receiving license approvals.

Major arguments made for lifting the Cuba travel ban altogether are that it abridges the rights of ordinary Americans to travel; it hinders efforts to influence conditions in Cuba and may be aiding Castro by helping restrict the flow of information; and Americans can travel to other countries with communist or authoritarian governments. Major arguments in opposition to lifting the Cuba travel ban are that more American travel would support Castro's rule by providing his government with potentially millions of dollars in hard currency; that there are legal provisions allowing travel to Cuba for humanitarian purposes that are used by thousands of Americans each year; and that the President should be free to restrict travel for foreign policy reasons.

In the 112th Congress, interest on the issue of Cuba travel and remittances continued. Legislation was introduced to roll back some of the easing of restrictions and some bills were introduced to further ease travel restrictions or lift them altogether, but ultimately none of the measures were enacted. To date in the 113th Congress, several legislative initiatives again have been introduced that would lift all travel restrictions: H.R. 871 (Rangel) would lift travel restrictions; H.R. 873 (Rangel) would lift restrictions on U.S. agricultural exports as well as travel restrictions; and H.R. 214 (Serrano) and H.R. 872 (Rangel) would lift the overall embargo on Cuba, including travel restrictions.

U.S. Agricultural Exports and Sanctions

U.S. commercial agricultural exports to Cuba have been allowed for more than a decade, but with numerous restrictions and licensing requirements. The 106th Congress passed the Trade Sanctions Reform and Export Enhancement Act of 2000 or TSRA (P.L. 106-387, Title IX) that allows for one-year export licenses for selling agricultural commodities to Cuba, although no U.S. government assistance, foreign assistance, export assistance, credits, or credit guarantees are available to finance such exports. TSRA also denies exporters access to U.S. private commercial financing or credit; all transactions must be conducted in cash in advance or with financing from third countries.

Cuba has purchased over $4.2 billion in products from the United States since 2001, largely agricultural products. U.S. exports to Cuba rose from about $7 million in 2001 to $404 million in 2004 and to a high of $712 million in 2008, far higher than in previous years, in part because of the rise in food prices and because of Cuba's increased food needs in the aftermath of several hurricanes and tropical storms that severely damaged the country's agricultural sector. From 2002 through 2010, the United States was the largest supplier of food and agricultural products to Cuba, but in 2011 Brazil's agricultural exports to Cuba superseded those of the United States.[45]

U.S. exports to Cuba declined considerably from 2009 through 2011, but began to increase again in 2012 (see Figure 5). In 2009, U.S. exports amounted to $533 million (25% lower than the previous year), and in 2010 U.S. exports declined further to $368 million (a 31% drop). In 2011, U.S. exports to Cuba declined by just 1.2% to $363 million.[46] Among the reasons for the decline, analysts cite Cuba's shortage of hard currency; credits and other arrangements offered by other governments to purchase their countries' products; and Cuba's perception that its efforts to motivate U.S. companies, organizations, local and state officials, and Members of Congress to push for change in U.S. sanctions policy toward Cuba have been ineffective.[47] In 2012, the level of U.S. exports to Cuba rose to $465 million, a 28% increase over the previous year, but still lower than export levels to Cuba in 2008 and 2009. Part of the increase in 2012 can be attributed to an increase in Cuba's import needs because of damage to the agricultural sector in eastern Cuba caused by Hurricane Sandy in October. Looking at the composition of U.S. exports to Cuba in recent years, the leading products have been cereals (especially corn and to a lesser extent wheat), poultry and pork, soybeans, prepared animal feed, and edible vegetables.

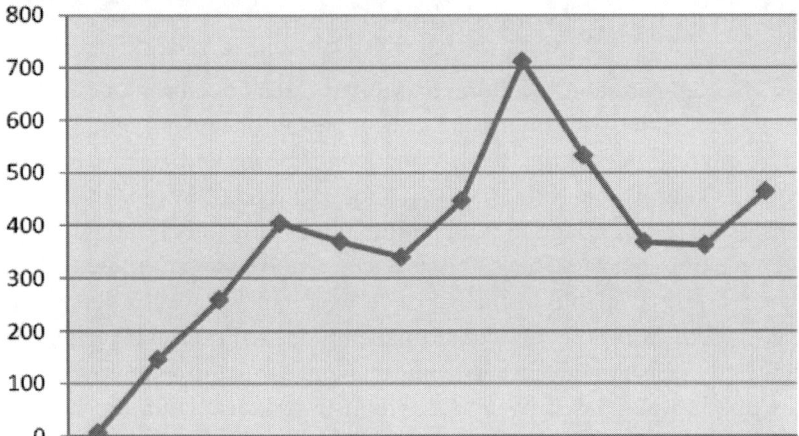

	2001	2002	2003	2004	2005	2006	2007	2008	2009	2010	2011	2012
U.S. $ mil.	7	146	259	404	369	340	447	712	533	368	363	465

Source: Created by CRS using information from the Global Trade Atlas, which uses data from the U.S. Department of Commerce.

Figure 5. U.S. Exports to Cuba, 2001-2012; (U.S. $ millions).

Over the past several years, there have been legislative efforts to further ease restrictions on agricultural exports to Cuba. For FY2010 and FY2011, Congress included a provision in omnibus appropriations measures (Division C, Section 619 of P.L. 111-117, and continued by reference in Division B, Section 1101 of P.L. 112-10) that temporarily overturned OFAC's 2005 clarification that TSRA's requirement for "payment of cash in advance" meant that the payment for the agricultural goods had to be received prior to the shipment of the goods from the port at which they were loaded in the United States. U.S. agricultural exporters and some Members of Congress had objected that OFAC's 2005 action constituted a new sanction that violated the intent of TSRA, and jeopardized million in agricultural sales. The legislative provisions cited above redefined payment of cash in advance for FY2010 and FY2011 to mean that payment was to be received before the transfer of title to, and control of, the exported items to the Cuban purchaser. This essentially meant that payment could occur before a shipment was offloaded in Cuba, rather than before an export shipment left a U.S. port.

For FY2012, Congress did not include a similar provision in the Consolidated Appropriations Act, 2012 (P.L. 112-74). While the Senate Appropriations Committee-approved version of the FY2012 Financial Services appropriations measure (S. 1573) had a provision that would have continued the definition of "payment of cash in advance" utilized in FY2010 and FY2011, this was not included in the final omnibus legislation. The Senate bill also contained another Cuba provision that would have prohibited restrictions on direct transfers from a Cuban financial institution to a U.S. financial institution in payment for licensed exports to Cuba. This provision as well was not included in the omnibus appropriations legislation.

To date in the 113th Congress, one bill has been introduced that includes measures to facilitate the export of U.S. agricultural products to Cuba, H.R. 873 (Rangel), while two other bills that would lift the overall embargo, H.R. 214 (Serrano) and H.R. 872 (Rangel), would lift restrictions and licensing requirements on U.S. agricultural exports to Cuba.

The U.S. International Trade Commission (USITC) produced a study in 2007 analyzing the effects of both U.S. government financing restrictions for agricultural exports to Cuba and U.S. travel restrictions on the level of U.S. agricultural sales to Cuba.[48] At the time of the study, the U.S. share of various Cuban agricultural imports was estimated to range from 0-99% depending on the commodity. If U.S. financing restrictions were lifted, the study estimated that the U.S. share of Cuban agricultural, fish, and forest products imports would rise to between one-half and two-thirds. According to the study, if travel restrictions for all U.S. citizens were lifted, the influx of U.S. tourists would be significant in the short term and would boost demand for imported agricultural products, particularly high-end products for the tourist sector. If both financing and travel restrictions were lifted, the study found that the largest gains in U.S. exports to Cuba would be for fresh fruits and vegetables, milk powder, processed foods, wheat, and dry beans.

In 2009, the USITC issued a working paper that updated the agency's 2007 study on U.S. agricultural sales to Cuba. The update concluded that if U.S. restrictions on financing and travel were lifted in 2008, U.S. agricultural exports to Cuba would have increased between $216 million and $478 million and the U.S. share of Cuba's agricultural imports would have increased from 38% to between 49% and 64%.[49] Among the U.S. agricultural products that would have benefited the most were wheat, rice, beef, pork, processed foods, and fish products.

In general, some groups favor further easing restrictions on agricultural exports to Cuba. U.S. agribusiness companies that support the removal of restrictions on agricultural exports to Cuba believe that U.S. farmers are unable to capitalize on a market so close to the United States. Those who support the lifting of financing restrictions contend such an action would help smaller U.S. companies increase their exports to Cuba more rapidly. Opponents of further easing restrictions on agricultural exports to Cuba maintain that U.S. policy does not deny such sales to Cuba, as evidenced by the large amount of sales since 2001. In particular, some agricultural producers that export to Cuba support continuation of the provision requiring payment of cash in advance because it ensures that they will be paid.

U.S. Funding to Support Democracy and Human Rights

Since 1996, the United States has provided assistance—through the U.S. Agency for International Development (USAID), the State Department, and the National Endowment for Democracy (NED)—to increase the flow of information on democracy, human rights, and free enterprise to Cuba.

USAID and State Department efforts are largely funded through Economic Support Funds (ESF) in the annual foreign operations appropriations bill. From FY1996-FY2012, Congress appropriated some $225 million in funding for Cuba democracy efforts. In recent years, this included $45.3 million for FY2008 and $20 million in each fiscal year from FY2009 through FY2012. The Administration's FY2013 request was for $15 million. Generally, as provided in appropriations measures, ESF has to be obligated within two years. According to the Government Accountability Office (GAO), USAID has received the majority of this funding while the State Department has also received funding since FY2004.

USAID's Cuba program has supported a variety of U.S.-based non-governmental organizations with the goals of promoting a rapid, peaceful transition to democracy, helping develop civil society, and building solidarity with Cuba's human rights activists. USAID maintains on its website that current USAID program partners are: Foundation of Human Rights in Cuba,

$3.4 million (2011-2014); Grupo de Apoyo a la Democracia, $3 million (2012-2015); International Relief and Development, $3.5 million (2011-2014); International Republican Institute, $3 million (2012-2015); Loyala University, $3 million (2010-2013); National Democratic Institute, $2.3 million (2011-2014); New America Foundation, $4.3 million (2012-2015); and Pan-American Development Foundation, $3.9 million (2011-2014). (See USAID's Cuba program website at: http://www.usaid.gov/where-we-work/latin-american-and-caribbean/cuba/our-work.)

NED is not a U.S. government agency but an independent nongovernmental organization that receives U.S. government funding. Its Cuba program is funded by the organization's regular appropriations by Congress as well as by funding from the State Department.

FY2011. The Administration again requested $20 million in ESF for FY2011 to support democracy and human rights projects. According to the Administration's request, the assistance would focus on providing humanitarian assistance to prisoners of conscience and their families, strengthening civil society, supporting issue-based civic action movements and coalitions, and promoting fundamental freedoms, especially freedom of expression and freedom of the press. The Senate version of the State Department and Foreign Operations appropriations measure, S. 3676, reported by the Senate Appropriations Committee on July 29, 2010 (S.Rept. 111-237), would have provided that $2 million of the ESF appropriated for Cuba be transferred and merged with funds for the National Endowment for Democracy for democracy programs in Cuba. Congress did not complete action on FY2011 appropriations until April 2011 when it approved a full-year appropriations measure (P.L. 112-10). In August 2011, the Administration made known its FY2011 foreign aid allocations by country, which included the full $20 million for Cuba democracy assistance that had been requested.

As notified to Congress in April 2012, of the $20 million, USAID would administer $8.9 million, the State Department's Bureau of Western Hemisphere Affairs would administer $1.6 million, and the State Department's Bureau of Democracy, Human Rights, and Labor (DRL) would administer $9.5 million, of which $4 million would be transferred to the National Endowment for Democracy. In terms of programs for the $20 million, $12.43 million would for democracy, civil society and media programs; $4.7 million would support human rights initiatives; and $2.87 million will be for program support.

FY2012. The Administration once again requested $20 million in ESF for FY2012 with the promotion of democratic principles as the core goal of assistance, and Congress supported the full amount in the conference report to the FY2012 Consolidated Appropriations Act (H.Rept. 112- 331 to H.R. 2055, P.L. 112-74). The budget request stated that there was an increased effort to manage programs more transparently, focus efforts on Cuba, and widen the scope of the civic groups receiving supports. According to the Administration's request, U.S. assistance would strengthen a range of independent elements of Cuban civil society, including associations and labor groups, marginalized groups, youth, legal associations, and women's networks. The programs would be designed to increase the capacity for community involvement of civil society organizations and networking among the groups. The program would also support Cuban efforts to document human rights violations, provide humanitarian assistance to political prisoners and their families, and build leadership skills of civil society leaders. Finally, the budget request maintained that U.S. assistance also would support the dissemination of information regarding market economies and economic rights.

The Senate Appropriations Committee-reported version of the FY2012 Department of State, Foreign Operations, and Related Programs Appropriations bill, S. 1601 (S.Rept. 112-85), would have provided $15 million in ESF for Cuba ($5 million less than the request), including humanitarian and democracy assistance, support for economic reform, private sector initiatives, and human rights. In its report to the bill, the committee maintained that it expected that funds would be made available, and programs carried out, in a transparent manner. The committee also would have directed that the USAID Administrator provide regular updates to the committee on the number of Cubans who receive assistance and the types of assistance. In contrast to the Senate bill, a draft House Appropriations Committee bill and report (marked up by the Subcommittee on State, Foreign Operations, and Relations Programs on July 27, 2011) would have recommended $20 million in ESF for Cuba (the full Administration's request), and would have directed that the funds be used only for democracy-building, and not for business promotion, economic reform, social development or other purposes expressly authorized by Section 109(a) of the Cuban Liberty and Democratic Solidarity Act of 1996 (P.L. 104-114). (See the draft committee report, available at http://appropriations.house.gov/UploadedFiles/FY12-SFOPSCombinedReport-CSBA.pdf.)

FY2013. For FY2013, the Administration requested $15 million for human rights and democracy programs for Cuba. According to the request, "U.S. assistance will continue to support human rights and civil society initiatives that promote basic freedoms, particularly freedom of expression. Programs will continue to provide humanitarian assistance to prisoners of conscience and their families, as well as strengthen independent Cuban civil society, and promote the flow of uncensored information to, from, and within the island."[50]

The Senate Appropriations Committee-reported version of the FY2013 State Department, Foreign Operations, and Related Programs Appropriations Act, S. 3241 (S.Rept. 112-172), would have provided $15 million in ESF for Cuba (the same as the Administration's request), including "for humanitarian assistance, support for economic reform, private sector initiatives, democracy, and human rights." In contrast, the House Appropriations Committee-reported version of the bill, H.R. 5857 (H.Rept. 112-94), would have provided $20 million in ESF ($5 million more than the Administration's request), but would transfer and merge the aid with funds available to the National Endowment for Democracy "to promote democracy and strengthen civil society in Cuba." The report to the House bill maintained that assistance "shall not be used for business promotion, economic reform, social development, or other purposes not expressly authorized by section 109(a)" of the Cuban Liberty and Democratic Solidarity Act (P.L. 104-114).

Congress did not complete action on FY2013 appropriations before the beginning of the fiscal year, but in September 2012, it approved a continuing resolution (H.J.Res. 117, P.L. 112-175) that continued FY2013 funding through March 27, 2013, at the same rate for projects and activities in FY2012, plus an across-the-board increase of 0.612%, although specific country accounts were left to the discretion of responsible agencies. On March 21, 2013, Congress completed action on full-year FY2013 appropriations with the approval of H.R. 933. This continues FY2013 funding for Cuba democracy programs, although the amount that ultimately will be allocated is uncertain because of the effects of sequestration set forth in the Budget Control Act of 2011 (P.L. 112-25), as amended by the American Taxpayer Relief Act (P.L. 112-240).[51]

National Endowment for Democracy. Until FY2008, NED's democratization assistance for Cuba had been funded largely through the annual Commerce, Justice, and State (CJS) appropriations measure, but is now funded through the State Department, Foreign Operations and Related Agencies appropriations measure.

According to NED, its Cuba funding in recent years has been as follows: $1.4 million in FY2008 for 11 projects; $1.5 million in FY2009 for 10 projects; $2.4 million in FY2010 for 15 projects; and $1.65 million in FY2011 for 13 projects. NED reported in its 2011 annual report (the most recent available) that it is funding the following organizations and projects for Cuba in that fiscal year: Afro-Cuban Alliance Inc; Center for a Free Cuba; Committee for Free Trade Unionism; Cuban Democratic Directorate; CubaNet News Inc.; Grupo Internacional para la Responsibilidad Social Corporativa en Cuba; Instituto Político para La Libertad Perú; National Democratic Institute for International Affairs; Observatorio Cubano de Derechos Humanos; People in Need; People in Peril Association; Civic Education; and Rule of Law.[52]

Oversight of U.S. Democracy Assistance to Cuba

GAO has issued several reports since 2006 examining USAID and State Department democracy programs for Cuba. In 2006, GAO issued a report examining programs from 1996 through 2005, and concluded that the U.S. program had significant problems and needed better management and oversight. According to GAO, internal controls, for both the awarding of Cuba program grants and oversight of grantees, "do not provide adequate assurance that the funds are being used properly and that grantees are in compliance with applicable law and regulations."[53] Investigative news reports on the program maintained that high shipping costs and lax oversight had diminished its effectiveness.[54]

GAO issued a second report in 2008 examining USAID's Cuba democracy program.[55] The report lauded the steps that USAID had taken since 2006 to address problems with its Cuba program and improve oversight of the assistance. These included awarding all grants competitively since 2006, hiring more staff for the program office since January 2008, and contracting for financial services in April 2008 to enhance oversight of grantees. The GAO report also noted that USAID had worked to strengthen program oversight through pre-award and follow-up reviews, improving grantee internal controls and implementation plans, and providing guidance and monitoring about permitted types of assistance and cost sharing. The 2008 GAO report also maintained, however, that USAID had not staffed the Cuba program to the level needed for effective grant oversight. GAO recommended that USAID (1) ensure that its Cuba program office is staffed at the level that is needed to fully implement planned monitoring activities; and (2) periodically assess the Cuba program's overall efforts to address and reduce grantee risks, especially regarding internal controls, procurement practices, expenditures, and compliance with laws and regulations.

More recently, in January 2013, GAO issued its third report on Cuba democracy programs.[56] The report concluded that USAID had improved its performance and financial monitoring of implementing partners' use of program funds, but found that the State Department's financial monitoring had gaps. Both agencies were reported to be taking steps to improve financial monitoring. GAO recommended that the Secretary of State take two actions to strengthen the agency's ability to monitor the use of Cuba democracy program funds: use a risk-based approach for program audits that considers specific indicators for program partners; and obtain sufficient information to approve implementing partners' use of subpartners.

Imprisonment of USAID Subcontractor since December 2009

As noted earlier, USAID subcontractor Alan Gross has been imprisoned in Cuba since December 2009 for his work on a Cuba democracy project designed to provide Cuba's Jewish community with communication equipment for wireless Internet connectivity. In March 2011, he was convicted by a Cuban court in March 2011 and sentenced to 15 years in prison. Gross has remained in prison despite numerous calls for his release on humanitarian grounds by Members of Congress, the Obama Administration, and many other religious and human rights groups. As noted above, his continued imprisonment is a key impediment to an improvement in U.S.-Cuban relations. Over the past year, Cuba has been linking the release of Alan Gross to the release of the so-called "Cuban five" (see text box below), who were convicted in the United States for espionage in 2001. The United States rejects such linkage, maintaining there is no equivalence between the cases.[57]

Gross was working as a USAID subcontractor for Development Alternatives Inc. (DAI), a Bethesda-based company that had received a contract from USAID to help support Cuban civil society organizations. As part of the project, Gross installed broadband Internet connections for three Jewish communities in the cities of Havana, Camagüey, and Santiago de Cuba. He was arrested on December 4, 2009, at Jose Martí International Airport in Havana when he was planning to leave the country after his fifth trip to Cuba under his subcontract with DAI. According to a statement at the time by DAI, Gross "was working with a peaceful, non-dissident civic group—a religious and cultural group recognized by the Cuban government—to improve its ability to communicate with its members across the island and overseas."[58]

After 14 months in prison, a Cuban court in Havana officially charged Gross on February 4, 2011, with "actions against the independence and territorial integrity of the state" pursuant to Article 91 of Cuba's Penal Code. After a two-day trial in March 2011, Gross was convicted and sentenced to 15 years in prison. Gross's lawyer had asked for the Cuban government to release

Gross as a humanitarian gesture, maintaining that his health continued to deteriorate and noting that his elderly mother had been recently diagnosed with lung cancer, and his daughter was recovering from cancer treatment. Cuba's Supreme Court heard arguments for Gross's appeal on July 22, 2011, but the court rejected the appeal on August 5, 2011.

There had been some hope in April 2012 that Cuba would positively respond to a humanitarian request by Alan Gross to visit his elderly sick mother in the United States for a period of two weeks, but this did not occur. In contrast, a U.S. federal judge in Florida granted René González, one of the so-called "Cuban five" convicted in 2001, the right to visit his dying brother in Cuba for two weeks.

In September 2012, Judy Gross, the wife of Alan Gross, expressed concern in media reports about the health of her husband, who has lost more than 100 pounds while in prison, and fears that he would not survive continued imprisonment. In early October 2012, Judy Gross expressed concern that her husband could have cancer. In November 2012, the Cuban government maintained that Gross was in normal health and that a biopsy on a lesion showed that he did not have cancer. Gross's lawyer has called for an independent medical examination by a doctor of Gross's choosing.[59]

On November 16, 2012, Alan and Judy Gross filed a suit in U.S. District Court against the U.S. government and his employer, DAI, alleging that they "failed to disclose adequately to Mr. Gross, both before and after he began traveling to Cuba, the material risks that he faced due to his participation in the project."[60] In late November 2012, Judy Gross urged President Obama to give the case top priority and to designate a special envoy to meet with the Cuban government for her husband's release.[61]

In the more than three years since Gross has been imprisoned, numerous U.S. officials and Members of Congress have raised the issue of Alan Gross's detention with the Cuban government and have called for his release. On December 3, 2012, the third anniversary of Gross's imprisonment, the State Department issued a statement again calling for his release, and asked the Cuban government to grant Gross's request to travel to the United States to visit his gravely ill 90-year old mother. On December 5, 2012, the Senate approved S.Res. 609 (Moran) by voice vote, marking the first congressional vote on the issue since Gross's detention. With 31 cosponsors, the resolution called for the immediate and unconditional release of Gross, and urged the Cuban government in the meantime to provide all appropriate diagnostic and medical treatment to address the full range of medical issues facing Mr. Gross and to allow him to choose a doctor to provide him with an independent medical assessment. A seven-member U.S. congressional delegation visited Cuba in February 2013, and again raised the issue of Gross's release. Cuban officials have called for talks with the United States aimed at resolving the Alan Gross case and continued to link his release to the case of the "Cuban five."

Cuban Five

In June 2001, five members of the so-called "Wasp Network" originally arrested in September 1998 were convicted on espionage charges by a U.S. Federal Court in Miami. Sentences handed down for the "Cuban five" in December 2001 ranged from 15 years to life in prison for three of the five. The group of five Cuban intelligence agents—Gerardo Hernández, Ramón Labañino, Antonio Guerrero, Fernando González, and René González— penetrated Cuban exile groups and tried to infiltrate U.S. military bases. The Cuban government vowed to work for the return of the "Cuban five" who have been dubbed "Heroes of the Republic" by Cuba's National Assembly. In December 2008, Cuban President Raúl Castro offered to exchange some imprisoned Cuban political dissidents for the "Cuban five," an offer that was rejected by the State Department, which maintained that the dissidents should be released immediately without any conditions.

In June 2009, the U.S. Supreme Court chose not to hear an appeal of the case of the "Cuban five" in which their lawyers were asking for a new trial outside Miami before an unbiased jury. Later in 2009, however, sentences for three of the five were reduced: Antonio Guerrero had his life sentence reduced to almost 22 years; Ramón Labañino had his life sentence reduced to 30 years; and Fernando González had his 19-year sentence reduced to 18 years. Gerardo Hernández, convicted of murder conspiracy for his role in the 1996 Brothers to the Rescue shootdown, is serving two life sentences.

René González, who received a 15-year sentence, was released from prison in early October 2011, but is serving three years of probation; a judge has ruled that he must serve it in the United States. Cuba had asked for González to be returned to Cuba upon his release from prison so that he could be reunited with his wife and family for humanitarian reasons. In late March 2012, González was allowed by a federal judge in Florida to visit his dying brother in Cuba for a period of two weeks, after which he returned to the United States.

Radio and TV Marti

U.S.-government sponsored radio and television broadcasting to Cuba—Radio and TV Martí— began in 1985 and 1990 respectively. According to the *Broadcasting Board of Governors (BBG) FY2013 Budget Request*, Radio and TV Martí "inform and engage the people of Cuba by providing a reliable and credible source of news and information." The BBG's Office of Cuba Broadcasting "uses a mix of media, including shortwave, medium wave, direct-to-home satellite, flash drives, and DVDs to help reach audiences in Cuba."[62]

Until October 1999, U.S.-government funded international broadcasting programs had been a primary function of the United States Information Agency (USIA). When USIA was abolished and its functions were merged into the Department of State at the beginning of FY2000, the BBG became an independent agency that included such entities as the Voice of America (VOA), Radio Free Europe/Radio Liberty (RFE/RL), Radio Free Asia, and the Office of Cuba Broadcasting (OCB), which manages Radio and TV Marti. OCB is headquartered in Miami, FL. Legislation in the 104th Congress (P.L. 104-134) required the relocation of OCB from Washington, DC, to south Florida. The move began in 1996 and was completed in 1998.

Radio Martí broadcasts on short and medium wave (AM) channels for 24 hours six days per week, and for 18 hours one day per week utilizing transmission facilities in Marathon, FL, and Greenville, NC, according to the BBG. It also transmits to Cuba 24 hours daily through Hispasat satellite television and the internet.

TV Martí programming has been broadcast through multiple transmission methods over the years. From its beginning in 1990 until July 2005, it was broadcast via an aerostat (blimp) from facilities in Cudjoe Key, Florida for four and one-half hours daily, but the aerostat was destroyed by Hurricane Dennis. From mid-2004 until 2006, TV Martí programming was transmitted for several hours once a week via an airborne platform known as Commando Solo operated by the Department of Defense utilizing a C-130 aircraft. In August 2006, OCB began to use contracted private aircraft to transmit prerecorded TV Martí broadcasts six days weekly, and by late October 2006 the OCB inaugurated an aircraft-broadcasting platform known as AeroMartí with the capability of transmitting live broadcasts. OCB currently uses two privately contracted airplanes for AeroMartí to transmit broadcasts two and one-half hours for five days weekly. Broadcasts are also transmitted via the internet and satellite television. In September 2011, the BBG awarded a contract to a Maryland firm to design and operate a text messaging system that can distribute up to 24,000 messages per week from OCB broadcasters to mobile phone users in Cuba, including the use of techniques to circumvent censorship.[63]

Funding for Cuba Broadcasting

From FY1984 through FY2012, Congress appropriated about $696 million for broadcasting to Cuba. In recent years, funding amounted to $28.416 million in FY2011 and an estimated $28.062 million in FY2012. The FY2013 request was for $23.594 million.

FY2012. The Administration requested $28.475 million for Cuba broadcasting in FY2012. The Senate Appropriations Committee version of the FY2012 State Department, Foreign Operations, and Related Programs Appropriations measure, S. 1601 (S.Rept. 112-85), recommended $28.181 million in funding for Cuba broadcasting, $294,000 less than the request. In contrast, a draft House Appropriations Committee report and bill (marked up by the House Appropriations Committee's Subcommittee on State, Foreign Operations, and Related Programs on July 27, 2011) recommended $30.175 million for Cuba broadcasting, $1.7 million more than the request. (See the draft committee report, available at http://appropriations.house.gov /UploadedFiles/ FY12-SFOPSCombinedReport-CSBA.pdf.) In final FY2012 appropriations action in the FY2012 Consolidated Appropriations Act (H.R. 2055, P.L. 112-74), Congress approved full funding of the Administration's $28.475 million request for broadcasting to Cuba. The BBG's estimate for FY2012 funding is $28.062 million.

FY2013. The Administration requested $23.594 million for Cuba broadcasting in FY2013, almost $4.5 million lower than FY2012 funding. According to the BBG's budget request, program reductions are possible because of OCB's planned streamlining in the planning and execution of news coverage and reliance on additional technical support from the BBG's International Broadcasting Bureau.

The Senate Appropriations Committee-reported FY2013 State Department, Foreign Operations, and Related Programs Appropriations Act, S. 3241 (S.Rept. 112-172), would have provided $23.4 million ($194,000 less than the Administration's request), while the House Appropriations Committee-reported bill, H.R. 5857 (H.Rept. 112-494), would have provided $28.062 million ($4.468 million more than the Administration's request and the same amount provided in FY2012). As already noted, Congress did not complete action on FY2013 appropriations, but in September 2012 it approved a continuing resolution (P.L. 112-175) that continued funding through March 27, 2013, at the same rate for projects and activities in FY2012 plus an across-the-board increase of 0.612%. On March 21, 2013, Congress completed action on full-year FY2013 appropriations with the approval of H.R. 933. The amount that ultimately will be allocated for Cuba broadcasting is uncertain because of the effects of sequestration set forth in the Budget Control Act of 2011 (P.L. 112-25), as amended by the American Taxpayer Relief Act (P.L. 112- 240).

Oversight of Radio and TV Martí

Both Radio and TV Martí have at times been the focus of controversies, including questions about adherence to broadcast standards. There have been various attempts over the years to cut funding for the programs, especially for TV Martí, which has not had much of an audience because of Cuban jamming efforts. In December 2006, press reports alleged significant problems in the OCB's operations, with claims of cronyism, patronage, and bias in its coverage.[64] In February 2007, the former director of TV Martí programming pled guilty in U.S. federal court to receiving more than $100,000 in kickbacks over a three-year period from a vendor receiving OCB contracts.[65]

From 1990 through 2008, there were numerous government studies and audits of the OCB, including investigations by the GAO, by a 1994 congressionally established Advisory Panel on Radio and TV Martí, by the State Department Office Inspector General (OIG), and by the combined State Department/BBG Office Inspector General.[66]

In January 2009, GAO issued a report asserting that the best available research suggests that Radio and TV Martí's audience is small, and cited telephone surveys since 2003 showing that less than 2% of respondents reported tuning in to Radio or TV Martí during the past week. With regard to TV Martí viewership, according to the report, all of the IBB's telephone surveys since 2003 show that less than 1% of respondents said that they had watched TV Martí during the past week. According to the GAO report, the IBB surveys show that there was no increase in reported TV Martí viewership following the beginning of AeroMartí and DirecTV satellite broadcasting in 2006.The GAO report also cited concerns with adherence to relevant domestic laws and international standards, including the domestic dissemination of OCB programming, inappropriate advertisements during OCB programming, and TV Martí's interference with Cuban broadcasts.[67] GAO testified on its report in a hearing held by the House Subcommittee on International Organizations,

Human Rights, and Oversight of the Committee on Foreign Affairs on June 17, 2009.

In April 2010, the Senate Foreign Relations Committee majority issued a staff report that concluded that Radio and TV Martí "continue to fail in their efforts to influence Cuban society, politics, and policy." The report cited problems with adherence to broadcast standards, audience size, and Cuban government jamming. Among its recommendations, the report called for the IBB to move the Office of Cuba Broadcasting back to Washington and integrate it fully into the Voice of America.[68]

In December 2011, GAO issued a report examining the extent to which the BBG's strategic plan for broadcasting required by the conference report to the FY2010 Consolidated Appropriations measure (H.Rept. 111-366 to H.R. 3288/P.L. 111-117) met the requirements established in the legislation. The BBG strategic plan was required to include (1) an analysis of the current situation in Cuba and an allocation of resources consistent with the relative priority of broadcasting to Cuba as determined by the annual Language Service Review and other factors; (2) the estimated audience sizes in Cuba for Radio and TV Martí and the sources and relative reliability of the data; (3) the cost of any and all types of TV transmission and the effectiveness of each in increasing such audience size; (4) the principal obstacles to increasing audience size; (5) an analysis of other options for disseminating news and information to Cuba and the cost effectiveness of each option; and (6) an analysis of the program efficiencies and effectiveness that can be achieved through shared resources and cost saving opportunities in radio and television production between Radio and TV Martí and the Voice of America. GAO found that the BBG's strategic plan lacked key information. Of the six requirements set forth in the conference report, the GAO found that the BBG's strategic plan fully addressed item (4) regarding the principal obstacles to increasing audience sized, but only partially addressed the other five items. The GAO report stated that the BBG can develop and provide more information to Congress, including an analysis of the cost savings opportunities of sharing resources between Radio and TV Martí and the Voice of America's Latin America Division.[69]

In May 2012, a controversy occurred involving an editorial by OCB Director Carlos García-Pérez in which he strongly criticized Cuban Cardinal Jaime Ortega and referred to the Cardinal as a "lackey" of the Cuban government.[70] The strong language was criticized by several Members of Congress, who called for the Administration to reject the comments against Cardinal Ortega.[71] The editorial raised significant questions about the editorial policy of OCB as well as OCB's adherence to broadcast standards.[72] Such an editorial, authored by the director of OCB, could lead one to conclude that the views articulated were those of the U.S. government. BBG's Director of Communications and External Affairs Lynne Weil maintained that such "editorials, unless otherwise stated, represent the views of the broadcasters only and not necessarily those of the U.S. government."[73]

Terrorism Issues[74]

Cuba was added to the State Department's list of states sponsoring international terrorism in 1982 (pursuant to Section 6(j) of the Export Administration Act of 1979; P.L. 96-72; 50 USC appendix 2504(j)) because of its alleged ties to international terrorism and support for terrorist groups in Latin America, and it has remained on the list since that time. Cuba had a long history of supporting revolutionary movements and governments in Latin America and Africa, but in 1992, Fidel Castro said that his country's support for insurgents abroad was a thing of the past. Cuba's change in policy was in large part due to the breakup of the Soviet Union, which resulted in the loss of billions of dollars in annual subsidies to Cuba, and led to substantial Cuban economic decline.

Critics of retaining Cuba on the terrorism list maintain that it is a holdover from the Cold War. They argue that domestic political considerations keep Cuba on the terrorism list and maintain that Cuba's presence on the list diverts U.S. attention from struggles against serious terrorist threats. Those who support keeping Cuba on the terrorism list argue that there is ample evidence that Cuba supports terrorism. They point to the government's history of supporting terrorist acts and armed insurgencies in Latin America and Africa. They point to the government's continued hosting of members of foreign terrorist organizations and U.S. fugitives from justice.

The State Department's *Country Reports on Terrorism 2011* report (issued July 31, 2012) maintained that "current and former members of Basque Fatherland and Liberty (ETA) continued to reside in Cuba," and that "press reporting indicated that the Cuban government provided medical care and political assistance" to members of the Revolutionary Armed Forces of Colombia (FARC). At the same time, the report maintained that there "was no indication that the Cuban government provided weapons or paramilitary training for either ETA or the FARC." With regard to ETA, the State Department reported that three suspected ETA members were arrested in

Venezuela after sailing there from Cuba and were deported back to Cuba in September 2011– one of the men, Jose Ignacio Echarte, is believed to have ties to the FARC and is a fugitive from Spain, which has requested his extradition.[75]

In more recent developments in 2012, Cuba has been playing a role in hosting talks between the FARC and Colombian government of President Juan Manuel Santos. Conversations began in Cuba with the FARC in early 2012, while formal peace talks began in Norway in October 2012 and then moved to Havana beginning in November 2012.[76]

Another issue noted in the 2011 terrorism report is that the Cuban government continues "to permit fugitives wanted in the United States to reside in Cuba," and provides such support as housing, food ration books, and medical care. In the 112th Congress, legislation was introduced, H.Res. 226 (King), that would have called for the immediate extradition or rendering of all fugitives from justice who are receiving safe harbor in Cuba in order to escape prosecution or confinement for criminal offenses committed in the United States.

As set forth in Section 6(j) of the Export Administration Act, a country's retention on the list may be rescinded in two ways. The first option is for the President to submit a report to Congress certifying that there has been a fundamental change in the leadership and policies of the government and that the government is not supporting acts of international terrorism and is providing assurances that it will not support such acts in the future. The second option is for the President to submit a report to Congress, at least 45 days in advance justifying the rescission and certifying that the government has not provided any support for international terrorism during the preceding six-months, and has provided assurances that it will not support such acts in the future.

A press report in February 2013 claimed that high ranking State Department officials concluded that Cuba should not be on the state sponsors of terrorism list, but State Department officials contend that the report was incorrect and that there are no current plans to remove Cuba from the list.[77] The State Department conducts an annual review to see whether a country should be on the list.

Some observers maintain that Cuba's role in facilitating Colombia's peace talks could ultimately be a factor in removing Cuba from the list.

Migration Issues[78]

Cuba and the United States reached two migration accords in 1994 and 1995 designed to stem the mass exodus of Cubans attempting to reach the United States by boat. On the minds of U.S. policymakers was the 1980 Mariel boatlift in which 125,000 Cubans fled to the United States with the approval of Cuban officials. In response to Fidel Castro's threat to unleash another Mariel, U.S. officials reiterated U.S. resolve not to allow another exodus. Amid escalating numbers of fleeing Cubans, on August 19, 1994, President Clinton abruptly changed U.S. migration policy, under which Cubans attempting to flee their homeland were allowed into the United States, and announced that the U.S. Coast Guard and Navy would take Cubans rescued at sea to the U.S. naval base at Guantanamo Bay, Cuba. Despite the change in policy, Cubans continued fleeing in large numbers.

As a result, in early September 1994, Cuba and the United States began talks that culminated in a September 9, 1994, bilateral agreement to stem the flow of Cubans fleeing to the United States by boat. In the agreement, the United States and Cuba agreed to facilitate safe, legal, and orderly Cuban migration to the United States, consistent with a 1984 migration agreement. The United States agreed to ensure that total legal Cuban migration to the United States would be a minimum of 20,000 each year, not including immediate relatives of U.S. citizens.

In May 1995, the United States reached another accord with Cuba under which the United States would parole the more than 30,000 Cubans housed at Guantanamo into the United States, but would intercept future Cuban migrants attempting to enter the United States by sea and would return them to Cuba. The two countries would cooperate jointly in the effort. Both countries also pledged to ensure that no action would be taken against those migrants returned to Cuba as a consequence of their attempt to immigrate illegally. On January 31, 1996, the Department of Defense announced that the last of some 32,000 Cubans intercepted at sea and housed at Guantanamo had left the U.S. Naval Station, most having been paroled into the United States.

Since the 1995 migration accord, the U.S. Coast Guard has interdicted thousands of Cubans at sea and returned them to their country, while those deemed at risk for persecution have been transferred to Guantanamo and then found asylum in a third country or eventually the United States. Those Cubans who reach shore are allowed to apply for permanent resident status in one year, pursuant to the Cuban Adjustment Act of 1966 (P.L. 89-732). In short, most interdictions, even in U.S. coastal waters, result in a return to Cuba, while those Cubans who touch shore are allowed to stay in the United States. This so-called "wet foot/dry foot" policy has been criticized by some as encouraging Cubans to risk their lives in order to make it to the United States and as encouraging alien smuggling. Others maintain that U.S. policy should welcome those migrants fleeing communist Cuba whether or not they are able to make it to land.

The number of Cubans interdicted at sea by the U.S. Coast Guard rose from 666 in FY2002 to a high of 2,868 in FY2007. In the three subsequent years, maritime interdictions declined significantly to 422 by FY2010 (see Figure 6). Major reasons for the decline were reported to include the U.S. economic downturn, more efficient coastal patrolling, and more aggressive prosecution of migrant smugglers by both the United States and Cuba.[79]

In FY2011 and FY2012, however, the number of Cubans interdicted by the Coast Guard increased respectively to 985 and 1,275.[80] Speculation on the reasons for the increase include Cuba's deteriorating economic and political situation; the Coast Guard's more efficient methods of interdiction; and the easing of the economic situation in the United States, making it easier for the payment of fees to migrant smugglers.[81] For FY2013, the number of Cubans interdicted was 311 as of March 21, 2013, according to Coast Guard statistics.

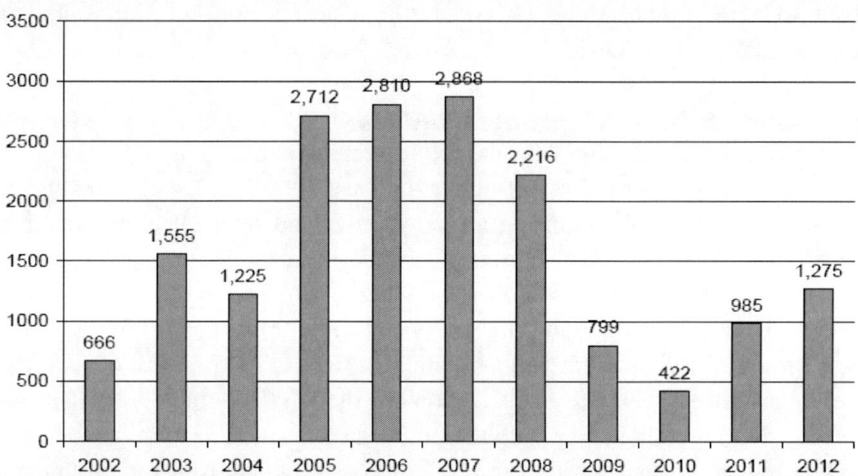

Source: CRS presentation of United States Coast Guard data. United States Coast Guard, Alien Migrant Interdiction, "Coast Guard Migrant Interdictions—Fiscal Year 1982-Present," October 23, 2012; current statistics available at http://www.uscg.mil/hq/cg5/cg531/AMIO/FlowStats/FY.asp.

Figure 6. Maritime Interdiction of Cubans, FY2002-FY2012.

Despite the U.S. Coast Guard's maritime interdiction program, thousands of unauthorized Cubans reach the United States each year, either by boat or at land ports of entry. U.S. Border Patrol apprehensions (largely coastal Florida) of unauthorized Cubans were 910 in FY2009, 712 in FY2010, and 959 in FY2011. These statistics are significantly lower than the FY2005-FY2008 period when Border Patrol apprehensions of Cubans averaged over 3,700 each year.[82]

Arrival of unauthorized Cubans at U.S. ports of entry—the majority from Mexico—averaged over 7,400 each year from FY2009-FY2011, although it showed a slightly rising trend from 7,053 in FY2009 to 7,798 in FY2011. Comparatively, however, the statistics are much lower than the average of almost 11,000 Cubans arriving at ports of entry annually from FY2005-FY2008.[83] In October 2008, Mexico and Cuba negotiated a migration accord in an attempt to curb the irregular flow of migrants through Mexico.[84] The agreement calls for Mexico to accept all unauthorized immigrants detained by Mexican authorities. Semi-annual U.S.-Cuban talks alternating between Cuba and the United States were held on the implementation of the 1994/1995 migration accords until they were suspended by the United States in 2004. The Obama Administration re-started the talks in 2009, and there were four rounds of talks until January 2011. In addition to migration issues, the talks became a forum to raise other issues of concern, including, for U.S. officials, the imprisonment of Alan Gross.

Cuba Alters Its Policy Regarding Exit Permits

In October 2012, the Cuban government announced that it would be updating its migration policy, effective January 14, 2013, by eliminating the long-standing policy of requiring an exit permit and letter of invitation from abroad for Cubans to travel abroad. Cubans would be able to travel abroad with just an updated passport and a visa issued by the country of destination, if required. Under the change in policy, Cubans could travel abroad for up to two years without forgoing their rights as Cuban citizens. The practice of requiring an exit permit has been extremely unpopular in Cuba and the government had been considering doing away with the practice for some time.

On its face, Cuba's action can be viewed as a human rights improvement, but how significant that improvement is will depend on how the law is implemented. The Cuban government said that it would fight against "brain drain," and that the new policy would not apply to scientists, athletes, and other professionals. In early January 2013, however, the Cuban government announced that the new travel policy would also apply to health care professionals, including doctors. It remains to be seen whether the government will allow political dissidents to leave the country and return.

When the new policy went into effect on January 14, 2013, thousands of Cubans lined up at government migration offices and travel agencies. Travel under the new policy requires an updated passport as well as any visas required by the receiving countries. While most countries require visas for Cubans, several Caribbean countries do not. Ecuador, which had not required a visa for Cubans visiting up to 90 days, announced on January 15, 2013, that it would require Cubans who wanted to visit to provide a letter of invitation from a legal resident in Ecuador with a commitment to cover expenses for the visiting Cuban(s).[85]

A U.S. State Department spokesman said that it welcomes any changes that would allow Cubans to depart from and return to their country freely. According to the State Department, Cuba's announced change is consistent with the Universal Declaration of Human Rights in that everyone should have the rights to leave any country, including their own, and return.[86] At the same time, however, Assistant Secretary of State for Western Hemisphere Affairs Roberta Jacobson cautioned that it is uncertain yet how the changes are to be implemented. She raised questions regarding whether Cuba would impose some controls on passports and whether everyone would be free to travel.[87] As noted above, Internet blogger Yoani Sánchez, who had been denied an exit permit for several years, received a new passport under the new policy and in February 2013 began a multi-nation trip that brought her to the United States in mid-March 2013. A number of dissidents, however, including those political prisoners who have been released on parole, have been restricted from traveling abroad.

In light of Cuba's new travel policy, some analysts have raised the question as to whether the United States should review its policy toward Cuban migrants, as set forth in the Cuban Adjustment Act of 1966 (P.L. 89-732), in which those Cubans arriving in the United States are allowed to apply for permanent resident status in one year.[88]

Anti-Drug Cooperation

Cuba is not a major producer or consumer of illicit drugs, but its extensive shoreline and geographic location make it susceptible to narcotics smuggling operations. Drugs that enter the Cuban market are largely the result of onshore wash-ups from smuggling by high-speed boats moving drugs from Jamaica to the Bahamas, Haiti, and the United States or by small aircraft from clandestine airfields in Jamaica. For a number of years, Cuban officials have expressed concerns over the use of their waters and airspace for drug transit and about increased domestic drug use. The Cuban government has taken a number of measures to deal with the drug problem, including legislation to stiffen penalties for traffickers, increased training for counternarcotics personnel, and cooperation with a number of countries on anti-drug efforts.

According to the State Department's 2013 *International Narcotics Control Strategy Report (INCSR)*, issued March 12, 2013, Cuba has a number of anti-drug-related agreements in place with other countries, including 35 judicial agreements regarding judicial proceedings and extradition and 35 bilateral counterdrug agreements. Since 1999, Cuba's Operation Hatchet has focused on maritime and air interdiction and the recovery of narcotics washed up on Cuban shores. As reported in the *INCSR*, Cuba seized 9.1 metric tons of illegal narcotics in 2011, and 2.1 metric tons in 2012 (through October), with the overwhelming majority consisting of wash ups. Since 2003, Cuba has aggressively pursued an internal enforcement and investigation program against its incipient drug market with an effective nationwide drug prevention and awareness campaign.

Over the years, there have been varying levels of U.S.-Cuban cooperation on anti-drug efforts. In 1996, Cuban authorities cooperated with the United States in the seizure of 6.6 tons of cocaine aboard the Miami-bound *Limerick*, a Honduran-flag ship. Cuba turned over the cocaine to the United States and cooperated fully in the investigation and subsequent prosecution of two defendants in the case in the United States. Cooperation has increased since 1999 when U.S. and Cuban officials met in Havana to discuss ways of improving anti-drug cooperation. Cuba accepted an upgrading of the communications link between the Cuban Border Guard and the U.S. Coast Guard as well as the stationing of a U.S. Coast Guard Drug Interdiction Specialist (DIS) at the U.S. Interests Section in Havana. The Coast Guard official was posted to the U.S. Interests Section in September 2000, and since that time, coordination has increased.

In the 2013 *INCSR*, the State Department reported that Cuba maintained a significant level of anti-drug cooperation with the United States in 2012. The Coast Guard shares tactical information related to narcotics trafficking on a case by case basis, and responds to Cuban information on vessels transiting through Cuban territorial seas suspected of smuggling. Bilateral cooperation led to multiple at-sea interdictions in 2012. The Cuban Border Guard reported 31 real-time reports of "go-fast" narcotics trafficking events in 2012 to the

U.S. Coast Guard, and its e-mail and phone notifications have increased in timeliness, quantity, and quality, according to the *INCSR*, occasionally including photographs of suspected vessels involved in narcotics trafficking.

Cuba maintains that it wants to cooperate with the United States to combat drug trafficking, and on various occasions has called for a bilateral anti-drug cooperation agreement with the United States.[89] In the 2011 *INCSR* (issued in March 2011) the State Department acknowledged that Cuba had presented the U.S. government with a draft bilateral accord for counternarcotics cooperation that is still under review. According to the State Department in the *INCSR*: "Structured appropriately, such an accord could advance the counternarcotics efforts undertaken by both countries." The report maintained that greater cooperation among the United States, Cuba, and its international partners—especially in the area of real-time tactical information-sharing and improved tactics, techniques, and procedures—would likely lead to increased interdictions and disruptions of illegal trafficking. These positive U.S. statements regarding a potential bilateral anti-drug cooperation agreement and greater multilateral cooperation in the region with Cuba were reiterated in the 2012 and 2013 *INCSRs*.

At a February 1, 2012, hearing before the Senate Caucus on International Narcotics Control on U.S.-Caribbean security cooperation, Caucus Chairman Senator Dianne Feinstein stated that "this limited cooperation we do have between our Coast Guard and Cuban authorities has been very useful, and I hope we can find ways to increase our counternarcotics cooperation with Cuba."[90]

The caucus released a report on September 13, 2012, in which Senator Feinstein recommended that the Obama Administration consider taking four steps to increase U.S. collaboration with Cuba on counternarcotics: 1) expand the U.S. Coast Guard and law enforcement presence at the U.S. Interests Section in Havana; 2) establish protocols for direct ship-to-ship communication between the U.S. Coast Guard and the Cuban Border Guard; 3) negotiate a bilateral counternarcotics agreement with Cuba; and 4) allow for Cuba's participation in the U.S.- Caribbean Security Dialogue.[91]

Cuba's Offshore Oil Development

Cuba is working toward potential development of its offshore oil resources, but it suffered setbacks in 2012 when three attempts by foreign oil companies drilling wells were unsuccessful. While the country has proven oil reserves of just 0.1 billion barrels, the U.S. Geological Survey estimates that offshore reserves in the North Cuba Basin could contain an additional 4.6 billion barrels of undiscovered technically recoverable crude oil. If oil is found, some experts estimate that it would take at least three to five years before production would begin.

While it is unclear whether offshore oil production could result in Cuba becoming a net oil exporter, it could reduce Cuba's current dependence on Venezuela for oil supplies. As noted above, Venezuela provides Cuba with some 100,000 barrels of oil per day. In 2011, Cuba produced about 55,000 barrels of oil per day on its own, with most production occurring onshore, and consumed just over 170,000, according to the U.S. Energy Information Administration.[92]

Cuba has had seven offshore deepwater oil projects involving nine foreign companies in 22 exploration blocs. (See Figure 7 for a map of Cuba's offshore oil blocks.) The Spanish oil company Repsol, in a consortium with Norway's Statoil and India's Oil and Natural Gas Corporation, began offshore exploratory drilling in late January 2012, using an oil rig known as the Scarabeo-9 (owned by an Italian oil services provider, Saipem, a subsidiary of the Italian oil company ENI). On May 18, 2012, however, Repsol announced that its exploratory well came up dry, and the company subsequently announced in late May that it would likely leave Cuba. Subsequently, in late May 2012, the Scarabeo-9 oil rig was used by the Malaysian company Petronas in cooperation with the Russian company Gazprom to explore for oil in a block off the coast of western Cuba. On August 6, 2012, however, Cuba announced that that the well was found not to be commercially viable because of its compact geological formation. In early September 2012, the Venezuelan oil company, PdVSA, announced that it had started exploring for oil off the coast of western Cuba, but on November 2, 2012, Cuba announced that the well was not commercially viable. In addition to these projects, Cuba has three additional offshore projects with foreign oil companies—PetroVietnam, Sonangol (Angola), and ONGC (India).

As a result of the three unsuccessful wells, the Scarabeo-9 oil rig left Cuba on November 14, 2012, reportedly headed to West Africa. Some oil experts maintain that it could be years before companies decide to return to drill again in Cuba's offshore deepwaters.[93] Most observers maintain that the failure to discover oil in the three wells drilled by the Scrrabeo-9 oil rig in 2012 is a significant setback for the Cuban government's efforts to develop its deepwater offshore hydrocarbon resources.[94]

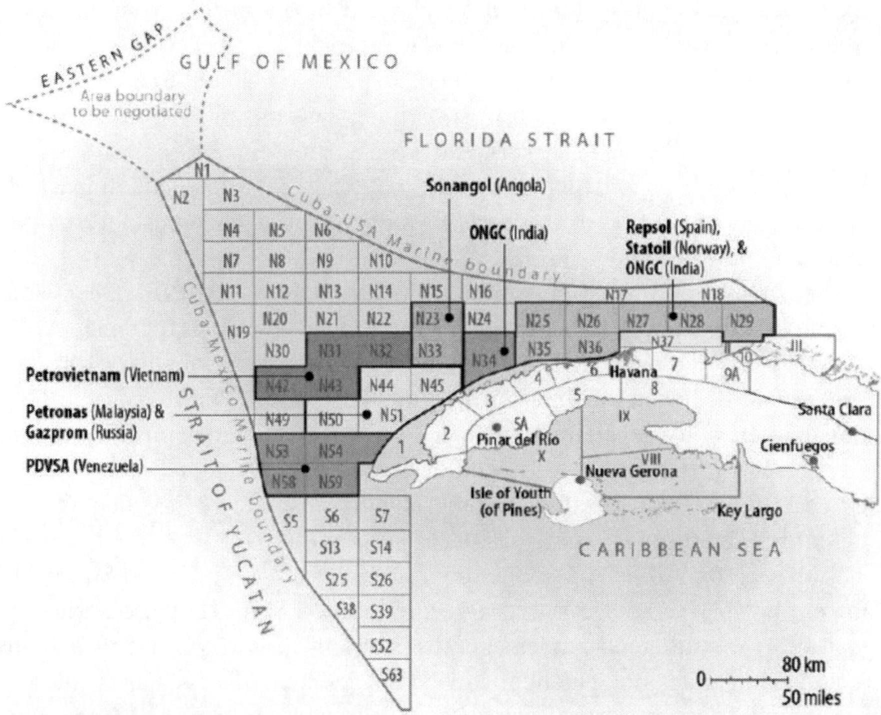

Source: Originally adapted by CRS from Jorge R. Piñon, presentation given at the Inter-American Dialogue, Washington, DC, October 8, 2010. Subsequently updated by CRS.

Notes: Petrobras (Brazil) signed an agreement for exploration of block N37 in October 2008, but announced its withdrawal in March 2011. The China National Petroleum Corporation (CNPC) had been reported in the past to be negotiating several offshore blocks, N19-N22 and N30.

Figure 7. Cuba's Offshore Oil Blocks.

It should be noted that a Russian company, Zarubezhneft, announced in December 2012 that it had begun drilling an exploratory well in a north coastal block (in shallow waters, not deepwater exploration) east of Havana off Cayo Coco, a Cuban tourist resort area. The company is using an oil rig known as the Songa Mercur operated by Songa Offshore, a Norwegian oil rig company. Drilling is expected to be completed in June 2013.[95]

In the aftermath of the Deepwater Horizon oil spill in the Gulf of Mexico, some Members of Congress and others expressed concern about Cuba's development of its deepwater petroleum reserves so close to the United States. They are concerned about oil spill risks and about the status of disaster preparedness and coordination with the United States in the event of an oil spill. Dealing with these challenges is made more difficult because of the long-standing poor state of relations between Cuba and the United States. If an oil spill did occur in the waters northwest of Cuba, currents in the Florida Straits could carry the oil to U.S. waters and coastal areas in Florida, although a number of factors would determine the potential environmental impact. If significant amounts of oil did reach U.S. waters, marine and coastal resources in southern Florida could be at risk.

The final report of the National Commission on the BP Deepwater Horizon Oil Spill and Offshore Drilling, issued in January 2011, maintained that since Mexico already drills in the Gulf of Mexico and Cuba has expressed an interest in deepwater drilling in the Gulf of Mexico, that it is in the U.S. national interest to negotiate with these countries to agree on a common, rigorous set of standards, a system of regulatory oversight, and operator adherence to an effective safety culture, along with protocols to cooperate on containment and response strategies in case of a spill.[96]

With regard to disaster response coordination, while the United States and Cuba are not parties to a bilateral agreement on oil spills, both countries are signatories to multilateral agreements that commit the two parties to prepare for and cooperate on potential oil spills. Under the auspices of the International Maritime Organization (IMO), the United States and Cuba have participated in several regional meetings (including Mexico in November 2011; Bahamas in December 2011; Curacao in February 2012; Jamaica in April 2012; and Mexico in August 2012) regarding oil spill prevention, preparedness, and response that have allowed information sharing among nations, including the United States and Cuba.

U.S. oil spill mitigation companies can be licensed by the Treasury and Commerce Departments to provide support and equipment in the event of an oil spill. One such example is a Florida-based company, Clean Caribbean & Americas, which has had licenses to be involved in Cuba since 2001. In addition, the U.S. Coast Guard has obtained licenses from Treasury and Commerce that allow it "to broadly engage in preparedness and response activities, and positions" the agency "to direct an immediate response in the event of a catastrophic oil spill."[97] Some energy and policy analysts, however, have called for the Administration to ease regulatory restrictions on private companies for the transfer of U.S. equipment and personnel to Cuba needed to prevent and combat a spill if it occurs.

Interest in Cuba's offshore oil development was strong in the 112th Congress, particularly over concerns about a potential oil spill, with three congressional hearings held and eight legislative initiatives introduced taking different approaches, none of which were enacted. The various policy approaches included: sanctioning foreign companies investing in or supporting Cuba's oil development; requiring the Secretary of the Interior to make recommendations on a joint contingency plan with Mexico, Cuba, and the Bahamas to ensure an adequate response to oil spills; and authorizing U.S. companies to engage in oil spill prevention and clean-up activities in Cuba's offshore oil sector as well as broader exploration and extraction activities.

LEGISLATIVE INITIATIVES IN THE 113TH CONGRESS

H.R. 214 (Serrano). Cuba Reconciliation Act. Would lift the trade embargo on Cuba. Introduced January 4, 2013; referred to the Committee on Foreign Affairs, and in addition to the Committees on Ways and Means, Energy and Commerce, Financial Services, the Judiciary, Oversight and Government Reform, and Agriculture.

H.R. 215 (Serrano). Baseball Diplomacy Act. Would waive certain prohibitions with respect to nationals of Cuba coming to the United States to play organized professional baseball. Introduced January 4, 2013; referred to the Committee on Foreign Affairs, and in addition to the Committee on the Judiciary.

H.R. 778 (Issa). Would modify a 1998 prohibition (Section 211 of Division A, Title II, P.L. 105- 277) by U.S. courts of certain rights relating to certain marks, trade names, or commercial names. The 1998 prohibition or sanction prevents trademark registrations and renewals from Cuban or foreign nations that were used in connection with a business or assets in Cuba that were confiscated, without the consent of the original owner. The bill would apply a fix so that the sanction would apply to all nationals and would bring the sanction into compliance with a 2002 World Trade Organization dispute settlement ruling. Introduced February 15, 2013; referred to the Committee on the Judiciary.

H.R. 871 (Range!). Would allow travel between the United States and Cuba. Introduced February 27, 2013; referred to the Committee on Foreign Affairs.

H.R. 872 (Range!). Would lift the trade embargo on Cuba. Introduced February 27, 2013; referred to the Committee on Foreign Affairs, and in addition to the Committees on Ways and Means, Energy and Commerce, the

Judiciary, Financial Services, Oversight and Government Reform, and Agriculture.

H.R. 873 (Range!). Would facilitate the export of U.S. agricultural products to Cuba, remove impediments to the export of medical devices and medicines to Cuba, allow travel to Cuba by U.S. legal residents, and establish an agricultural export promotion program with respect to Cuba. Introduced February 27, 2013; referred to the Committee on Foreign Affairs, and in addition to the Committees on Ways and Means, the Judiciary, Agriculture, and Financial Services.

H.Res. 121 (Hastings, FL). Would honor Cuban blogger Yoani Sánchez "for her ongoing efforts to challenge political, economic, and social oppression by the Castro regime." Introduced March 15, 2013; referred to the Committee on Foreign Affairs and in addition to the Committee on the Judiciary.

APPENDIX A. SELECTED EXECUTIVE BRANCH REPORTS AND WEB PAGES

U.S. Relations with Cuba, Fact Sheet,
State Department *Date:* June 21, 2012
Full Text: http://www.state.gov/r/pa/ei/bgn/2886.htm

Congressional Budget Justification for Foreign Operations FY2013, Annex: Regional Perspectives (pp. 768-769 of pdf), State Department
Date: April 3, 2012
Full Text: http://www.state.gov/documents/organization/185015.pdf

Country Reports on Human Rights Practices 2011, Cuba,
State Department
Date: May 24, 2012
Full Text: http://www.state.gov/j/drl/rls/hrrpt/humanrightsreport/index.htm?dynamic_load_id= 186505

Country Reports on Terrorism 2011
(State Sponsors of Terrorism chapter),
State Department *Date:* July 31, 2012
Full Text: http://www.state.gov/j/ct/rls/crt/2011/195547.htm

Cuba Country Page,
State Department *Full Text:* http://www.state.gov/p/wha/ci/cu/

Cuba Country Page, U.S. Agency for International Development
Full Text: http://www.usaid.gov/where-we-work/latin-american-and-caribbean/cuba/our-work

Cuba Sanctions, Treasury Department
Full Text: http://www.treasury.gov/resource-center/sanctions/Programs/pages/cuba.aspx

Cuba: What You Need to Know About U.S. Sanctions Against Cuba,
Treasury Department, Office of Foreign Assets Control
Date: January 24, 2012
Full Text: http://www.treasury.gov/resource-center/sanctions/Programs/Documents/cuba.pdf

International Religious Freedom Report, 2011, Cuba,
State Department *Date:* July 30, 2012
Full Text: http://www.state.gov/j/drl/rls/irf/religiousfreedom/index.htm?dlid=192965#wrapper

International Narcotics Control Strategy Report 2013, Vol. I, Cuba (pp. 138-140), State Department
Date: March 2013
Full Text: http://www.state.gov/documents/organization/204265.pdf

Trafficking in Persons Report 2012 (Cuba, pp. 133-134 of pdf),
State Department *Date:* June 19, 2012
Full Text: http://www.state.gov/documents/organization/192594.pdf

End Notes

[1] Portions of this background are drawn from U.S. Department of State, "Background Note: Cuba," April 28, 2011. For further background, see *Cuba, A Country Study*, ed. Rex A. Hudson, Federal Research Division, Library of Congress, (Washington, DC: U.S. Government Printing Office, 2002); "Country Profile: Cuba," Federal Research Division, Library of Congress, September 2006, available at http://lcweb2.loc.gov/frd/cs/profiles/Cuba.pdf; *Cuba, A Short History*, ed. Leslie Bethell (Cambridge University Press, 1993);

and Hugh Thomas, *Cuba: The Pursuit of Freedom*, (New York, Harper & Row, Publishers, 1971).

[2] U.S. Senator Orville Platt introduced an amendment to an army appropriation bill that was approved by both houses and enacted into law in 1901.

[3] For more on Cuba's political succession, see CRS Report RS22742, *Cuba's Political Succession: From Fidel to Raúl Castro*. For background discussion of potential Cuban political scenarios envisioned in the aftermath of Fidel Castro's stepping down from power in 2006, see CRS Report RL33622, *Cuba's Future Political Scenarios and U.S. Policy Approaches*.

[4] Daniel Dombey, Richard Lapper, and Andrew Ward, "A Family Business, Cuban-Americans Look Beyond the Havana Handover," Financial Times, February 27, 2008.

[5] Juan O. Tamayo, "Cuban Communists OK Term Limits for Party and Government Officials," *Miami Herald*, January 29, 2012, and "Cuba's Communists Meet to Update Party, Not Much Buzz on Street," *Miami Herald*, January 28, 2012; Patricia Grogg, "Cuba: Party Aims for Efficient, Inclusive Socialism," *Inter Press Service*, February 1, 2012.

[6] "Castro Dynasty Capped at 59 Years," Latin American Weekly Report, February 28, 2013; Damien Cave and Victoria Burnett, "As Castro Era Drifts to Close, A New Face Steps in at No. 2," *New York Times*, February 28, 2013; Marc Frank, "Castro Successor Lacks Charisma But Is Experienced Manager," *Reuters*, February 26, 2013.

[7] Comisión Cubana de Derechos Humanos y Reconciliación Nacional (CCDHRN), "Lista Parcial de Sancionados o Procesados por Motivos Políticos o Político-Sociales," January 22, 2013.

[8] Amnesty International, "Cuban Journalist Named Prisoner of Conscience," January 30, 2013; and "Cuban Activists Talk About Lack of Basic Freedoms, 10 Years on from Mass Crackdown," March 18, 2013.

[9] Amnesty International, "Cuba, Further Information: Prisoner of Conscience on Hunger Strike: Calixto Ramón Martínez Arias," March 14, 2013.

[10] CCDHRN, "Cuba: Algunos Actos de Represión Política en el Mes de Diciembre de 2012," January 3, 2013, available at http://www.procubalibre.org/informes/pdf/OVERVIEW_DICIEMBRE_2012.pdf.

[11] CCDHRN, "Cuba: Algunos Actos de Represión Política en el Mes de Febrero de 2013," March 4, 2013, available at http://www.cubanet.org/wp-content/uploads/2013/03/OVERVIEW_FEBRERO_ 2013.pdf.

[12] Amnesty International, *Routine Repression, Political Short-Term Detentions and Harassment in Cuba,* March 2012.

[13] The project collected thousands of signatures supporting a national plebiscite for political reform in accordance with a provision of the Cuban Constitution.

[14] Peter Finn, "Account of Cuban Crash Adds to Mystery," *Washington Post*, March 6, 2013; "I Only Wish They Were Nightmares, and Not Memories," (Editorial, Opinion) *Washington Post*, March 6, 2013.

[15] "Cuba, Aron Medig Reitera en una Reunión con la Hija de Oswaldo Payá que No Recuerda Nada del Accidente," *European Press*, March 13, 2013

[16] U.S. Department of State, Daily Press Briefing, March 28, 2013.

[17] Amnesty International, "Urgent Action, Cuban Man Targeted for Government Criticism," November 15, 2012, and "Cuban Activists Released," November 28, 2012.

[18] The term "SATS" is a Scandinavian term used in the theater to describe the moment of concentration just before the theatrical action takes place. Yoani Sanchez, "Who is Antonio Rodiles and What is 'State of SATS?," Estado de SATS, January 22, 2013, available at

http://www.estadodesats.com/2013/01/who-is-antonio-rodiles-and-what-is-state-of-sats-yoani-sanchez.html

[19] Amnesty International, "Threats and Assault on Activists," November 30, 2012.

[20] Committee to Protect Journalists, "After the Black Spring, Cuba's New Repression," July 6, 2011.

[21] Information and statistics were drawn from several sources: U.S. Department of State, "Background Note: Cuba," November 7, 2011; Economist Intelligence Unit, "Cuba Country Report," February 2013; Oficina Nacional de Estadísticas, "Anuario Estadístico de Cuba, 2011"; and Marc Frank, "Cuban Sugar Harvest Revs Up, But a Bit Behind Schedule," February 4, 2013.

[22] World Bank, World Development Report 2012, p. 402.

[23] "Cuba Country Report," *Economist Intelligence Unit (EIU)*, February 2013.

[24] Peter Orsi, "Economy Czar Says Cuba to Authorize More Areas of Small Businesses, Private Employment," *Associated Press*, December 13, 2012; "Cuba's Private Sector Continues to Expand Under Economic Reform," *Xinhua News Agency*, July 2, 2012.

[25] Marc Frank, "Cuba Cuts State Payroll, Private Sector Jobs Grow 23 Pct in 2012," *Reuters News*, December 27, 2012.

[26] *Lineamientos de la Política Económica y Social Del Partido y la Revolución*, VI Congreso del Partido Comunista de Cuba, approved April 18, 2011, available at http://thecubaneconomy.com/wp-content/uploads/2011/05/Lineamientos-de-la-Pol%C3%ADtica-econ%C3%B3mica-y-Social-del-Partido-y-la-Revoluci%C3%B3n-Aprobado-el-18-de-abril-de-2011.-VI-Congreso-del-PCC.pdf. In addition, the Cuban Communist Party published a report comparing the original 291 guidelines to the final 313 guidelines and how they changed. See *Información Sobre el Resultado del Debate de los Lineamientos de la Política Económica y Social Del Partido y la Revolución*, VI Congreso del Partido Comunista de Cuba, May 2011, available at http://www.cubadebate.cu/wp-content/uploads/2011/05/ tabloide_debate_lineamientos.pdf.

[27] Oscar Espinosa Chepe, *Cambios en Cuba: Pocos, Limitados y Tardios*, Havana Cuba, February 2011, available at http://reconciliacioncubana.files.wordpress.com/2011/03/cambios-en-cuba.pdf.

[28] U.S. Department of State, "Country Reports on Human Rights Practices for 2011, Cuba," May 24, 2012.

[29] Marc Frank, "Cuba Growing Less Food than 5 Ys Ago Despite Agriculture Reforms," *Reuters*, August 31, 2012 and "Cuba Sees Economy Growing 3.1 Pct in 2012, Below Forecast," *Reuters*, December 3, 2012.

[30] "Castro Declares He Had 'a Good Visit" with Russia's Medvedev," *Agence France Presse*, February 23, 2013.

[31] For further background, see section on "Cuba and the OAS" in archived CRS Report R40193, *Cuba: Issues for the 111th Congress*, by Mark P. Sullivan; also see CRS Report R42639, *Organization of American States: Background and Issues for Congress*, by Peter J. Meyer.

[32] U.N. OCHA, "Cuba, Plan of Action, Response to Needs Arising from Hurricane Sandy," November 2012, available at http://reliefweb.int/sites/reliefweb.int/files/resources/CUBA%20Action%20Plan%20FINAL.pdf.

[33] For additional background, see archived CRS Report RL30386, *Cuba-U.S. Relations: Chronology of Key Events 1959-1999*, by Mark P. Sullivan.

[34] U.S. Department of State, "Commission for Assistance to a Free Cuba," White House Fact Sheet, December 8, 2003.

[35] See the Commission's May 2004 report, available at http://pdf.usaid.gov/pdf_docs /PCAAB 192. pdf

[36] U.S. Department of State, Secretary of State Condoleezza Rice, "Statement on Cuba's Transition," February 24, 2008.

[37] White House, "President Bush Delivers Remarks on Cuba," March 7, 2008.

[38] U.S. Department of State, "The Path to Freedom Countering Repression and Strengthening Civil Society in Cuba," Testimony by Roberta S. Jacobson, Assistant Secretary, Bureau of Western Hemisphere Affairs, Senate Foreign Relations Committee, Subcommittee on the Western Hemisphere, June 7, 2012, available at: http://www.state.gov/p/wha/rls/rm /2012/191935.htm

[39] White House, "Remarks by the President at the Summit of the Americas Opening Ceremony," April 17, 2009.

[40] U.S. Department of State, "The Path to Freedom Countering Repression and Strengthening Civil Society in Cuba," Testimony by Roberta S. Jacobson, Assistant Secretary, Bureau of Western Hemisphere Affairs, Senate Foreign Relations Committee, Subcommittee on the Western Hemisphere, June 7, 2012.

[41] U.S. Department of State, "Scores of Cuban Democracy Activists Detained on the Eve of Human Rights Day," Press Statement, December 10, 2012.

[42] For additional information, see CRS Report RL31139, *Cuba: U.S. Restrictions on Travel and Remittances*, by Mark P. Sullivan.

[43] U.S. Department of the Treasury, OFAC, "Comprehensive Guidelines for License Applications to Engage in Travel-Related Transactions Involving Cuba," Revised May 10, 2012.

[44] Damien Cave, "Licensing Rules Slow Tours to Cuba," *New York Times*, September 16, 2012; Paul Haven, "U.S. Travel Outfits Say Rules for Legal Travel to Cuba Getting Tighter," *Associated Press*, September 13, 2012.

[45] Global Trade Atlas, derived by looking at reporting partners' exports to Cuba.

[46] Department of Commerce statistics, as presented by Global Trade Atlas.

[47] Juan Tamayo, "Big Drop in U.S. Agricultural Sales to Cuba," *Miami Herald*, July 29, 2010; Marc Frank, "U.S. Food Sales to Cuba Continued Decline in 2011," *Reuters News*, February 22, 2012; U.S.-Cuba Trade and Economic Council, Inc. "Economic Eye on Cuba," October 2012.

[48] USITC, *U.S. Agricultural Sales to Cuba: Certain Economic Effects of U.S. Restrictions,* USITC Publication 3932, July 2007, available at http://www.usitc.gov/ext_relations/news_ release/2007/er0719ee1.htm.

[49] USITC, *U.S. Agricultural Sales to Cuba: Certain Economic Effects of U.S. Restrictions, An Update,* Office of Industries Working Paper, by Jonathan R. Coleman, No. ID-22, June 2009, available at http://www.usitc.gov/publications/332/working_papers/ID-22.pdf

[50] U.S. Department of State, *Congressional Budget Justification for Foreign Operations, FY2013, Annex: Regional Perspectives*, April 3, 2012, p. 768.

[51] See CRS Report R42994, *The Budget Control Act, Sequestration, and the Foreign Affairs Budget: Background and Possible Impacts*, by Susan B. Epstein.

[52] See more on the projects available from NED's website at http://www.ned.org/publications /annual-reports/2011-annual-report/latin-america-and-the-caribbean/cuba.

[53] U.S. Government Accountability Office (GAO), U.S. Democracy Assistance for Cuba Needs Better Management and Oversight, GAO-07-147, November 2006.

[54] Oscar Corral, "Federal Program to Help Democracy in Cuba Falls Short of Mark," *Miami Herald*, November 14, 2006, and "Is U.S. Aid Reaching Castro Foes?" *Miami Herald*, November 15, 2006.

[55] U.S. GAO, *Foreign Assistance: Continued Efforts Needed to Strengthen USAID's Oversight of U.S. Democracy Assistance for Cuba*, GAO-09-165, November 2008.

[56] U.S. GAO, *Cuba Democracy Assistance, USAID's Program is Improved, but State Could Better Monitor Its Implementing Partners*, GAO-13-285, January 2013.

[57] U.S. Department of State, Daily Press Briefing, May 11, 2012, available at http://www.state.gov/r/pa/prs/dpb/2012/ 05/189753.htm.

[58] Development Alternatives Inc. "Updated Statement from DAI President and CEO Dr. James Boomgard Regarding Detainee in Cuba," January 7, 2010, available at http://dai.com/news-publications/news/updated-statement-daipresident-and-ceo-dr-james-boomgard-regarding-detainee

[59] "Cuba Says Imprisoned American in "Normal" Health," *Agence France Presse*, November 28, 2102.

[60] David Adams, "American Contractor Jailed in Cuba Sues U.S., Employer for $60 Million," *Reuters*, November 16, 2012.

[61] "Judy Gross Urges Action from Obama to Get Husband Released," *EFE News Service,* November 30, 2012.

[62] See the full text of the Broadcasting Board of Governors FY2013 budget request, available at http://www.bbg.gov/ wp-content/media/2012/02/FY-2013-BBG-Congressional-Budget-Request-FINAL-2-9-12-Small.pdf.

[63] See information on the BBG contract available, awarded September 23, 2011, by the BBG at https://www.fbo.gov/ index?s=opportunity&mode=form&id=98ec01c39584ddd3631450c952eb2eea&tab=core&_cview=1.

[64] Oscar Corral, "Radio, TV Martí Face Another Government Audit," *Miami Herald*, December 18, 2006, and "Problems Dog Broadcaster," *Miami Herald*, December 19, 2006.

[65] Jay Weaver, "TV Martí Executive Admits Taking Kickbacks," *Miami Herald*, February 14, 2007.

[66] See the following reports and audits from 1990 through 2008: U.S. General Accounting Office (GAO), *Broadcasts to Cuba, TV Marti Surveys are Flawed*, GAO/NSIAD-90-252, August 1990; U.S. GAO, *TV Marti, Costs and Compliance with Broadcast Standards and International Agreements*, GAO/NSIAD-92-199, May 1992; U.S. GAO, Letter to Hon. Howard L. Berman and Hon. John F. Kerry regarding Radio Marti broadcast standards, GAO/NSIAD-93-126R, February 17, 1993; Advisory Panel on Radio and TV Marti, *Report of the Advisory Panel on Radio and TV Marti*, Three Volumes, March 1994; U.S. GAO, *Radio Marti, Program Review Processes Need Strengthening*, GAO/NSIAD94-265, September 1994; U.S. GAO, *U.S. Information Agency, Issues Related to Reinvention Planning in the Office of Cuba Broadcasting*, GAO/NSIAD-96-110, May 1996; U.S. Department of State, Office of the Inspector General, *Review of Policies and Procedures for Ensuring that Radio Marti Broadcasts Adhere to Applicable Requirements*, 99- IB-010, June 1999; U.S. Department of State and the Broadcasting Board of Governors, Office of Inspector General, *Review of the Effectiveness and Implementation of Office of Cuba Broadcasting's New Program Initiatives*, Report No. IBO-A-03-01, January 2003, and *Report of Inspection, Office of Cuba Broadcasting,* Report No. ISP-IB 07-35, June 2007; and U.S. GAO, *Broadcasting to Cuba, Weaknesses in Contracting Practices Reduced Visibility into Selected Award Decisions*, GAO-08-764, July 2008.

[67] U.S. GAO, *Broadcasting to Cuba, Actions Are Needed to Improve Strategy and Operations*, GAO-09-127, January 2009.

[68] U.S. Congress, Senate Committee on Foreign Relations, *Cuba: Immediate Action Is Needed To Ensure the Survivability of Radio and TV Marti*, committee print, 111th Cong., 2nd sess., April 29, 2010, S.Prt. 111-46 (Washington: GPO, 2010).

[69] U.S. Government Accountability Office, *Broadcasting Board of Governors Should Provide Additional Information to Congress Regarding Broadcasting to Cuba,* December 13, 2011, available at http://www.gao.gov/assets/590/ 586869.pdf.

[70] William Booth, "U.S. Broadcaster Call Archbishop a Castro "Lackey," Washington Post, May 6, 2012. The editorial no longer appears on the website of Radio/TV Martí, but is available at http://porcubaparacuba.blogspot.com/2012/05/ editorial-de-radio-y-tv-marti-acerca.html.

[71] See a letter to Secretary of State Hillary Rodham Clinton, a member of the Broadcasting Board of Governors, from five Members of Congress, May 8, 2012, available at http://www.democracyinamericas.org/pdfs/ortegaletter.pdf.

[72] Authorization legislation establishing both Radio and TV Martí require broadcasting to Cuba to be in accordance with all Voice of America standards to ensure the broadcast of programs which are objective, accurate, balanced, and which present a variety of views (§3(b) of P.L. 98-111, as amended, and §243(b) of P.L. 101-246, as amended).

[73] William Booth, "U.S. Broadcaster Calls Archbishop a Castro "Lackey," *Washington Post*, May 6, 2012.

[74] For background information, see archived CRS Report RL32251, *Cuba and the State Sponsors of Terrorism List*, by Mark P. Sullivan, August 22, 2006.

[75] "Spain Requests Extradition of ETA Suspect from Cuba," *Agence France Presse*, October 14, 2011.

[76] See CRS Report R42982, *Peace Talks in Colombia*, by June S. Beittel.

[77] Bryan Bender, "Talk Grows of Taking Cuba Off Terror List," *Boston Globe*, February 21, 2013; U.S. Department of State, Daily Press Briefing, February 21, 2013.

[78] For additional background on migration issues through mid-2009, see CRS Report R40566, *Cuban Migration to the United States: Policy and Trends*, by Ruth Ellen Wasem.

[79] Alfonso Chardy and Juan Tamayo, "Exodus of Cubans Slowing," *Miami Herald*, October 6, 2010.

[80] U.S. Coast Guard, Alien Migrant Interdiction, Coast Guard Office of Law Enforcement, "Total Interdictions, Fiscal Year 1982 to Present," October 23, 2012.

[81] Alfonso Chardy and Juan O. Tamayo, "Illegal Cuban Migration, After Years of Decline, Is Up Again," Miami Herald, *Miami Herald*, October 8, 2011; Alfonso Chardy and Juan O. Tamayo, "Number of Cubans Trying to Enter U.S. Increases," *Miami Herald*, June 17, 2012.

[82] U.S. Department of Homeland Security, Office of Immigration Statistics, "Apprehensions by the U.S. Border Patrol: 2005-2010," July 2011; FY2011 statistics provided to CRS by U.S. Border Patrol.

[83] Statistics on arrivals of unauthorized Cubans at U.S. ports of entry provided to CRS from U.S. Customs and Border Protection, August 24, 2012. Also see CRS Report R40566, *Cuban Migration to the United States: Policy and Trends*, by Ruth Ellen Wasem.

[84] Diego Cevallos, "Migration: More and More Cubans Entering U.S. Through Mexico," *Inter Press Service News Agency*, June 17, 2008.

[85] Republic of Ecuador, Ministry of Foreign Relations, Commerce, and Integration, "Ecuador Requerirá Carta de Invitación par Ingreso de Ciudadanos Cubanos," Press Release, January 15, 2013.

[86] U.S. Department of State, "Daily Press Briefing," October 16, 2012.

87 U.S. Department of State, Assistant Secretary for Western Hemisphere Affairs Roberta S. Jacobson, "Preview of the Pathways for Prosperity/America's Competiveness Forum in Cali, Colombia, October 23-25," October 18, 2012.

88 David Adams and Tom Brown, "Cuban Perks Under Scrutiny in U.S. Immigration Reform," *Reuters News*, February 8, 2013; Stephen Johnson "Recommendations for the New Administration: Interests, Policies, and Challenges in the Americas," Center for Strategic and International Studies, November 21, 2012; and Philip Peters, "Migration Policy Reform: Cuba Gets Started, U.S. Should Follow," Lexington Institute, December 2012.

89 On March 12, 2002, Cuba's Ministry of Foreign Affairs and the Cuban Interests Section in Washington delivered three diplomatic notes to the U.S. Interests Section in Havana and the State Department in Washington proposing agreements on drug interdiction, terrorism, and migration issues. See "Statement from the Ministry of Foreign Affairs: Prominent Drug Trafficker Arrested in our Country," Information Office, Cuban Interests Section, March 17, 2002; "Cuba Offers to Sign Anti-Drug Pact," *Miami Herald,* April 8, 2006.

90 "Senate Caucus on International Narcotics Control Holds Hearing on Drug-Related Violence in the Caribbean and U.S. Security Assistance Through the Caribbean Basin Security Initiative," *CQ Congressional Transcripts*, February 1, 2012.

91 U.S. Congress, Senate United States Senate Caucus on International Narcotics Control, *Preventing a Security Crisis in the Caribbean*, 112th Cong., 2nd sess., September 2012, pp. 38-40, available at http://www.feinstein.senate.gov/ public/index.cfm/files/serve/?File_id= 90bb66bc-3371-4898-8415-fbfc31c0ed24.

92 U.S. Energy Information Administration, "Country Analysis Briefs: Caribbean," May 1, 2012.

93 Jeff Franks, "Drilling Rig Leave Cuba, Taking Oil Hopes With It," Reuters News, November 14, 2012, Peter Orsi, "Cuba Says 3rd Deep-Water Oil Well Sunk This Year Not Commercially Viable," *AP Newswire*, November 2, 2012.

94 "Cuba Offshore Oil Search Fails for a Third Time," *Agence France Presse*, November 2, 2012; "PdVSA Has Third Dry Well in Cuba Deepwater Exploration: Report," *Platts Commodity News*, November 2, 2012.

95 "Russia's Zarubezhneft Drills Exploration Well in Cuban Offshore Block L," *Platts Commodity News*, December 19, 2012.

96 National Commission on the BP Deepwater Horizon Oil Spill and Offshore Drilling, *Deepwater, The Gulf Oil Disaster and the Future of Offshore Drilling,* Report to the President, p. 254 and p. 300. See the full text of the report at http://www.oilspillcommission.gov/sites/default/files/documents/DEEPWATER_ReporttothePresident_FINAL.pdf.

97 U.S. Congress, House Committee on Transportation and Infrastructure, Subcommittee on Coast Guard and Maritime Transportation, *Offshore Drilling in Cuba and the Bahamas: The U.S. Coast Guard's Oil Spill Readiness and Response Planning*, 112th Cong., 2nd sess., January 30, 2012, 112-70 (Washington: GPO, 2012). Testimony of Rear Admiral William Baumgartner, Commander, Seventh District, and Rear Admiral Cari Thomas, Director of Policy Response, U.S. Coast Guard, available at http://republicans.transportation.house.gov /Media/file/TestimonyCGMT/2012-01-30- BaumgartnerThomas.pdf.

In: Cuba
Editor: Gustavo Nogueira

ISBN: 978-1-62618-970-6

Chapter 2

CUBA'S OFFSHORE OIL DEVELOPMENT: BACKGROUND AND U.S. POLICY CONSIDERATIONS*

Neelesh Nerurkar and Mark P. Sullivan

SUMMARY

Cuba is moving toward development of its offshore oil resources. While the country has proven oil reserves of just 0.1 billion barrels, the U.S. Geological Survey estimates that offshore reserves in the North Cuba Basin could contain an additional 4.6 billion barrels of undiscovered technically recoverable crude oil. The Spanish oil company Repsol, in a consortium with Norway's Statoil and India's Oil and Natural Gas Corporation, is expected to begin offshore exploratory drilling in early 2012, and several other companies are considering such drilling. At present, Cuba has six offshore projects with foreign oil companies. If oil is found, some experts estimate that it would take at least three to five years before production would begin.

In the aftermath of the Deepwater Horizon oil spill, some Members of Congress and others have expressed concern about Cuba's development of its deepwater petroleum reserves so close to the United States. They are concerned about oil spill risks and about the status of preparedness and

---

* This is an edited, reformatted and augmented version of the Congressional Research Service Publication, CRS Report for Congress R41522, dated November 28, 2011.

coordination in the event of an oil spill. Dealing with these challenges is made more difficult because of the long-standing poor state of relations between Cuba and the United States. If an oil spill did occur in the waters northwest of Cuba, currents in the Florida Straits could carry the oil to U.S. waters and coastal areas in Florida, although a number of factors would determine the potential environmental impact. If significant amounts of oil did reach U.S. waters, marine and coastal resources in southern Florida could be at risk.

The Obama Administration has been making efforts to prepare for a potential oil spill in Cuban waters that could affect the United States. This has included: updating oil spill area contingency plans covering Florida and developing a broader offshore drilling response plan; engaging with Repsol over its oil spill response plans (including plans to inspect the oil rig that Repsol will use); and licensing U.S. companies to provide personnel and export equipment needed for oil spill preparedness and response. Some energy and policy analysts have called for the Administration to ease regulatory restrictions on the transfer of U.S. equipment and personnel to Cuba for oil spill preparedness and response. Some have also called for direct U.S.-Cuban government cooperation to minimize potential oil spill damage, looking at U.S. cooperation with Mexico as a potential model as well as information sharing and cooperation through multilateral channels under the auspices of the International Maritime Organization. In contrast, some policy groups call for the United States to focus on preventing Cuba from engaging in offshore oil exploration altogether.

In the 112th Congress, five legislative initiatives have been introduced taking varying approaches toward Cuba's offshore oil development, and there have been two oversight hearings. H.R. 372 would authorize the Secretary of the Interior to deny oil leases and permits to those companies that engage in activities with the government of any foreign country subject to any U.S. government sanction or embargo. S. 405, among its provisions, would require the development of oil spill response plans for nondomestic oil spills in the Gulf of Mexico, including recommendations for a joint contingency plan with Mexico, Cuba, and the Bahamas.

H.R. 2047 would impose visa restrictions on foreign nationals and economic sanctions on companies that help facilitate the development of Cuba's offshore petroleum resources. S. 1836 and H.R. 3393 would provide that foreign offshore oil developers would be liable for damages from oil spills that enter U.S. waters.

INTRODUCTION

Long dependent on oil imports, Cuba has invited foreign companies to explore for and produce petroleum in its north offshore region, which could hold 4.6 billion barrels of technically recoverable oil according to U.S. Geological Survey (USGS) estimates.

One of those companies, Spain-based Repsol, is expected to start exploratory drilling in early 2012. A number of other companies, all government-owned national oil companies except for Repsol, are also considering exploratory offshore drilling in Cuban waters. Repsol's exploratory drilling in Cuban waters reportedly would occur approximately 55 to 60 miles south of Key West, Florida.[1]

Cuba's offshore development so close to the United States raises implications for U.S. policy focusing on oil spill risks and the status of U.S.-Cuban cooperation on preparedness and response in the case of a major oil spill. The Deepwater Horizon oil spill in the U.S. Gulf of Mexico heightened concerns about oil spill risks and raised the potential of U.S.-Cuban engagement regarding a potential oil spill in Cuban waters. However, the prospects for addressing these concerns are complicated by longstanding U.S. policy to isolate communist Cuba.

This report first examines Cuba's oil sector, including current production and consumption levels. It then looks at Cuba's offshore development, including the Repsol project, other offshore projects involving state-owed foreign oil companies, and the outlook for Cuba's offshore oil production.

The report then analyzes considerations for the United States raised by Cuba's offshore oil development, examining oil spill risks and environmental dangers if spilled oil reaches U.S. waters, the status of disaster coordination between the United States and Cuba, and potential approaches on the issue.

The report then examines the debate over broader U.S. involvement in Cuba's offshore oil development, and touches on two outstanding boundary issues related to Cuba's offshore oil development. Finally, the report examines legislative initiatives that have been advanced to deal with Cuba's offshore oil development.

CUBA'S OIL SECTOR

Current Situation

Cuba currently has proven oil reserves of 0.1 billion barrels and natural gas reserves of 2.5 trillion cubic feet.[2] These are located on shore or near shore, and were the focus of oil exploration and production until recently. The USGS estimates that the offshore North Cuba Basin could contain an additional 4.6 billion barrels of undiscovered technically recoverable crude oil resources, as well as 0.9 billion barrels of natural gas liquids and 9.8 trillion cubic feet of natural gas.[3,4] More than 70% of that oil may be in a portion of the North Cuba Basin stretching from about 70 miles west of the west end of the island for about 300 miles eastward in a narrow band known as the North Cuba Foreland Basin (see **Figure 1**). Separately, Cuban officials claimed in 2008 that Cuban offshore resources could be as much as 20 billion barrels of undiscovered crude, but in April 2011 Cuban officials lowered those estimates to 5 to 9 billion barrels.[5]

Cuba produced 51 thousand barrels of oil a day (Kb/d) in 2010 from the onshore or shallow, near shore fields. The output is mostly heavy, sour (sulfur-rich) crude that requires advanced refining capacity to process.[6] Cuba currently accesses offshore fields located near its northern coast through horizontal drilling from onshore rigs. Canadian companies Peberco and Sherritt developed near-shore assets from onshore block 7 (see **Figure 2**), but the Cuban government terminated that lease in 2009.

Cuba consumed 165 Kb/d of oil in 2009, down from 225 Kb/d two decades ago. Cuban domestic production increased and consumption fell after the Soviet Union curtailed its support for Cuba in the early 1990s. Most of Cuba's oil today is used for power generation, with relatively small amounts used for transportation. This implies net imports of 114 Kb/d. This comes from Venezuela, which has stepped into the former Soviet Union's role as a patron of the Cuban government. According to the official agreement between the two nations, Venezuela provides Cuba with oil at indexed prices and with long-term financing for up to 40% of oil imports at subsidized interest rates.[7] Cuba compensates Venezuela at least in part through offering medical and education services, including sending doctors to Venezuela.

According to the U.S. Energy Information Administration, Cuba currently has about 300 Kb/d of simple crude refining capability. However, not all of this is currently producing and Cuba has a limited amount of additional complex capacity to process the heavy sour crudes it produces. A significant amount of the oil going into power generation is burned directly as crude instead of as refined products, which can damage power plants. Of Cuba's imports, roughly 60% are refined products, mostly distillate and residual fuel oil. The rest is crude oil.[8]

Petroleos de Venezuela S.A. (PdVSA), Venezuela's state-owned national oil company (NOC), is helping *Unión Cuba Petróleo* (Cupet), Cuba's NOC, to expand and upgrade Cuba's refining capacity. Their Cuvenpetrol joint venture brought online the previously defunct Cienfuegos refinery in 2007, and they are pursing further expansion there with the assistance of the China National Petroleum Corporation (CNPC) and Chinese lenders.[9] Renovations at the Hermanos Diaz refinery and construction of a new refinery at the port of Matanzas are also planned. The upgrades may help Cuba process more of its own heavy crudes, which could be especially useful if production increases, as well as for processing crude imported from Venezuela.

(Three areas comprising the North Cuba Basin assessed by the USGS)

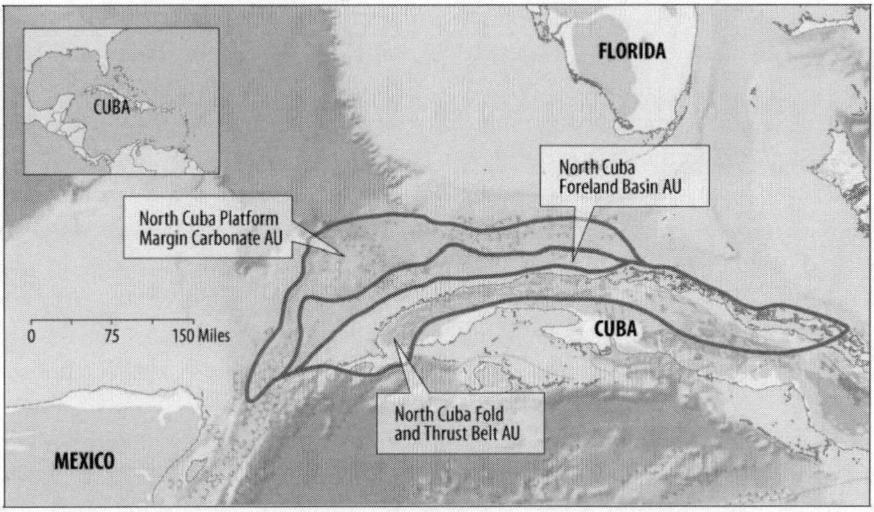

Source: U.S. Geological Survey, "Assessment of Undiscovered Oil and Gas Resources of the North Cuba Basin, 2004," (February 2005). http://walrus.wr.usgs.gov/infobank/programs/html/factsheets/pdfs/2005_3009.pdf. Adapted by CRS.

Note: "AU" = Assessment Unit.

Figure 1. North Cuba Basin.

Offshore Development

The Repsol Project

Repsol YPF, a publicly traded oil company based in Spain, is expected to begin drilling an offshore exploratory well in Cuba's exclusive economic zone (EEZ) in early 2012. The project, called the Jagüey prospect, reportedly is about 55 to 60 miles south of Key West, Florida, as noted above. This is not Repsol's first offshore exploration venture in Cuba. It drilled Cuba's only prior deepwater well, Yamagua-1, in 2004 in offshore block 27, roughly 20 miles northeast of Havana.[10] Repsol discovered petroleum resources, but deemed them commercially insufficient to justify producing.[11]

In its current project, Repsol leads a consortium which also includes Norway's NOC, Statoil, and India's NOC, the Oil and Natural Gas Corporation (ONGC).[12] Repsol has a 40% stake in the venture, with the other two partners each holding a 30% stake. The consortium has rights to six exploration blocks located off Cuba's northern shore (see **Figure 2**).

Repsol has collected seismic data and now awaits arrival of offshore oil rig Scarabeo-9, which it has contracted to carry out exploratory drilling from its owner, Italian oil services provider, Saipem.[13] Scarabeo-9 was built at a shipyard in Yantai, Shandong province, China. According to reports, the only major U.S. made component in the rig is the blowout preventer (which is the type of equipment that failed during the Deepwater Horizon oil spill).[14] The rig moved to Singapore, where its marine and drilling systems were completed. Originally expected to be completed in September 2009, Scarabeo-9 was delayed several times. The rig is now expected to arrive in Cuba in late December 2011 or early January 2012.[15] (As discussed below, U.S. officials from the Coast Guard and Department of the Interior will inspect the rig before it enters Cuban waters. See "Oil Spill Preparedness and Response" below.)

Repsol has committed to Cuban authorities to drill one exploratory well, and may add additional wells depending on its results.[16] Scarabeo-9 may drill additional wells for other companies with Cuban offshore exploration and production licenses, and Malaysia's NOC Petronas is reportedly next in line, according to Repsol officials. According to Cuban officials, there are plans for five wells to be drilled.[17]

Other Offshore Projects

Other foreign companies have five other lease agreements for offshore blocks in Cuba, and at least one more is being negotiated. Lease holders are conducting seismic surveys, and may be preparing for exploratory drilling. Apart from Repsol, the companies are all state-owned. Some of the NOCs' governments, including Brazil, Russia, and China, have recently made loans to Cuba to support development of infrastructure as well as energy, minerals, and agriculture sectors.[18]

Separate from its consortium with Repsol, ONGC contracted for two additional blocks in 2006 (see **Figure 2**). It may be preparing to move from seismic analysis to exploratory drilling as it has already started soliciting bids for necessary equipment.[19] Malaysia's NOC, Petronas, has partnered with Russian NOC Gazprom, in a contract on four blocks off the western coast of Cuba. (Gazprom and Petronas have also partnered to develop the Badra field in Iraq.[20]) They are studying seismic data and could begin drilling in 2012.[21] Vietnam's NOC, PetroVietnam, holds contracts for four offshore blocks west of Cuba.[22] PetroVietnam may partner with Russian NOC Zarubezhneft, which has separate contracts for onshore and near shore blocks. Venezuela's NOC, PdVSA, has a license to explore four western offshore blocks. Finally, Angola's NOC, Sonangol, signed an agreement to operate two offshore blocks in December 2010.[23]

Chinese NOC, CNPC, is in negotiations for Cuban offshore blocks.[24] Chinese companies have never previously drilled off Cuba's coast, though CNPC does provide some onshore drilling services. (Even Scarabeo-9, though it was built in China, is neither owned nor leased by a Chinese company.) As mentioned above, CNPC is also helping Cuba refurbish its Cienfuegos refinery.

Petrobras, Brazil's NOC, had signed an agreement in 2008 for offshore block N37, off Cuba's northern coast.[25] Based on seismic data it collected as well as other company priorities, Petrobras decided to relinquish its contract in March 2011. Company statements indicated that it would rather focus on oil prospects in Brazil.[26]

Outlook for Cuba's Offshore Production

Exploratory drilling by Repsol and others will provide more information on the potential for Cuban output. If oil is found, some experts estimate that companies would have to invest in developing production capacity for at least three to five years before production could begin.[27] However, production could be delayed due to a number of factors, such as the availability of offshore oil field development services. Development will take place at a slower rate than might otherwise be the case due to U.S. sanctions, which prohibit involvement from U.S. companies and prohibit use of equipment with more than 10% U.S.

content.[28] Once production starts, it will likely grow slowly over the course of years. For the foreseeable future, any incremental increase in Cuban production is likely to be small relative to the roughly 85 million barrel a day global oil market.

Some analysts have argued that Cuba could produce enough oil to become an oil exporter; however, this remains very speculative at this juncture. First, there remain uncertainties about when oil production could begin and at what rate it could be produced. Second, Cuba would need to offset the roughly 130 Kb/d of oil it currently imports to meet existing demand before becoming a net exporter. Third, current Cuban oil demand may grow, especially if the economy grows or the government loosens control over oil use as more domestic supply becomes available.

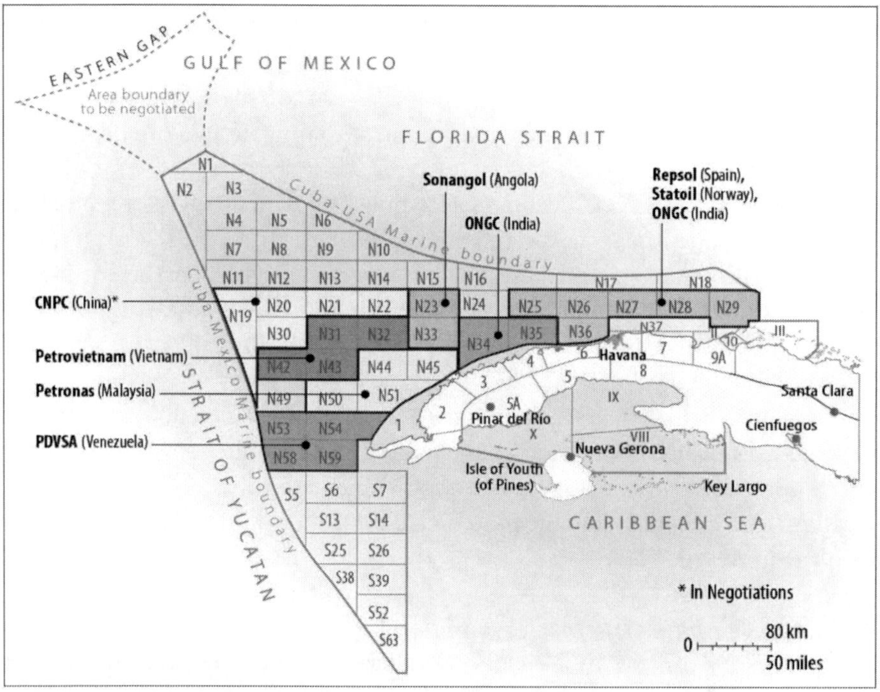

Source: Adapted by CRS from Jorge R. Piñon, Presentation given at the Inter-American Dialogue, Washington, DC, October 8, 2010.

Notes: Petronas took on Gazprom as a partner in its Cuba offshore project in November 2010. Petrobras (Brazil) signed an agreement for exploration of block N37 in October 2008, but announced its withdrawal in March 2011.

Figure 2. Cuba's Offshore Blocks.

Cuba is still likely to trade more oil—especially as refining capacity increases—but its net trade balance for oil may not necessarily shift to a significant oil export surplus. It depends on how much oil is found and developed and what happens to domestic Cuban demand. What is more certain is that lower net import needs may reduce Cuba's dependence on imports from Venezuela.

IMPLICATIONS AND CONSIDERATIONS FOR U.S. POLICY

Oil Spill Risks[29]

The Deepwater Horizon oil spill in the U.S. Gulf of Mexico heightened concerns over the potential of an oil spill in Cuban waters and the risk such a spill could affect Florida's waters and coastal areas.[30] As noted above, Repsol's current plans for drilling in Cuba fall within about 55 to 60 miles south of Key West, Florida. Were an oil spill to occur in these areas, it could have environmental impacts in the United States. Oil can be spilled from acute exploration and production accidents, through longer-term discharge from operations, or through transportation accidents, such as a tanker collision or pipeline rupture.

Risks of a Spill in Cuban Waters

The U.S. agency in charge of enforcing safety and environmental regulations on the U.S. Outer Continental Shelf, including oil spill response, is the Department of the Interior's Bureau of Safety and Environmental Enforcement (BSEE). In addition, several statutes, including the Clean Water Act and the Oil Pollution Act, establish a liability regime for oil spills. Offshore exploration and production operations in non-U.S. waters may not be governed by analogous regulations or fall under a liability structure that creates an incentive to minimize oil spills. Since the Repsol project is only the second deepwater well to be drilled in Cuba's EEZ, Cuban officials are in the process of developing and implementing up-to-date regulations to prevent offshore drilling accidents and contingency plans to address accidents if they do occur.[31] They have pledged to follow the highest international environmental and safety standards, and have expressed a strong willingness to cooperate with the United States and other countries on safety measures.[32] However, as the recent U.S. experience in the Gulf of Mexico illustrates, even the long-time existence of regulations and regulator may not always prevent an oil spill.

According to a 2008 American Petroleum Institute study of U.S. offshore oil spills, the largest cause of spilled oil is loss of well control or "blowouts" at offshore platforms.[33] Currently, only exploration wells are planned in Cuba. Their results will be analyzed before production wells and transportation infrastructure is considered. However, there have been major oil spills from exploratory wells in the past. Two of the largest accidental oil spills in world history resulted from blowouts at exploratory wells in the Gulf of Mexico – the Deepwater Horizon oil spill in the U.S. Gulf of Mexico and the 1979 Ixtoc oil spill in Mexico's section of the Gulf of Mexico.

It is difficult to assess the likelihood of a spill. According to Saipem, Scarabeo-9 is built to Norwegian standards, including extra equipment to shut off blown-out wells beyond what is required in the United States.[34] Repsol has significant offshore experience, including projects in the U.S. Gulf of Mexico. It has had issues with oil spills, which is not abnormal for an oil company.[35] (See "Oil Spill Preparedness and Response" below.) Among other Cuban lease holders, Statoil has extensive offshore experience, including projects in the U.S. Gulf of Mexico, and are generally seen as accomplished offshore operators. Petronas, ONGC, and PetroVietnam also have offshore experience. PdVSA does not, but its offshore project appears the furthest from seeing drilling activity among existing licenses. Cuban officials claim they are taking necessary regulatory precautions, including incorporating safety practices from the United Kingdom and the United States.[36]

Risks that Oil Spilled in Cuban Waters Reaches the United States

If an oil spill were to occur in the waters northwest of Cuba, currents in the Florida Straits could carry that oil to U.S. waters and coastal areas in southern and south eastern Florida.[37] However, any environmental impact to Florida would depend on many factors at the time of a spill, including size and location of the oil spill, ocean conditions in the area, prevailing wind direction and velocity, temperature of the water and the air, the type of oil spilled, and effectiveness of any cleanup efforts. The wide variety of factors render impossible a precise description of the environmental impact were an oil spill to occur in Cuban waters.

Even if prevailing winds and current conditions favored rapid transport of spilled oil to the Florida coastline, other factors would also affect the rate of spill dispersal and, in part, determine how much of the spill reached the U.S. coast. The physical and chemical characteristics of an oil spill change over time, a process known as "weathering." How much weathering takes place after a spill occurs would affect the nature of the oil and the degree of impact. How fast oil spreads depends on volume spilled and the viscosity of the oil.[38] As the spill spreads out, the lighter and more volatile components of the oil would evaporate at a rate that depends on water and air temperature, as well as wind speed and wave action.[39] Over time, and depending on waves and turbulence at the sea surface, the spill would start to break up, or disperse. Other factors, such as oxidation, biodegradation, interaction with sediments, all contribute to the changing character of an oil spill over time and during its transport by ocean currents and winds.[40]

Finally, the extent of any cleanup activities will influence how much of the spill persists in the environment. In general, the faster and more expansive the cleanup effort, the more likely it may limit damage to the environment. (See "Oil Spill Preparedness and Response" below for a discussion of U.S. policy related to preparedness and response in the event of an oil spill.)

Assets at Risk If Spilled Oil Reaches U.S. Waters

If significant quantities of oil did reach U.S. waters, risks to the marine and coastal resources of Southern Florida could be of particular concern. The coastal and ocean resources of the region provide recreational, commercial, and ecological benefits to both local communities and the nation.

One of the more vulnerable areas that could be at risk is the Florida Keys and adjacent areas. The Florida Keys National Marine Sanctuary includes state and national parks, wildlife refuges, ecological reserves, research areas, and sanctuary preservation areas. North of the Florida Keys are the Everglades and Biscayne National Parks. As one moves up Florida's east coast, barrier beaches backed by lagoons and wetlands dominate the geography. And then there are the densely populated areas of Miami-Dade, Broward, and Palm Beach Counties.

The Florida Keys and adjacent areas comprise diverse and interrelated marine systems. The Florida reef is the most extensive living coral reef in North American waters, stretching for 325 miles. Reefs, sea grass beds and mangroves in the region provide habitats for many marine animals, including a number of threatened and endangered species. These coral reefs and related coastal ecosystems are valuable because they provide protection from erosion and flooding, especially from severe storms such as hurricanes.

Depending on timing, size, and location, an oil spill can cause significant harm to individual organisms and entire populations in marine and coastal habitats.[41] Spills can cause impacts over a range of time scales, from days to years, or even decades for certain spills. Acute exposure to an oil spill can kill organisms or have non-lethal but debilitating affects on organism development, feeding, reproduction, or disease immunity. Ecosystems in which they exist can also be harmed.[42] Certain habitats in the area—such as coral reefs, mangrove swamps, and salt marshes—are especially vulnerable.[43] Long-term, chronic exposure, as occurs from continuous oil releases such as leaking pipelines, offshore production discharges, and non-point sources (e.g., urban runoff) can see impacts spread from sea life to the survival and reproductive success of marine birds and mammals.[44]

Southern Florida's natural resources are closely integrated with its economic interests. Southern Florida supports significant tourism as well as commercial and recreational fishing. Florida's tourism industry directly employs more than a million people. The 84 million tourists that visited Florida in 2008 spent around $65 billion.[45] The Deepwater Horizon spill illustrated that an oil spill can significantly harm the tourism industry of affected areas. A well-publicized oil spill can even weaken tourism in a nearby area, regardless of the actual threat to human health created by the spill.

Oil Spill Preparedness and Response

In light of oil spill concerns, there has been increased congressional and public interest on the status of oil spill preparedness and response and coordination between Cuba and the United States. A number of analysts and policy groups have been encouraging U.S.-Cuban engagement on the issue,[46] while some policy groups maintain that the United States should focus on preventing Cuba from engaging in offshore oil exploration altogether.[47]

The Obama Administration has been making efforts to prepare for a potential oil spill in Cuban waters that could affect the United States. The U.S. Coast Guard has been working with state, local, and other federal agencies to ensure that area contingency plans covering Florida are adequate. The National Oceanic and Atmospheric Administration (NOAA) in cooperation with the Department of the Interior's Bureau of Ocean Energy Management (BOEM) has run trajectory models in order to identify potential landfall areas along the U.S. coasts, information that is being used to enhance the area contingency plans.[48] Since March 2011, the Coast Guard's Seventh District in Miami has been working to develop an International Offshore Drilling Response Plan. As part of this effort, the Coast Guard hosted an inter-agency table top exercise in Miami on November 17, 2011, responding to a fictitious international spill off the coast of Florida.[49]

U.S. agencies have also engaged with officials from Repsol, which has provided information regarding its plans related to drilling and oil spill response. The company has offered U.S. agencies an opportunity to inspect the Scarabeo-9 oil rig. Both the Coast Guard and the Department of the Interior's Bureau of Safety and Environmental Enforcement (BSEE) are planning to inspect the rig before it enters Cuban waters. According to U.S. officials, Repsol maintains that it will adhere to U.S. regulations and the highest industry standards when conducting its exploratory drilling in Cuban waters.[50]

Currently the United States and Cuba are not parties to a bilateral agreement on oil spills. In the aftermath of the Deepwater Horizon spill, however, U.S. officials in Havana kept the Cuban government informed about the oil spill in working-level discussions. With Cuba's movement toward developing its offshore oil resources so close to the United States, some analysts have called for more institutionalized or formal U.S.-Cuban cooperation and planning to minimize potential damage from an oil spill. Given the comprehensive U.S. economic sanctions on Cuba, some analysts have called for the Administration to amend or rescind regulations that restrict the transfer of equipment, technology, and personnel that would be needed to combat an oil spill in Cuba.[51] Some energy analysts assert that foreign oil companies operating in Cuba need to have full access to technology and personnel in order to prevent or manage a spill.[52] Some maintain that the U.S. embargo has forced drillers to use second-hand equipment to avoid buying from U.S. companies.[53]

U.S. oil spill mitigation service companies can be licensed through the Treasury Department's Office of Foreign Assets Control (OFAC) and the Department of Commerce's Bureau of Industry and Security (BIS) to provide oil spill prevention and containment support to companies operating in Cuba. According to the Department of State, the United States expects any foreign oil company engaged in oil exploration activities in Cuba to have adequate safeguards in place to prevent oil spills and contingency plans to address a spill should it happen.[54]

Since 2001, a Florida-based company, Clean Caribbean & Americas (CCA), has received U.S. licenses to send technical advisers and trainers to assist foreign oil companies in Cuba to prepare to respond to a large oil spill. The actual material and equipment is stored in Fort Lauderdale and would be sent to Cuba by air and sea in the event of a major oil spill.[55] For a Tier 1 oil spill, one that is small and localized, foreign oil companies drilling offshore in

Cuba would maintain their own capabilities and equipment. For a Tier 2 oil spill, involving larger quantities of oil that could spread beyond the immediate vicinity where the spill took place, near shore oil operators and the Cuban government would supply equipment to help respond to the spill. A much larger Tier 3 oil spill, like a major tanker accident or an offshore well blowout, would require international assistance, like that provided by Clean Caribbean & Americas, which would move equipment into Cuba.[56] This type of oil spill response mechanism for large Tier 3 spills is a typical arrangement that has developed internationally over the past 30 years. CCA's President Paul Schuler maintains that involvement of Cuban and U.S. agencies in drills and exercises would enhance preparedness and response to a potential oil spill in Cuba.[57]

In October 2011 congressional testimony, Schuler stated that his company was jointly developing Repsol's oil spill contingency plan with a UK-based company, Oil Spill Response Ltd (ORSL), which has a large stockpile of air mobile equipment and has no restrictions on sending equipment and personnel to Cuba. According to Schuler, the Scarabeo-9 will have Tier 1 equipment on site to provide initial rapid response, while ORSL is sending equipment to Cuba to be place at the port of Mariel for Tier 2 spill reinforcement. In the event of a Tier 3 spill , both CCA and OSRL would mobilize resources if needed. However, if a spill similar to the Deepwater Horizon spill occurred, Schuler testified that he would expect the mobilization of significant other resources. He encouraged loosening up the licensing process so more U.S. companies and resources could be made available if needed.[58] Schuler also maintained during questioning that CCA was the only U.S. company at this juncture to have a license to export oil spill response equipment to Cuba, while two others are licensed to provide management and training services.

In May 2010, OFAC approved a license for the Texas-based International Association of Drilling Contractors (IADC) to travel to Cuba to discuss safety and mitigation of environmental hazards with Cuban authorities. After the meeting in August 2010, IADC President Lee Hunt maintained that the Cubans are eager to work with U.S. industry to ensure safer drilling.[59] In May 2011, IADC sponsored a conference in Trinidad and Tobago on the topic of improving oil industry environmental practices that featured a panel on Cuba's offshore drilling at the May 12-13, 2011, conference.[60]

In early November 2011, Wild Well Control (a Houston-based company) reportedly was granted a U.S. license to provide engineers and other specialists to companies experiencing a blowout in Cuba waters. The company is also seeking a license to allow it to provide firefighting equipment and well capping technology.[61]

In recent congressional hearings, BSEE Director Michael Bromwich maintained that the Department of Commerce is reviewing several applications for post-incident oil spill containment and cleanup as well as applications for a subsea well containment system and related equipment, such as submersible vehicles and subsea construction, dive support, and well intervention vehicles to prepare for and to operate in the event of a spill.[62] According to press reports in October 2011, the Houston-based Helix Energy Solutions Group has applied for U.S. licenses to export a capping stack into Cuban waters in the even of a massive oil spill.[63]

U.S.-Mexico Cooperation as a Potential Model

U.S. cooperation with the Mexican government on oil spills could serve as a potential model for U.S.-Cuban government engagement on disaster preparedness and coordination. The United States and Mexico negotiated a cooperation agreement in 1980 regarding pollution caused by oil and other hazardous substances. The agreement called for the two countries to establish a joint contingency plan in order to ensure an adequate response to spills.[64] The joint plan that was developed – known as Mexus Plan – sets forth standard operating procedures in case of pollution incidents that threaten the coastal waters or marine environment of the border zone of both countries. The plan lays out the organization of the response teams for each country, including the federal and state agencies involved. It provides for joint response teams to be formed and activated when needed, and provides for coordination, planning, and logistics of the joint response. The U.S. response team is coordinated by the Coast Guard's Assistant Commandant for Marine Safety and Environmental Protection.[65]

Following the model of U.S.-Mexican cooperation on oil spills could ensure optimal bilateral engagement with Cuba on oil spill contingency planning. Such a model would likely first entail the negotiation of a cooperation agreement on oil spills followed by the development of a joint contingency plan. Even before an agreement and plan are in place, initial discussions and dialogue on the issue could increase preparedness in the case of a spill. Once the agreement and joint plan are in place, regular meetings and periodic exercises could provide for the maintenance of the joint contingency plan.

As with U.S.-Mexican cooperation, the Coast Guard would likely play a leading coordinating role. Such Coast Guard cooperation with Cuba on oil spill preparedness and response would likely be made easier because of the Coast Guard's existing cooperation with Cuba on migration and drug trafficking issues.[66]

The final report of the National Commission on the BP Deepwater Horizon Oil Spill and Offshore Drilling, issued in January 2011, maintained that since Mexico already drills in the Gulf of Mexico and Cuba has expressed an interest in deepwater drilling in the Gulf of Mexico, that it is in the U.S. national interest to negotiate with these countries to agree on a common, rigorous set of standards, a system of regulatory oversight, and operator adherence to an effective safety culture, along with protocols to cooperate on containment and response strategies in case of a spill.[67] Mexican officials have also called for discussions between the three countries.[68]

Some energy analysts have also argued that the Bahamas should also be included in any movement in cooperation on oil spill response preparedness between Cuba and the United States since that country also is looking to eventually develop its deepwater oil and natural gas potential and because of the close location of many Bahamian islands to Cuba and the United States.[69]

As noted below, legislation has been introduced in the 112th Congress, S. 405 (Nelson), that, among its provisions, would require the Secretary of the Interior to work toward the development and implementation of oil spill response plans for spills in the eastern Gulf of Mexico. This would require recommendations on a joint contingency plan with Mexico, Cuba, and the Bahamas.

Cooperation through Multilateral Channels

Both Cuba and the United States are signatories to multilateral agreements that commit the two parties to prepare for and cooperate on potential oil spills. This includes the International Convention on Oil Pollution Preparedness, Response, and Cooperation (OPRC), which was adopted under the auspices of the International Maritime Organization (IMO) in 1990 and entered into force in 1995. The convention was adopted in response to a U.S. environmental initiative in the aftermath of the 1989 Exxon Valdez oil spill. Under the convention, parties are required to establish measures for dealing with pollution incidents, either nationally or in co-operation with other countries.[70] The IMO is given a central role under the convention in providing information services, education and training, and technical services and assistance.

Both Cuba and the United States are also parties to the Convention for the Protection and Development of the Marine Environment of the Wider Caribbean Region, known as the Cartagena Convention, which was adopted in

1983 and entered into force in 1986. The agreement includes a Protocol Concerning Co-operation in Combating Oil Spills in the Wider Caribbean Region. The protocol calls for an exchange of information among the signatories regarding contacts, laws, regulations, institutions, and operational procedures relating to the prevention of oil spill incidents and to the means of reducing and combating the harmful effects of oil spills. It also states that parties to the agreement should conclude appropriate bilateral or multilateral subregional arrangements as necessary to facilitate implementation. It obligates each party to assist other parties in response to an oil spill incident according to these arrangements.[71]

Short of direct U.S.-Cuban bilateral engagement on oil spill preparedness and coordination, these two multilateral agreements could provide a mechanism for some U.S.-Cuban cooperation on oil spills. For example, in order to implement the Cartagena Agreement's protocol on oil spill cooperation in the Caribbean, the IMO maintains a regional activity center in Curaçao, Netherlands Antilles, known as the Regional Marine Pollution Emergency Information and Training Center for the Wider Caribbean (RAC/REMPEITC-Caribe). The Center's objective is to strengthen the operational effectiveness of the Cartagena Agreement and OPRC through the provision of technical services, training activities, information sharing, and exercises.[72] The United States and Cuba could work through the IMO and its regional center in Curacao to engage on oil spill preparedness and coordination.

As noted above, the IMO sent a technical mission to Cuba in June 2010 to evaluate the Cuba's preparedness to respond to the Deepwater Horizon oil spill. The mission made several recommendations for Cuba to improve its national contingency plan to respond to oil spills, including the development of a training plan and increased cooperation with the IMO's regional training center in Curaçao (such as attending meeting, participating in projects, and receiving IMO assistance through this regional institution).[73]

In December 2011, U.S. officials will be participating in a regional seminar in the Bahamas sponsored by the IMO that focuses on pollution related to offshore oil exploration. Officials from Cuba, the Bahamas, Mexico, and Jamaica have been invited to participate in the seminar entitled "Regional OPRC Seminar on National Plans for Marine Pollution Preparedness and Response Related to Offshore Units and Regional Cooperation" According to BSEE Director Bromwich, "the seminar will provide a valuable opportunity for participating countries to lean about other nations' plans for emergency well control and oil spill response." Bromwich also maintained that such a multilateral approach is the most effective means of safeguarding U.S. interests and that the Administration intends to continue to pursue such multilateral engagements in the Gulf of Mexico.[74]

Debate over U.S. Investment in Cuba's Energy Sector

Since the United States imposed comprehensive economic sanctions on Cuba in the early 1960s, most financial transactions with Cuba have been prohibited, including U.S. investment in Cuba's offshore energy sector. The Cuban Assets Control Regulations (CACR, found at 31 CFR 515), first issued by the Treasury Department in 1963, lay out a comprehensive set of economic sanctions against Cuba, including a prohibition on most financial transactions. The CACR have been amended many times over the years to reflect changes in policy and remain in force today. The Cuban Liberty and Democratic Solidarity Act of 1996 (P.L. 104-114), enacted in the aftermath of Cuba's shooting down of two U.S. civilian planes in February 1996, codified the Cuban embargo, including all the restrictions under the CACR. The codification is especially significant because of its long-lasting effect on U.S. policy toward Cuba. The executive branch is prohibited from lifting the economic embargo until certain democratic conditions are met. The CACR still provides the executive branch with the ability to modify the embargo restrictions, but the President cannot suspend or completely terminate the Cuban embargo regulations without first determining that a transition government or democratically-elected government is in power in Cuba.[75]

Some U.S. business and policy groups have called on Congress and the Administration to allow U.S. oil companies to become involved in Cuba's offshore oil development. Several legislative initiatives were introduced in the 111th Congress (S. 774, H.R. 1918, and S. 1517) that would have specifically authorized such activities and amended U.S. law to allow for travel for such activities (see "Legislative Initiatives" below). A major business argument in favor of U.S. involvement in Cuba's offshore energy sector is that U.S. failure to enter into the Cuban market completely hands over potential investment opportunities to foreign competitors.[76] As mentioned above, national oil companies from Russia, China, Venezuela, and elsewhere have been investing in Cuba's energy industry. In a 2009 report, the Brookings Institution offered several additional reasons for U.S. involvement in Cuba's offshore development. The report maintains: that it would help reduce Cuba's dependence on Venezuela for its oil imports; that it would increase U.S.

influence in Cuba if U.S. companies had a significant presence in the county; that U.S. companies have the expertise to develop Cuba's offshore oil and gas in a safe and responsible manner; and that it is preferable to have U.S. companies involved because they have higher standards of transparency than some foreign oil companies.[77]

On the opposite side of the policy debate, a number of policy groups and Members of Congress oppose engagement with Cuba, including U.S. investment in Cuba's offshore energy development. A legislative initiative introduced in the 111th Congress, H.R. 5620, would have gone further by imposing visa restrictions and economic sanctions on foreign companies and their executives who help facilitate the development of Cuba's petroleum resources. The bill asserted that offshore drilling by or under the authorization of the Cuban government poses a "serious economic and environmental threat to the United States" because of the damage that an oil spill could cause. Opponents of U.S. support for Cuba's offshore oil development also argue that such involvement would provide an economic lifeline to the Cuban government and thus prolong the continuation of the communist regime. They maintain that if Cuba reaped substantial economic benefits from offshore oil development, it could reduce societal pressure on Cuba to enact market-oriented economic reforms. Some who oppose U.S. involvement in Cuba's energy development contend that while Cuba might have substantial amounts of oil offshore, it will take years to develop. They maintain that the Cuban government is using the enticement of potential oil profits to break down the U.S. economic embargo on Cuba.[78]

Boundary Issues

There are two boundary issues related to Cuba's development of its offshore hydrocarbon resources. The first involves a 1977 bilateral agreement that delineated a maritime boundary between Cuba and the United States in the Straits of Florida and eastern Gulf of Mexico. The second involves an undelineated section of the Gulf of Mexico known as the eastern gap with claims by the United States, Mexico, and Cuba. (See **Figure 2**, which shows both the maritime boundary between the United States and Cuba and the eastern gap area.)

When the United States and Cuba negotiated the 1977 maritime boundary agreement, U.S. policymakers viewed it as important to avoid maritime enforcement problems and to establish an agreed limit for fisheries and continental shelf activities (such as exploitation of hydrocarbon resources). Both countries, which have opposing coasts ranging from between 77 and 90 miles apart, agreed to the provisional application of the agreement pending permanent entry into force following the exchange of instruments of ratification. While the boundary agreement was submitted to the U.S. Senate in January 1979 for its advice and consent to ratification, and the Senate Foreign Relations Committee subsequently reported the treaty favorably in August 1980, the Senate has not ratified it. According to the Department of State, final action has been deferred because of the political relations between Cuba and the United States, not because of any stated objection to the boundary.[79] Nevertheless, Cuba and the United States have exchanged diplomatic notes every two years extending the provisional application of the agreement for a two-year period. The most recent exchange of notes occurred May 20, 2010, with an effective date of January 5, 2010. As noted in State Department testimony to the Senate Foreign Relations Committee in June 1980, the provisional application of the agreement falls under the President's authority to establish boundaries, pending the full Senate's consideration of the treaty.[80] The treaty itself, in Article V, included a provision stating the parties agreed to apply the terms of the agreement provisionally, and according to the Department of State, this "constituted an executive agreement within the body of the treaty."[81]

Some Members of Congress have called on the Administration to rescind the provisional application of the 1977 boundary agreement with the view that it would likely curtail Cuba's offshore oil development. U.S. withdrawal from the agreement, however, would have no practical effect on Cuba's offshore oil development. According to then-National Security Adviser James Jones in late September 2010, withdrawal from the agreement would have no discernable effect on the Cuban government and could create further boundary claim disputes for the United States.[82]

The eastern gap—an undelineated area of the Gulf of Mexico beyond the 200-mile exclusive economic zones of Cuba, Mexico, and the United States—could potentially hold large amounts of oil, although to date there is little hard data to confirm this. The demarcation of the area is open for negotiations among the three countries, but will likely await an improvement in relations between Cuba and the United States.[83] A potential model for these negotiations is a treaty signed in 2000 between the United States and Mexico for a western gap in the Gulf of Mexico.[84] Negotiations involving three countries, however, would likely be more complicated than a single bilateral agreement with Mexico. In May 2009, Cuba made a submission to the U.N. Commission on the Limits of the Continental Shelf (CLCS) regarding the eastern gap, but all three states— Cuba, Mexico, and the United States—maintained that the submission did not prejudice the final delimitation of the outer continental shelf agreed to by these states.[85]

LEGISLATIVE INITIATIVES AND OVERSIGHT

Legislative initiatives in the 111th Congress, none of which received consideration, focused on two approaches toward Cuba's offshore oil development. The first approach would have allowed for U.S. investment in Cuba's offshore energy development, while the second approach would have imposed sanctions on individuals and foreign companies that helped the development of Cuba's offshore petroleum resources.

In the 112th Congress, the five legislative initiatives introduced to date also different approaches to Cuba's offshore oil development, but would not include U.S. investment in Cuba's offshore energy development. One approach, as reflected by H.R. 372 and H.R. 2047, would allow for the sanctioning of companies involved in Cuba's offshore oil development if the companies also wanted to conduct hydrocarbon operations in U.S. offshore waters. Another approach, reflected by S. 405, would impose requirements on companies conducting hydrocarbon operations off the coast of Cuba if the companies also wanted leases for oil and gas development in U.S. waters, and would also require the development and implementation of oil spill response plans for nondomestic oil spills in the Gulf of Mexico, including a joint contingency plan with Mexico, Cuba, and the Bahamas. A third approach, reflected by S. 1836 and H.R. 3393, would provide that Americans affected by oil spills that originate in foreign waters could hold those responsible accountable and seek compensation for damages.

111th Congress

In the 111th Congress, legislative initiatives reflected two contrasting policy approaches toward Cuba's development of its offshore oil reserves. One approach would have allowed for U.S. involvement in Cuba's offshore oil sector, while the other approach would have imposed sanctions on foreign companies and individuals who assisted the development of Cuba's petroleum resources.

Reflecting the first approach, S. 774 (Dorgan), H.R. 1918 (Flake), and S. 1517 (Murkowski) would have authorized U.S. companies to work with Cuba for the exploration and extraction of oil, and to export without license all necessary equipment to Cuba. The bills would have amended the Trade Sanctions Reform and Export Enhancement Act of 2000 or TSRA (P.L. 106-387, Title IX) to provide for a general license for travel by persons engaging in hydrocarbon exploration and extraction activities. H.R. 1918 would have gone further and allowed for the importation of hydrocarbon resources from Cuba. In addition to these initiatives that specifically would have authorized involvement in Cuba's offshore energy sector, several other broader legislative initiatives in the 111th Congress that would have lifted all economic sanctions on Cuba by default would have allowed for U.S. investment in Cuba's energy sector.

In contrast, reflecting the second approach, H.R. 5620 (Ros-Lehtinen), the Caribbean Coral Reef Protection Act of 2010, would have imposed visa restrictions and economic sanctions on foreign nationals who helped facilitate the development of Cuba's petroleum resources. The initiative would have amended the Cuban Liberty and Democratic Solidarity Act of 1996 (P.L. 104-114) to exclude from the United States certain aliens (and their spouses, minor children, or agents) whose companies invested $1 million or more that contributed to the ability of Cuba to develop its offshore petroleum resources. The bill also would have provided for the imposition of sanctions if the President determined that a person had made an investment on or after January 10, 2005 of $1 million or more (or any combination of investments that equaled or exceeded $1 million or more in any 12-month period) that contributed to the enhancement of the Cuba's ability to develop its offshore petroleum resources. If such a determination were made, the President would have been required to propose two or more sanctions from a menu of sanctions listed in the bill.

112th Congress

Interest in Cuba's offshore oil development has continued in the 112th Congress as foreign oil companies have moved forward with plans to begin exploratory drilling. To date, five legislative initiatives have been introduced taking different approaches, and two congressional hearings have been held examining the issue.

H.R. 372 (Buchanan), introduced January 26, 2011, would amend the Outer Continental Shelf Lands Act to authorize the Secretary of the Interior to deny oil and gas leases and permits "to persons who engage in activities with the government of any foreign country that is subject to any sanction or an embargo" by the U.S. government. The intent of the legislation is to provide a disincentive to companies involved, or contemplating becoming involved, in Cuba's oil development, although the scope of the legislation is much broader and could affect other oil companies, including U.S. companies, not involved in Cuba. Because the bill does not define "sanction," the term could be used to refer to such U.S. restrictions as export controls or limits on foreign assistance. With this use of the term, many countries worldwide could be construed as being subject to a U.S. sanction, and as a result, any energy company that engages in activities with one of these countries could be denied an oil and gas lease in the United States under the proposed legislation.

S. 405 (Bill Nelson), the Gulf Stream Protection Act of 2011, introduced February 17, 2011, would require a company that is conducting oil or gas operations off the coasts of Cuba to submit an oil response plan for their Cuba operations and demonstrate sufficient resources to respond to a worst case scenario if the company wanted to lease drilling rights in the United States. The bill would also require the Secretary of the Interior to carry out an oil spill risk analysis and planning process for the development and implementation of oil spill response plans for nondomestic oil spills in the Gulf of Mexico. The Secretary of the Interior would be required, among other things, to include recommendations for Congress on a joint contingency plan with the countries of Mexico, Cuba, and the Bahamas to ensure an adequate response to oil spills located in the eastern Gulf of Mexico.

H.R. 2047 (Ros-Lehtinen), the Caribbean Coral Reef Protection Act of 2011 (identical to a bill introduced in the 111th Congress and noted above), was introduced May 26, 2011, and would impose visa restrictions on foreign nationals and economic sanctions on companies that help facilitate the development of Cuba's offshore petroleum resources. The bill would exclude from the United States aliens who invest $1 million or more that contributes to the enhancement of the ability of Cuba to develop its offshore oil resources. It would also require the imposition of sanctions (two or more from a menu of listed sanctions) if the President determined that a person had made an investment of $1 million on or after January 10, 2005, that contributed to Cuba's offshore oil development.

Both H.R. 3393 (Rivera) and S. 1836 (Menendez), would amend the Oil Pollution Act of 1990 to clarify that the Act applies to oil spills by foreign offshore units that occur in water beyond the exclusive economic zone of the United States. H.R. 3393 would amend the Act to provide that if the spill involves a foreign offshore unit in a country that has been designated by the Secretary of State as a state sponsor of terrorism, then the liability limit would be three times the liability limit (the current limit is the total of all removal costs plus $75 million). In contrast, S. 1836 would amend the Act to provided that there would be "no limitation on liability… for any incident involving a foreign offshore unit in which oil is discharged and enters or poses a substantial threat to enter the navigable waters or the exclusive economic zone." S. 1836 also would remove the $1 billion cap on the use of the Oil Spill Liability Trust Fund for oil spills by a foreign offshore unit that enters or poses a substantial threat to enter the navigable waters or the exclusive economic zone. H.R. 3393 would also amend the Federal Water Pollution Control Act to allow to make the owners of foreign offshore oil units subject to civil penalties if a discharge reaches the navigable waters of the United States, adjoining shorelines, or waters of the contiguous zones. The bill would also multiply each maximum penalty by three if the foreign offshore unit is located, in whole or in part, in the territorial sea or on the continental shelf of a foreign country that is a state sponsor of terrorism.

Oversight Hearings

Two congressional oversight hearings have been held in the 112th Congress on the issue of Cuba's offshore oil development and the implications for the United States. On October 18, 2011, the Senate Energy and Natural Resources Committee hearing held a hearing on Outer Continental Oil Spill Response Capabilities featuring officials from the U.S. Coast Guard and the Department of the Interior's Bureau of Safety and Environmental Enforcement as well as private witnesses from Florida International University and Clean Caribbean & Americas (a Florida company that is helping Repsol develop an oil spill contingency plan).[86]

On November 2, 2011, the House Natural Resources Committee, Subcommittee on Energy and Mineral Resources, held a hearing on North American offshore energy that also featured Coast Guard and Department of the Interior officials, as well as private witnesses from Cuba Democracy Advocates, Florida International University, the Environmental Defense Fund, and the Rule of Law Committee for the Oceans.[87]

CONCLUSION

Concern over Cuba's offshore oil development is likely to continue, especially if exploratory drilling begins as anticipated in early 2012. An oil spill in Cuban waters potentially could carry oil to U.S. waters and coastal areas in Florida, and potentially could threaten marine and coastal resources. The U.S. government has licensed some companies to provide oil spill prevention and containment support to companies operating in Cuba and is reviewing additional licenses. U.S. officials have also engaged with Repsol, and will be inspecting the oil rig before it enters Cuban waters. Policymakers may want to review whether U.S.-Cuban government engagement is warranted in order to maximize preparedness and response in the event of a major spill. Legislative initiatives have been introduced in the 112th Congress reflecting contrasting approaches toward Cuba's offshore development.

End Notes

[1] Karen Boman, "Repsol to Begin Offshore Cuba Drilling Later This Year," *Rigzone*, May 20, 2011.

[2] Unless otherwise noted, data on oil volumes in this report come from the Energy Information Administration's *International Energy Statistics,* see http://tonto.eia.doe.gov/cfapps/ipdbproject/IEDIndex3.cfm.

[3] These are risk-weighted mean estimates. There are uncertainties around such estimates that the USGS addresses by providing various risk weighted estimates with different degrees of certainty. Further, technically recoverable estimates do not take into account if the reserves are commercially attractive to produce. For more details see the following source and footnote 4. Christopher J. Schenk et al., *Assessment of Undiscovered Oil and Gas Resources of the North Cuba Basin, Cuba, 2004*, U.S. Department of the Interior, U.S. Geological Survey, World Assessment of Oil and Gas Fact Sheet, February 2005. http://walrus.wr.usgs.gov/infobank/programs/html/factsheets/pdfs/2005_3009.pdf.

[4] For an explanation of reserves and resources terms and concepts, please see CRS Report R40872, *U.S. Fossil Fuel Resources: Terminology, Reporting, and Summary*, by Gene Whitney, Carl E. Behrens, and Carol Glover.

[5] Leslie Moore Mira, "Cuba lowers its resource estimate to 9 billion barrels: official," *Platts Commodity News*, April 5, 2011.

[6] Energy Information Administration (EIA), "Country Analysis Brief: Caribbean," U.S. Department of Energy. November 2009, http://www.eia.doe.gov/emeu/cabs/Caribbean/OilProduction.html.

[7] Bureau of Western Hemisphere Affairs, "Background Notes: Cuba," U.S. Department of State, March 25, 2010.

[8] Imports data is for 2007, the most recent available figures from EIA.

[9] "CNPC Secures Cuban oil contract," *China Economic Review*, November 24, 2010.

[10] "Exploration and Production: Operations," Repsol YPF, 2005 (accessed 11/8/2010), http://www.repsol.com/es_en/corporacion/accionistas-e-

inversores/inf_economicofinanciera/informes_financieros/HTML/AreasNegocio/04/default.aspx?Pagina=16.

[11] Repsol, "Global Presence: Cuba," Repsol, April 30, 2010. http://www.repsol.com/es_en/corporacion/conocer-repsol/ presencia-global/cuba.aspx.

[12] Statoil is also looking to explore for oil in the Bahamas, where it has partnered with the Bahamas Petroleum Company. However, following the Deepwater Horizon oil spill, the Bahamian government suspended the consideration process for all oil exploration and drilling applications until the country has stringent environmental protocols in place to mitigate against a catastrophic oil well leak.

[13] Saipem is a subsidiary of publicly traded Italian oil major ENI S.p.a.

[14] Daniel Wallis, "Cuban oil rig set to cause waves in Washington," *Reuters*, May 17, 2011.

[15] David Goodhue, "Sen. Nelson Targeting Cuban Oil Exploration Near Florida," *Miami Herald*, November 21, 2011.

[16] Jeff Franks, "Repsol moving ahead with Cuba oil plans," *Reuters*, April 5, 2011.

[17] Carlos Batista, "Cuba to drill five new oil wells by 2013," *AFP*, April 5, 2011.

[18] Note that Brazil's national oil company, Petrobras, has subsequently withdrawn from its offshore lease. Kate Joynes- Burgess, "Russia Comes to Cuba's Aid with Economic Deal," *IHS Global Insight Daily Analysis*, July 20, 2009. Bert Wilkinson, "Caribbean: China Consolidates Influence as U.S. Frets," *Inter Press News Service*, May 28, 2009. Daniel McCleary, "Brazil to Loan $300M to Cuba to Refurbish Port of Mariel," *Dow Jones International*, July 13, 2009.

[19] ONGC Videsh Limited, "Tender for Supply of Sub-sea Well Heads and Large OD Casting Pipes for Block-N34 and N-35 of Cuba Off-shore," press release, February 8, 2010, http://www.ongcvidesh.com/TenderFiles/31.pdf.

[20] "Gazprom Neft Heads for Cuba," *International Oil Daily, Energy Intelligence Group*, October 6, 2010.

[21] Gazprom has taken a 30% stake in the blocks originally contracted just to Petronas in a 2007 agreement with the Cuban government. "Gazprom Takes State In Cuban Offshore Blocks," *Rigzone*, November 16, 2010.

[22] PetroVietnam, "E&P Worldwide—Caribbean & South America," PetroVietnam, November 11, 2010; http://pvep.com.vn/Default.aspx?pageid=122&action=view&flash=cuba.

[23] "Sonangol Signs Deal With Cuban Oil Company," *Angola Press Agency*, December 20, 2010.

[24] Linda Hutchinson-Jafar, "China's CNPC in talks for possible Cuba oil block," July 13, 2011.

[25] Rosa Tania Valdes, "Lula, Fidel Castro hold 'emotional' meeting," *Reuters*, February 24, 2010.

[26] Marc Frank, "Petrobras has relinquished Cuba oil block -official," *Reuters*, March 10, 2011.

[27] Jorge Piñon, *Cuba's Energy Crisis: Part III*, Cuba Transition Project, part of the Institute for Cuban and Cuban- American Studies at University of Miami, January 26, 2006, http://ctp.iccas.miami.edu/FOCUS_Web/Issue72.htm.

[28] See 15 CFR 734.4, which sets forth the 10% de minimis U.S. content provision in the Export Administration Regulations.

[29] This section uses research and analysis from CRS Specialists Peter Folger, Jonathon Ramsuer, and Harold Upton.

[30] For background on the Deepwater Horizon Spill itself, see CRS Report R41262, *Deepwater Horizon Oil Spill: Selected Issues for Congress*, coordinated by Curry L. Hagerty and Jonathan L. Ramseur.

[31] The International Maritime Organization (IMO) sent a technical assistance mission to Cuba in June 2010 to evaluate the level of preparation to respond to the Deepwater Horizon oil spill. The mission made several recommendations for Cuba to improve its national contingency plan, including the development of a training plan. See IMO, "Cuba, Misión de Asesoría Técnica," June 5-13, 2010, prepared by Klaus Essig. Also see the testimony of Jorge R. Piñon, Florida International University, before the Senate Committee on Energy and

Natural Resources, October 18, 2011, available at: http://energy.senate.gov/public/_files/PinonTestimony10182011.pdf.

[32] Testimony of Daniel J. Whittle, Environmental Defense Fund, before the House Committee on Natural Resources, Subcommittee on Energy and Mineral Resources, November 2, 2011, available at: http://naturalresources.house.gov/UploadedFiles/WhittleTestimony11.02.11.pdf.

[33] The Department of the Interior defines a "loss of well control" as "uncontrolled flow of formation or other fluids, including flow to an exposed formation (an underground blowout) or at the surface (a surface blowout), flow through a diverter, or uncontrolled flow resulting from a failure of surface equipment or procedures". Also see Dagmar Schmidt Etkin, "Analysis of U.S. Oil Spillage," American Petroleum Institute, August 2009. http://www.api.org/ehs/water/ spills/upload/356-Final.pdf.

[34] Construction of the rig was originally ordered by Norwegian firm Frigstad, but the contract was later transferred to Saipem. See more details on Scarabeo 9's specification at Saipem's website, http://www.snamprogetti.it/media_gallery/ brochure/Scarabeo9.pdf.

[35] Repsol, "Corporate Responsibility 2009," Repsol, April 26, 2010. http://www.repsol.com/es_en/corporacion/responsabilidad-corporativa/informe-responsabilidad-corporativa/default.aspx. Note that Repsol, along with U.S. firm Pride of North America, is currently under investigation by a Spanish court for an offshore oil spill in the Mediterranean. Repsol officials have described the spill as "a minor one-time incident which was solved and cleaned up within days." (Martin Robert, "Spain court probes Repsol oil spillage: report," *Reuters*, July 2, 2010.)

[36] Desiree Connor, "Cuba says safety a priority in offshore oil plan," *Reuters*, May 12, 2011.

[37] Waters in the Florida Straits between Cuba and Florida move eastward from the Gulf of Mexico into the Atlantic Ocean, feeding the Gulf Stream. This is the Florida Current, which stretches east and north through the Florida Straits and up the western side of the North Atlantic.

[38] International Tanker Owners Pollution Federation Limited (ITOPF), *Fate of Marine Oil Spills*, Technical Information Paper No.2, United Kingdom, 2002, http://www.itopf.com/_assets/documents/tip2.pdf.

[39] Ibid. Refined petroleum products, such as kerosene and gasoline, might evaporate completely. Heavier oils, or the heavier components of crude oil, may not undergo much evaporation; however, they may clump together and sink.

[40] Ibid.

[41] National Research Council (NRC), *Oil in the Sea III: Inputs, Fates, and Effects*, National Academies of Science, p. 4.

[42] Ibid., p. 127. These "sub-lethal" effects can occur at concentrations that are several orders of magnitude lower than concentrations that cause death.

[43] Ibid., p. 120.

[44] Ibid., p. 134. However, due to the increasing complexity of factors over time, studies on chronic effects are often met with debate and some controversy.

[45] These are 2008 figures provided by 'Visit Florida,' the state's official tourism marketing corporation. http://media.visitflorida.org/research.php.

[46] For example, see *As Cuba Plans to Drill in the Gulf of Mexico, U.S. Policy Poses Needless Risks to Our National Interest*, Center for Democracy in the Americas, February 2011, available at http://democracyinamericas.org/pdfs/ Cuba_Drilling_and_US_Policy.pdf.

[47] For example, see the testimony of Mauricio Claver-Carone, Cuba Democracy Advocates, before the House Committee on Natural Resources, Subcommittee on Energy and Mineral Resources, November 2, 2011, available at: http://naturalresources.house.gov/UploadedFiles/ClaverCaroneTestimony11.02.11.pdf.

[48] Testimony of Michael R Bromwich, Director, BSEE, Department of the Interior, before the Senate Committee on Energy and Natural Resources, October 18, 2011, available at: http://energy.senate.gov/public/_files/BromwichTestimony10182011.pdf.

[49] Lesley Clark, "U.S. Wary of Cuba's Drilling Plans, The Chief of the Miami Coast Guard Office Says His Agency Is Reviewing Response Scenarios for a Possible Spill Out of Cuba," *Miami Herald*, October 1, 2010; William Gibson, "Coast Guard Braces for Potential Cuban Oil Spill," *South Florida Sun-Sentinel*, March 23, 2011; Testimony of Vice Admiral Brian M. Salerno, U.S. Coast Guard, before the Senate Committee on Energy and Natural Resources, October 18, 2011, available at: http://energy.senate.gov/public/_files/VADMSoralcubastatement.pdf; "Coast Guard Prepares for International Offshore Drilling Close to Our Shores," *gCaptain.com*, November 19, 2011.

[50] Bromwich, op cit.

[51] Jorge R. Piñon and Robert L. Muse, "Coping with the Next Oil Spill: Why U.S.-Cuba Environmental Cooperation is Critical," *U.S. Cuba Relations at Brookings*, Issue Brief No. 2, May 2010.

[52] Clifford Krauss, "Cuba's Oil Plans for Deep Waters Raise Concerns," *New York Times*, September 30, 2010.

[53] "U.S. Embargo May Hinder Oil Spill Response in Cuba Waters: Driller," *Platts Commodity News*, May 1, 2011.

[54] U.S. Department of State, "Cuba: Oil Exploration, Question at the July 16, 2010 Daily Press Briefing," July 19, 2010. OFAC licenses cover travel and any financial transactions while BIS licenses cover the export of commodities.

[55] Telephone conversation with Paul A. Schuler, President, Clean Caribbean & Americas (CCA), November 3, 2010. For further background on the work of CCA in Latin America and the Caribbean, see its website at http://www.cleancaribbean.org/cgi-bin/loadAll.cgi?toget=2index.

[56] For an explanation of the tiered oil spill response categories, see International Petroleum Industry Environmental Conservation Association (IPIECA), "Guide to Tiered Preparedness and Response," IPIECA Report Series Vol. 14, 2007.

[57] Telephone conversation with Paul A. Schuler, November 3, 2010. Also see "Florida Firm Ready to Clean Up in Event of Cuba Oil Spill," *CubaNews*, December 2010, pp. 14-15.

[58] Testimony of Paul Schuler, President & CEO, Clean Caribbean & Americas, before the Senate Committee on Energy and Natural Resources, October 18, 2011, available at: http://energy.senate.gov/public/_files/SchulerTestimony10182011.pdf.

[59] Monica Hatcher, "Cuba Drilling Poses Spill Issue Group Says Trade Embargo Could Hinder a Response by the U.S.," *Houston Chronicle*, September 5, 2010. For further background on IADC, see http://www.iadc.org/.

[60] "U.S. Gives Permission for Cubans to Attend Drilling Conference," *Platts Commodity News*, April 28, 2011; See the agenda of the IADC conference, available at http://www.iadc.org/conferences/Environment_2011/.

[61] Gary Gentile, "Wild Well Control Seeks License to Provide Spill Equipment for Cuban Ops," *Platts Commodity News*, November 4, 2011.

[62] Bromwich, op cit.

[63] Gary Gentile, "Helix Applies for U.S. Licenses to Export Spill Equipment to Cuba," *Platts Commodity News*, October 31, 2011.

[64] U.S. Department of State, "Mexico, Pollution: Marine Environment, Agreement signed at Mexico City, July 24, 1980," TIAS, 10021.

[65] United States Coast Guard, "Mexus Plan, The Joint Contingency Plan Between the United Mexican States and the United States of America Regarding Pollution of the Marine Environment by Discharges of Hydrocarbons and Other Hazardous Substance," February 25, 2000.

[66] For background on U.S. cooperation with Cuba on migration and drug trafficking, see CRS Report R41617, *Cuba: Issues for the 112th Congress*.

[67] National Commission on the BP Deepwater Horizon Oil Spill and Offshore Drilling, *Deepwater, The Gulf Oil Disaster and the Future of Offshore Drilling,* Report to the President, p. 254 and p. 300. See the full text of the report at http://www.

oilspillcommission.gov/sites/default/files/documents/DEEPWATER_ReporttothePresident_FINAL.pdf.

[68] Tom Doggett, "U.S. fears Cuba oil drilling, Mexico suggests talks," April 20, 2011.

[69] Jorge R. Piñon, "Joint Spill Planning," *Oil & Gas Journal*, March 7, 2011.

[70] U.S. Congress, Senate, "International Convention on Oil Pollution Preparedness, Response, and Co-operation, 1990," 102d Congress, 1st Session, Treaty Doc. 102-11, August 1, 1991 (Washington: GPO).

[71] U.S. Department of State, "Marine Pollution, Wider Caribbean Region, Convention between the United States of America and Other Governments, Cartagena, March 14, 1983," TIAS 11085.

[72] See the website of the IMO's regional Caribbean center at http://cep.unep.org/racrempeitc.

[73] IMO, "Cuba, Misión de Asesoría Técnica," June 5-13, 2010, prepared by Klaus Essig, pp. 41-42.

[74] Bromwich, op cit.

[75] For background, see U.S. Government Accountability Office, U.S. Embargo on Cuba" Recent Regulatory Changes and Potential Presidential or Congressional Actions," September 17, 2009; and Dianne E. Rennack and Mark P. Sullivan, *U.S.-Cuban Relations: An Analytic Compendium of U.S. Policies, Laws, & Regulations*, The Atlantic Council, Washington, DC, March 2005.

[76] Jake Colvin, "The Case for Business," in *9 Ways for US to Talk to Cuba & For Cuba to Talk to US*, The Center for Democracy in the Americas, Washington, DC, 2009.

[77] "Cuba: A New Policy of Critical and Constructive Engagement, Report of the Brookings Project on U.S. Policy Toward a Cuba in Transition," Brookings Institution, April 2009.

[78] Frank Calzón, "Search for Oil Won't Cure the Economy," *Miami Herald*, October 1, 2010.

[79] U.S. Department of State, Bureau of Oceans and International Environmental and Scientific Affairs, "Limits in the Seas, No. 110, Maritime Boundary: Cuba–United States," February 21, 1990.

[80] U.S. Congress, Senate Committee on Foreign Relations, *Three Treaties Establishing Maritime Boundaries Between the United States and Mexico, Venezuela, and Cuba*, (to accompany Execs. F, G & H, 96-1), 96th Congress, 2nd session., August 5, 1980, Executive Rept. No. 96-49, p. 19.

[81] Ibid., p. 26. Also for a discussion of the provisional application of treaties, see U.S. Congress, Senate Committee on Foreign Relations, *Treaties and Other International Agreements: The Role of the United States Senate*, committee print, prepared by the Congressional Research Service, 106th Congress, 2nd session, January 2001, S. Prt. 106-71 (Washington: GPO, 2001), pp. 113-114.

[82] Lesley Clark and Sara Kennedy, "Cuba Ready to Drill for Oil Deeper Than BP," *Miami Herald*, September 30, 2010.

[83] Jorge R. Piñon and Jonathan Benjamin-Alvarado, "Extracting Cuba's Oil and Gas: Challenges and Opportunities," in *Cuba's Energy Future*, ed. by Jonathan Benjamin-Alvarado, Brookings Institution Press, Washington, DC, 2010, p. 31.

[84] The Senate Committee on Foreign Relations reported the treaty favorably in September 2000, and the full Senate agreed to the resolution of advice and consent to ratification on October 18, 2000. See U.S. Senate, *Treaty with Mexico on Delimitation of Continental Shelf*, 106th Congress, 2nd sess., July 27, 2000, Treaty Doc. 106-39; and U.S. Congress, Senate Committee on Foreign Relations, *Treaty with Mexico on Delimitation of the Continental Shelf*, 106th Congress, 2nd session, September 29, 2000, Exec. Rept. 106-19.

[85] The role of the CLCS is to facilitate the implementation of the U.N Convention on the Law of the Sea with regard to the establishment of the outer limits of the continental shelf beyond 200 nautical miles. The Commission considers data and other material submitted by coastal states concerning the outer limits of the continental shelf and makes recommendations to coastal states on such matters, but without prejudice to the question of delimitation of the

continental shelf between states with opposite or adjacent coasts. See the homepage of the CLCS, at http://www.un.org/ Depts/los/clcs_new/clcs_home.htm.

[86] For hearing testimony, see the website of the Senate Committee on Energy and Natural Resources, available at: http://energy.senate.gov/public/index.cfm?FuseAction=Hearings. Hearing&Hearing_ID=f37547ef-039b-373a-dc68- 113595376178.

[87] For testimony, see the website of the November 2, 2011, hearing of the House Natural Resources Committee, Energy and Mineral Resources Subcommittee, available at: http://naturalresources.house.gov/Calendar/EventSingle.aspx?EventID=260052.

In: Cuba
Editor: Gustavo Nogueira

ISBN: 978-1-62618-970-6

Chapter 3

CUBA: U.S. RESTRICTIONS ON TRAVEL AND REMITTANCES[*]

Mark P. Sullivan

SUMMARY

Restrictions on travel to Cuba have been a key and often contentious component in U.S. efforts to isolate Cuba's communist government since the early 1960s.

Under the George W. Bush Administration, restrictions on travel and on private remittances to Cuba were tightened. In March 2003, the Administration eliminated travel for people-to-people educational exchanges unrelated to academic coursework.

In June 2004, the Administration further restricted family and educational travel, eliminated the category of fully-hosted travel, and restricted remittances so that they could only be sent to the remitter's immediate family. Initially there was mixed reaction to the Administration's June 2004 tightening of Cuba travel and remittance restrictions, but opposition to the policy grew, especially within the Cuban American community regarding the restrictions on family travel and remittances.

---

[*] This is an edited, reformatted and augmented version of Congressional Research Service, Publication No. RL31139, dated November 9, 2012.

Obama Administration Policy

Under the Obama Administration, Congress took action in March 2009 by including two provisions in the FY2009 omnibus appropriations measure (P.L. 111-8) that eased restrictions on family travel and travel related to marketing and sale of agricultural and medical goods to Cuba. Subsequently, in April 2009, President Obama announced that his Administration would go further and allow unlimited family travel and remittances. Regulations implementing these changes were issued in September 2009. The new regulations also included the authorization of general licenses for travel transactions for telecommunications-related sales and for attendance at professional meetings related to commercial telecommunications.

In January 2011, the Obama Administration announced policy changes further easing restrictions on travel and remittances. The measures (1) increase purposeful travel to Cuba related to religious, educational, and people-to-people exchanges; (2) allow any U.S. person to send remittances to non-family members in Cuba and make it easier for religious institutions to send remittances for religious activities; and (3) permit all U.S. international airports to apply to provide services to licensed charter flights. These new measures, with the exception of the expansion of eligible airports, are similar to policies that were undertaken by the Clinton Administration in 1999, but subsequently curtailed by the Bush Administration in 2003-2004.

Legislative Action in the 112th Congress

In the first session of the 112th Congress, there were several attempts aimed at rolling back the Obama Administration's actions easing restrictions on travel and remittances. The House Appropriations Committee version of the FY2012 Financial Services and General Government Appropriations bill, H.R. 2434, would have rolled back President Obama's easing of restrictions on remittances and family travel; efforts to include the provision in an FY2012 "megabus" appropriations measure, H.R. 2055, were unsuccessful. (Notably in the second session, neither the House nor Senate Appropriations Committee-reported versions of the FY2013 Financial Services and General Government Appropriations measure, H.R. 6020 and S. 3301 respectively, have provisions regarding U.S. restrictions on travel or remittances to Cuba.) Among other measures, H.R. 2583, the FY2012 Foreign Relations Authorization Act, would require enforcement of travel regulations as in effect on January 19, 2009, and

H.R. 2831 would amend the Cuban Adjustment Act of 1966 in an attempt to curb frequent travel to Cuba by Cubans who have recently immigrated to the United States.

Several initiatives were also introduced in the 112th Congress that would further ease or lift such restrictions altogether. H.R. 1886 would prohibit restrictions on travel to Cuba. H.R. 1888, in addition to removing some restrictions on the export of U.S. agricultural products to Cuba, would also prohibit Cuba travel restrictions. Two initiatives that would lift the overall Cuba embargo, H.R. 255 and H.R. 1887, also would lift restrictions on travel and remittances to Cuba. H.R. 380 would prohibit funding to enforce restrictions on travel for educational activities in Cuba.

RECENT DEVELOPMENTS

In September 2012, press reports cited a slowdown in the Treasury Department's approval of licenses for people-to-people travel since the agency had issued new guidelines in May 2012. By early October 2012, however, companies conducting the people-to-people travel reported that they were once again receiving license approvals. (See "Developments in 2012" below.)

On June 7, 2012, the Senate Foreign Relations Committee, Subcommittee on Western Hemisphere, Peace Corps, and Global Narcotics, held a hearing on Cuba's human rights situation in which Assistant Secretary of State for Western Hemisphere Affairs Roberta Jacobson expressed strong U.S. support for democracy and human rights activists in Cuba and defended the Obama's Administration policy on travel and remittances. (See "Developments in 2012" below.)

On May 31, 2012, the House Committee on the Judiciary, Subcommittee on Immigration on Policy Enforcement, held a hearing on H.R. 2831 (Rivera), a bill that would amend the Cuban Adjustment Act of 1966 (P.L. 89-732) by providing that an alien from Cuba would be ineligible for adjustment to permanent resident status if he or she returns to Cuba before becoming a U.S. citizen. The intent of the bill is to curb travel to Cuba by those who have recently emigrated from Cuba. (See "Legislative Initiatives in the 112th Congress" below.)

On May 10, 2012, Treasury Department's Office of Foreign Assets Control (OFAC) tightened restrictions on people-to-people travel by making changes to its license guidelines. The revised guidelines require an organization applying for a people-to-people license to describe how the travel "would enhance contact with the Cuban people, and/or support civil society in Cuba, and/or promote the Cuban people's independence from Cuban authorities." (See "Chronology of Cuba Travel Restrictions" below.)

OVERVIEW OF THE U.S. RESTRICTIONS

Since the United States imposed a comprehensive trade embargo against Cuba in the early 1960s, there have been numerous policy changes to restrictions on travel to Cuba. The embargo regulations do not ban travel itself, but place restrictions on any financial transactions related to travel to Cuba, which effectively result in a travel ban. Accordingly, from 1963 until 1977, travel to Cuba was effectively banned under the Cuban Assets Control Regulations (CACR) issued by the Treasury Department's Office of Foreign Assets Control (OFAC) to implement the embargo. In 1977, the Carter Administration made changes to the regulations that essentially lifted the travel ban. In 1982, the Reagan Administration made other changes to the CACR that once again restricted travel to Cuba, but allowed for travel-related transactions by certain categories of travelers. Under the Clinton Administration, there were several changes to the Treasury Department regulations, with some at first tightening the restrictions, and others later loosening the restrictions.

Under the George W. Bush Administration, the travel regulations were tightened significantly, with additional restrictions on family visits, educational travel, and travel for those involved in amateur and semi-professional international sports federation competitions. In addition, the categories of fully-hosted travel and people-to-people educational exchanges unrelated to academic coursework were eliminated as permissible travel to Cuba. The Bush Administration also cracked down on those traveling to Cuba illegally, further restricted religious travel by changing licensing guidelines for such travel, and suspended the licenses of several travel service providers in Florida for license violations.

Under the Obama Administration, Congress took action in March 2009 (P.L. 111-8) to ease restrictions on travel by Cuban Americans to visit their family in Cuba and on travel related to the marketing and sale of agricultural and medical goods to Cuba. In April 2009, President Obama went even further by announcing that all restrictions on family travel and on remittances to family members in Cuba would be lifted, and on September 3, 2009, the Treasury Department issued regulations implementing these policy changes. In

January 2011, President Obama took further action to ease restrictions on travel and remittances to Cuba by providing new general licenses for travel involving educational and religious activities and restoring a specific license authorizing travel for people-to-people exchanges. The Administration also restored a general license for any U.S. person to send remittances to Cuba (up to $500 per quarter) and created a general license for remittances to religious organizations. Finally, the Administration also expanded the U.S. airports eligible to provide services to flights to and from Cuba. In most respects, with the exception of the expansion of eligible airports, these new measures appear to be similar to policies that were undertaken by the Clinton Administration in 1999 but were subsequently curtailed by the Bush Administration in 2003 and 2004.

The President has the authority to ease restrictions on travel to Cuba. For example, the President could choose to authorize travel to Cuba under a general license for all eligible categories of travel. Lifting all the restrictions on travel, however, would require legislative action. This is because of the codification of the embargo in Section 102(h) of the Cuban Liberty and Democratic Solidarity Act of 1996 (P.L. 104-114); that act conditions the lifting of the embargo, including the travel restrictions, on the fulfillment of certain democratic conditions in Cuba. Although the Administration retains flexibility through licensing authority to ease travel restrictions, the President may not lift all restrictions on travel as set forth in the CACR. Moreover, a provision in the Trade Sanctions Reform and Export Enhancement Act of 2000 (§910(b) of P.L. 106-387, Title IX) prevents the Administration from licensing travel for tourist activities, and defines such activities as any activity not expressly authorized in the 12 broad categories of travel set forth in the CACR regulations. This legislative provision essentially circumscribes the authority of the executive branch to issue travel licenses for activities beyond the broad categories of travel allowed, and would have to be amended, superseded by new legislation, or repealed in order to expand categories of travel to Cuba or lift travel restrictions altogether.

June 2004 Tightening of Travel and Remittance Restrictions

There was mixed reaction to the Bush Administration's June 2004 tightening of Cuba travel and remittance restrictions, including within the Cuban American community. President Bush maintained that such restrictions would "prevent the regime from exploiting hard currency of tourists and remittances to Cubans to prop up their repressive regime."[1] Supporters of the tightened restrictions argued that both educational and family travel to Cuba had become fronts for tourist travel. Tightening up on such travel, they argued, would deny the regime dollars that help maintain its repressive control. (According to the Commission for Assistance for a Free Cuba, some 125,000 family visits to Cuba in 2003 resulted in about $96 million in hard currency for the government.)[2] Another argument made by some supporters of the tightened restrictions was that the limiting of family travel to once every three years would help ensure that such travel was limited to family emergencies. Along these lines, some argued that limiting family travel would make travelers more sensitive to political repression on the island and highlights that Cuban Americans are political refugees, not economic immigrants. Some supporters of the additional remittance restrictions argued that the Bush Administration demonstrated a continuation of the compassionate policy of supporting the Cuban people by not cutting the level of remittances allowed, $300 per quarter. They emphasized that the Administration only took action to ensure that the remittances would be restricted to immediate family members and not benefit certain members of the Cuban government and Cuban Communist Party.

Opponents of the tightened travel and remittance restrictions made a number of policy arguments. They maintained that the restrictions were anti-family and violated the basic principle of family reunification. Some in the Cuban American community argued that the policy of restricting family visits was inhumane and only resulted in more suffering for Cuban families. They especially opposed the additional restrictions that did not allow travel to visit cousins, aunts, uncles, and more-distant relatives. Another argument opposing restrictions on travel and private remittances was that the steps would have no effect on reducing repression in Cuba or weakening the government's instruments of repression. Opponents of the tightened restrictions maintained that the new restrictions were opposed by several prominent Cuban dissidents, including Oswaldo Payá of the Varela Project and Elizardo Sanchez of the Cuban Commission for Human Rights and National Reconciliation. Miriam Leiva, one of the founders of the Ladies in White human rights group, maintained that the policy punished dissidents and their families; she compared the U.S. restrictions to the situation faced by Cubans, who cannot travel without permission from the Cuban government.[3] Former political prisoner Oscar Espinosa Chepe, released from prison in December 2004, called the U.S. policy "absurd," maintaining that "what we need is to create space for dialogue."[4]

There were also concerns that the new restrictions were drafted without considering the full consequences of their implementation. For example, the elimination of the category of fully-hosted travel raised concerns about the status some 70 U.S. students receiving full scholarships at the Latin American School of Medicine in Havana. The school has more than 3,000 students from 23 countries and consists of a six-month pre-med program and a six-year medical school program. Members of the Congressional Black Caucus, who were instrumental in the establishment of the scholarship program for U.S. students, expressed concern that the students could have been forced to abandon their medical education because of the new OFAC regulations. As a result of these concerns, OFAC ultimately licensed the medical students to continue their studies and engage in travel-related transactions.

In the aftermath of the Bush Administration's tightening of travel restrictions, there was increased opposition to the policy and several groups were established opposing the Administration's actions. A group known as ENCASA, the Emergency Network of Cuban American Scholars and Artists for Change in Cuba Policy, launched a media campaign in 2006 opposing the travel restrictions.[5] In June 2006, another group of some 450 scholars known as the Emergency Coalition to Defend Educational Travel (ECDET) filed suit in U.S. federal court in Washington against the Treasury Department, maintaining that travel restrictions violated academic freedom.[6] (On November 4, 2008, the U.S. Court of Appeals for the District of Columbia found that the travel restrictions do not violate the right to academic freedom.)[7]

With regard to family travel, a group in Miami, the Association of Christian Women in Defense of the Cuban Family, organized several protests against the tightened family travel restrictions.[8] In March 2008, Cuban Americans living in Vermont filed a complaint in U.S. federal court in Burlington, VT, that U.S. restrictions on family travel to Cuba violate their civil rights. Affiliates of the American Civil Liberties Union of Florida, Massachusetts, and Vermont subsequently filed a brief in support of the complaint. Human Rights Watch maintained that the U.S. travel policies inflicted harm on Cuban families and undermined the freedom of movement of hundreds of thousands of Cuban Americans.[9] In a 2005 report, Human Rights Watch cited numerous cases of family hardships after the tightened family travel restrictions went into effect, including the inability to visit children, sick or dying parents, or to attend funerals.[10]

A 2007 Florida International University poll examining attitudes of the Cuban American community in South Florida showed that about 64% of respondents wanted to return to the less restrictive policies on travel and remittances that were in place in 2003. Moreover, 55.2% of respondents supported allowing unrestricted travel overall, not just family travel.[11]

Easing of Restrictions in 2009

The tightening of family travel restrictions became an issue during the 2008 presidential campaign with candidate Barack Obama pledging to lift restrictions for family travel and remittances to Cuba.

With the election of Obama, the 111th Congress moved to ease family travel restrictions in March 2009 by approving two provisions that eased sanctions on travel to Cuba in FY2009 omnibus appropriations legislation (P.L. 111-8). Unlike the Bush Administration, the Obama Administration did not threaten to veto such legislation easing Cuba sanctions. This marked the first congressional action easing Cuba sanctions in almost a decade.

In the first provision, as implemented by the Treasury Department, family travel was again allowed once every 12 months under a general license to visit a close relative for an unlimited length of stay, and the limit for daily expenditure allowed by family travelers became the same as for other authorized travelers to Cuba (the State Department maximum per diem rate for Havana).

The definition of "close relative" was expanded to mean any individual related to the traveler by blood, marriage, or adoption who is no more than three generations removed from that person.

The second provision in the omnibus measure required a general license for travel related to the marketing and sale of agricultural and medical goods to Cuba. The Treasury Department's Office of Foreign Assets Control ultimately issued regulations implementing this omnibus provision on September 3, 2009. The regulations require a written report at least 14 days before departure identifying both the traveler and the producer or distributor and describing the purpose and scope of such travel. Another written report is required within 14 days of return from Cuba describing the activities conducted, the persons met, and the expenses incurred. The regulations also require that such travelers under this provision be regularly employed by a producer or distributor of the agricultural commodities or medical products or an entity duly appointed to represent such a producer or distributor. The activity schedules for such travelers cannot include free time, travel, or recreation in excess of that consistent with a full work schedule.

Going even further, the Obama Administration announced several significant measures to ease U.S. sanctions on Cuba in April 2009. Fulfilling a campaign pledge, President Obama announced that all restrictions on family travel and on remittances to family members in Cuba would be lifted. This significantly superseded the action taken by Congress in March that had essentially reverted family travel restrictions to as they had been before they were tightened in 2004. Under the new policy announced by the Administration in April, there are no limitations on the frequency or duration of family visits (which would still be covered under a general license) and the 44-pound limitation on accompanied baggage was removed. Family travelers are allowed to spend the same as allowed for other travelers, up to the State Department's maximum per diem rate for Havana (which varies, but currently is $166).[12] With regard to family remittances, the previous limitation of no more than $300 per quarter was removed with no restriction on the amount or frequency of the remittances. Authorized travelers were again authorized to carry up to $3,000 in remittances.[13] Regulations for the above policy changes were issued by the Treasury and Commerce Departments on September 3, 2009.

Easing of Restrictions in 2011

On January 14, 2011, the Obama Administration announced a series of policy changes further easing restrictions on travel and remittances to Cuba that had been rumored in the second half of 2010. The changes are designed to make it easier to engage in educational, religious, and other types of people-to-people travel and allow all Americans to send remittances to Cuba. The changes are similar to policy that was in place from 1999 under the Clinton Administration through mid-2004 under the Bush Administration. President Obama directed the Secretaries of State, Treasury, and Homeland Security to amend regulations and policies "in order to continue efforts reach out to the Cuban people in support of their desire to freely determine their country's future."[14] The Administration maintains that the policy changes will increase people-to-people contact, help strengthen Cuban civil society, and make Cuban people less dependent on the Cuban state.[15] The changes are being taken at the same time that the Cuban government is laying off thousands of state workers and increasing private enterprise through an expansion of the authorized categories for self-employment.

According to the White House announcement, the policy changes would be enacted through modifications to existing regulations. This occurred on January 28, 2011, when the Departments of the Treasury and Homeland Security published changes to the regulations in the *Federal Register*.[16]

The measures (1) increase purposeful travel to Cuba related to religious, educational, and journalistic activities (general licenses are now authorized for certain types of educational and religious travel; people-to-people travel exchanges are authorized via a specific license); (2) allow any U.S. person to send remittances to non-family members in Cuba and make it easier for religious institutions to send remittances for religious activities (general licenses are now authorized for both); and (3) allow all U.S. international airports to apply to provide services to licensed charter flights to and from Cuba. In most respects, these new measures appear to be similar to policies that were undertaken by the Clinton Administration in 1999, but were subsequently curtailed by the Bush Administration in 2003 and 2004. An exception is the expansion of airports to service licensed flights to and from Cuba. While the new travel regulations immediately went into effect for those categories of travel falling under a general license category, OFAC delayed processing applications for new travel categories requiring a specific license (such as people-to-people exchanges) until it updated and issued guidelines.[17] These ultimately were issued in April 2011: *Comprehensive Guidelines for License Applications to Engage in Travel-related Transactions Involving Cuba*.[18]

- **Purposeful Travel**. With regard to purposeful travel, the policy changes allowed religious organizations to sponsor religious travel to Cuba under a general license as opposed to the previous requirement for a specific license for such travel. Restrictions on educational travel were eased in several ways: educational travel for academic credit is now allowed under a general license (instead of a specific license as previously required); students are now allowed to participate through academic institutions other than their own; and instructor support is allowed from adjunct and part-time staff. Academic institutions will also be allowed to apply for specific licenses to sponsor or cosponsor academic seminars, conferences, and workshops related to Cuba and allow faculty, staff, and students to attend. People-to-people exchanges, under the auspices of an organization that sponsors and organizes such programs, are now allowed under a specific license (such activities previously had been allowed from 1999-2003).

- **Remittances.** The policy changes restored a general license category available for any U.S. person to send up to $500 in remittances per quarter to non-family members in Cuba (but not to senior Cuban government officials or senior members of the Cuban Communist Party) to support private economic activity, among other purposes. A general license also was created for remittances to religious institutions in Cuba in support of religious activities.
- **U.S. Airports**. The policy changes expanded the number of eligible airports in the United States authorized to serve licensed charter flights to and from Cuba. The Clinton Administration had expanded airports eligible to service license charter flights beyond that of Miami International Airport to international airports in Los Angeles and New York (JFK) in 1999, but the January 2011 policy change allows all U.S. international airports to apply to provide services for chartered flights to and from Cuba under certain conditions. The airport would need to have adequate customs and immigration capabilities, and a licensed travel service provider would need to have expressed an interest in providing service to and from Cuba from the airport.

By early July 2011, OFAC confirmed that it had approved the first licenses for U.S. people-topeople organizations to bring U.S. visitors to Cuba, and the first such trips began in August 2011.[19] On July 25, 2011, however, prior to the trips beginning, OFAC issued an advisory maintaining that misstatements in the media had suggested that U.S. policy now allows for virtually unrestricted group travel to Cuba, and reaffirmed that travel conducted by people-topeople travel groups licensed for travel to Cuba must "certify that all participants will have a full-time schedule of educational exchange activities that will result in meaningful interaction between the travelers and individuals in Cuba." The advisory stated that authorized activities by people-to-people groups are not "tourist activities," and pointed out that the Trade Sanctions Reform and Export Enhancement Act of 2000 prohibits OFAC from licensing transactions for tourist activities.[20]

As a result of the OFAC advisory, a luxury travel company, Abercrombie & Kent, announced in early August 2011 that it was suspending all its Cuba bookings "until it can ensure it is fully compliant" with the OFAC guidance. Abercrombie & Kent does not have an OFAC license, but reportedly had arranged to travel pursuant to the license of a group known as the Foundation for Caribbean Studies.[21]

The Department of Homeland Security, U.S. Customs and Border Protection (CBP) subsequently announced its approval of 12 additional airports eligible to provide passenger air service between the United States and Cuba, bringing the total number of airports approved to 15. The newly authorized airports are Atlanta, Baltimore-Washington (BWI), Chicago O'Hare, Dallas-Fort Worth, Fort Lauderdale-Hollywood, Houston, New Orleans, Oakland (CA), Pittsburgh, San Juan (Puerto Rico), Southwest Florida International Airport (Fort Myers), and Tampa.[22]

Policy groups in favor of increased U.S. engagement with Cuba largely praised the Administration's action as a significant step forward in reforming U.S.-Cuban relations and as an important means to expand the flow of information and ideas to Cuba and to increase the income of Cubans working in the expanding private sector. Perhaps more surprisingly, the Miami-based Cuban American National Foundation (CANF) strongly supported the Administration's policy changes. According to CANF President Francisco "Pepe" Hernández: "A greater ability to send remittances in conjunction with increased contact and communication with those on the island will help to break the chains of dependency that the Castro regime has used to oppress those inside Cuba."[23]

In contrast, policy groups opposed to easing U.S. sanctions have criticized the Administration, maintaining that the policy changes will help prop up Cuba's repressive government when it is most vulnerable because of the difficult economic situation. Opponents of the policy changes argue that sending dollars via increased travel by Americans and increased remittances will actually help the Cuban government maintain in place its repressive policies. They also argue that easing the restrictions on travel and remittances will not bring about respect for human rights in Cuba.

The Cuban government characterized the U.S. policy changes as positive, but maintained that they were limited in scope and did not alter policy toward Cuba. A statement by Cuba's Ministry of Foreign Affairs maintained that the policy changes did not restore the right to travel to Cuba for all American citizens, and that the United States should lift the blockade (embargo) and the prohibition on travel to Cuba if it is interested in expanding and facilitating contacts between Cubans and Americans.[24]

In the first session of the 112th Congress, there were several attempts aimed at rolling back the Obama Administration's actions easing restrictions on travel and remittances, including a provision originating in the House Appropriation Committee's version of the FY2012 Financial Services and General Government appropriations measure, H.R. 2434. The White House had threatened to veto the bill if it contained the provision, and stood firm when congressional leaders were considering including the provision in a "megabus" FY2012 appropriations bill, H.R. 2055. Ultimately congressional leaders agreed not to include the provision in the appropriations measure (P.L. 112-74). (As a political tradeoff, however, congressional leaders dropped another Cuba provision that would have continued, for the third consecutive year, to clarify the definition of "payment of cash in advance" for U.S. agricultural and medical exports to Cuba so that the payment was due upon delivery in Cuba as opposed to being due before the goods left U.S. ports.) (See "Legislative Initiatives in the 112th Congress" below.)

Developments in 2012

In 2012, some Members of Congress expressed concerns about people-to-people travel that appeared to be focusing on tourist activities rather than on purposeful travel. In response, the Treasury Department issued an announcement in March 2012 warning about misleading advertising regarding some people-to-people trips that could lead to OFAC investigating the organization conducting the trips. The announcement maintained that licenses could be revoked and that organizations may be issued a civil penalty up to $65,000 per violation.[25] OFAC followed up this announcement in May 2012 by revising its people-to-people license guidelines. The revised guidelines reflect similar language to the March announcement and also require an organization applying for a people-to-people license to describe how the travel "would enhance contact with the Cuban people, and/or support civil society in Cuba, and/or promote the Cuban people's independence from Cuban authorities."[26] (For more details, see "Chronology of Cuba Travel Restrictions" below.)

On April 27, 2012, a suspicious fire destroyed the Coral Gables, FL, office of Airline Brokers, a travel agency specializing in flights to Cuba. The Coral Gables fire department subsequently determined the fire to be caused by arson. The Federal Bureau of Investigation and the Bureau of Alcohol, Tobacco, Firearms, and Explosives became involved in the investigation of the fire.[27]

In June 7, 2012, congressional testimony, Assistant Secretary of State for Western Hemisphere Affairs Roberta Jacobson set forth a clear-cut description of U.S. policy toward Cuba in which she expressed strong U.S. support for democracy and human rights activists in Cuba and defended the Obama's Administration policy on travel and remittances. The Assistant Secretary asserted that "the Obama Administration's priority is to empower Cubans to freely determine their own future." She maintained that "the most effective tool we have for doing that is building connections between the Cuban and American people, in order to give Cubans the support and tools they need to move forward independent of their government." The Assistant Secretary maintained that "the Administration's travel, remittance and people-to-people policies are helping Cubans by providing alternative sources of information, taking advantage of emerging opportunities for self-employment and private property, and strengthening civil society."[28]

In September 2012, various press reports cited a slowdown in the Treasury Department's approval or reapproval of licenses for people-to-people travel since the agency had issued new guidelines in May (described above). Companies conducting such programs complained that the delay in the licenses was forcing them to cancel trips and even to lay off staff.[29] By early October 2012, however, companies conducting the people-to-people travel maintained that they were once again receiving license approvals.

Chronology of Cuba Travel Restrictions

1960—In the first trade restrictions on Cuba after the rise to power of Fidel Castro, President Eisenhower placed most U.S. exports to Cuba under validated license controls, except for nonsubsidized food, medicines, and medical supplies. The action did not include restrictions on travel.

1962/1963—In February 1962, President Kennedy imposed a trade embargo on Cuba because of the Castro government's ties to the Soviet Union. Pursuant to the President's directive, the Department of the Treasury's Office of Foreign Assets Control (OFAC) issued the Cuban Import Regulations. On July 9, 1963, OFAC issued a more comprehensive set of prohibitions, the Cuban Assets Control Regulations, which effectively banned travel by prohibiting any transactions with Cuba.

1977—In March, the Carter Administration announced the lifting of restrictions on U.S. travel to Cuba that had been in place since the early 1960s. The Carter Administration lifted the travel ban by issuing a general license for travel-related transactions for those visiting Cuba. Direct flights were also allowed.

1982—In April, the Reagan Administration reimposed restrictions on travel to Cuba, although it allowed for certain categories of travel, including travel by U.S. government officials, employees of news or filmmaking organizations, persons engaging in professional research, or persons visiting their close relatives. It did not allow for ordinary tourist or business travel that had been allowed since the Carter Administration's 1977 action.

1984—On June 28, the Supreme Court, in a 5-4 decision in the case of *Regan v. Wald*, rejected a challenge to the ban on travel to Cuba and asserted the executive branch's right to impose travel restrictions for national security reasons.

1993—The Clinton Administration, in June 1993, slightly amended restrictions on U.S. travel to Cuba. Two additional categories of travel were allowed: travel to Cuba "for clearly defined educational or religious activities"; and travel "for activities of recognized human rights organizations." In both categories, travelers were required to apply for a specific license from OFAC.

1994—In August, President Clinton announced several measures against the Cuban government in response to an escalation in the number of Cubans fleeing to the United States. Among these measures, the Administration tightened travel restrictions by prohibiting family visits under a general license, and allowing specific licenses for family visits only "when extreme hardship is demonstrated in cases involving extreme humanitarian need" such as terminal illness or severe medical emergency. Such visits required a specific license from OFAC. In addition, professional researchers were required to apply for a specific license, whereas since 1982 they had been able to travel freely under a general license. (*Federal Register*, August 30, 1994, pp. 44884-44886.)

1995—In October, President Clinton announced measures to ease some U.S. restrictions on travel and other activities with Cuba, with the overall objective of promoting democracy and the free flow of ideas. The new measures included authorizing general licenses for transactions relating to travel to Cuba for Cuban Americans making yearly visits to close relatives in "circumstances that demonstrate extreme humanitarian need." This reversed the August 1994 action that required specific licenses. However, those traveling for this purpose more than once in a 12-month period would need to apply to OFAC for a specific license. In addition, the new measures allowed for specific licenses for free-lance journalists traveling to Cuba. (*Federal Register*, October 20, 1995, pp. 54194-54198.)

1996—On February 26, following the shootdown of two U.S. civilian planes two days earlier by Cuban fighter jets, President Clinton took several measures against Cuba, including the indefinite suspension of charter flights between Cuba and the United States. Qualified licensed travelers could go to Cuba, provided their flights were routed through third countries.

1998—On March 20, following Pope John Paul II's January trip to Cuba, President Clinton announced several changes in U.S. policy toward Cuba, including the resumption of licensing for direct charter flights to Cuba. On July 2, OFAC issued licenses to nine air charter companies to provide direct passenger flights from Miami International Airport to Havana's José Martí International Airport.

1999—On January 5, President Clinton announced several measures to support the Cuban people that were intended to augment changes implemented in March 1998. Among the measures introduced was the expansion of direct passenger charter flights from additional U.S. cities other than Miami. In August, the State Department announced that direct flights to Cuba would be allowed from New York and Los Angeles. In addition, President Clinton also announced in January 1999 that measures would be taken to increase people-to-people exchanges. As a result, on May 13, 1999, OFAC issued a number of changes to the Cuba embargo regulations that effectively loosened restrictions on certain categories of travelers to Cuba. Travel for professional research became possible under a general license, and travel for a wide range of educational, religious, sports competition, and other activities became possible with specific licenses authorized by OFAC on a case-by-case basis. In addition, those traveling to Cuba to visit a close family member under either a general or specific license only needed to "demonstrate humanitarian need," as opposed to "extreme humanitarian need" that had been required since 1995. (*Federal Register*, May 13, 1999, pp. 25808-25820.)

2000—In October, Congress approved and the President signed the Trade Sanctions Reform and Export Enhancement Act of 2000 (Title IX of P.L. 106-387), which included a provision that prohibited travel-related transactions for "tourist activities," which as set forth in Section 910(b)(2) of the act are defined as any activity not authorized or referenced in the existing travel regulations (31 CFR 515.560, paragraphs (1) through (12)). The congressional action appeared to circumscribe the authority of the OFAC to issue specific travel licenses on a case-by-case basis that do not fit neatly within the categories of travel already allowed by the regulations.

2001—On July 12, OFAC published regulations pursuant to the provisions of the Trade Sanctions and Export Enhancement Act of 2000 (Title IX of P.L. 106-387) that prohibited travel-related transactions for "tourist activities." (*Federal Register*, July 12, 2001, pp. 36683-36688.) On July 13, 2001, President Bush announced that he had asked the Treasury Department to enhance and expand the capabilities of OFAC to prevent, among other things, "unlicensed and excessive travel."

2003—On January 29, OFAC published proposed enforcement guidelines (as an appendix to 31 CFR Part 501) for all its economic sanctions programs and additional guidelines (as an appendix to 31 CFR Part 515) for the Cuba sanctions program. The general guidelines provided a procedural framework for OFAC's enforcement of economic sanctions, while the Cuba-specific guidelines consist of penalties for different embargo violations. (*Federal Register*, January 29, 2003, pp. 4422-4429.)

On March 24, 2003, OFAC announced that the Cuba travel regulations were being amended to ease travel to Cuba for those visiting close relatives. (*Federal Register*, March 24, 2003, pp. 14141-14148.) Travel was permitted to visit relatives to within three degrees of relationship of the traveler and was not restricted to travel in circumstances of humanitarian need. The new regulations also increased the amount a traveler may carry, up to $3,000 (compared to $300 previously), although the limit of $300 per quarter destined for each household remained. Finally, the regulations were tightened for certain types of educational travel. People-to-people educational exchanges unrelated to academic coursework were no longer allowed. Some groups lauded the restriction of these educational exchanges because they believed they had become an opportunity for unrestricted travel; others criticized the Bush Administration's decision to restrict the second largest category of travel to Cuba in which ordinary people were able to travel and exchange with their counterparts on the island.

On October 10, 2003, President Bush instructed the Department of Homeland Security, as part of a broader initiative on Cuba, to increase inspections of travelers and shipments to and from Cuba in order to more strictly enforce the trade and travel embargo.

2004—On February 26, President Bush ordered the Department of Homeland Security to expand its policing of the waters between Florida and Cuba with the objective of stopping pleasure boating traffic. (*Federal Register*, March 1, 2004, pp. 9315-9517.)

On June 16, 2004, OFAC published changes to the CACR implementing the President's directives to implement certain recommendations of the Commission for Assistance to a Free Cuba. The new regulations tightened travel restrictions in several ways. Fully-hosted travel was eliminated as a legal category of permissible travel. Family visits were restricted to one trip every three years under a specific license to visit only immediate family (grandparents, grandchildren, parents, siblings, spouses, and children) for a period not to exceed 14 days. The daily amount of money that family visitors could spend while in Cuba was reduced from the State Department per diem rate for Havana (then $179) to $50. Specific licenses for visiting non-Cuban nationals in Cuba (such as a student) were limited to when the family member visited was in "exigent circumstances." The general license for amateur or semi-professional athletic teams to travel to Cuba to engage in sports competitions was eliminated; such travel now required a specific license. (*Federal Register*, June 16, 2004, pp. 33768-33774)

Specific licenses for educational activities were further restricted in several ways: the institutional licenses were restricted to undergraduate and graduate institutions, while the category of educational exchanges sponsored by secondary schools was eliminated; the duration of institutional licenses was shortened from two to one year; three types of licensed educational activities—structural education programs in Cuba offered as part of a course at the licensed institution; formal courses of study offered at a Cuban academic institution; and teaching at a Cuban academic institution—were required to be no shorter than 10 weeks.

The new regulations also further restricted sending cash remittances to Cuba. Quarterly remittances of $300 could still be sent, but were restricted to members of the remitter's immediate family and could not be remitted to certain government officials and certain members of the Cuban Communist Party. The regulations were also changed to reduce the amount of remittances that authorized travelers may carry to Cuba, from $3,000 to $300. This reversed OFAC's March 2003 changes to the regulations that had increased the amount that authorized travelers could carry to $3,000.

On June 22, 2004, the Department of Commerce's Bureau of Industry and Security (BIS) published regulations related to the recommendations of the Commission for Assistance to a Free Cuba. The new regulations placed new limits on gift parcels sent to Cuba and personal baggage of travelers going to Cuba. Gift parcels could no longer contain items such as seeds, clothing, personal hygiene items, veterinary medicines and supplies, fishing equipment and supplies, and soap-making equipment. Baggage was limited to 44 pounds. (*Federal Register*, pp. 34565-34567)

On July 8, 2004, the U.S. Coast Guard published regulations requiring U.S. vessels less than 100 meters to have a Coast Guard permit to enter Cuban territorial waters. (*Federal Register*, pp. 41367-41374)

2005—On March 31, OFAC made changes to its guidelines for license applications related to religious travel. According to the guidelines, specific licenses issued under CFR 515.566(b) for religious organizations only authorized up to 25 individuals to travel to Cuba no more than once per calendar quarter. The specific licenses under this section would not be valid for more than one year. (OFAC, *Comprehensive Guidelines for License Applications to Engage in Travel-related Transactions Involving Cuba*, Revised September 2004, p. 40, the relevant paragraph was updated March 31, 2005).

2009—On March 11, President Obama signed into law the Omnibus Appropriations Act, 2009 (P.L. 111-8), with two provisions easing restrictions on travel to Cuba.

Section 620 of Division D amended the Trade Sanctions Reform and Export Enhancement Act of 2000 (TSRA) to require the Secretary of the Treasury to issue regulations for travel to, from, or within Cuba under a general license for the marketing and sale of agricultural and medical goods, meaning that there would be no requirement to obtain special permission from OFAC. Such travel had required a specific license from OFAC, issued on a case-by-case basis. OFAC maintained that it would issue regulations in the coming weeks, although a letter from Secretary of the Treasury Timothy Geithner published in the *Congressional Record* stated that the new regulations "would provide that the representatives of only a narrow class of businesses would be eligible, under a new general license, to travel to market and sell agricultural and medical goods." The Secretary also maintained that "any business using the general license would be required to provide both advance written notice outlining the purpose and scope of the planned travel and, upon return, a report outlining the activities conducted, including the persons with whom they met, the expenses incurred, and business conducted in Cuba." (*Congressional Record*, March 10, 2009, p. S2933.)

Section 621 of Division D prohibited funds from being used to administer, implement, or enforce family travel restrictions that were imposed by the Bush Administration in June 2004. OFAC implemented this provision by reinstating a general license for family travel as it existed prior to the Bush Administration's tightening of restrictions in June 2004. As implemented by OFAC, travel was allowed once every 12 months to visit a close relative for an unlimited length of stay, and the limit for daily expenditure allowed by family travelers became the same as for other authorized travelers to Cuba (State Department maximum per diem rate for Havana in effect when the travel takes place.) The new general license also expanded the definition of "close relative" to mean any individual related to the traveler by blood, marriage, or adoption who was no more than three generations removed from that person.

On April 13, 2009, President Obama directed that all restrictions on family travel and on remittances to family members in Cuba be lifted. The Administration also announced measures to expand the scope of eligible humanitarian donations through gift parcels and to increase telecommunications links with Cuba. (See the White House fact sheet available at http://www.whitehouse.gov/the_press_office/Fact-Sheet-Reaching -out-to-the-Cuban-people/.)

On September 3, 2009, OFAC issued amendments to the Cuban Assets Control Regulations implementing President Obama's policy changes with regard to family travel, remittances, and greater telecommunications links with Cuba. The amendments also included new categories of travel under general licenses, including travel for the marketing and sale of agricultural and medical goods (implementing the legislative provision approved in March 2009 described above) and travel for telecommunications providers and those attending professional meetings for commercial telecommunications transactions. (*Federal Register*, September 8, 2009, pp. 46000- 46007.) On the same day, the Department of Commerce's Bureau of Industry and Security issued amendments to the Export Administration Regulations that expanded the value and list of eligible item that may be included in gift parcels to Cuba and removed the previous weight limit of 44 pounds for accompanied baggage to Cuba. (*Federal Register*, September 8, 2009, pp. 45985- 45990.)

2011—On January 14, the White House announced that President Obama had directed the Secretaries of State, Treasury, and Homeland Security to make changes to regulations and policies to (1) increase purposeful travel to Cuba related to religious, educational, and journalistic activities; (2) allow any U.S. person to send remittances to non-family members in Cuba and make it easier for religious institutions to send remittances for religious activities; and (3) allow all U.S. international airports to provide services to licensed charter flights to and from Cuba. (See the White House statement at http://www.whitehouse.gov/the-press-office/2011/01/14/ reaching-out-cuban-people.)

On January 28, 2011, OFAC issued changes to the CACR implementing the revised policy announced by the President on January 14 and designed to increase purposeful travel and ease restrictions on remittances to non-family members in Cuba and to religious institutions for religious activities (*Federal Register*, January 28, 2011, pp. 5072-5078). On the same day, the Department of Homeland Security (DHS), U.S. Customs and Border Protection (CBP), issued changes to DHS regulations to allow additional international airports in the United States to request approval of CBP to process authorized flights between the United States and Cuba (*Federal Register*, January 28, 2011, pp. 5058-5061).

On April 21, 2011, OFAC issued revised guidelines for travel license applications reflecting the policy changes set forth in January 2011. (U.S. Department of the Treasury, OFAC, *Comprehensive Guidelines for License Applications to Engage in Travel-related Transactions Involving Cuba,* Revised April 19, 2011.)

On July 25, 2011, OFAC issued an advisory reaffirming that travel conducted by people-topeople travel groups licensed for travel to Cuba must "certify that all participants will have a full-time schedule of educational exchange activities that will result in meaningful interaction between the travelers and individuals in Cuba." (U.S. Department of the Treasury, OFAC, "Cuba Travel Advisory," July 25, 2011)

2012—On March 9, 2012, OFAC published an announcement regarding advertising for peopleto-people travel, noting that all advertisements must state the name of the licensed organization conducting the travel and that the organization must use the name under which their OFAC travel was licensed unless the group requests and receives a license amendment from OFAC to use an alternative name. The announcement also stated that advertising that appeared to suggest that the people-to-people trips were focused on activities that travelers may undertake off hours (after their daily full-time schedule of people-to-people activities) may give an incorrect impression and prompt OFAC to contact the licensed organization and conduct an investigation. It maintained that people-to-people organizations that failed to meet requirements of their licenses may have their licenses revoked or be issued a civil penalty up to $65,000 per violation. (U.S. Department of the Treasury, OFAC, "Advertising Educational Exchange Travel to Cuba for People-to-People Contact," March 9, 2012, available at http://www.treasury.gov /resourcecenter/sanctions/Programs/Pages/cuba_ppl_notice.aspx.)

On May 10, 2012, OFAC tightened restrictions on people-to-people travel by making changes to its license guidelines. The revised guidelines reflect similar language to the March 2012 announcement described above regarding advertising. The revised guidelines also require an organization applying for a people-to-people license to describe how the travel "would enhance contact with the Cuban people, and/or support civil society in Cuba, and/or promote the Cuban people's independence from Cuban authorities." Just as in 2011, the guidelines require applicants to certify that the predominant portion of activities engaged in will not be with prohibited Cuban government or Cuban Communist Party officials (as defined in 31 CFR 515.337 and 31 CFR

515.338), but the changes in May 2012 require that the sample itinerary for the proposed travel needs to specify how meetings with such officials advance purposeful travel by enhancing contact with the Cuban people, supporting civil society, or promoting independence from Cuban authorities (U.S. Department of the Treasury, OFAC, *Comprehensive Guidelines for License Applications to Engage in Travel-Related Transactions Involving Cuba*, Revised May 10, 2012).

CURRENT PERMISSIBLE TRAVEL TO CUBA

At present, certain categories of travelers may travel to Cuba under a general license, which means that there is no need to obtain special permission from OFAC. Nevertheless, those individuals traveling under a general license must be able to document that their travel qualifies under a general license, and must keep records for a period of five years after the travel transactions take place. Those eligible for travel under a general license include those visiting close relatives in Cuba; full-time journalists; full-time professionals conducting professional research (of a noncommercial, academic nature) or attending conferences sponsored by international professional organizations or associations; faculty, staff, and students of accredited U.S. graduate and undergraduate degree-making institutions engaged in one of several categories of educational activities in Cuba; members and staff of religious organizations engaged in a full-time program of in religious activities; and travel related to licensed sales of agricultural, medical, and telecommunications products.

In addition, a wide variety of travelers engaging in educational, religious, and humanitarian activities and people-to-people exchanges may be eligible for specific licenses. Applications for specific licenses are reviewed and granted by OFAC on a case-by-case basis. Some specific licenses may authorize multiple trips to Cuba over an extended period of time. Applicants for specific license have to wait for OFAC to issue the license prior to engaging in travel-related transactions.

The travel regulations can be found at 31 CFR 515.560, which references other sections of the Cuban Assets Control Regulations (CACR) for travel-related transaction licensing criteria. In addition, OFAC publishes Comprehensive Guidelines for License Applications to Engage in Travel-related Transactions Involving Cuba, which were most recently revised in May 2012, and a List of Authorized Providers of Air, Travel and Remittance Forwarding Services to Cuba, most recently revised in October 2012.[30]

The U.S. government does not collect data on the number of Americans traveling to Cuba,[31] although the Department of Transportation (DOT) collects statistics on the number of passengers on direct flights from the United States to Cuba. According to statistics from the DOT's database, there were almost 196,000 passengers on direct flights from the United States to Cuba in 2009, while the number rose to almost 242,000 in 2010, and to over 348,000 in 2011.[32] These statistics, however, do not reflect Americans that travel to Cuba through other countries, such as Canada, Mexico, or other Latin American and Caribbean nations. In addition, it is likely that some Americans (especially Cuban Americans visiting family) travel to Cuba multiple times within a one-year period so that overall number of visitors from the United States could be somewhat inflated. Various press estimates maintain that the overall number of Americans traveling to Cuba in 2011 was some 400,000, with the majority Cuban Americans visiting family.[33]

General License Categories

- **Family Visits.** Persons subject to the jurisdiction of the United States and persons traveling with them who share a common dwelling as a family visiting a close relative who is a national of Cuba or who is a U.S. government employee assigned to the U.S. Interests Section in Havana without limits on the duration or frequency of visits (31 CFR 515.561(a)). A close relative is defined as any individual related to the traveler by blood, marriage, or adoption who is no more than three generations removed from the traveler or from a common ancestor with the traveler (31 CFR 515.339).
- **Official Government Business.** Officials of the U.S. government, foreign governments, and certain intergovernmental organizations traveling on official business (31 CFR 515.562).
- **Journalistic Activity.** Persons regularly employed as journalists by a news reporting organization or by persons regularly employed as supporting broadcast or technical personnel (31 CFR 515.563(a)).
- **Professional Research and Meetings.** Full-time professionals conducting professional research in their areas (provided that the research is of a noncommercial, academic nature; that the research comprises a full work schedule in Cuba; and that the research has a substantial likelihood of public dissemination) or attending professional meetings or conferences in Cuba organized by an

international professional organization, institution, or association that regularly sponsors meetings or conferences in other countries (31 CFR 515.564(a)(1) and 515.564(a)(2)). A new category for professional meetings for commercial telecommunications transactions was added in September 2009 (31 CFR 515.564(a)(3)).

- **Educational Activities.** Accredited U.S. graduate and undergraduate degree-making institutions, including faculty, staff, and students involved in (1) participation in a structured educational program in Cuba as part of a course offered for credit by the sponsoring U.S. academic institution; (2) noncommercial academic research in Cuba specifically related to Cuba for the purpose of obtaining a graduate degree; (3) participation in a formal course of study at a Cuban academic institution, provided the formal course of study in Cuba will be accepted for credit toward the student's graduate or undergraduate degree; (4) teaching at a Cuban academic institution by an individual regularly employed in a teaching capacity at the sponsoring U.S. academic institution, provided that the teaching activities are related to an academic program at the Cuban institution and that the duration of the teaching will be no short than 10 weeks; or (5) sponsorship of a Cuban scholar to teach or engage in other scholarly activity at the sponsoring U.S. academic institution (31 CFR 515.565(a)).
- **Religious Activities.** Religious organizations located in the United States, including members and staff of such organizations engaged in a full-time program of religious activities (31 CFR 515.566(a)).
- **Travel Related to Sales of Agricultural Commodities, Medicine, or Medical Devices.** Employees of a producer or distributor of agricultural or medical commodities or an entity representing such a firm. The regulation also sets forth a requirement for written reports, before and after the trip, to be submitted to OFAC describing the purpose and scope of the travel and the business activities conducted (31 CFR 515.533(e)).
- **Travel Related to Sales of Telecommunication Items.** Employees of a U.S. telecommunications services provider or an entity representing such a provider. The regulation also requires written reports to be submitted to OFAC before and after the trip, describing the purpose and scope of the travel and the business activities conducted (31 CFR 515.533(f)).

Specific License Categories

- **Family Visits.** Persons subject to the United States and persons travelling with them who share a common dwelling as a family visiting a close relative who is neither a national of Cuba nor a U.S. government employee assigned to the U.S. Interests Section in Havana. (31 CFR 515.561(b)).
- **Free-lance Journalists**. Travel directly incident to journalistic activities for a free-lance project upon submission of an adequate written application with required documentation (31 CFR 515.563(b)).
- **Professional research and professional meetings.** Travel directly incident to professional research and professional meetings that do not qualify for a general license (31 CFR 515.564(b)).
- **Academic Educational Activities.** Travel for individuals to engage in academic educational activities (noncommercial academic research; participation in a formal course of study at a Cuban academic institution; or teaching at a Cuban academic institution) that are not authorized by a general license for educational activities (31 CFR 515.565(b)(1)).
- **People-to People Exchanges.** Travel for organizations authorizing educational exchanges not involving academic study pursuant to a degree program when those exchanges take place under the auspices of an organization that sponsors and organizes such programs to promote people-to-people contact (31 CFR 515.565(b)(2)).
- **Academic Seminars, Conference, and Workshops.** Accredited U.S. graduate or undergraduate degree-granting academic institutions to sponsor or co-sponsor academic seminars, conference, and workshops related to Cuba or global issues involving Cuba, including attendance at such events by faculty, staff, and students of the licensed institution (31 CFR 515.565(b)(3)).
- **Religious Activities.** Travel for religious activities by individuals or organizations that do not qualify for a general license (31 CFR 515.566(b)).
- **Public Performances, Clinics, Workshops, Athletic and Other Competitions and Exhibitions.** Travel for organizations and individuals participating in a public performance, clinic, workshop, athletic or other competition, or exhibition in Cuba. The event must be open for attendance and, in relevant situations, participation by the

Cuban public. All U.S. profits after costs must be donated to an independent nongovernmental organization in Cuba or a U.S.-based charity with the objective, to the extent possible, of promoting people-to-people contact or otherwise benefiting the Cuban people. Such donations are not required for certain amateur or semi-professional athletic competitions held under the auspices of international sports federations. (31 CFR 515.567).

- **Support for the Cuban People.** Those traveling for activities in support of the Cuban people, such as activities of recognized human rights organizations, activities designed to promote a rapid, peaceful transition to democracy, and activities intended to strengthen civil society (31 CFR 515.574).
- **Humanitarian Projects.** Those involved in humanitarian projects in Cuba, such as medical and health-related projects, construction projects, intended to benefit legitimately independent civil society groups, environmental projects, projects involving non-formal educational training, within Cuba or off island, on topics including civil education, journalism, advocacy and organizing, adult literacy and vocational skills, community-based grass roots projects, projects suitable to the development of small-scale enterprise, projects related to agricultural and rural development that promote independent activity, and projects involving the donation of goods to meet basic human needs (31 CFR 515.575).
- **Activities of Private Foundations or Research or Educational Institutes.** Those involved in activities of private foundations or research or education institutes that have an established interest in international relations to collect information related to Cuba for noncommercial purposes (31 CFR 515.576).
- **Exportation, Importation, or Transmission of Information or Informational Materials.** Those involved in the importation, exportation, or transmission of informational materials, defined as publications, films posters, phonograph records, photographs, microfilms, microfiche, tapes, compact disks, CD ROMS, artworks, news wire feeds, and other informational and informational articles (31CFR 515.545(b)).
- **Exportation of Licensable Products**. Those involved in activities related to marketing, sales negotiation, accompanied delivery, or servicing of exports to Cuba authorized by the Department of Commerce and who are not already authorized under general licenses

for activities related to marketing and sales of agricultural and medical products or telecommunications services (31CFR 515.533(g)).

Current Restrictions on Remittances

According to a 2007 Government Accountability Office (GAO) report, no reliable data exist for cash remitted directly or indirectly from the United States to Cuba, although the report maintained that data from several sources showed that worldwide remittances to Cuba amounted to between $900 million and $1 billion.[34]

In more recent years, remittance flows to Cuba are estimated to have been $1.2 billion in 2010 and $1.3 billion in 2011, according to a study by Manuel Orozco at the Inter-American Dialogue.[35] Another study by the Havana Consulting Group LLC reports significantly higher amounts of remittances that have increased from $1.3 billion in 2007 to almost $2.3 billion in 2011. This study argues that remittances to Cuba have been boosted significantly by several factors, including the Obama Administration's lifting of restrictions on family travel and remittances, economic policy changes in Cuba (such allowing more private sector activities), and reduced wire-transfer costs.[36]

U.S. restrictions on such remittances are regulated by the Cuban Assets Control Regulations (CACR) and, just like restrictions on travel, have changed over time.

- **Family Remittances.** As noted above, President Obama announced in April 2009 that restrictions on remittances to family members in Cuba would be lifted. In September 2009, OFAC issued amendments to the CACR implementing the Administration policy changes on remittances. The current regulations authorize a general license for family remittances and remove the limitation on the amount and frequency of family remittances that persons 18 years of age or older may provide to close relatives in Cuba (31 CFR 515.570(a)). As with the travel-related transactions, a close relative is defined as any individual related to the remitter by blood, marriage, or adoption who is no more than three generations removed from the remitter or from a common ancestor with the remitter. The regulations still prohibit remittances to certain officials of the Cuban government and Cuban Communist party. The regulations also authorize two one-time $1,000 emigration-related remittances (31 CFR 515.570(e)), raised from $500 previously. Depository institutions no longer need a specific license for sending remittances to Cuba, although both depository institutions and other licensed remittance forwarders are required to collect information showing compliance with remittance provisions (31 CFR 515.572(a)(3)).

- **Non-Family Remittances.** In January 2011, the Obama Administration restored a general license category for any U.S. person to send remittances of up to $500 per quarter to non-family members in Cuba, including, but not limited to, remittances to support the development of private businesses. These remittances, however, cannot be provided to senior Cuban government officials or senior members of the Cuban Communist Party (31 CFR 515.570(b)).
- **Remittances to Religious Organizations.** In January 2011, the Administration created a general license for remittances to religious institutions in Cuba in support of religious activities (31 CFR 515.570(c)). Prior to this, such remittances were authorized by a specific license.
- **Remittances to U.S. Students in Cuba.** In January 2011, the Administration created a general license authorizing remittances to close relatives who are students in Cuba pursuant to a general or specific license authorizing certain educational activities (31 CFR 515.570(d)).
- **Remittances to Independent Non-governmental Entities and Individuals in Cuba.** Any person subject to U.S. jurisdiction may apply for a specific license to provide remittances to independent non-governmental entities in Cuba, including but not limited to pro-democracy groups and civil society groups, or members of such entities, or to individuals or independent non-governmental entities to support the development of private businesses, including small farms. The types of activities for which transfers will be considered include, but are not limited to, assistance for an independent farmers' cooperative in purchasing goods or support to an independent group in operating a nursing home for the elderly (31 CFR 515.570(g)(1)).
- **Carrying of Remittances to Cuba.** The amount of total authorized remittances that may be carried to Cuba (for all types of remittances) is $3,000 (31 CFR 515.560(c)(4)(i)).

Enforcement of Travel Restrictions: Civil Penalties

Beginning in April 2003, OFAC began making available a regular listing of civil penalties enforcement information for its sanctions programs, including violations of the Cuba travel regulations.[37] According to a Treasury Department spokesman, the information was being made available to make the process more transparent to the public. Under the Trading with the Enemy Act, the Secretary of the Treasury may impose civil fines up to $65,000 per violation of the Cuban Assets Control Regulations.[38] According to OFAC, typical individual penalties have been much lower (see Table 1). Penalties against companies are generally much larger.

Since April 2003, enforcement actions for the Cuba travel regulations have included penalties against the following companies: Metso Minerals, Zim American Israeli Shipping Company, Playboy Enterprises, Omega World Travel, Mr. Travel, Havanatur & Travel Service, American Airlines, Cuba Paquetes, MRP Group Inc., Air Jamaica, Trek Tours (Rhode Island), Premiere Travel of Ohio, Hialeah Gardens Immigration Agency, Only Believe Ministries (Ohio), the Salvation Army (Texas Division), Beau Rivage Resorts Inc. (Mississippi), E & J Gallo Winery (California), the Four Oaks Foundation (New York), Pioneer Valley Travel (Massachusetts), the International Bicycle Fund (Washington State), Augsburg College (Minnesota), the U.S./Cuba Labor Exchange (Michigan), Coda International Tours Inc. (Florida), Travelocity.com (Texas), American Express Company (Mexico), Lakes Community Credit Union (Michigan), Sonida International (New York), Journey Corporation Travel Management (New York), RMO Inc. (Colorado), Tours International America (California), Aerovacations Inc. (California), Agoda Company (Thailand), Center for Cross Cultural Study Inc. (Massachusetts), Priceline.com (Connecticut), Magic USA Tours (Florida), Philips Electronics of North America Corporation (New York), and First Incentive Travel (Florida).

Many other companies have received penalties for violating other aspects of the Cuba embargo regulations, including some that have been assessed multi-million dollar penalties.

In addition, the listing shows that from 2004-2010 numerous individuals had civil penalties assessed or reached informal settlements for alleged violations of various restrictions under the Cuban Assets Control Regulations. Since 2004, according to the information provided on OFAC's website, over

900 individuals either have been assessed a penalty or reached an informal settlement for violations of the Cuba regulations (not just travel-related restrictions) with more than $1.2 million in penalties.

The number of individuals penalized by OFAC has fallen considerably since 2006 as backlogged cases have been resolved (see Table 1). A total of 290 individuals were penalized in 2004, 579 in 2005, 21 in 2006, and 17 in 2007. In 2008, OFAC reported that 32 individuals were penalized, with the majority for the purchase of Cuban cigars over the Internet, while in 2009, three individuals were penalized, and just one individual in 2010. In 2011, and to date in 2012, no individuals have been for violating the CACR.

Table 1. Cuba Sanctions: OFAC Penalties of Individuals, 2004-2010

Penalties	2004	2005	2006	2007	2008	2009	2010
Total Number	290	579	21	17	32	3	1
Total Dollar Amount $	448,603	629,917	52,779	41,712	52,172	25,175	525
Average Dollar Amount $	1,547	1,088	2,513	2,454	1,630	8,392	525

Source: Information drawn from civil penalties and enforcement information provided on OFAC's website, available at http://www.treasury.gov/resource-center /sanctions/CivPen /Pages/civpen-index2.aspx.

ARGUMENTS FOR LIFTING CUBA TRAVEL RESTRICTIONS

Those who argue in favor of lifting restrictions on travel to Cuba contend that the travel ban hinders U.S. efforts to influence political and economic conditions in Cuba. They maintain that the best way to realize change in Cuba is to lift restrictions, allowing a flood of U.S. citizens to travel and engage in conversations with average Cubans. They point to the influence of person-to-person contact in Russia and Eastern European nations, which they argue ultimately helped lead to the fall of communism in the Soviet bloc. They maintain that restricting travel by ordinary Americans prevents interaction and information exchanges with ordinary Cubans, exchanges that can help break down the Cuban government's tight control and manipulation of news; that the current travel ban actually supports the Cuban government in its efforts to restrict information provided to the Cuban people; and that it in effect supports the Cuban government's totalitarian control over the Cuban nation.

A second argument made by those who want to lift travel restrictions is that the ban abridges the rights of ordinary Americans to travel. They contend that such restrictions on the right to travel subvert the first amendment right of free speech. They maintain that the U.S. government should not limit the categories of travelers who can visit Cuba or subject many prospective travelers to the requirement of applying for specific licenses, subject to denial, in order to engage in peopleto-people contact.

Those in favor of lifting the travel ban also argue that U.S. citizens can travel to other communist or authoritarian governments around the world, such as the People's Republic of China, Vietnam, Burma, and Iran. They point out that Americans could travel to the Soviet Union before its breakup. Supporters of changing travel policy toward Cuba argue that their proposals would still allow the President to prohibit such travel in times of war or armed hostilities, or if there were imminent danger to the health or safety of Americans. They argue that these conditions do not exist with regard to Cuba, and point to a May 1998 Defense Intelligence Agency report that concluded that "Cuba does not pose a significant military threat to the U.S. or to other countries in the region."[39]

Those arguing for lifting travel restrictions also point to human rights activists in Cuba who themselves argue for the lifting of such sanctions. According to the prominent Cuban human rights activist Elizardo Sanchez: "The more Americans on the streets of Cuban cities, the better for the cause of a more open society in Cuba."[40] Miriam Leiva, founder of the Ladies in White human rights group, and Oscar Espinosa Chepe, a formerly jailed independent economist on conditional release, support lifting travel restrictions for all Americans, maintaining that Americans could help Cuba's efforts for democracy by sharing simple conversation and sharing every-day experiences with Cubans.[41]

Supporters of lifting the travel ban maintain that such a move could be done without lifting the underlying U.S. embargo on trade and financial transactions with Cuba. They point to the 1977- 1982 period when the travel ban was essentially lifted, but the overall embargo remained in place.

Finally, some supporters of lifting the travel restrictions argue that the U.S. economy would benefit from increased demand for air and cruise travel, which reportedly would expand U.S. economic output. According to a report prepared for the Center for International Policy, a policy group that advocates lifting the embargo, U.S. economic output would expand by $1.18 billion-

$1.61 billion, with the creation of between 16,888 and 23,020 jobs if travel restrictions were lifted.[42]

ARGUMENTS FOR MAINTAINING CUBA TRAVEL RESTRICTIONS

Those favoring the continuation of restrictions on travel to Cuba point out that there are already significant provisions in U.S. law permitting Americans to travel there for legitimate reasons that support the Cuban people and not the Cuban government. They point out that thousands of Americans travel to Cuba legally under the various provisions of the Cuban embargo regulations, and that now Cuban Americans may visit close relatives without restrictions. Other categories of travel allowed include students, journalists, researchers, artists, musicians, and athletes.

A second argument made for maintaining restrictions on travel to Cuba is that lifting the travel ban entirely will open the floodgates to American tourist travel that will support Raúl Castro's rule by providing his government with millions in tourist receipts. Advocates of restricting travel oppose any loosening that could prolong the Castro regime by propping it up with increased income. In contrast to those supporting tourist travel, they believe that continued travel restrictions will help influence Cuba's policy. They argue that since the collapse of the Soviet Union and the loss of Soviet subsidies to Cuba, the travel and embargo regulations have contributed to the Cuban government's decision to cut the military's size and budget by half since 1989 and to introduce limited economic reforms. Lifting travel restrictions, they argue, would eliminate the U.S. leverage on Cuba to enact further reforms and to improve the human rights situation.

Those favoring the maintenance of travel restrictions argue that the reality of the human rights situation dispels the notion that American tourists would be engaging in exchanges with ordinary Cubans. They maintain that the thousands of European, Canadian, and other tourists who travel to Cuba each year largely stay in tourist hotels that are off-limits to most Cubans and thus have no discernible effect on the human rights situation in Cuba.

Some opposed to lifting travel restrictions argue that there should not be tourist travel as long as Cuba provides refuge to violent criminals who have escaped U.S. justice. The State Department maintains that more than 70 fugitives from U.S. justice are hiding out in Cuba, including convicted murderer Joanne Chesimard, who killed a New Jersey state trooper in 1973.

Finally, many opponents of legislation to lift the Cuba travel restrictions argue that the authority to impose such restrictions is an important foreign policy tool for the President. They point out that the President has the authority to restrict travel when it is in the national security or foreign policy interests of the United States, and has utilized that policy tool when needed. They point to past instances of restricting travel to Libya, Vietnam, and North Korea. With regard to Cuba, they point to the 1984 Supreme Court decision in the case of *Regan v. Wald* that upheld restrictions on travel to Cuba imposed by the Reagan Administration.

LEGISLATIVE INITIATIVES IN THE 112TH CONGRESS

As noted above, in the first session of the 112th Congress, there were several attempts aimed at rolling back the Obama Administration's actions easing restrictions on travel and remittances. Several initiatives were also introduced that would further ease or lift such restrictions altogether.

FAA Reauthorization

During consideration of the Federal Aviation Administration reauthorization bill, S. 223, in February 2011, an amendment was submitted, but never considered, S.Amdt. 61 (Rubio), that would have prohibited an expansion of flights to locations in countries that are state sponsors of terrorism (which includes Cuba).

FY2012 Financial Services and General Government Appropriations

The House Appropriations Committee reported its version of the FY2012 Financial Services and General Government Appropriations bill, H.R. 2434, on July 7, 2011, with a provision in Section 901 that would have rolled back the Obama Administration's actions easing restrictions on family travel and on remittances overall. (The Senate Appropriations Committee version of the measure, S. 1573, did not contain a similar provision.) The House provision had been offered as an amendment by Representative Mario Diaz-Balart that was agreed to by voice vote during the committee's June 24, 2011, markup of the measure. The provision would have repealed amendments to the Cuban Assets Control Regulations made since January 19, 2009, regarding family travel (31 CFR 515.561), carrying remittances (31 CFR 515.560(c)(4)(i)), and sending remittances to Cuba (31 CFR 515.570). According to the provision, such regulations would be restored and carried out as in effect on January 19, 2009, notwithstanding any guidelines, opinions, letters, presidential directives, or agency practices relating to such regulations that are issued or carried out after such date.

If the provision were to be enacted, family travel would be limited to once every three years for a period of up to 14 days and would require a specific license from the Treasury Department; licensed travelers would be allowed to carry just $300 in remittances compared to the $3,000 currently allowed; family remittances would be limited to $300 per quarter; non-family remittances restored by the Obama Administration, up to $500 per quarter, would not be allowed; and the general license for remittances to religious organizations would be eliminated, with such remittances permitted via specific license.

The White House's Statement of Administration Policy on H.R. 2434, issued July 13, 2011, stated that Administration opposed Section 901 because it would reverse the President's policy on family travel and remittances, and that the President's senior advisors would recommend a veto if the bill contained the provision. According to the statement, Section 901 "would undo the President's efforts to increase contact between divided Cuban families, undermine the enhancement of the Cuban people's economic independence and support for private sector activity in Cuba that come from increased remittances from family members, and therefore isolate the Cuban people and make them more dependent on Cuban authorities."[43]

A second Cuba amendment agreed to by voice vote during the markup of H.R. 2434 was offered by Representative Jeff Flake. The amendment made changes to the committee report to the bill (H.Rept. 112-136) and would require a report from OFAC on the current number of pending applications seeking specific licenses related to educational exchanges not involving academic study pursuant to a degree program under the auspices of an organization that sponsors and organizes such programs to promote people-to-people contact. The report also would require information on the number of these licenses that OFAC has approved to date, its plan for getting through the current queue of license applications, and its plan for expeditiously reviewing those applications in the future.

In November 2011, an attempt to include the Senate version of the Financial Services appropriations measure, S. 1573, in a "minibus" with two other full-year appropriations measures and a short-term continuing resolution failed in part because of disagreement over a Cuba provision that would have allowed direct transfers from a Cuban financial institution to a U.S. financial institution to pay for U.S. agricultural and medical exports to Cuba.

In December 2011, a legislative battle ensued over the Consolidated Appropriations Act, FY2012, H.R. 2055, a "megabus" bill that combined nine full-year appropriations measures, including the Financial Services and General Government bill. At issue was the potential inclusion of two Cuba provisions that had been in the House Appropriations Committee-approved version of the Financial Services bill, H.R. 2434: one described above that would roll back to January 2009 the Obama Administration's actions easing restrictions on family travel and on remittances; and the second a provision that would continue to clarify, for the third fiscal year in a row, the definition of "payment of cash in advance" for U.S. agricultural and medical exports to Cuba so that the payment was due upon delivery in Cuba as opposed to being due before the goods left U.S. ports. (The text of the two Cuba provisions was also included in Division C, Sections 632 and 634, of H.R. 3671, a new "megabus" bill introduced by House Republicans on December 14, 2011.)

Ultimately congressional leaders agreed to not include the two Cuba provisions in H.R. 2055 (H.Rept. 112-331) and the measure was approved by the House and Senate, respectively, on December 16 and 17, 2011, and signed into law on December 23, 2011 (P.L. 112-74). The White House reportedly had exerted strong pressure not to include the Cuba provision that would have rolled back the Administration's easing of restrictions on travel and remittances. Dropping the second provision on the definition of "payment of cash in advance" for U.S. agricultural and medical products appears to have been a political tradeoff made to compensate for the travel rollback provision being dropped.

FY2012 Foreign Relations Authorization Act

In other congressional action, on July 21, 2011, the House Committee on Foreign Affairs marked up H.R. 2583 (H.Rept. 112-223), the FY2012 Foreign Relations Authorization Act, with a provision (§1126 of the reported bill) that would require the President to fully enforce all U.S. regulations on travel to Cuba as in effect on January 19, 2009, and impose the corresponding penalties against individuals determined to be in violation of such regulations. The provision was added by an amendment offered by Representative David Rivera, approved 36-6, that had the intent of reinstating tighter travel restrictions as they existed under the Bush Administration in January 2009.

Amendments to the Cuban Adjustment Act

Two additional measures introduced in August 2011 would amend the Cuban Adjustment Act of 1966 (CAA, P.L. 89-732) in order to curb travel to Cuba by Cubans who have recently immigrated to the United States. Introduced on August 1, 2011, H.R. 2771 (Rivera) would amend the CAA to increase to five years the period during which a Cuban national must be physically present in the United States in order to qualify for adjustment of status to that of a permanent resident. The legislation also would provide that an alien shall be ineligible for adjustment to permanent resident status if the alien returns to Cuba after admission or parole into the United States before becoming a U.S. citizen. A more recent version, H.R. 2831 (Rivera), introduced August 30, 2011, just contains the provision maintaining that an alien from Cuba shall be ineligible for adjustment to permanent resident status under the CAA if he or she returns to Cuba before becoming a U.S. citizen. The House Committee on the Judiciary, Subcommittee on Immigration on Policy Enforcement, held a hearing on H.R. 2831 on May 31, 2012 (available at http://judiciary.house.gov/hearings/Hearings%202012/hear_ 05312012_3.html).

Initiatives to Ease Restrictions on Travel and Remittances

In contrast to measures aimed at rolling back the Obama Administration's polices easing travel and remittances to Cuba, several measures would ease of lift travel restrictions altogether. H.R. 1886 (Rangel) would prohibit restrictions on travel to Cuba. H.R. 1888 (Rangel), in addition to removing some restrictions on the export of U.S. agricultural products to Cuba, would also prohibit Cuba travel restrictions. Two initiatives that would lift the overall embargo on trade and restrictions on financial transaction with Cuba, H.R. 255 (Serrano) and H.R. 1887 (Rangel), would also lift restrictions on travel and remittances to Cuba. H.R. 380 (Lee) would provide that no funds made available to the Department of the Treasury may be used to implement, administer, or enforce regulations to require specific licenses for travel-related transactions directly related to educational activities in Cuba.

LEGISLATIVE ACTION FROM THE 106TH TO THE 111TH CONGRESS

Legislative Initiatives in the 111th Congress

The 111th Congress took action in March 2009 to ease restrictions on family travel and travel for the marketing and sale of agricultural and medical goods. The eased family travel restrictions were superseded by the Obama Administration's April 2009 action to allow unlimited family travel and remittances. At the same time, the Administration also eased restrictions for travel for telecommunications-related sales and for attendance at professional meetings related to commercial telecommunications. Numerous other bills introduced in the 111th Congress would have lifted or eased restrictions on travel and remittances to Cuba, but these restrictions were not considered. One House initiative, H.R. 4645 (Peterson), would have lifted all restrictions on travel to Cuba and also would have eased restrictions on the payment mechanisms for U.S. agricultural exports to Cuba. The House Agriculture Committee approved the measure, but no further action was taken on the bill.

First Session Action

On March 11, 2009, President Obama signed into law the Omnibus Appropriations Act, 2009 (P.L. 111-8), with two provisions easing restrictions on travel to Cuba. (The provisions were identical to provisions that had been included in the Senate Appropriations Committee version of the FY2009 Financial Services and General Government Appropriations bill in the 110th Congress, S. 3260.)

In the enacted bill, Section 620 of Division D, Financial Services and General Government Appropriations Act, 2009, amended the Trade Sanctions Reform and Export Enhancement Act of 2000 (TSRA) to require the Secretary of the Treasury to issue regulations for travel to, from, or within Cuba under a general license for the marketing and sale of agricultural and medical goods, meaning that there would be no requirement to obtain special permission from

OFAC. Such travel had required a specific license from OFAC, issued on a case-by-case basis. OFAC issued regulations implementing this provision on September 3, 2009.

Section 621 of Division D prohibited funds from being used to administer, implement, or enforce family travel restrictions that were imposed by the Bush Administration in June 2004. OFAC implemented this provision by reinstating a general license for family travel as it existed prior to the Bush Administration's tightening of restrictions in June 2004. As implemented by the Treasury Department, travel was allowed once every 12 months to visit a close relative for an unlimited length of stay, and the limit for daily expenditure allowed by family travelers became the same as for other authorized travelers to Cuba (the State Department maximum per diem rate for Havana). The new general license also expanded the definition of "close relative" to mean any individual related to the traveler by blood, marriage, or adoption who is no more than three generations removed from that person. This provision was superseded by the Obama Administration's further liberalization of family travel to Cuba announced in April 2009.

The joint explanatory statement to P.L. 111-8 also required the Department of the Treasury to prepare a report within 90 days on the steps that it is taking to assess OFAC's allocation of resources for investigating and penalizing violations of the Cuba embargo with respect to the numerous other sanctions programs it administers. As part of the report, the Treasury Department was directed to provide detailed information on OFAC's Cuba-related licensing on its enforcement of the Cuba embargo.

On November 19, 2009, the House Committee on Foreign Affairs held a hearing on U.S. restrictions on travel to Cuba entitled "Is it Time to Lift the Ban on Travel to Cuba?" that featured former U.S. government officials and other private witnesses.

Second Session Action

In the second session, the only legislative action related to Cuba travel restrictions occurred in the House Committee on Agriculture, and no subsequent action was taken. On March 11, 2010, the committee held a hearing to review U.S. agricultural sales to Cuba. At the hearing, there was discussion of recently introduced H.R. 4645 (Peterson), a measure that would remove restrictions on travel to Cuba and also remove some restrictions regarding payments for U.S. agricultural exports to Cuba. On June 30, 2010, the committee reported out H.R. 4645 by a vote of 25-20 (H.Rept. 111-653). The bill would have lifted all restrictions on travel to Cuba. It also included two provisions easing restrictions on the payment mechanisms for U.S. agricultural exports to Cuba. The House Committee on Foreign Affairs was scheduled to hold a markup of the bill on September 29, 2010, but postponed its consideration, and in the aftermath of the 2011 U.S. legislative elections, no further action was taken. An identical companion bill in the Senate, S. 3112 (Klobuchar), was introduced March 15, 2010, and referred to the Committee on Foreign Relations.

On April 29, 2010, the House Ways and Means Committee, Subcommittee on Trade, held a hearing on U.S.-Cuba policy that examined whether relaxing current Cuba travel and trade restrictions would advance U.S. economic objectives, as well as U.S. political and human rights goals in Cuba.

Additional Initiatives in the 111th Congress

Several other legislative initiatives were introduced in the 111th Congress that would have eased restrictions on travel to Cuba, but no action was taken on these measures. H.R. 874 (Delahunt)/S. 428 (Dorgan) and H.R. 1528 (Rangel) would have prohibited restrictions on travel to Cuba. H.R. 188 (Serrano), H.R. 1530 (Rangel), and H.R. 2272 (Rush) would have lifted the overall embargo on trade and financial transactions with Cuba, including travel restrictions. H.R. 1531 (Rangel)/S. 1089 (Baucus) would have facilitated the export of U.S. agricultural products to Cuba and also would have prohibited restrictions on travel to Cuba. H.R. 332 (Lee) would have eased restrictions on educational travel by providing that no funds made available to the Department of the Treasury may be used to implement, administer, or enforce regulations to require specific licenses for travel-related transactions directly related to educational activities in Cuba. S. 774 (Dorgan), H.R. 1918 (Flake), and S. 1517 (Murkowski) would have amended the Trade Sanctions Reform and Economic Enhancement Act of 2000 to require the Secretary of the Treasury to authorize travel to Cuba under a general license in connection to hydrocarbon exploration and extraction activities. In contrast, H.Con.Res. 132 (Tiahrt) would have called for the fulfillment of certain democratic conditions before the United States increases trade and tourism to Cuba.

Legislative Initiatives in the 110th Congress

In the 110th Congress, several House and Senate committee versions of appropriations bills had provisions that would have eased restrictions on travel to Cuba in various ways, but none of these provisions were included in final enacted legislation. Numerous other bills were introduced that would have eased restrictions on travel and remittance in various ways, but no action was taken on these measures.

First Session Action

In the first session of the 110th Congress, two Senate Appropriations Committee-reported versions of appropriations bills had provisions that would have eased restrictions on travel to Cuba for the marketing and sale of agricultural and medical goods, but ultimately these provisions were not included in the FY2008 Consolidated Appropriations Act (P.L. 110-161). The Senate version of the FY2008 Financial Services and General Government appropriations bill, reported July 19, 2007, H.R. 2829, had a provision in Section 620 that would eased such travel restrictions, while the Senate version of the FY2008 Agriculture appropriations bill, S. 1859, reported July 24, 2007, had such a provision in Section741.

Second Session Action

In the second session, several versions of House and Senate appropriations bills had provisions easing Cuba travel restrictions and other Cuba sanctions, but none of these were included in the FY2009 continuing resolution. The House Appropriations Committee approved its version of the Financial Services and General Government Appropriations bill for FY2009 on June 25, 2008, which contained provisions in Title VI that would have eased restrictions on the sale of U.S. agricultural exports to Cuba and on family travel to Cuba. The committee ultimately introduced and reported the bill, H.R. 7323, on December 10, 2008 (H.Rept. 110-920). With regard to family travel, Section 622 would have allowed for such travel once a year (instead of the current restriction of once every three years), while Section 623 would have expanded such travel by a person to visit an aunt, uncle, niece, nephew, or first cousin (instead of the current restriction limiting such travel to visit a spouse, child, grandchild, parent, grandparent, or sibling).

On July 14, 2008, the Senate Appropriations Committee reported its version of the FY2009 Financial Services and General Government Appropriations bill, S. 3260 (S.Rept. 110-417), which included provisions easing restrictions on family travel and on travel to Cuba relating to the commercial sale of agricultural and medical goods. With regard to family travel, Section 620 would have provided that no funds could be used to administer, implement, or enforce the Administration's June 2004 tightening of restrictions related to travel to visit relatives in Cuba. With regard to travel for agricultural or medical sales, Section 619 would have allowed for a general license for such travel instead of a specific license that requires permission from the Treasury Department.

On July 21, 2008, the Senate Appropriations Committee reported its version of the FY2009 Agriculture Appropriations bill, S. 3289 (S.Rept. 110-426), with a provision in Section 737 that would have eased restrictions on travel to Cuba for the sale of agricultural and medical goods. The provision would have allowed for a general license for such travel instead of a specific license that requires permission from the Treasury Department. The measure had been approved by the committee on July 17, 2008.

Additional Initiatives in the 110th Congress

A number of other initiatives introduced in the 110th Congress would have eased Cuba travel restrictions. H.R. 654 (Rangel), S. 721 (Enzi), and Section 254 of S. 554 (Dorgan) would prohibit the President from regulating or prohibiting travel to Cuba or any of the transactions incident to travel. Two bills that would lift overall economic sanctions—H.R. 217 (Serrano) and H.R. 624 (Rangel)—would also lift travel restrictions. H.R. 177 (Lee) would ease restrictions on educational travel to Cuba. H.R. 757 (Delahunt) would lift restrictions on family travel and the provision of remittances for family members in Cuba. H.R. 1026 (Moran, Jerry), which would facilitate the sale of U.S. agricultural products to Cuba, includes a provision that would provide for general license authority for travel-related transactions for people involved in agricultural sales and marketing activities or in the transportation of such sales. H.R. 2819 (Rangel) and S. 1673 (Baucus), which would ease restrictions on U.S. agricultural and medical exports to Cuba, would also lift restrictions on travel to Cuba. The Senate Committee on Finance held a hearing on S. 1673 on December 11, 2007.

Legislative Initiatives in the Aftermath of 2008 Hurricanes

In the aftermath of the Hurricanes Gustav and Ike that struck Cuba in late August and early September 2008, several legislative initiatives were introduced that would have temporarily eased U.S. embargo restrictions in several areas, including restrictions on family travel, remittances, the provision of gift parcels, and the sale of relief supplies to Cuba. On September 15, 2008, Senator Dodd offered S.Amdt. 5581 to the Department of Defense authorization bill (S. 3001) that would have, for a 180-day period: allowed unrestricted family travel; eased restrictions on remittances by removing the limit and allowing any American to send remittances to Cuba; expanded the list of allowable items that may be included in gift parcels; and allowed for unrestricted U.S. cash sales of food, medicines, and relief supplies to Cuba. The amendment was not considered, and therefore not part of the final bill.

In the House, two legislative initiatives were introduced in the aftermath of the hurricanes that would have temporarily eased restrictions in various ways. On September 16, 2008, Representative Flake introduced H.R. 6913, which would have prohibited any funds from going to the Department of Commerce to implement, administer, or enforce tightened restrictions on the contents of gift parcels to Cuba that were introduced in June 2004. On September 18, 2008, Representative Delahunt introduced H.R. 6962, the Humanitarian Relief to Cuba Act, which would have, for a 180-day period: allowed unrestricted family travel; eased restrictions on remittances by removing the limit and allowing any American to send remittances to Cuba; and expanded the list of allowable items that may be included in gift parcels.

Legislative Initiatives in the 109th Congress

In the 109th Congress, several amendments to FY2006 and FY2007 appropriations bills that would have eased Cuba travel restrictions in various ways and restrictions on sending gift parcels to Cuba were defeated. Several bills were introduced that would have lifted or eased restrictions on travel and the provision of remittances to Cuba, but no action was taken on these measures.

First Session Action

On June 30, 2005, the House rejected three amendments easing Cuba sanctions to H.R. 3058, the FY2006 Transportation, Treasury, Housing and Urban Development, Judiciary, District of Columbia, and Independent Agencies Appropriations Act. The amendments failed during House floor consideration: H.Amdt. 420 (Davis) on family travel, by a vote of 208-211; H.Amdt. 422 (Lee) on educational travel, by a vote of 187-233; and H.Amdt. 424 (Rangel) on the overall embargo, by a vote of 169-250. An additional amendment on religious travel, H.Amdt. 421 (Flake), was withdrawn, and an amendment on family travel by members of the U.S. military, H.Amdt. 419 (Flake), was ruled out of order for constituting legislation in an appropriations bill. The introduction of H.Amdt. 419 was prompted by the case of a U.S.

military member who served in Iraq, Sgt. Carlos Lazo, who was prohibited from visiting his two sons in Cuba because he last visited there in 2003.

During June 29, 2005, Senate consideration of H.R. 2361, the FY2006 Interior, Environment, and Related Agencies Appropriations Act, the Senate rejected (60-35; a two-thirds majority vote was required) a motion to suspend the rules with respect to S.Amdt. 1059 (Dorgan), which would have allowed travel to Cuba under a general license for the purpose of visiting a member of the person's immediate family for humanitarian reasons. The amendment was then ruled out of order. Its introduction had also been prompted by the case of Sgt. Carlos Lazo who wanted to visit his sons in Cuba, one of whom was gravely sick.

On June 15, 2005, the House rejected (210-216) H.Amdt. 270 (Flake) to H.R. 2862, the FY2006 Science, State, Justice, Commerce, and Related Agencies Appropriations Act. The amendment would have prohibited the use of funds to implement, administer, or enforce June 2004 tightened restrictions on sending gift parcels to Cuba. H.Amdt. 269 (McDermott), which would have prohibited the use of funds in the bill to prosecute any individual for travel to Cuba, was offered but subsequently withdrawn.

During April 6, 2005, Senate floor consideration of the FY2006 and FY2007 Foreign Affairs Authorization Act, S. 600, the Senate considered S.Amdt. 281 (Baucus) and a second-degree amendment, S.Amdt. 282 (Craig) that would have facilitated the sale of U.S. agricultural products to Cuba. The language of the amendments consisted of the provisions of S. 328 (Craig), the Agricultural Export Facilitation Act of 2005, which included a provision for a general license for travel transactions related to the marketing and sale of agricultural products, as opposed to the current requirement of a specific license for such travel transactions. Neither action on the amendments nor on S. 600 was completed.

Second Session Action

On June 14, 2006, the House rejected two amendments to the FY2007 Transportation/Treasury appropriation bill, H.R. 5576 that would have eased Cuba travel restrictions. H.Amdt. 1050 (Rangel), rejected by a vote of 183-245, would have prohibited funds from being used to implement the overall economic embargo of Cuba. H.Amdt. 1051 (Lee), rejected by a vote of 187-236, would have prohibited funds from being used to implement the Administration's June 2004 tightening of restrictions on educational travel to Cuba. An additional Cuba amendment, H.Amdt. 1032 (Flake), would have prohibited the use of funds to amend regulations relating to travel for religious activities in Cuba; it was withdrawn from consideration.

In other action, on June 22, 2006, the Senate Appropriations Committee reported its version of the FY2007 Agriculture appropriations bill, H.R. 5384 (S.Rept. 109-266), which contained a provision (§755) liberalizing travel to Cuba related to the sale of agricultural and medical goods. The provision would have provided for such travel under a general license, instead of under a specific license as currently allowed, issued on a case-by-case basis by the Treasury Department. Final action on the appropriations measure was not completed by the end of the 109th Congress. Similar Senate provisions in FY2004 and FY2005 agricultural appropriations bills were stripped out of the final enacted measures.

Additional Initiatives in the 109th Congress

A number of other legislative initiatives were introduced in the 109th Congress that would have eased restrictions on travel and remittances to Cuba. Two bills—S. 894 (Enzi) and H.R. 1814 (Flake)—would have specifically lifted overall restrictions on travel to Cuba. H.R. 2617 (Davis) would have prohibited any additional restrictions on per diem allowances, family visits to Cuba, remittances, and accompanied baggage beyond those that were in effect on June 15, 2004. H.R. 3064 (Lee) would have prohibited the use of funds available to the Department of the Treasury to implement regulations from June 2004 that tightened restrictions on travel to Cuba for educational activities. H.Con.Res. 206 (Serrano), introduced in the aftermath of Hurricane Dennis that struck Cuba in July 2005 (causing 16 deaths and significant damage), would have expressed the sense of Congress that the President should temporarily suspend restrictions on remittances, gift parcels, and family travel to Cuba to allow Cuban-Americans to assist their relatives.

Two bills—H.R. 208 (Serrano) and H.R. 579 (Paul)—would have lifted the overall embargo on trade and financial transactions with Cuba, including restrictions on travel and remittances to Cuba.

Finally, two identical bills dealing with easing restrictions on exporting agricultural commodities to Cuba—H.R. 719 (Moran of Kansas) and S. 328 (Craig)—included provisions that would have provided for a general license for travel transactions related to the marketing and sale of agricultural products, as opposed to the current requirement of a specific license for such travel transactions.

Legislative Initiatives in the 108th Congress[44]

In the 108th Congress, several FY2004 and FY2005 appropriations bills had provisions that would have eased Cuba travel restrictions in various ways, but ultimately these provisions were not included in final appropriations measures. The Administration had threatened to veto legislation if it contained provisions weakening Cuba sanctions. In addition, several bills in the 108th Congress were introduced that specifically would have lifted or eased restrictions on travel to Cuba, but no action was taken on these measures.

First Session Action

Since action on FY2003 Treasury Department appropriations was not completed before the end of the 107th Congress, the 108th Congress faced early action on it and other unfinished FY2003 appropriations measures. The final version of the FY2003 omnibus appropriations measure, H.J.Res. 2 (P.L. 108-7), which included Treasury Department appropriations, did not include provisions affecting restrictions on travel to Cuba. The White House had threatened to veto the measure if it contained provisions weakening the embargo. While the Senate version did not include the Senate Appropriations Committee provision from the 107th Congress that would have eased travel restrictions by prohibiting any funding for enforcing the Cuba travel regulations, it did include a provision (contained in Division J, Section 124) that would have expedited action on travel applications for travel by OFAC within 90 days of receipt. Ultimately, however, the Senate provision was dropped in the conference report (H.Rept. 108-10) on the omnibus measure.

Both the House and Senate versions of the FY2004 Transportation-Treasury appropriations bill, H.R. 2989, had nearly identical provisions that would have prevented funds from being used to administer or enforce restrictions on travel or travel-related transactions. But the provisions were dropped in the conference report to the FY2004 Consolidated Appropriations Act, P.L. 108-199 (H.R. 2673, H.Rept. 108-401, filed November 25, 2003), which incorporated seven regular appropriations acts, including Transportation-Treasury appropriations. The conference also dropped two Cuba provisions from the House version of H.R. 2989 that would have eased restrictions on remittances and on people-to-people educational exchanges. The White House again threatened to veto any legislation that would weaken economic sanctions against Cuba.

The House provisions had been approved during September 9, 2003, House floor consideration of the H.R. 2989: H.Amdt. 375 (Flake), approved by a vote of 227-188, would have prevented funds from enforcing travel restrictions (§745 of the House version); H.Amdt. 377 (Delahunt), approved by a vote of 222-196, would have prevented funds from enforcing restrictions on remittances (§746); and H.Amdt. 382 (Davis), approved by a vote of 246-173, would have prohibited funds from being used to eliminate the travel category of people-to-people educational exchanges (§749).

During Senate floor consideration of H.R. 2989 on October 23, 2003, the Senate approved by voice vote S.Amdt. 1900 (Dorgan), nearly identical to the Flake amendment noted above that would have prevented funds from being used to administer or enforce restrictions on travel or travel-related transactions (§643 of the Senate version). A motion to table the Dorgan amendment was defeated by a vote of 59-36. The Senate approved the bill by a vote of 91-3. The only difference between the Senate and House language was that the Dorgan amendment, as amended by S.Amdt. 1901 (Craig), provided that the section would take effect one day after enactment of the bill.

In other action, the conference on the FY2004 Consolidated Appropriations Act, P.L. 108-199 (H.R. 2673), also dropped a provision in the Senate version of the FY2004 agriculture appropriations bill that would have allowed travel to Cuba under a general license for travel related to the sale of agricultural and medical goods. On July 17, 2003, the Senate Appropriations Committee approved its version of the FY2004 agriculture appropriations bill, S. 1427, that included a provision (§760) allowing travel to Cuba under a general license (which does not require applying to the Treasury Department) for travel related to the commercial sale of agricultural and medical goods. The Senate included this provision when it approved H.R. 2673 on November 6, 2003. The House-passed version of the bill, H.R. 2673, had no such provision. At present, such travel to Cuba is allowed with OFAC's approval of a specific license. In early June 2003, the Treasury Department rejected an application to travel to Cuba for organizers of a second U.S. food and agribusiness fair in Havana.[45] The first such trade fair, held in September 2002, featured some 288 exhibitors from more than 30 states and resulted in millions in U.S. agricultural sales to Cuba.[46]

Second Session Action

Several FY2005 appropriations measures had provisions that would have eased Cuba sanctions, but these were dropped in the FY2005 omnibus appropriations measure (H.R. 4818, H.Rept. 108- 792).

The House-passed version of the FY2005 Commerce, Justice, and State appropriations bill, H.R. 4754, approved July 8, 2004 (397-18), included a provision (§801) that would have prohibited funds from being used to implement, administer, or enforce recent amendments to the Cuba embargo regulations that tightened restrictions on gift parcels and baggage taken by individuals for travel to Cuba. The provision was added by a Flake amendment, H.Amdt. 647, approved by a vote of 221-194 on July 7, 2004. The Senate version of the bill, S. 2809, as reported out of committee, did not include such a provision.

Both the House-approved version of the FY2005 Transportation/Treasury appropriations bill, H.R. 5025, and the Senate Appropriations Committee version of the bill, S. 2806, had provisions that would have eased Cuba sanctions in various ways. In its statement of policy on H.R. 5025, the Administration indicated that the President would veto the measure if it contained provisions weakening Cuba sanctions.

The House-passed version of H.R. 5025 had three provisions that would have eased Cuba sanctions. During floor consideration on September 21, 2004, by a vote of 225-174, the House approved a Davis (of Florida) amendment (H.Amdt. 769), which provided that no funds could be used to administer, implement, or enforce the Bush Administration's June 2004 tightening of restrictions on visiting relatives in Cuba. On September 22, 2004, the House approved two additional Cuba amendments by voice vote, a Waters amendment (H.Amdt. 770) that would have prohibited funds from being used to implement any sanction imposed on private commercial sales of agricultural commodities or medicine or medical supplies to Cuba and a Lee amendment (H.Amdt. 771) that would have prohibited funds from being used to implement, administer, or enforce the Bush Administration's June 2004 tightening of restrictions on travel for educational activities. The House also rejected a Rangel amendment (H.Amdt. 772) on September 22, 2004, by a vote of 225-188 that would have more broadly prohibited funds from being used to implement, administer, or enforce the economic embargo of Cuba. During September 15, 2004, House floor consideration of H.R. 5025, Representative Jeff Flake announced his intention not to offer an amendment, as he had for the past three years, which would have prohibited funds from being used to administer or enforce restrictions on travel or travel-related transactions.

The Senate version of the FY2005 Transportation/Treasury appropriations bill, S. 2806, as reported out of the Senate Appropriations Committee (S.Rept. 108-342) on September 15, 2004, had a provision (§222) that would have prohibited funds from administering or enforcing restrictions on Cuba travel or travel-related transactions. That provision, which was proposed by Senator

Byron Dorgan, was unanimously approved by the Subcommittee on Transportation, Treasury, and General Government on September 9, 2004.

The Senate version of the FY2005 Agriculture Appropriation bill, S. 2803, as reported by the Senate Appropriations Committee (S.Rept. 108-340), had a provision (§776) that would have directed the Secretary of the Treasury to promulgate regulations allowing for travel to Cuba under a "general license" when it was related to the commercial sale of agricultural and medical products. The House-passed version of the bill, H.R. 4766, had no such provision. In its statement of policy on the bill, the Administration stated that the President would veto the measure if it contained a provision weakening Cuba sanctions.

Additional Initiatives in the 108th Congress

Among other initiatives introduced in the 108th Congress, but not acted upon, two bills would specifically have lifted restrictions on travel to Cuba: S. 950 (Enzi), introduced April 30, 2003, and H.R. 2071 (Flake), introduced May 13, 2003. H.R. 3422 (Serrano), introduced October 30, 2003, would, among other provisions, have lifted restrictions on travel to Cuba. Three broad legislative initiatives were introduced that would have lifted all Cuba embargo restrictions, including those on travel: H.R. 188 (Serrano), introduced January 7, 2003, S. 403 (Baucus), introduced February 13, 2003, and H.R. 1698 (Paul), introduced April 9, 2003. Another initiative, S. 2449 (Baucus)/H.R. 4457 (Otter), introduced respectively on May 19 and 20, 2004, would have required yearly congressional approval for the renewal of trade and travel restrictions with respect to Cuba. Finally, H.R. 4678 (Davis of Florida), introduced June 24, 2004, in the aftermath of the President's tightening of Cuba sanctions, would have barred certain additional restrictions on travel and remittances to Cuba.

Legislative Initiatives in the 107th Congress[47]

In the 107th Congress, although various measures were introduced that would have eliminated or eased restrictions on travel to Cuba and the House voted in both the first and second sessions to prohibit spending to administer the travel regulations, no legislative action was completed by the end of the second session.

First Session Action

During July 25, 2001, floor action on H.R. 2590, the FY2002 Treasury Department appropriations bill, the House approved an amendment that would prohibit spending for administering Treasury Department regulations restricting travel to Cuba. H.Amdt. 241, offered by Representative Flake (which amended H.Amdt. 240 offered by Representative Smith), would prohibit funding to administer the Cuban Assets Control Regulations (administered by OFAC) with respect to any travel or travel-related transaction. The amendment was approved by a vote of 240 to 186, compared to a vote of 232-186 for a similar amendment in last year's Treasury Department appropriations bill.

The Senate version of H.R. 2590, approved September 19, 2001, did not include any provision regarding U.S. restrictions on travel to Cuba, and the provision was not included in the House-Senate conference on the bill (H.Rept. 107-253). During Senate floor debate, Senator Byron Dorgan noted that he had intended to offer an amendment on the issue, but that he decided not to because he did not want to slow passage of the bill. He indicated that he would support the House provision during conference, but ultimately, however, the House-Senate conference report on the bill did not include the Cuba provision. In light of the changed congressional priorities in the aftermath of the September 11 attacks on New York and Washington, conference negotiators reportedly did not want to slow passage of the bill with any controversial provisions. The Bush Administration had threatened to veto the Treasury bill if it included the Cuba travel provision.

Second Session Action

The Cuba travel issue received further consideration in the second session of the 107th Congress. A bipartisan House Cuba working group of 40 Representatives vowed as one of its goals to work for a lifting of travel restrictions. On February 11, 2002, the Senate Appropriations Committee's Subcommittee on Treasury and General Government held a hearing on the issue, featuring Administration and outside witnesses.

The travel issue was part of debate during consideration of the FY2003 Treasury Department appropriations bill (H.R. 5120 and S. 2740). Secretary of State Colin Powell and Secretary of the Treasury Paul O'Neill said they would recommend that the President veto legislation that includes a loosening of restrictions on travel to Cuba (or a weakening of restrictions on private financing for U.S. agricultural exports to Cuba).[48] The White House also stated that President Bush would veto such legislation.[49]

In July 23, 2002, floor action on H.R. 5120, the House approved three Cuba sanctions amendments, including one on the easing of travel restrictions offered by Representative Jeff Flake. The House approved the Flake travel amendment (H.Amdt. 552), by a vote of 262-167 that would provide that no funds could be used to administer or enforce the Treasury Department regulations with respect to travel to Cuba. The Flake amendment would not prevent the issuance of general or specific licenses for travel to Cuba. Some observers raised the question of whether the effect of this amendment would be limited since the underlying embargo regulations restricting travel would remain unchanged; enforcement action against violations of the relevant embargo regulations could potentially take place in future years when the Treasury Department appropriations measure did not include the funding limitations on enforcing the travel restrictions.[50]

During consideration of H.R. 5120, the House also rejected two Cuba amendments. A Rangel amendment (H.Amdt. 555), rejected by a vote of 204-226, would have prevented any funds in the bill from being used to implement, administer, or enforce the overall economic embargo of Cuba, which includes travel. A Goss amendment (H.Amdt. 551), rejected by a vote of 182-247, would have provided that any limitation on the use of funds to administer or enforce regulations restricting travel to Cuba or travel-related transactions would only apply after the President certified to Congress that certain conditions were met regarding biological weapons and terrorism.[51] The rule for the bill's consideration, H.Res. 488 (H.Rept. 107-585), had provided that the Goss amendment would not be subject to amendment.

The House subsequently passed H.R. 5120 on July 24, 2002, by a vote of 308-121, with the three Cuba amendments, including the Flake Cuba travel amendment.

The Senate version of the Treasury Department appropriations measure, S. 2740, as reported by the Senate Committee on Appropriations on July 17, 2002 (S.Rept. 107-212), included a provision, in Section 516, that was similar, although not identical, to the Flake amendment described above. It provided that no funds may be used to enforce the Treasury Department regulations with respect to any travel or travel-related transactions, but would not prevent OFAC from issuing general and specific licenses for travel to Cuba. In addition, Section 124 of the Senate bill stipulated that no Treasury Department funds for "Departmental Offices, Salaries, and Expenses" may be used by OFAC, until OFAC has certain procedures in place to expedite license applications for travel to Cuba.

Congress did not complete action on the FY2003 Treasury Department appropriations measure before the end of the 107th Congress, so action was deferred until the 108th Congress.

Additional Legislative Initiatives in the 107th Congress

Several other initiatives were introduced in the 107th Congress that would have eased U.S. restrictions on travel to Cuba, but no action was taken on these measures.

- H.R. 5022 (Flake), introduced June 26, 2002, would have lifted all restrictions on travel to Cuba.
- Several broad bills would have lifted all sanctions on trade, financial transactions, and travel to Cuba: H.R. 174 (Serrano), the Cuban Reconciliation Act, introduced January 3, 2001, and identical bills S. 400 (Baucus) and H.R. 798 (Rangel), the Free Trade with Cuba Act, introduced February 27 and 28, 2001, respectively.
- S. 1017 (Dodd) and H.R. 2138 (Serrano), the Bridges to the Cuban People Act of 2001, introduced June 12, 2001, would, among other provisions, have removed all restrictions on travel to Cuba by U.S. nationals or lawful permanent resident aliens.
- Several bills would, among other provisions, have repealed the travel restrictions imposed in the 106th Congress by the Trade Sanctions Reform and Export Enhancement Act of 2000 (P.L. 106-387, Title IX, Section 910). These include identical bills S. 402 (Baucus) and H.R. 797 (Rangel), the Cuban Humanitarian Trade Act of 2001, introduced February 27 and 28, 2001; S. 171 (Dorgan), introduced January 24, 2001; and S. 239 (Hagel), the Cuba Food and Medicine Access Act of 2001, introduced February 1, 2001.

Legislative Initiatives in the 106th Congress

The only action completed by the 106th Congress relating to Cuba travel involved a tightening of travel restrictions. The final version of the FY2001 agriculture appropriations measure (P.L. 106- 387, Title IX, Trade Sanctions Reform and Export Enhancement Act of 2000) included a provision that restricts travel to Cuba to those categories of non-tourist travel already allowed by the Treasury Department regulations. Section 910 of the law provides that neither general nor specific licenses for travel to Cuba can be provided for activities that do not fit into the 12 categories expressly authorized in the Cuban Assets Control Regulations, Section 515.560 (a) of Title 31, CFR, paragraphs (1) through (12).

As noted in the law, the Secretary of the Treasury may not authorize travel-related transactions "for travel to, from, or within Cuba for "tourist activities," which are defined as any activity that is not expressly authorized in the 12 categories of the regulations. The provision prevents the Administration from loosening the travel restrictions to allow tourist travel. This, in effect, strengthens restrictions on travel to Cuba and somewhat circumscribes the authority of OFAC to issue specific travel licenses on a case-by-case basis under Section 515.560 (b) of Title 31, CFR. OFAC in the past has utilized that section to provide specific licenses for activities that do not fit neatly within the categories of travel set forth in 515.560 (a), including such travel for medical evacuations of Americans legally in Cuba and for U.S. contractors servicing the needs of the U.S. Interests Section. (Regulations implementing the provision of the law were issued by OFAC on July 12, 2001.)

In other legislative action, the Senate considered the issue of travel to Cuba in June 30, 1999 floor action on the FY2000 Foreign Operations Appropriations bill, S. 1234. An amendment was introduced by Senator Christopher Dodd that would have terminated regulations or prohibitions on travel to Cuba and on transactions related to such travel in most instances.[52] The Senate defeated the amendment by tabling it in a 55-43 vote on June 30, 1999. On November 10, 1999, Senator Dodd introduced identical language as S. 1919, the Freedom to Travel to Cuba Act of 2000, but no action was taken on the bill.

The House took up the issue of travel to Cuba when it considered H.R. 4871, the Treasury Department appropriations bill, on July 20, 2000. A Sanford amendment was approved (232-186) to prohibit funds in the bill from being used to administer or enforce the Cuban Assets Control Regulations with respect to any travel or travel-related transaction. Subsequently, the language of the amendment was dropped from a new version of the FY2001 Treasury Department appropriations bill, H.R. 4985, introduced on July 26. H.R. 4985 was appended to the conference report on the Legislative Branch appropriations bill—H.R. 4516, H.Rept. 106-796—in an attempt to bypass Senate debate on its version of the Treasury appropriations bill, S. 2900. The Senate initially rejected this conference report on September 20, 2000, by a vote of 28-69, but later agreed to the report, 58-37, on October 12. The House had agreed to the conference report earlier, on September 14, 2000, by a vote of 212-209.

End Notes

[1] President George W. Bush, "Remarks After Meeting with the Commission for Assistance for a Free Cuba," U.S. Department of State, May 6, 2004.

[2] Commission for Assistance to a Free Cuba, Report to the President, May 2004, p. 37.

[3] Miriam Leiva, "Whose Country Is It, Anyway?" May 24, 2004, http://Salon.com; and "Why Deal with North Korea and Not Cuba," *Miami Herald*, March 1, 2008.

[4] David Adams, "Dissidents Say It's Time to Open Talks," *St Petersburg Times*, December 18, 2006.

[5] Oscar Corral, "Scholars, Artists Rip Embargo," *Miami Herald*, April 26, 2006.

[6] "Cuba's Campus Attrition," *CQ Weekly*, July 24, 2006; also see ECDET's website available at http://www.ecdet.org/.

[7] Jack Chang, "Court Upholds Limits on Student Trips to Cuba," *Miami Herald*, November 5, 2008.

[8] Laura Morales, "Protesters Call for Family-Friendly Cuban Travel," *Miami Herald*, August 27, 2006.

[9] Human Rights Watch, *World Report 2008*, January 2008.

[10] Human Rights Watch, Families Torn Apart, *The High Cost of U.S. and Cuban Travel Restrictions*, October 2005.

[11] "2007 FIU Cuba Poll," Institute for Public Opinion Research and Cuban Research Institute, Florida International University.

[12] See: U.S. Department of State, Foreign Per Diem Rates by Location, available at http://aoprals.state.gov/web920/per_diem.asp

[13] White House, "Fact Sheet: Reaching Out to the Cuban People," April 13, 2009.

[14] White House, Office of the Press Secretary, "Reaching Out to the Cuban People," January 14, 2011, available at http://www.whitehouse.gov/the-press-office/2011/01/14/reaching-out-cuban-people

[15] Mary Beth Sheridan, "Obama Loosens Travel Restrictions to Cuba," *Washington Post*, January 15, 2011.

[16] Department of the Treasury, "Cuban Assets Control Regulations," Vol. 76, No. 19 *Federal Register* 5072-5078, January 28, 2011; Department of Homeland Security, "Airports of Entry or Departure for Flights to and from Cuba," Vol. 76, No. 19 *Federal Register* 5058-5061, January 28, 2011.

[17] CRS correspondence with the Treasury Department, March 17, 2011.

[18] The guidelines were subsequently revised again in May 2012. See: U.S. Department of the Treasury, Office of Foreign Assets Control, *Comprehensive Guidelines for License Applications to Engage in Travel-related Transactions Involving Cuba*, May 10, 2012, available at http://www.treasury.gov/resource-center/sanctions/Programs/Documents/cuba_tr_app.pdf

[19] Peter Orsi, "U.S. Licensing Travel Operators to Start Up Legal Cuba Trips, Treasury Department Says," *Associated Press*, July 1, 2011; Mimi Whitefield, "People-to-People Tours to Cuba Take Off Thursday," *Miami Herald*, August 10, 2011; and Jeff Franks, "Purposeful Cuba Trips Resume," *Chicago Tribune*, August 18, 2011. Also see the following online resource: *Organizations Sponsoring People-to-People Travel to Cuba*, Latin America Working Group Education Fund, available at http://www.lawg.org/storage/documents/people2people.pdf

[20] U.S. Department of the Treasury, OFAC, "Cuba Travel Advisory," July 25, 2011, available at http://www.treasury.gov/resource-center/sanctions/Programs/Documents/cuba_trav_adv.pdf

[21] See the Cuba travel website of Abercrombie & Kent at http://www.abercrombiekent.com /travel-destinations/latinamerica-luxury-travel/Cuba/, and the website of the Foundation for Caribbean Studies at http://www.caribbeanstudies.org/Home.html. Also see Juan Tamayo, " U.S. Company Suspends Cuba Tours," *Miami Herald*, August 7, 2011.

[22] U.S. Department of Homeland Security, Customs and Border Protection, "Approved U.S. Ports of Entry for Flights to and from Cuba," June 21, 2011.

[23] Cuban American National Foundation, Press Release, "Cuban American National Foundation Supports New Cuba Policy Measures," January 14, 2011.

[24] Republic of Cuba, Ministry of Foreign Affairs, "Statement by the Ministry of Foreign Affairs," January 14, 2011.

[25] U.S. Department of the Treasury, OFAC, "Advertising Educational Exchange Travel to Cuba for People-to-People Contact," March 9, 2012, available at http://www.treasury.gov /resourcecenter /sanctions/Programs/Pages/cuba_ppl_notice.aspx

[26] U.S. Department of the Treasury, OFAC, "Comprehensive Guidelines for License Applications to Engage in Travel-Related Transactions Involving Cuba," Revised May 10, 2012, available at http://www.treasury.gov/resourcecenter/sanctions/Programs/Documents /cuba_ tr_app.pdf

[27] Federal Bureau of Investigation, Miami Division, "FBI Release Photographs of a Vehicle of Interest in Connection with Coral Gables Fire," Press Release, June 7, 2012.

[28] Testimony of Roberta S. Jacobson, Assistant Secretary of State for Western Hemisphere Affairs, Senate Foreign Relations Committee, Subcommittee on Western Hemisphere, Peace Corps, and Global Narcotics, at a hearing entitled "The Path to Freedom: Countering Repression and Strengthening Civil Society," June 7, 2012, available at http://www.state.gov/p/wha/rls/rm/2012/191935.htm

[29] Damien Cave, "Licensing Rules Slow Tours to Cuba," *New York Times*, September 16, 2012; Paul Haven, "U.S. Travel Outfits Say Rules for Legal Travel to Cuba Getting Tighter," *Associated Press*, September 13, 2012.

[30] See OFAC's web page on Cuba Sanctions, available at http://www.treasury.gov/resource-center/sanctions/Programs/ pages/cuba.aspx. OFAC's *Comprehensive Guidelines for License Applications to Engage in Travel-Related Transactions Involving Cuba*, revised May 10, 2012, is available at http://www.treasury.gov/resource-center/sanctions/ Programs/Documents/cuba_tr_app.pdf; and OFAC's *List of Authorized Providers of Air, Travel and Remittance Forwarding Services to Cuba*, revised October 23, 2012, is available at http://www.treasury.gov/resource-center/ sanctions/Programs/Documents/cuba_tsp.pdf.

[31] According to a 2007 Government Accountability Office (GAO) report, there are no reliable estimates of total U.S. travel to Cuba because U.S. and Cuban government data are incomplete and cover different populations. See U.S. GAO, Economic Sanctions: Agencies Face Competing Priorities in Enforcing the U.S. Embargo on Cuba, GAO-08-80, November 30, 2007, pp. 31-33. Nevertheless, in May 2004, the inter-agency Commission for Assistance to a Free Cuba estimated that some 160,000-200,000 legal and illegal travelers visited Cuba from the United States annually over the past decade. The Commission maintained that the largest category of legal travel to Cuba consisted of Cuban Americans visiting their families, accounting for 125,000 out of 160,000 total U.S. visitors to Cuba in 2003. See Commission for Assistance to Free Cuba, Report to the President, May 2004, pp. 28 and 36. A July 2007 U.S. International Trade Commission (ITC) report estimated that about 171,000 Americans traveled to Cuba in 2005, and concluded that lifting travel restrictions would result in U.S. travel increasing to between 550,000 and 1 million. U.S.

ITC, *U.S. Agricultural Sales to Cuba: Certain Economic Effects of U.S. Restrictions*, USITC Publication 3932, July 2007, pp. xi and 3-14 to 3-17.

[32] Air carrier statistics are derived from a "TranStats" database maintained by the U.S. Department of Transportation, Research and Innovative Technology Administration, Bureau of Transportation Statistics, available at http://www.transtats.bts.gov/homepage.asp

[33] For example, see Nick Miroff, "U.S. Trade to Cuba Grows as Restrictions Are Eased," *NPR, All Things Considered,* February 6, 2012; David Harrison, "Cuban Refugees Shouldn't Return, Lawmaker Says," *CQ Today*, May 31, 2012.

[34] U.S. Government Accountability Office, *Economic Sanctions: Agencies Face Competing Priorities in Enforcing the U.S. Embargo on Cuba,* GAO-08-80, November 30, 2007, p. 34. In 2004, the Commission for Assistance to a Free Cuba estimated that U.S. cash remittances to Cuba amounted to an estimated $400 million-$800 million per year, although the report also noted that some estimates were as high as $1 billion annually. See Commission for Assistance to a Free Cuba, Report to the President, May 2004, p. 34.

[35] Manuel Orozco, "The Market for Money Transfers: Ranking of Remittance Service Providers in Latin America and the Caribbean," Inter-American Dialogue, October 24, 2012, available at http://www.thedialogue.org/uploads/Remittances_and_Development/LACRemittance MarketCompeititionandScorecard _shortermemo_2012.pdf

[36] Emilio Morales and Joseph L. Scarpaci, "Opening Up on Both Shorelines Helps Increase Remittances Sent to Cuba in 2011 by About 20%," The Havana Consulting Group LLC, March 12, 2012.

[37] See OFAC's website for information on civil enforcement, available at http://www.treasury.gov/resource-center/ sanctions/CivPen/Pages/civpen-index2.aspx.

[38] Department of the Treasury, Office of Foreign Assets Control, "Economic Sanctions Enforcement Guidelines," 74 *Federal Register* 57593-57608, November 9, 2009.

[39] Defense Intelligence Agency. Report on Cuban Threat to U.S. National Security. May 6, 1998.

[40] Congressional Record, July 25, 2001, p. H4599.

[41] William Gibson, "Support Mounts for Cuba Travel," *South Florida Sun-Sentinel*, April 2, 2009; and "Cuban Dissidents Hail U.S. Lifting of Travel Limits," *EFE News Service*, April 13, 2009.

[42] *The Impact on the U.S. Economy of Lifting Restrictions on Travel to Cuba*, The Brattle Group, Washington, DC. Prepared by Dorothy Robyn, James D. Reitzes, and Bryan Church. July 15, 2002.

[43] Executive Office of the President, Office of Management and Budget, Statement of Administration Policy, H.R. 2434 – Financial Services and General Government Appropriations Act, 2012, July 13, 2011.

[44] For a complete listing and discussion of all Cuba bills in the 108th Congress, see CRS Report RL31740, *Cuba: Issues for the 108th Congress*, by Mark P. Sullivan.

[45] Nancy San Martin, "U.S. Pulls Plug on Cuba Expo," *Miami Herald*, June 18, 2003.

[46] Nancy San Martin, "U.S. Official Dampens Trade-Show Enthusiasm with Talks of Cuban Credit," *Miami Herald*, September 29, 2002.

[47] For a complete listing and discussion of all Cuba bills in the 107th Congress, see CRS Report RL30806, *Cuba: Issues for the 107th Congress*, by Mark P. Sullivan and Maureen Taft-Morales.

[48] U.S. Department of State, International Information Programs, Washington File, "Bush Administration Opposes Legislative Efforts to Amend Cuba Policy," July 16, 2002.

[49] White House, Press Briefing by Ari Fleischer, July 24, 2002.

[50] "House Approves Limits on Treasury Enforcement of Cuba Embargo," *Inside U.S. Trade*, July 26, 2002.

[51] For further information on the issues of biological weapons and terrorism as they relate to Cuba, see CRS Report RL30806, *Cuba: Issues for the 107th Congress*, by Mark P. Sullivan and Maureen Taft-Morales.

[52] The Dodd amendment allowed for travel restrictions to be imposed if the United States is at war with Cuba, if armed hostilities are in progress, or when threats to physical safety or public health exist. Under current law, the Secretary of State has the same authority to restrict travel (22 USC 211a).

In: Cuba
Editor: Gustavo Nogueira
ISBN: 978-1-62618-970-6

Chapter 4

CUBA DEMOCRACY ASSISTANCE: USAID'S PROGRAM IS IMPROVED, BUT STATE COULD BETTER MONITOR ITS IMPLEMENTING PARTNERS*

United States Government Accountability Office

WHY GAO DID THIS STUDY

Since 1996, Congress has appropriated $205 million to USAID and State to support democracy assistance for Cuba. Because of Cuban government restrictions, conditions in Cuba pose security risks to the implementing partners— primarily NGOs—and subpartners that provide U.S. assistance.

For this report, GAO (1) identified current assistance, implementing partners, subpartners, and beneficiaries; (2) reviewed USAID's and State's efforts to implement the program in accordance with U.S. laws and regulations and to address program risks; and (3) examined USAID's and State's monitoring of the use of program funds. This report is a publicly releasable version of a Sensitive But Unclassified Report that GAO issued in December 2012.

* This is an edited, reformatted and augmented version of the Highlights of GAO-13-285, a report to the Chairman, Committee on Foreign Relations, U.S. Senate, dated January 2013.

To address these objectives, GAO analyzed program activities and funding, relevant laws and regulations, and practices for monitoring the use of funds. GAO also conducted performance and financial reviews of a nongeneralizable sample of six implementing partners—representing about 60 percent of USAID and State funding for awards and contracts active in fiscal year 2011—and 11 subpartners.

What GAO Recommends

GAO is recommending that State take steps to improve its financial monitoring of implementing partners and provide clear guidance for approving subpartners. State concurred with GAO's recommendations and cited steps they are taking to address them.

What GAO Found

The U.S. Agency for International Development (USAID) and Department of State (State) provide democracy assistance for Cuba aimed at developing civil society and promoting freedom of information. Typical program beneficiaries include Cuban community leaders, independent journalists, women, youths, and marginalized groups. USAID receives the majority of funding allocated for this assistance, although State has received 32 percent of funding since 2004. In recent years, both USAID and State have provided more funding for program implementation to for-profit and nongovernmental organizations (NGO) with a worldwide or regional focus than to universities and to NGOs that focus only on Cuba. All types of implementing partners, but worldwide or regional organizations in particular, used subpartners to implement program activities under 21 of the 29 awards and contracts that GAO reviewed.

USAID and State legal officials view the Cuba democracy program's authorizing legislation as allowing the agencies discretion in determining the types of activities that can be funded with program assistance. Agency officials added that the agencies ensure that program activities directly relate to democracy promotion as broadly illustrated in related program legislation. The officials stated that organizations are expected to work with agency program officers to determine what activities are permitted or appropriate. In addition, they said that program partners and subpartners are expected to spend U.S. government funds consistent with U.S. laws, and that requirements in primary award agreements generally flow down to any subpartners.

USAID has improved its performance and financial monitoring of implementing partners' use of program funds by implementing new policies and hiring contractors to improve monitoring and evaluation and to conduct financial internal controls reviews, but GAO found gaps in State's financial monitoring. While GAO found some gaps in implementing partners' performance planning and reporting, both agencies are taking steps to improve performance monitoring. For financial monitoring, USAID performs financial internal controls reviews of its implementing partners with the assistance of an external auditor. Since 2008, USAID has used a risk-based approach to determine the coverage and frequency of the 30 reviews the auditor has conducted, which have identified weaknesses in implementing partners' financial management, procurement, and internal controls. However, because of resource constraints, State did not perform financial internal controls reviews for more than two-thirds of its implementing partners during fiscal years 2010 through 2012. State procured an external financial auditor in September 2012 that plans to review more than half of State's implementing partners, and has taken steps toward implementing a risk-based approach for scheduling these reviews. Federal regulations generally require agencies to approve the use of subpartners. GAO found that USAID issued specific guidance in 2011 to its implementing partners on requirements for subpartner approval. While State told GAO it has similar requirements, State's requirements are not clearly specified in its written guidance. As a result, State was not provided with the information it would have needed to approve at least 91 subawards and subcontracts that were obligated under eight awards.

ABBREVIATIONS

DRL	Bureau of Democracy, Human Rights, and Labor
LAC	Latin American and Caribbean Bureau
M/E	Monitoring and evaluation
NED	National Endowment for Democracy
NGO	Nongovernmental organization
OMB	Office of Management and Budget
OTI	Office of Transition Initiatives
Partner	Implementing partner

RFA	Request for Application
RFP	Request for Proposal
State	Department of State
USAID	United States Agency for International Development
USINT	U.S. Interests Section in Havana, Cuba
WHA	Bureau of Western Hemisphere Affairs

January 25, 2013

The Honorable John F. Kerry
Chairman
Committee on Foreign Relations
United States Senate

Dear Mr. Chairman:

The U.S. Agency for International Development (USAID) and the Department of State (State) provide assistance to increase respect for human rights and fundamental freedoms, strengthen independent Cuban civil society, and foster greater access to independent information on the island.[1] Cubans have lived under an authoritarian regime for more than 50 years; in that time, Cubans have faced continuous restrictions in citizens' rights to organize, express themselves freely, and participate in political life without fear of being harassed and repressed by the Cuban government. Conditions in Cuba pose risks and challenges for the delivery of program assistance on the island, which is largely provided by implementing partners (partners) and their subpartners.[2] In fiscal years 1996 through 2011, Congress appropriated $205 million for Cuba democracy assistance, with about 90 percent of program funding provided since fiscal year 2004. While USAID has received most of this funding, State has also received program funding allocations since fiscal year 2004.

In November 2006 and November 2008, we reported problems with USAID's management and oversight of its Cuba program, such as oversight weaknesses that increased the risk of grantees' improper use of grant funds and noncompliance with U.S. laws and regulations.[3] We made several recommendations to help USAID strengthen its program management and monitoring, including recommendations to improve the timeliness of preaward reviews, and to ensure partners develop approved implementation plans for awards. USAID has reported taking several steps to address the recommendations. However, security risks and challenges related to the delivery of program assistance persist.

To respond to your request, this report focuses on USAID's and State's efforts to provide democracy assistance for Cuba. We (1) identified the types and amounts of assistance that USAID and State have provided, as well as characteristics of their partners, subpartners, and program beneficiaries; (2) reviewed USAID's and State's efforts to implement the program in accordance with U.S. laws and regulations and to address program risks; and (3) examined USAID's and State's monitoring of the use of program funds. This report is a publicly releasable version of a prior GAO report, issued in December 2012, that USAID and State had designated Sensitive But Unclassified. Although the information provided in this report is more limited in scope, it addresses the same questions and uses the same overall methodology as the sensitive report.

In conducting this review, we identified all awards and contracts funded from 1996 to 2012 and analyzed selected characteristics of all 22 USAID and State partners that received program funding from fiscal years 2007 to 2009 appropriations and were active in fiscal year 2011. These partners received 29 awards and contracts during that period. We reviewed relevant U.S. laws and regulations and agency and departmental policies and procedures, and we interviewed USAID and State officials regarding program implementation and related risks. We also analyzed reported activities, assistance delivered, and management and internal controls[4] for a nonprobability,[5] nongeneralizable sample of six USAID and State partners and 11 of their subpartners, to assess performance and financial monitoring and oversight of their awards and contracts. We selected at least one partner from each of the four USAID and State bureaus and offices implementing program assistance. The six selected partners were among the top 15 recipients of program funding awarded in fiscal years 2007 through 2010 and represented about 60 percent of funding for awards active in fiscal year 2011. These six partners received 10 USAID and State awards and contracts over the time period we reviewed—five from USAID and five from State. We reviewed and analyzed performance and financial documentation and data and conducted interviews with USAID and State officials and with representatives of partners and subpartners in our sample. In addition, we conducted fieldwork in Miami, Florida, where we interviewed representatives and reviewed documentation at local partners and subpartners, and at the U.S. Interests Section (USINT) in Havana, Cuba, where we interviewed U.S. officials and observed activities at post.[6] Appendix I provides further details on our scope and methodology.

We conducted this performance audit from September 2011 to January 2013 in accordance with generally accepted government auditing standards. Those standards require that we plan and perform the audit to obtain sufficient, appropriate evidence to provide a reasonable basis for our findings and conclusions based on our audit objectives. We believe that the evidence obtained provides a reasonable basis for our findings and conclusions based on our audit objectives.

BACKGROUND

Challenges Affecting Cuba Democracy Assistance

Conditions in Cuba continue to pose substantial challenges for U.S. assistance. Cuba is a Communist state that restricts nearly all political dissent on the island; tactics for suppressing dissent in Cuba include surveillance, arbitrary arrests, detentions, travel restrictions, exile, criminal prosecutions, and loss of employment. Furthermore, there is no free press in Cuba, and independent journalists and activists are harassed and imprisoned. The Cuban government substantially restricts and controls the flow of information, limiting access to the Internet, cell phones, radio antennas, and other items, and restricting their use through high costs, punitive laws, and the threat of confiscation. Moreover, the government routinely jams all external, non-Cuban broadcasts, including the U.S. government-supported Radio and TV Martí broadcasts.

The United States, which maintains an embargo on most trade with Cuba, does not have diplomatic relations with the Cuban government. Consequently, USAID does not work cooperatively or collaboratively with Cuban government agencies, as it does in most other countries receiving U.S. democracy assistance. USAID does not have staff in Cuba, and State does not have staff dedicated to the Cuba democracy program in Havana.[7] USAID and State program staff have been unable to obtain visas to visit Cuba over the past decade, which poses challenges for program implementation, monitoring, and evaluation. In addition, Cuban law prohibits citizens from cooperating with U.S. democracy assistance activities. In December 2009, a subcontractor working for one of USAID's partners was arrested in Cuba while delivering computer equipment to provide Internet access to Jewish communities on the island. He was subsequently sentenced to 15 years in prison for "acts against the independence or the territorial integrity of the state."

Roles and Responsibilities for Implementing Cuba Democracy Assistance

Several USAID and State bureaus and offices implement Cuba democracy assistance efforts, including soliciting proposals, competitively awarding funds, and monitoring program implementation.

- USAID:
  - **The Latin American and Caribbean Bureau (LAC), Office of Cuban Affairs**, is chiefly responsible for implementing Cuba democracy assistance efforts.
  - **The Management Bureau** has various offices that also assist in overseeing program awards and contracts.
  - **The Office of Transition Initiatives (OTI)** oversaw implementation of a single contract from fiscal years 2009 through 2012. OTI's Cuba program efforts were envisioned from their inception to be temporary, as is typically the case with OTI's programs, which generally aim to provide short-term assistance.
- State:
  - **The Bureau of Democracy, Human Rights, and Labor (DRL)** is responsible for managing and overseeing the majority of State's Cuba democracy assistance program activities.
  - **The Bureau of Western Hemisphere Affairs (WHA)**, including the U.S. Interests Section in Havana, Cuba, (USINT), also manages and oversees Cuba program activities.,
  - **The Bureau of Administration** assists DRL and WHA in the financial management and oversight of Cuba democracy assistance awards.[8]

Reporting on Cuba democracy assistance in 2006,[9] we found that USINT delivered some assistance to independent groups and individuals in Cuba, including assistance provided by USAID- and State-funded awardees. Because of heightened security concerns, USINT no longer has a role in implementing assistance for USAID and State/DRL partners. However, USINT continues to provide information on conditions in Cuba, facilitates and assists with State/WHA training courses, and supports civil society in Cuba.

Table 1 outlines key USAID and State roles and responsibilities for providing U.S. democracy assistance for Cuba.

Table 1. Key USAID and State Roles and Responsibilities for Cuba Democracy Assistance

Responsible entity	Staff roles/responsibilities
USAID	
LAC, Office of Cuban Affairs	A Director, four Program Managers, a Budget Analyst, and an Administrative Assistant share full- time responsibilities for program management and oversight. These responsibilities include program design, implementation, and monitoring as well as budgeting and financial management.
Management Bureau	Various staff members assist in program oversight, including closing out awards and contracts, conducting risk assessments, and auditing programs.
OTI	A Country Representative dedicated full-time to Cuba and a part-time Deputy Team Leader shared responsibilities for all aspects of the temporary program's management and oversight through the end of program efforts in fiscal year 2012.
State	
DRL	Two Program Officers share part-time responsibilities for Cuba program planning, implementation, monitoring and evaluation. A Foreign Affairs Officer also provides part-time assistance, primarily on policy issues related to Cuba.
WHA	About a dozen staff members each spend about 5 to 10 percent of their time on Cuba program management and oversight.
USINT, Havana	Several staff, dedicated part-time to providing information on conditions in Cuba, facilitate and assist with State/WHA training courses and support civil society in Cuba.
Bureau of Administration	Various staff members are responsible for financial oversight and providing financial management guidance to DRL and WHA Cuba program partners.

Source: GAO analysis of USAID, State, and other records.

In addition, the National Endowment for Democracy (NED), an independent, nongovernmental organization, funds programs to promote democracy in Cuba through both direct congressional appropriations and with funding that it receives through State/DRL.[10]

Program partners—such as nongovernmental organizations, universities, and development contractors—and subpartners also have roles and responsibilities for Cuba democracy program assistance. USAID and State provide funding for partners through various mechanisms, including grants and cooperative agreements (together referred to as "awards" in this report) and contracts. Partners may also award funding to subpartners to assist them in implementing program efforts. Subpartners may include consultants, subcontractors, subgrantees, and recipients of grants under contract.

USAID AND STATE FOCUS ON PROMOTING CUBAN CIVIL SOCIETY AND ACCESS TO INFORMATION; MOST FUNDING GOES TO WORLDWIDE OR REGIONAL PARTNERS AND THEIR SUBPARTNERS

USAID and State support democracy for Cuba by providing awards and contracts to partners with objectives related to developing civil society and promoting freedom of information. USAID receives the majority of funding allocated for this assistance, although State has received 32 percent of funding since it started taking part in the program in 2004. Since 2008, USAID and State have awarded more funds to larger organizations with a worldwide or regional presence than to the other two categories of typical awardees: universities, and smaller organizations that focus only on Cuba. Under 21 of the 29 recent awards and contracts that we reviewed, partners used subpartners to implement program activities and obligated about 40 percent of the funding associated with these awards and contracts to subpartners. Worldwide or regional organizations provided more than 90 percent of the funding provided to subpartners.

Program Focuses on Promoting Civil Society and Access to Information

USAID's and State's democracy assistance efforts for Cuba generally focus on developing an independent civil society and promoting freedom of information in Cuba. The overall goal and guiding principle of U.S. democracy assistance for Cuba is to improve the effectiveness of citizens to participate in activities affecting their lives and to increase access to information.[11] Efforts to develop Cuban civil society include training in organizational and community development, leadership, and advocacy.

Related material assistance may include the provision of books, pamphlets, movies, music, and other materials that promote democratic values. In addition, efforts to promote freedom of information have included the following, among other activities:

- information technology training for Cuban nationals, ranging from basic computing to blogging;
- journalism training;
- support for independent publications; and
- provision of material assistance.

USAID and State officials noted that in recent years, program efforts have included a greater focus on information technology, particularly on supporting independent bloggers and developing social networking platforms on the island.

Several partners we reviewed received funding to support international solidarity activities, although agency and partner officials indicated that the program recently has reduced its focus on off-island activities to foster support for democracy in Cuba. Activities of this type that State/DRL funded in fiscal years 2009 through 2011 included the following:

- an essay contest for Latin American youths related to Cuba Solidarity Day, and
- exhibits and documentaries presented outside Cuba for the purpose of bringing awareness globally to Cuban human rights issues and civil society development.

In recent years, USAID and State had few awards or contracts focused solely on such humanitarian assistance as assistance for political prisoners and their families, according to agency officials. USINT officials noted that humanitarian assistance has declined along with a decrease in the number of political prisoners in Cuba.[12] Officials added that USINT itself no longer provides any humanitarian assistance on the island.

To broaden reach and impact, Cuba democracy assistance efforts have expanded beyond a focus on traditional activists to include groups such as poor and rural communities, religious organizations, small businesses, and information technology enthusiasts. Typical program beneficiaries also include Cuban community leaders, independent journalists, independent bloggers, women, and youths.

Table 2 summarizes information on recent program assistance and target beneficiaries.

Table 2. Cuba Democracy Assistance Provided from Fiscal Years 2009 to 2011

Purpose of assistance	Examples of assistance provided	Target beneficiaries
Development of independent civil society	• Training on topics such as democracy building, economic rights, free markets, leadership and advocacy, organizational and community development, conducting small projects, and human rights abuse response and documentation • Books, pamphlets, and other materials to promote democratic values • Support for the development of independent social networking platforms • Technical and material assistance to increase awareness of freedom of expression.	Community leaders Cuban youths and students, girls and women, poor and rural communities Other groups
Promotion of freedom of information	• Information technology training, such as basic computing, Internet, and blogging • Journalism training • Digital newspaper publication to disseminate work of independent journalists • Development and distribution of media content for youth • Digital photo contest • Shortwave independent radio station • Material assistance	Cuban public Independent Cuban journalists Cuban youths
Humanitarian assistance	• Food products, over-the-counter medication, basic clothing • Financial aid for family members of political prisoners • Material assistance for civil society organizations targeting political prisoners and their families	Political prisoners and their families

Source: GAO analysis of USAID, State, and other records.

USAID Has Received the Majority of Allocations, but State's Portion Has Grown in Recent Years

In fiscal years 1996 through 2011, Congress appropriated $205 million for Cuba democracy assistance, appropriating 87 percent of these funds since 2004. Increased funding for Cuba democracy assistance was recommended by the interagency Commission for Assistance to a Free Cuba, which was established by President George W. Bush in 2003.[13] Program funding, which peaked in 2008 with appropriations totaling $44.4 million, has ranged between $15 and $20 million per year during fiscal years 2009 through 2012. For fiscal year 2013, USAID and State reduced their combined funding request to $15 million, citing operational challenges to assistance efforts in Cuba.[14]

In fiscal years 1996 through 2011, $138.2 million of Cuba democracy funds were allocated to USAID and $52.3 million were allocated to State (see figure 1).[15]

When the Cuba democracy program began in 1996, USAID was the only agency involved and USAID/LAC was the only programming bureau.

- USAID/LAC has received the largest total amount of program funding and has continued to receive the largest annual amount, averaging $12.1 million annually since fiscal year 2004.
- USAID/OTI received program funding totaling $14.3 million from the appropriations for fiscal years 2007 through 2010.[16]

State has received 32 percent of Cuba democracy funding since fiscal year 2004.

- State/DRL has received an average of $5.8 million annually since fiscal year 2004.[17]
- State/WHA has received an average of $1.4 million annually since fiscal year 2008.

USAID and State Have Provided Most Recent Awards and Contracts to Worldwide or Regional Organizations

USAID and State have awarded funding for Cuba democracy assistance to three categories of partners: (1) Cuba-specific nongovernmental organizations (NGO), (2) worldwide or regional organizations, and (3) universities.[18]

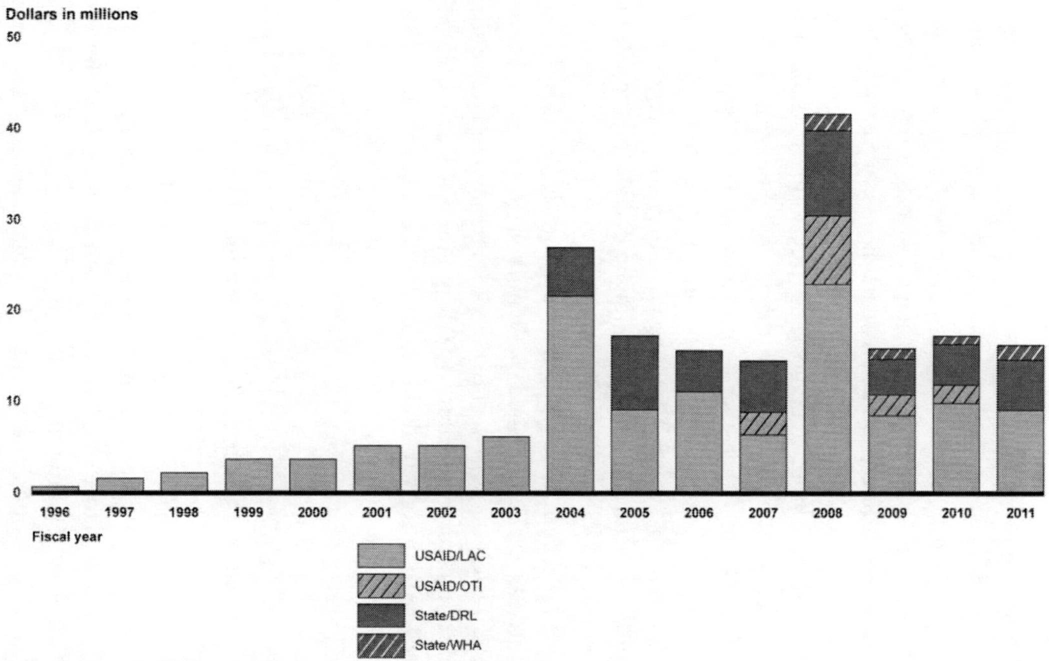

Source: GAO analysis of USAID and State data.

Note: The fiscal year 2012 appropriations for Cuba democracy funding have not yet been allocated between the USAID and State bureaus. Since 1984, through its congressional appropriation, NED has also provided democracy assistance to Cuba ranging from $0 to $1.4 million annually from 2006 to 2010. In addition, since 2005, State/DRL has transferred funding to NED for Cuba democracy assistance that has totaled $14.4 million and has ranged from $0 to $4 million annually.

Figure 1. Cuba Democracy Funding Allocations to USAID and State from Appropriations for Fiscal Years 1996-2011.

USAID's and State's awards and contracts tended to share certain characteristics, such as their broad objectives and amounts awarded, depending on the type of partner.

- *Objectives*. USAID and State awards and contracts to Cuba-specific NGOs and to worldwide or regional organizations have generally funded similar types of program activities, such as efforts to provide training and material assistance on the island. Awards to universities have tended to have different objectives. In the early years of the Cuba program, awards to universities funded activities such as research on how to promote a democratic transition in Cuba and scholarships to study at universities in the United States. Since the mid-2000s, after finding that the Cuban government would not provide exit visas for Cuban students to study in the United States, USAID and State have awarded funding to universities largely for programs to provide distance learning training to Cubans on the island or courses at universities in other Latin American countries.
- *Amount of award or contract*. USAID's awards and contracts in fiscal years 1996 through 2012 averaged $1.9 million for Cuba-specific NGOs, and $2.1 million for worldwide or regional organizations. State's awards and contracts averaged $0.8 million for Cuba-specific NGOs and $0.9 million for worldwide or regional organizations. Both USAID's and State's awards and contracts to universities averaged $0.8 million.

In fiscal years 1996 through 2012, USAID and State had a combined total of 111 awards and contracts to 51 partners representing all three types of organizations (see figure 2). Many of the awards were concentrated among certain partners, with 25 of these partners receiving multiple awards from USAID, State, or both. For example, one partner received a combined 11 awards from USAID and State, more than any other partner, and 10 of the 51 partners received 67 percent of total funding to partners.[19]

Since fiscal year 2008, regional or worldwide organizations have had more active USAID and State awards and contracts each year, and have received more funding, than Cuba-specific NGOs or universities. Prior to 2008, Cuba-specific NGOs had more active USAID awards than the other categories of recipients in most years. However, the program's partners have consistently included worldwide or regional organizations, some of which have a history of working on Cuba issues (see figure 3). For example, for awards that began in fiscal years 1996 through 2007, Cuba-specific NGOs received 48 percent of award funding, worldwide or regional organizations received 43 percent, and universities received 9 percent. In contrast, for awards made since fiscal year 2008, worldwide or regional organizations received 74 percent of award and contract funding, while Cuba-specific NGOs received almost 17 percent, and universities received almost 10 percent. As we previously reported, this greater use of worldwide or regional organizations, which began in 2008, reflected more formal requirements for submitting proposals and USAID's decision to fund awards and contracts that incorporate capacity building for subpartners as an important element.[20]

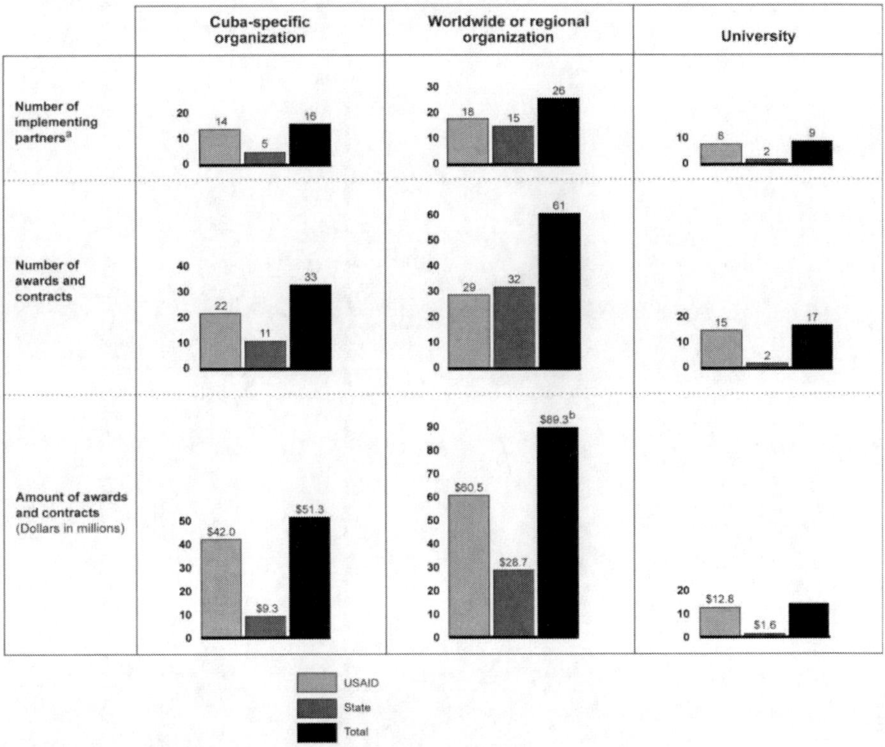

Source: GAO analysis of USAID and State records.

[a] The sum of USAID's and State's implementing partners exceeds the total for each type of organization because 11 implementing partners, across the different types of organizations, received awards from both USAID and State.

[b] Difference in sum of subtotals and total is due to rounding.

Figure 2. Characteristics of USAID and State Cuba Democracy Assistance Implementing Partners and Their Awards and Contracts, Fiscal Years 1996-2012.

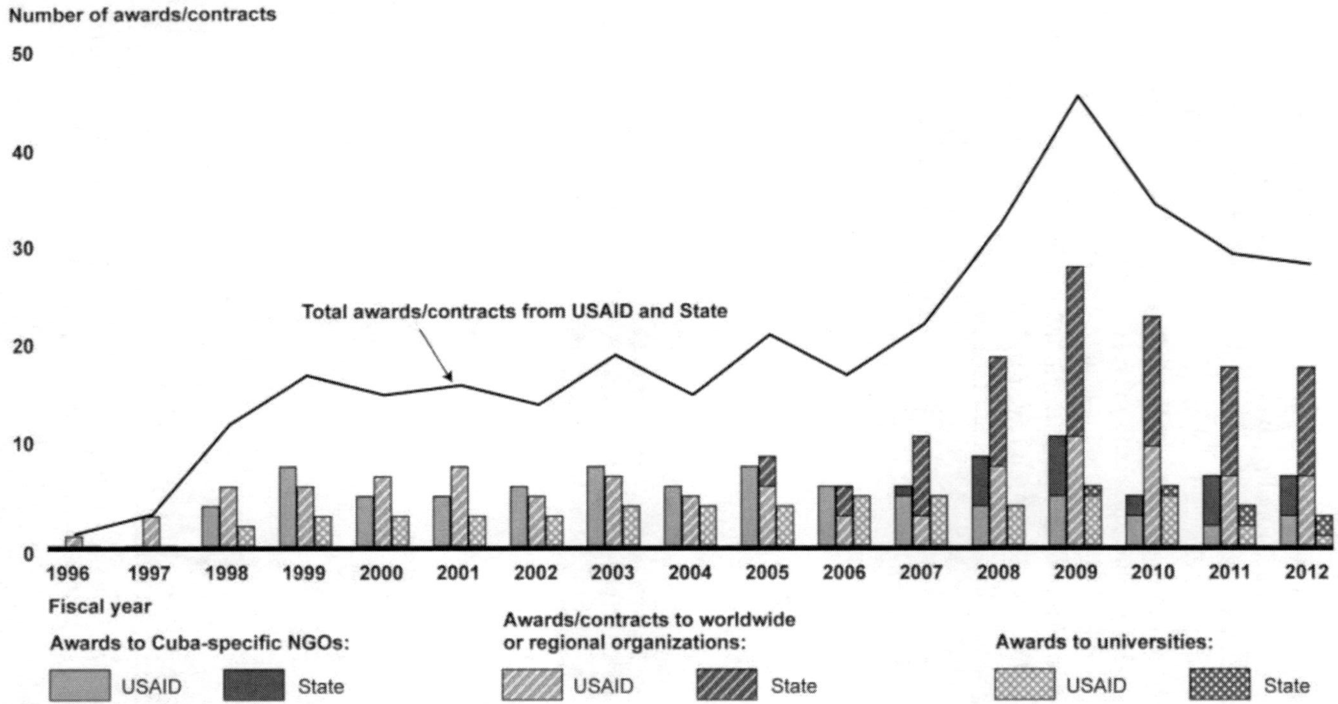

Source: GAO analysis of USAID and State data.

Figure 3. Number of Active USAID and State Awards and Contracts for Cuba Democracy Assistance, by Type of Organization, Fiscal Years 1996-2012.

Most Partners Use Subpartners to Help Implement Program Activities

Many partners, and worldwide or regional organizations in particular, use subpartners to help carry out their Cuba democracy assistance work. We reviewed 29 recent awards and contracts to determine the extent to which partners use subpartners to implement program activities.[21] We found that partners used subpartners under 21 of the 29 awards and contracts, obligating about 40 percent of the funding under these awards and contracts to subpartners. On average, partners that used subpartners under an award or contract had 12 subpartners.

However, the numbers of subpartners under each of the 21 awards varied:

- Four awards had one subpartner.
- Seven awards had between two and nine subpartners.
- Ten awards and contracts had more than 10 and up to 38 subpartners.

The purposes of subawards and subcontracts also varied greatly. Many subawards and subcontracts were for discrete activities, such as to conduct workshops. Other subawards and subcontracts covered an array of tasks, such as content development and instruction for a distance learning course or development, training, and support for civil society networks. Accordingly, subpartners included different types of non-profit and for-profit organizations as well as individuals who worked as consultants that provided the skills necessary to implement the varying activities.

Furthermore, the amount of funding that went to subpartners ranged from less than $5,000 to several hundred thousand dollars. For six of the 21 awards and contracts with subpartners, the majority of program funding was obligated to subpartners. In such cases, subpartners generally performed all or most of the programmatic functions under the overall award or contract, while the partners' main functions were to provide strategic direction of the overall award or contract and to perform management functions such as reporting to the agency and overseeing their subpartners.

Worldwide or regional organizations were more likely to use subpartners than were the other categories of organizations. In total, 93 percent of the subawards and subcontracts were awarded by worldwide or regional organizations. Also, on average, worldwide or regional organizations had 12 subpartners for each of their awards or contracts, while Cuba-specific NGOs had three and universities had five. Correspondingly, five of the six partners that obligated the majority of their funding to subpartners were worldwide or regional organizations.

Agencies Exercise Discretion under Authorizing Legislation in Funding Cuba Program Activities

USAID and State legal officials view the Cuba program's authorizing legislation as providing the purposes for which foreign assistance funds may be used and allowing discretion to determine which program activities will be funded. The officials stated that they view the types of activities listed in section 109(a) of the Helms-Burton Act as illustrating, not limiting, the types of program assistance that the agencies can provide. Specific authority for Cuba democracy assistance activities was provided in section 1705 of the Cuban Democracy Act of 1992 and in section 109(a) of the Helms-Burton Act in 1996.[22] Section 1705 authorizes the donation of food for NGOs and individuals in Cuba; exports of medicines and medical supplies, instruments, and equipment; and assistance to appropriate NGOs to support efforts by individuals and organizations to promote nonviolent democratic change in Cuba. Section 109(a) authorizes assistance and other support that may be provided, such as published informational matter for independent democratic groups, humanitarian assistance for victims of political repression and their families, and support for democratic and human rights groups.

USAID and State legal officials said that the agencies ensure that program activities directly relate to democracy promotion as broadly illustrated in related program legislation. For example, the officials noted that the types of activities that fit within the scope of "democracy promotion," as that term has been broadly defined in various foreign assistance appropriations,[23] would be the types of activities eligible for funding under section 109(a) of the Helms-Burton Act. They added that, while the agencies have not compiled a list of activities that will be approved or not approved for funding under the Cuba program, proposed or approved activities are set forth in agency congressional notifications and listed in individual requests for proposals or applications and in award agreements and contracts. In addition, they said that organizations are expected to work with agency program officers to determine what activities are permitted or appropriate, and whether Department of Treasury and Commerce authorizations, as required, already exist for delivery of various types of assistance or whether the organization must instead apply for a license.[24] Furthermore, they noted that program partners and subpartners, including subpartners based in other countries, are expected to spend U.S. government funds consistent with U.S. laws and that requirements in primary award agreements and contracts generally flow down to any subpartners.

USAID HAS IMPROVED ITS MONITORING OF PARTNERS; STATE'S MONITORING DOES NOT ENSURE PROGRAM FUNDS ARE USED AS INTENDED

Since 2008, USAID has worked to improve performance and financial monitoring of its Cuba program partners.[25] However, we found gaps in State's financial monitoring efforts. For performance monitoring, we found some deficiencies in the performance planning and reporting conducted by USAID's and State's partners in our nongeneralizable sample, but both agencies are taking steps to improve their performance monitoring. For financial monitoring, USAID has hired an external auditor to perform financial internal controls reviews of its partners, and has used a risk- based approach considering criteria such as award value and prior issues identified to determine the coverage and frequency of the 30 reviews the auditor has conducted. These reviews have identified financial management, procurement, and internal controls weaknesses that USAID has taken steps to address. While State conducted no financial internal controls reviews for at least two-thirds of its partners between fiscal year 2010 and 2012, State recently hired an external auditor to perform such reviews starting in fiscal year 2013 and has taken steps to implement a risk-based approach to prioritize the scheduling of its reviews. Specifically, State plans to complete reviews for three-quarters of State/DRL's partners and none of State/WHA's partners. In addition, in accordance with federal regulations, the agencies approve partner requests to award funding to specific subpartners. In June 2011, USAID provided specific written guidance to its partners on what USAID requires for approval of subpartners. State has provided limited written guidance on approval to some partners, which does not clearly inform partners of the specific types of information State requires for approval. As a result, State was not provided with the detailed information that officials told us would have been required for State to have approved 91 subawards and subcontracts that were obligated under eight of its recent awards.

USAID and State Are Taking Steps to Improve Their Performance Monitoring to Address Weaknesses in Some Partners' Performance Planning and Reporting

USAID Continues to Take Steps to Improve Performance Monitoring; USAID's Partners' Had Some Weaknesses in Performance Planning

USAID has taken steps to improve its ability to monitor its Cuba program partners' performance, by working with them to improve their performance planning and reporting.[26] USAID has numerous requirements for partners' performance planning and reporting, the key elements of which are summarized below.

- **Performance Planning:** USAID directs its Cuba program partners to establish monitoring and evaluation (M/E) plans that include certain specific characteristics.[27] USAID works with partners to include more detailed information on indicators in their M/E plans. USAID/LAC also required its one contractor to perform data quality assessments on its performance data.
- **Performance Reporting:** USAID requires awardees to submit progress reports on a quarterly basis and requires contractors to submit monthly and annual progress reports, among others.[28] USAID uses information in these performance reports to track the progress of individual awards and contracts and to track the progress of the overall Cuba program. According to USAID officials, USAID first reviews the reporting to compare it against targets set in partners' M/E plans. In addition, USAID analyzes and aggregates the information reported by partners to track performance for USAID's Cuba program and to report to State's Office of U.S. Foreign Assistance Resources on government-wide performance.

We reviewed the M/E plans and progress reports for the five USAID awards or contracts in our nongeneralizable sample, which began in fiscal year 2008 or 2009 (see table 3).[29] We found some weaknesses in the partners' M/E plans but found detailed reporting against indicators in the progress reports we reviewed. For example, our analysis indicates that all M/E plans we reviewed included clearly defined indicators for program activities.

However, not all partners specified targets and data collection methods for each indicator in their M/E plans. Establishing targets for indicators during the planning stage is important because targets form the standard against which actual results will be compared and assessed. Specifying data collection methods for each indicator enables the agency to determine whether it will be realistic for the partner to measure performance for that indicator in a timely manner. One partner included no information on data collection methods in their M/E plan, while another partner only included general information on their planned data collection methods, such as by stating that the data would be collected by subpartners.

Based on review of the partners' progress reports, we found that all partners in our sample reported to USAID on progress through quantitative updates against each indicator, allowing USAID to gauge the specific progress made during each reporting period.[30] One partner also reported progress for each individual subpartner, including by reporting the number of each subpartner's beneficiaries disaggregated by target group. Partners' progress reports also provided narrative information describing program activities, challenges encountered, and planned activities for the next reporting period.

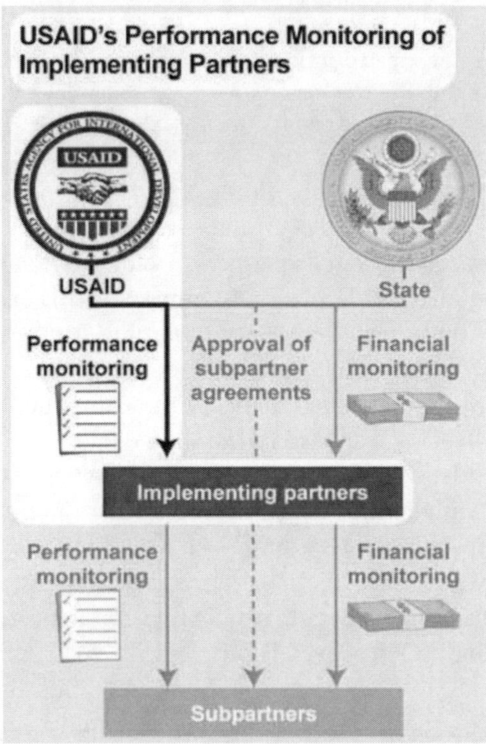

Source: GAO analysis of USAID and State data; Nova Development (clip art); and Corel Graphics (logos).

Table 3. Characteristics of Partners' Performance Planning and Reporting for Selected USAID Awards and Contracts

Partner	Awarding agency and bureau/ office	Performance Planning: M/E plans			Performance reporting: progress reports
		Indicators clearly defined	Target set for each indicator	Data collection methods identified for each indicator	Updates on progress clearly reported for each indicator
Partner #1	USAID/OTI	•	°	◒	•
Partner #2	USAID/LAC	•	◒	•	•
Partner #3	USAID/LAC	•	°	°	•
Partner #4	USAID/LAC	•	•	•	•
Partner #5	USAID/LAC	•	•	•	•

Legend: • = Yes; ◒ = Partially; ° = No.

Source: GAO analysis of USAID and partner records.

Note: Each row in the table represents a separate award or contract. M/E plans for our sample partners were created in fiscal year 2008 or 2009 when their awards and contracts began. The progress reports we reviewed were from fiscal year 2011, except for Partner #2 progress reports from fiscal year 2009.

Although we found some gaps in these partners' performance planning, USAID/LAC has been working to improve the quality of performance information that it receives from its partners,[31] with a particular emphasis since 2010 on improving their M/E plans. To improve M/E plans and partners' reporting based on those plans, in 2010, according to USAID, USAID/LAC conducted in-depth assessments of each of its partners' M/E plans to determine whether they included indicator tracking tables, definitions of indicators, data collection responsibilities, data quality limitations, and other key information. Also, in September 2010, USAID/LAC hired an M/E contractor to work with each of its partners to further improve and standardize their performance management systems. The M/E contractor has worked with partners to identify and track the most appropriate indicators, including any applicable standardized indicators that USAID/LAC can aggregate across the partners to determine its own overall progress. In 2011, this M/E contractor also provided training to each partner and helped them to improve their M/E plans, for example by specifying quarterly targets and data collection plans for each indicator. According to M/E contractor representatives, partners' performance planning has improved, although additional improvement is needed in the quality of some partners' data. In fiscal year 2013, the M/E contractor plans to perform data quality audits of the partners.[32]

State Is Taking Steps to Improve Performance Monitoring to Address Its Partners' Weaknesses in Performance Planning and Reporting

State has also made some recent improvements to performance monitoring of its Cuba program, in the areas of both performance planning and reporting. State's requirements for performance planning and reporting include:

- **Performance Planning:** State/DRL and State/WHA have provided different requirements for prospective partners regarding elements of M/E plans.
  - **State/DRL.** In 2010, State/DRL increased the level of requirements for prospective partners' M/E plans through the request for proposal (RFP) it issued that year.[33] Previously, State/DRL required that prospective partners submit an M/E plan but did not specify characteristics that the M/E plan should include. The RFP issued in 2010 specified that M/E plans should include a baseline and target for each indicator and data collection methods and sources, among other characteristics. In addition, the RFP referenced M/E guidance available on State/DRL's website that included more details on how to develop an effective M/E plan.
  - **State/WHA.** For the State/WHA award in our sample, the RFP issued in fiscal year 2010 required the prospective partner to submit an M/E plan outlining performance indicators, sources and means for verification, risks and assumptions for goals and objectives, and expected results and activities. For WHA's most recent RFPs issued in fiscal year 2012, State/WHA also included additional guidance for prospective partners' M/E plans, for example, including by defining indicators and providing a sample M/E plan. In addition, State/WHA further clarified that all indicators in M/E plans must include measurable, numerical targets.
- **Performance Reporting:** Both State bureaus require their partners to submit quarterly progress reports.[34] According to State/DRL and State/WHA officials, they review partners' quarterly reports against the partners' planned performance to confirm that the awards are making progress toward established targets and that activities align

with the award's objectives.[35] State officials then analyze the quarterly progress reports to aggregate progress for each of its bureaus with USAID's progress to be able to report government-wide performance for the Cuba democracy program to State's Office of U.S. Foreign Assistance Resources.

State's partners that we reviewed ranged in the amount and kind of detail they included in their M/E plans as well as in their progress reports. For the State awards in our sample, we found that State/WHA's partner had the most clearly defined M/E plan (see table 4). This partner's M/E plan included specific and clearly defined indicators, targets against which the partner could measure its performance, and clear plans for data collection. For the four State/DRL awards, the partners included indicators in their M/E plans but did not define them.[36]

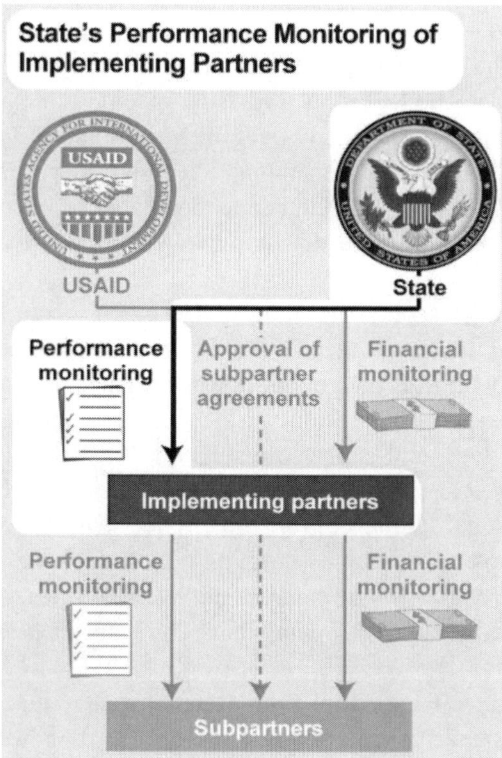

Source: GAO analysis of USAID and State data; Nova Development (clip art); and Corel Graphics (logos).

Table 4. Characteristics of Partners' Performance Planning and Reporting for Selected State Awards

Partner	Awarding agency and bureau/ office	Performance Planning: M/E plans			Performance reporting: progress reports Updates on progress
		Indicators clearly defined	Target set for each indicator	Data collection methods identified for each indicator	Updates on progress clearly reported for each indicator
Partner #3[a]	State/DRL	◒	°	•	°
Partner #3	State/DRL	◒	◒	•	°
Partner #3	State/DRL	◒	◒	◒	◒
Partner #5	State/DRL	◒	•	◒	◒
Partner #6	State/WHA	•	•	•	•

Legend: • = Yes; ◒ = Partially; ° = No.

Source: GAO analysis of State and partner records.

Notes: Each row in the table represents a separate award. M/E plans for our sample partners were created in fiscal year 2009 or 2010 when their awards and contracts began. The progress reports we reviewed were from fiscal year 2011, except for the reports for Partner #3's third award, which were from fiscal year 2012.

[a] One State/DRL partner (Partner #3) received three awards during the period of our sample, and we reviewed that partner's planning and reporting for all three awards.

We also found that one partner with three State/DRL awards did not set clear targets for a number of its indicators. In addition, two partners identified some data collection methods in their M/E plans, but did not clearly identify which methods would be used to collect data for each individual indicator.

We also found that, for State/WHA's award, progress reports included detailed reporting against each indicator, as well as additional qualitative and quantitative information on overall progress including survey results and statistics. On the other hand, reporting for the State/DRL awards lacked such detail. For example, for one State/DRL award, the partner tracked its performance in quarterly reports for only 3 of the more than 10 indicators in its M/E plan. For three other awards, the partner did not aggregate or track performance against any specific indicators in their progress reports. While not reporting on progress against specific indicators, these partners generally reported anecdotally on the topics covered in the indicators or scattered some performance data throughout their reports.

In September 2012, State/DRL awarded a contract to a firm specializing in M/E, which could address such gaps in its partners' performance planning and reporting.[37] In the area of performance planning, State/DRL has directed the M/E contractor to provide training and technical assistance to its partners to improve their M/E plans, such as to ensure they include information on data collection methods. In addition, State/DRL directed the M/E contractor to develop indicators for all State/DRL partners to report on that meet or surpass data quality standards. This should allow State/DRL to more easily aggregate information on the overall performance of its Cuba program partners.

Partners Generally Monitored Subpartner Performance and Used Various Methods to Monitor Beneficiaries

The partners in our sample had various policies, procedures, and mechanisms in place for monitoring subpartner performance, which they compiled information on to report to the agencies.[38] We found that partners generally required subpartners to report on their activities quarterly or monthly, and at the end of a subaward or subcontract. Some partners also required subpartners to submit trip reports after any travel to Cuba.[39] Other monitoring practices cited by partners included site visits to subpartners and frequent communication via phone, email, or in-person.

Because of security concerns and limited on-site monitoring in Cuba, partners and subpartners use a variety of methods to verify the delivery of assistance to Cuban beneficiaries. Representatives from USAID's M/E contractor indicated that partners have had difficulty collecting and reporting data because of Cuban beneficiaries' reluctance to maintain and provide specific information in writing (e.g., timesheets, attendance sheets, or other documents naming beneficiaries). However, the M/E contractor has found that partners and subpartners often communicate with beneficiaries though various means. Similarly, according to representatives of partners and subpartners we interviewed,[40] some delivery verification methods they used included the following:

- having future travelers ask beneficiaries how they used assistance provided previously; and

- observing beneficiaries' use of assistance through remote or indirect means—for example, through articles published online that demonstrate that beneficiaries received training.

Partners generally aggregate information obtained from such methods in their progress reports to the agencies. However, in certain instances, the monitoring methods selected have limited subpartners' ability to track and report detailed information. For example, one subpartner reported on an indicator, the number of signatures on petitions, by providing data that it obtained over the phone (instead of through document reviews that would prevent double counting of signatures and allow for other data quality checks). In addition, USAID's M/E contractor has found cases in which data could not be transmitted in a timely manner, preventing reporting on activities in the quarter when they were implemented. As a result, both the partners and the agencies can have difficulty knowing the exact numbers and identities of beneficiaries in Cuba.

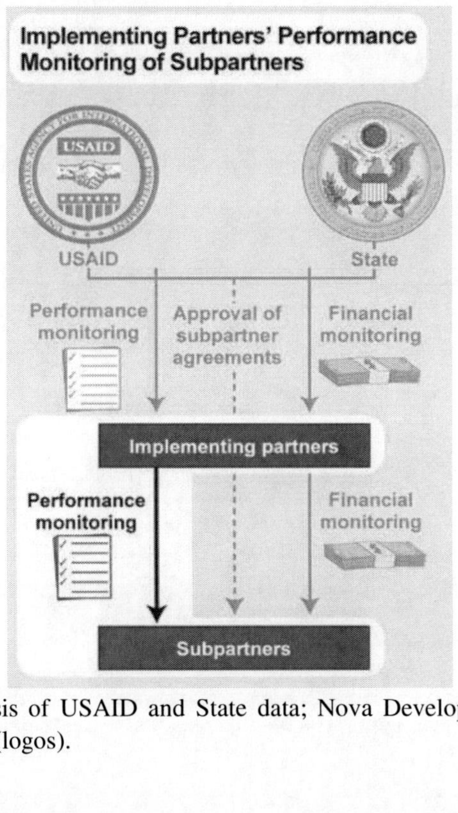

Source: GAO analysis of USAID and State data; Nova Development (clip art); and Corel Graphics (logos).

USAID Has Improved Financial Monitoring of Partners; State Has Not Consistently Conducted Financial Internal Controls Reviews

USAID Has Bolstered Financial Monitoring of Its Partners

Since 2008, USAID has made improvements to financial monitoring of its Cuba program partners.[41] In April 2008, USAID/LAC hired an external auditor to perform financial internal controls reviews of its partners to ensure that they have appropriate internal controls and to review selected transactions under the program to ensure that they are allowable, allocable, and reasonable.[42] Since 2008, this auditor has conducted 30 audits across 13 of the 16 partners USAID funded during fiscal years 2007 through 2010. These audits are in addition to audits performed by USAID's Management Bureau and its Inspector General.[43] Across its different auditing entities, USAID has a goal of reviewing each partner approximately once every 6 months. Other risk-based factors are considered in the scheduling and sequencing of reviews, such as preaward reviews, prior audit findings, and period of performance.

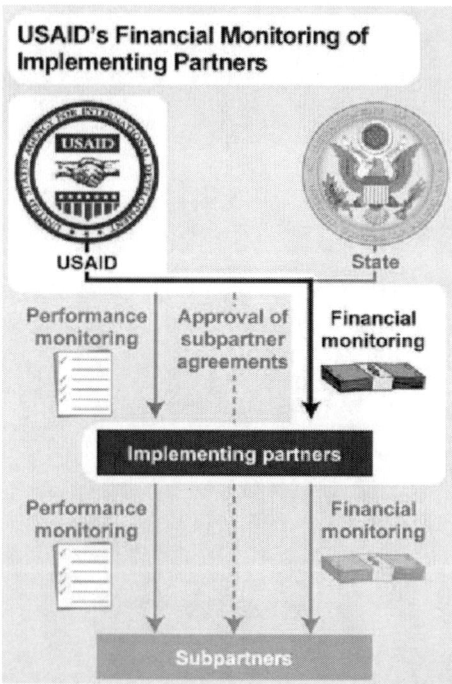

Source: GAO analysis of USAID and State data; Nova Development (clip art); and Corel Graphics (logos).

Through December 2011, the external auditor had found 50 instances of unsupported costs, such as insufficient documentation and lack of authorization, and 15 instances of excessive costs charged to an award or contract, such as charging incorrect rates or expenses not allocable to the award or contract. In sum, the external auditor questioned 11 percent of the charges made to USAID/LAC during the external auditor's periods of review.[44] In addition, the external auditor found inadequacies in the following three main areas:

- the financial management systems at 11 partners,
- procurement standards at 8 partners, and
- internal controls at 8 partners.

Two of the external auditor's most common specific findings were that (1) partners did not properly complete their quarterly financial reports, and (2) partners did not perform a cost-price analysis before procuring a subpartner or equipment to ensure that it was procured at a fair price.[45] Specifically, the external auditor found that four partners did not provide any documentation of a cost-price analysis. In addition, the auditor found that another four partners had insufficiently completed or documented cost-price analyses, either by performing them verbally but not documenting them or by having unorganized or unexplained documentation of the cost-price analysis, limiting the external auditor's ability to confirm the reasonableness of the costs in question.

As a result of the external auditor's findings, USAID/LAC has provided training to partners and, according to USAID/LAC's external auditor, partners have made improvements. First, USAID/LAC asked the external auditor to provide briefings to the Cuba program partners at their December 2008 and March 2011 quarterly meetings on topics such as unallowable expenses, internal control standards, and procurement regulations. In addition, our review of audit reports issued from 2008 through 2011 showed that the external auditor found fewer inadequacies at some partners that had previously been audited. An official with USAID's external auditor who is responsible for these audits noted that recent reviews have found that the partners have improved their financial management capacity.

USAID/OTI used other processes to regularly monitor the financial performance of its partner under the Cuba democracy program. According to USAID/OTI, USAID/OTI staff worked closely with their Cuba program partner to plan future expenditures and reviewed documentation related to individual subpartners to determine each subpartner's real costs. In addition,

USAID/OTI officials maintained a database that the partner updated on a weekly basis, allowing USAID/OTI to monitor all expenditures weekly.

Based on our financial internal controls reviews of the five USAID awards and contracts in our nongeneralizable sample (see table 5 in appendix I),[46] we found that partners' internal controls included (1) policies to prevent the commingling of U.S. government funds, such as unique accounting codes to identify awards and separate bank accounts for U.S. government funds; (2) policy manuals to instruct employees on the proper use of U.S. government funds received through grants and contracts; and (3) procedures to segregate incompatible financial duties. We found that USAID/LAC was overcharged for some overhead and labor expenses by one of USAID/LAC's partners in our sample.[47]

State Has Not Consistently Performed Financial Internal Controls Reviews of Its Partners but Plans to Bolster Its Monitoring of Some Awards

State has not consistently conducted ongoing, in-depth financial monitoring of its Cuba program partners. State's Bureau of Administration is responsible for conducting financial internal controls reviews of State/DRL's and State/WHA's partners.[48] However, the Bureau of Administration has conducted financial internal controls reviews of less than one-third of partners active since fiscal year 2010. For State/DRL awards, the Bureau of Administration conducted one review in each fiscal year for 2010 and 2011 and four in July 2012. It conducted no reviews of State/WHA partners.[49] According to State officials, the Bureau of Administration attempts to conduct financial internal controls reviews at least once during the course of each DRL and WHA award but has not done so for many of its awards because of staffing turnover and constraints in the Bureau of Administration.

In September 2012, State/DRL awarded funding to an external auditor to perform financial internal controls reviews in fiscal year 2013. During this fiscal year, State intends for the auditor to perform one review of three-quarters of State/DRL's partners and no reviews of State/WHA's partners.[50] State provided documentation to us showing that, in October 2012, State/DRL worked with the external auditor to develop a preliminary plan to select the ordering of its partners to be reviewed using a risk- based approach that considered criteria such as the value of awards, any prior financial compliance issues identified, and the partners' internal administrative capacity.

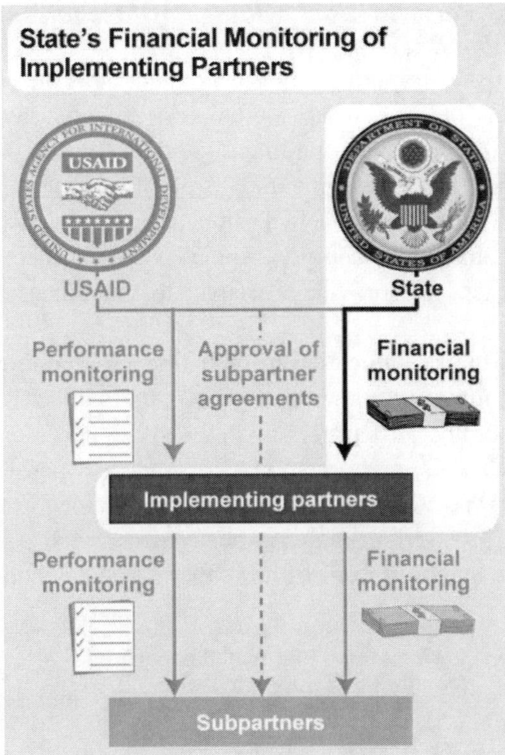

Source: GAO analysis of USAID and State data; Nova Development (clip art); and Corel Graphics (logos).

We conducted financial internal controls reviews on three partners with five State awards in our nongeneralizable sample (see table 5 in appendix I).[51] These three partners had been recently reviewed by USAID/LAC's external auditor, because they have also been USAID/LAC awardees, and had made internal control improvements in response to the auditor's findings.[52]

Similar to our review of USAID's partners, we found that partners had internal control mechanisms in place, including (1) policies to prevent the commingling of U.S. government funds, (2) policy manuals to instruct employees on the proper use of U.S. government funds, and (3) procedures to segregate incompatible financial duties. However, State has had eight partners with nine State/DRL and State/WHA awards active in fiscal years 2011 or 2012 that have received no financial internal controls reviews during the course of their awards, through either State's Bureau of Administration or USAID's external auditor.

USAID and State Partners Used a Range of Mechanisms for Financial Monitoring of Subpartners

Our review of the six partners in our nongeneralizable sample found that partners had written policies and procedures for financial monitoring of their subpartners' use of program funding.[53] For example, partners had risk assessment processes to determine the level of monitoring required for a certain subpartner, depending on that subpartner's capacity and the type of subaward or subcontract. In addition, some of the partners required certain types of subpartners to provide receipts to document 100 percent of expenses.[54]

To test the partners' application of their financial monitoring policies and procedures, we conducted reviews of 11 subpartners under the six partners in our sample.[55] Generally, all partners maintained the necessary documentation (i.e., receipts, timesheets, authorizations) to support expenses incurred at the subpartner level. We found that partners maintained varying levels of documentation on cost-price analyses performed and that one partner had incomplete documentation for one of its subpartner's expenditures.

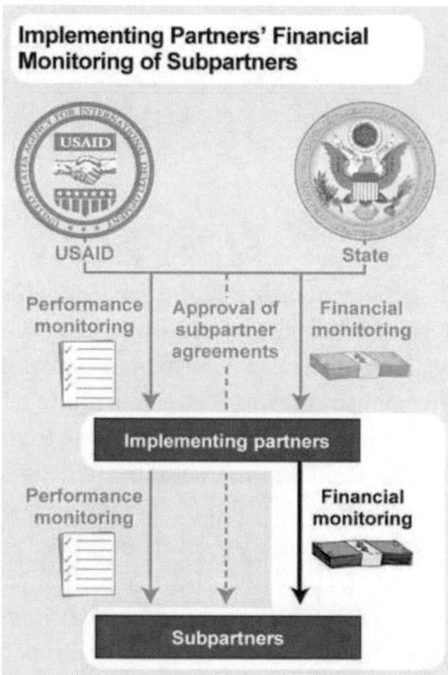

Source: GAO analysis of USAID and State data; Nova Development (clip art); and Corel Graphics (logos).

- For three of the five subpartners with fixed-price subcontracts in our sample,[56] documentation supporting the partners' cost-price analyses included (1) the actual amounts paid for similar services to subpartners on previous awards, (2) price quotes to procure supplies and equipment from various vendors, or (3) surveys demonstrating the market value of labor paid for different labor categories to substantiate that the amounts were within industry standards. For two of the five subpartners with fixed-price subcontracts within our sample, the partners documented that they believed the price of the subcontract to be fair based on the partner's prior experience.[57]
- One subpartner of a USAID/LAC partner submitted its receipts in a foreign language that staff at the partner could not read and provided little explanation of the receipts.

Agencies Have No Direct Relationships with Subpartners except to Approve Their Funding, and State Does Not Provide Clear Guidance on Subpartner Approval Requirements

Agencies Have No Direct Relationships with Subpartners Other Than Approving Partner Requests for Their Funding

USAID and State have no direct relationships with their partners' subpartners. Partners are responsible for all oversight of their subpartners and for reporting to the agencies any updates and problems related to the subpartners' work, such as through any quarterly reports and site visits. However, the agencies are generally required to approve any partner requests to award funding to subpartners.[58]

USAID Provides Specific Guidance to Its Partners on Requirements for Subpartner Approval

USAID has provided guidance to its Cuba program partners on what is required for approval of subpartners, both during the preaward phase and during the course of the award.

- **Preaward phase.** According to USAID officials, subpartners can be considered pre-approved during the preaward phase if they are described in award proposals, in accordance with the standard provisions that are referenced in partners' awards. USAID officials indicated that some of USAID's partners in the past had interpreted the term "described" to include any reference in a proposal to a subpartner's activities, even if the proposal did not specify that the work would be completed by a subpartner or provide specific information about the subpartner. As a result, USAID learned—through reviews conducted since May 2009 by its external auditor and its own subsequent reviews—that some partners had subcontracts and subawards USAID had not approved. In response, in June 2011, USAID added guidance to its requests for applications (RFA) on the type of information that partners must submit in order to receive prior approval for all types of subpartners.[59] This information includes, for example,

- the name of the proposed subpartner,
- a description of the work to be performed by the subpartner under the award,
- the total estimated cost to be paid to the subpartner, and
- a detailed, line-item budget.

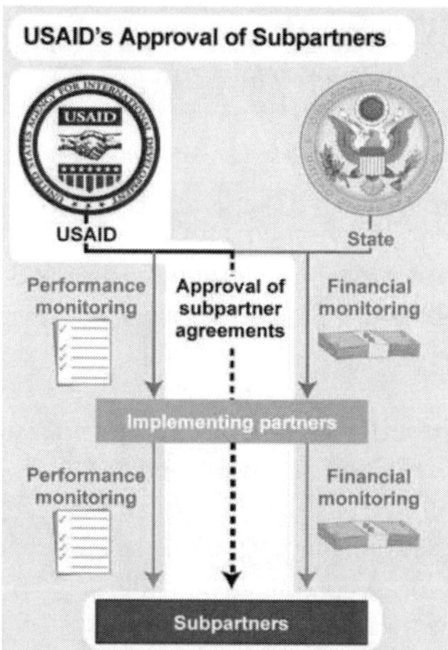

Source: GAO analysis of USAID and State data; Nova Development (clip art); and Corel Graphics (logos).

In some cases, USAID accepts this information orally if the partner is concerned that the leaking of this information could compromise the security of individual consultants who travel to Cuba. To provide further clarity to partners on whether or not USAID considers subpartners as approved during the preaward phase, USAID stated in its RFAs issued in fiscal year 2011 that a subaward or subcontract is not considered approved until the USAID Agreement Officer in the Management Bureau signs a letter approving it.

- **Award phase.** For USAID approval during the award phase, the partner must submit all of the same information as required during the preaward phase, as well as a copy of the proposed agreement with the subpartner and documentation of the process through which the subaward or subcontract was procured. [60] In May 2011, to further its understanding of the work to be performed by its subpartners, USAID/LAC set up a technical evaluation committee. For ongoing awards, partners submit proposed subpartners for approval to the committee. The committee may ask for information on proposed subpartner activities, among other things, to ensure that the programmatic content of the work to be performed by subpartners fits in the scope of the overall award.

State Does Not Provide Clear Guidance to Its Partners on Subpartner Approval Requirements

State's Bureau of Administration is responsible for approving subawards and subcontracts under State/DRL and State/WHA awards, and has requirements similar to USAID's. However, State does not clearly inform its Cuba program partners of these requirements in written guidance and, as a result, some partners have not provided the required information.

According to language in State's Standard Terms and Conditions,[61] prior written approval is required for any subawards or subcontracts unless they are described in the application and funded in the approved award.[62] Specifically, State officials told us that the Bureau of Administration requires certain information and documentation to approve subpartners, including

- the name of the subpartner organization,
- a copy of the draft agreement with the subpartner,
- the amount of and budget for the agreement,
- the period of performance, and
- a description of the subpartner's role.

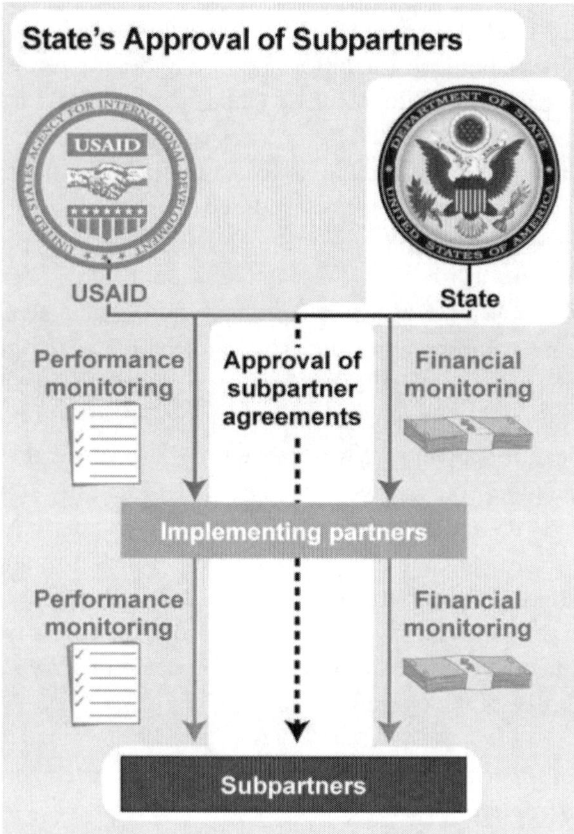

Source: GAO analysis of USAID and State data; Nova Development (clip art); and Corel Graphics (logos).

According to State officials, State's requirements are currently the same whether a partner obtains preapproval during the preaward phase or during the course of the award. They added that, for consultants under State/DRL awards, the Bureau of Administration does not require their names because of security and sensitivity concerns. Instead, according to officials, the Bureau of Administration requires information on the amount of the consultancy contract, the consultant's budget, and a description of the consultant's role and qualifications.[63] Nevertheless, the detailed information that State told us is required to preapprove subawards, subcontracts, and consultants is not specified in written guidance to all partners. For example, in State's handbook for grant recipients, State informs recipients to provide in award proposals details on any subpartners, but does not specify the type of information to provide. In addition, based on our review of fiscal year 2010 and 2011 award documents, we found that in cases when State/DRL is made aware of a prospective partner's intention to have a subpartner through review of their proposal, State/DRL included a requirement in the partner's award to provide a copy of the agreement with that specific subpartner to State within 10 days of its execution. However, State/DRL omitted this requirement from awards for which it was not clear the partner intended to use a subpartner. As a result, we found that many partners only received the broad written guidance in State's Standard Terms and Conditions and recipient handbook.

Based on our analysis of the use of subpartners under recent State awards,[64] we found that State had sufficient information to preapprove the subawards and subcontracts under State/WHA's awards. However, State did not have the detailed information that, according to State officials, would be required to approve 91 subawards and subcontracts to which partners obligated funding under 8 State/DRL awards.

According to representatives of one of the State/DRL partners, they assumed that State was aware of their subawards and subcontracts and considered them preapproved because the partners' proposals had referenced the types of work to be performed under the award by the subpartners. We found, however, that the partners' proposals only provided general information for all proposed subpartners, such as an estimated total cost aggregated for all subawards and subcontracts. The proposals did not specify information about individual proposed subawards or subcontracts, such as proposed periods of performance, a description of the work to be performed, or copies of draft subpartner agreements. Officials said that State has provided training to grants officers over time to ensure greater consistency in the application of preapproval requirements. However, we interviewed two partners with ongoing State/DRL awards, and both were still unaware of the information required for subpartner approval.

CONCLUSION

USAID has been implementing Cuba democracy assistance efforts since 1996, and State's role in the program has increased since it began providing assistance several years later, in 2004. More than $200 million has been provided for these efforts over the past 15 years, with recent growth in the use of worldwide and regional organizations that often use subpartners to help implement program activities. Despite ongoing challenges stemming from the difficult operating environment in Cuba, since our 2006 and 2008 reports, USAID has taken steps to improve its performance and financial monitoring of Cuba democracy program awards. While State has also taken initial steps to improve performance monitoring of its Cuba program awards, we found that State's financial monitoring was lacking in certain areas.

For performance monitoring of Cuba program partners, both USAID and State have required partners to submit program planning and reporting documents that the agencies use to monitor their partners' implementation of program activities and progress toward program goals. Although we found some gaps in these efforts, such as instances in which partners did not identify targets in performance plans, lacked clearly defined indicators, or did not report on established indicators, both agencies are taking steps to improve performance monitoring of their partners. Specifically, since 2010, USAID has used an external contractor to enhance its Cuba program monitoring and evaluation efforts. In September 2012, State hired an organization for a similar purpose, with work on this effort slated to begin in fiscal year 2013.

To enhance financial monitoring, in 2008 USAID hired an external auditor to perform financial internal controls reviews of its Cuba program partners and used a risk-based approach to determine how often each partner should be reviewed to enable more efficient and effective reviews, with resources focused on areas of greater risk.

Such an approach considers key factors such as the value of awards, coverage, previously identified deficiencies, award type, and the frequency of the reviews that will be needed. USAID's auditor conducted 30 audits through fiscal year 2012, which identified questionable charges and weaknesses in partners' financial management, procurement standards, and internal controls. State did not conduct financial internal controls reviews for more than two-thirds of its awards during fiscal years 2010 through 2012, although State recently awarded funding to an external auditor for this purpose and has taken steps toward implementing a risk-based approach for these reviews. However, because these actions were taken recently, State's ability to ensure that funds are being spent as intended remains unknown until it has completed these reviews. Moreover, unlike USAID, State has not provided clear guidance to its partners regarding requirements for subpartner approval.

As a result, State lacks complete and accurate information on its partners' use of subpartners to implement program efforts. Without adequate information on program subpartners, State has limited ability to fully understand and assess its partners' use of program funds.

RECOMMENDATIONS FOR EXECUTIVE ACTION

To strengthen State's ability to monitor the use of Cuba democracy program funds, we recommend that the Secretary of State take the following two actions:

- To enhance financial oversight, use a risk-based approach for program audits, including those conducted by an external auditor, that considers, among other factors, specific indicators—such as value of awards, prior deficiencies, oversight coverage, and frequency—for each of State's Cuba program partners.
- To obtain sufficient information to approve implementing partners' use of subpartners, provide clear guidance to implementing partners regarding requirements for approval of the use of subpartners, and monitor implementing partners to ensure that they adhere to these requirements.

AGENCY COMMENTS AND OUR EVALUATION

We provided a draft of this report to USAID and State for review and comment. USAID noted that it is a challenge to implement assistance programs in countries where USAID does not have dedicated staff in-country, and cited their ongoing commitment to ensuring that Cuba democracy assistance programs managed by USAID receive appropriate management and oversight to minimize waste and mismanagement and maximize impact on the ground in Cuba. USAID highlighted steps that the agency has taken to improve program management, such as dedicating additional resources to conduct financial audits and monitoring of awardees, and conducting pre-award audits on organizations with limited or no experience managing USAID- funded projects. USAID expressed appreciation for our recognition of the agency's program improvements.

State concurred with both of our recommendations and noted relevant actions it has taken or plans to take. Regarding our recommendation to enhance financial oversight through using a risk-based approach for Cuba program audits, State noted that the external auditor that State/DRL recently procured to audit some of its partners has taken steps to implement a risk-based approach. State further noted that the department is evaluating staffing in its Bureau of Administration and audit requirements to be able to address program oversight needs not covered by this external auditor. State also concurred with our recommendation to obtain sufficient information to approve implementing partners' use of subpartners. State said that it plans to hold meetings with awardees to discuss award requirements and provide an orientation on resources and technical support available to Cuba program awardees.

USAID and State also provided technical comments that we have incorporated, as appropriate.

Sincerely yours,

David Gootnick, Director
International Affairs and Trade

APPENDIX I. OBJECTIVES, SCOPE, AND METHODOLOGY

This report (1) identifies the types and amounts of democracy assistance that the United States Agency for International Development (USAID) and the Department of State (State) have provided to Cuba and characteristics of implementing partners, subpartners, and program beneficiaries; (2) reviews USAID's and State's efforts to implement the program in accordance with U.S. laws and regulations and to address program risks; and (3) examines USAID's and State's monitoring of the use of program funds. This report is a publicly releasable version of a prior GAO report, issued in December 2012, that USAID and State had designated Sensitive But Unclassified.

To identify types and amounts of democracy assistance, and characteristics of implementing partners (partners), subpartners, and beneficiaries, we reviewed congressional notifications, agency and partner documents and data on program awards and funding, copies of award agreements and contracts and any modifications, partner interim and final reports, and other key documents and data. We interviewed agency officials and partner representatives to corroborate information and data obtained. We also discussed Cuba democracy assistance with officials at the National Endowment for Democracy.[65] We reviewed amounts of assistance and characteristics of partners that received program funding in fiscal years 1996 through 2012. To test the reliability of funding data, we compiled lists of all funding that went to each partner and sent these lists to USAID and State for verification. In addition, we reviewed the use of subpartners under all of USAID's and State's 29 awards and contracts that were active in fiscal year 2011, which were funded with appropriations from fiscal years 2007 through 2009. These awards and contracts were awarded to 22 partners. We obtained information and data from each of the partners on their use of subpartners under the respective awards and contracts. For the purposes of our review, we defined subpartners as recipients of subawards, including subgrants, grants under contract, subcontracts, or consultants. To test the reliability of subpartner data, we compared information obtained from partners to agency information, and interviewed agency and partner officials regarding any discrepancies. We determined data on program funding and on the use of subpartners were sufficiently reliable for the purposes of this report.

To review USAID's and State's efforts to ensure that program implementation is consistent with U.S. laws and regulations, and to provide guidance to partners and subpartners regarding program risks, we reviewed relevant U.S. laws and regulations and agency and departmental policies and procedures. We also interviewed USAID and State legal and program officials regarding program implementation and related risks. For selected partners and subpartners, we reviewed award agreements; contracts; and partner guidance, policies, and procedures regarding program security risks and travel security and safety. We also interviewed representatives of partners and subpartners regarding program security risks and traveler safety and security measures.

We analyzed reported activities and assistance delivered, and management and internal controls for a nonprobability,[66] nongeneralizable sample of six USAID and State partners and 11 of their subpartners, in order to assess performance and financial monitoring and oversight of their awards and contracts. While the results of our analysis of these six partners are not generalizable to the population, we selected this nonprobability sample to be generally reflective of other partners in the population and to cover a large proportion of the overall dollar value of aid. We selected at least one partner from each of the four USAID and State bureaus and offices implementing program assistance—USAID's Bureau of Latin American and Caribbean Affairs (LAC); the Office of Transition Initiatives (OTI) within USAID's Bureau for Democracy, Conflict, and Humanitarian Assistance; State's Bureau of Democracy, Human Rights, and Labor (DRL); and State's Bureau of Western Hemisphere Affairs (WHA). While there were in total 29 partners from which we selected, the six selected partners were among the top 15 recipients of program funding awarded in fiscal years 2007 through 2010, and represented about 60 percent of funding for awards active in fiscal year 2011.

Other factors that we considered in selecting partners included the timing of the awards and contracts—we selected partners with awards or contracts active in fiscal year 2011—and other strategic factors, such as the type of activity planned under the award or contract and whether the partner had ongoing or new program activities planned for fiscal year 2012 and beyond. We judgmentally selected two subpartners for further review under each of the partners, except for the one partner that only used one subpartner. To select subpartners, we considered factors such as funding received, whether the subpartner had recent activity in fiscal year 2011, the type of activity implemented, and other strategic factors. These subpartners are not generalizable to the population, but provided additional program context and examples for the purposes of our review. Table 5 provides additional information on the partners in our sample.

To examine USAID's and State's monitoring of the use of program funds, we reviewed and analyzed performance and financial documentation and data and conducted interviews with USAID and State officials, as well as representatives of partners and subpartners in our sample. In addition, we conducted fieldwork in Miami, Florida—where we interviewed representatives and reviewed documentation at local partners and subpartners—and at the U.S. Interests Section (USINT) in Havana, Cuba—where we interviewed U.S. officials and partner representatives, and observed WHA-funded democracy assistance activities at post. We examined USAID's and State's program operational plans and performance progress reports; agency and partner policy and procedure manuals and program guidance; agendas and information presented at quarterly partner meetings; partner and subpartner award agreements and contracts; partner implementation, monitoring and evaluation (M/E) plans; partner and subpartner interim and final performance and financial reports; and audits of partner activities, among other documents and data.

To review partners' M/E plans, we assessed whether each plan had three of the basic elements of M/E plans, as described in USAID and State guidance: (1) clearly defined indicators, (2) targets set for each indicator, and (3) data collection methods specified for each indicator. We selected these criteria since they were common elements that M/E plans should have, according to USAID and State guidance. Because we focus on assessing the agencies' and partners' abilities to monitor, not to evaluate, the Cuba democracy program, we did not select criteria for assessing any portions of the M/E plans related to evaluation. We analyzed each M/E plan to determine the extent to which the plan incorporated each element. For example, we determined that an M/E plan partially met the criterion for clearly defined indicators if the plan had indicators but did not provide definitions for the indicators. For targets, we assessed a partner's M/E plan as having partially met the criterion if there were relevant targets specified for some but not all indicators. For data collection methods, we assessed the partner's M/E plan as having partially met the criterion if data collection methods were only described for some indicators or if the plan generally described data collection methods, but did specify which methods pertained to each indicator.

To review partners' progress reports, we assessed whether progress was clearly reported against each indicator identified in the M/E plans. We determined that a partner's progress report met this criterion if the report included specific updates on progress for each indicator, with any progress for quantitative indicators (e.g., number of beneficiaries) reported in numeric form. We assessed a partner's progress report as partially having met the criterion if it reported progress in this specific way on some but not all indicators. We assessed a partner's progress report as not having met the criterion if the partner did not clearly report progress on any of its indicators.

Table 5. Selected USAID and State Implementing Partners

Implementing partner	Awarding agency and bureau/office for each award/contract	Total obligation to award/ contract	Number and type of subpartners reviewed per award/contract
Partner #1	USAID/OTI	$11,600,000	Two subgrantees
Partner #2	USAID/LAC	$6,857,817	Two subgrantees
Partner #3	USAID/LAC	$3,300,000	One subcontractor and one subgrantee
	State/DRL	$591,000	—
	State/DRL	$500,000	—
	State/DRL	$837,053	—
Partner #4	USAID/LAC	$3,700,000	Two subcontractors
Partner #5	USAID/LAC	$3,000,000	One subgrantee
	State/DRL	$591,000	One subcontractor
Partner #6	State/WHA	$821,232	One subcontractor

Source: GAO analysis of USAID, State, and implementing partner records.

Additionally, we interviewed officials from the two organizations contracted by USAID/LAC to conduct performance and financial reviews of partners. We reviewed partner and subpartner internal controls and related residual fiscal accountability risk, and also performed walk- throughs of their disbursement processes and reviewed invoices and other supporting documentation. We primarily focused our review on compliance with internal controls standards relating to monitoring of program funds and to reviewing certain control activities.[67] We performed selected expenditure testing at each partner and subpartner in our sample, when applicable, to identify potential internal controls or financial management issues. We also reviewed our previous reports and interviewed experts to identify lessons learned and to better understand challenges related to providing democracy assistance for Cuba.

We conducted this performance audit from September 2011 to January 2013 in accordance with generally accepted government auditing standards. Those standards require that we plan and perform the audit to obtain sufficient, appropriate evidence to provide a reasonable basis for our findings and conclusions based on our audit objectives. We believe that the evidence obtained provides a reasonable basis for our findings and conclusions based on our audit objectives.

End Notes

[1] The Cuban Democracy Act of 1992 (Pub. L. No. 102-484, Div. A, Tit. XVII) and the Cuban Liberty and Democratic Solidarity (LIBERTAD) Act of 1996 (Pub. L. No. 104-114, commonly known as the Helms-Burton Act) authorize assistance and other support for individuals and independent nongovernmental organizations (NGOs) to promote peaceful, nonviolent democratic change in Cuba through various types of democracy-building efforts. Appropriations are provided through the annual Department of State, Foreign Operations, and Related Programs Appropriations Act, generally in the Economic Support Fund account.

[2] For the purposes of our review, partners include contractors, grantees, and recipients of cooperative agreements. Awardees include recipients of grants and cooperative agreements. Subpartners include recipients of subawards (subgrants, grants under contract, consultants, etc.) and subcontracts. Program beneficiaries are the ultimate targets of program assistance, including human rights and democracy advocates on the island, as well as Cuban civil society actors.

[3] See GAO, *Foreign Assistance: U.S. Democracy Assistance for Cuba Needs Better Management and Oversight,* GAO-07-147 *(Washington, D.C.: Nov. 15, 2006); and GAO, Foreign Assistance: Continued Efforts Needed to Strengthen USAID's Oversight of U.S. Democracy Assistance for Cuba,* GAO-09-165 *(Washington, D.C.: Nov. 24, 2008). These reports focused primarily on assistance provided by USAID's Latin American and Caribbean Bureau. A classified version of the November 2006 report was published in May 2007.*

[4] Internal controls comprise the plans, methods, and procedures used to meet missions, goals, and objectives and, in doing so, support performance-based management.

[5] A nonprobability sample is not randomly selected, but rather is selected based on the use of a set of key criteria; also, it is not a large enough sample for findings to be generalizable to the full population of partners.

[6] The United States does not have an embassy in Cuba, but USINT provides services including consular services and public diplomacy.

[7] However, staff at USINT have a role in overseeing program efforts that are funded by State's Bureau of Western Hemisphere Affairs.

[8] The Office of Acquisitions Management is the primary office within the Bureau of Administration involved in these functions. In addition, there are other bureaus and offices that have administrative and oversight functions for State's Cuba program.

[9] GAO-07-147.

[10] NED was not included in the scope of this review, because it is not a U.S. government agency and because USAID and State provide the majority of Cuba democracy assistance.

[11] The overall goal of the Cuba program is laid out in USAID and State's most recent performance plan report for fiscal year 2011. This report also notes that the Cuba program is focused on advancing democracy and improving human rights conditions by supporting efforts to promote democratic reforms in Cuba, and promoting increased respect for human rights and fundamental freedoms in Cuba. State officials noted that State/DRL Cuba program efforts focus on civil society and human rights, including freedom of information as well as workers' and women's rights.

[12] USINT officials noted that, during a crackdown in 2003 known as the "Black Spring," the Cuban government arrested 75 dissidents. All of the dissidents have since been released, with many exiled to Spain. According to the USINT officials, in the past 2 to 3 years the Cuban government has shifted away from sentencing convicted dissidents to long prison terms; instead, the government now is generally detaining dissidents for shorter periods of time. Officials added that as of January 2012 there were several dozen political prisoners in Cuba, but that it is difficult to ascertain exact numbers as there are no official statistics.

[13] President Bush established the Commission for Assistance to a Free Cuba in October 2003 to identify (1) ways in which the U.S. government could hasten the end of the Castro dictatorship and (2) U.S. programs to assist the Cuban people during a transition to democracy. The commission's July 2006 report recommended providing $80 million over 2 years to increase support for Cuban civil society, disseminate uncensored information to Cuba, expand international awareness of conditions in Cuba, and help realize a democratic transition in Cuba. The report also recommended subsequent annual funding of at least $20 million until the end of the Castro regime.

[14] State officials stated that they do not believe this reduction will affect program efficacy.

[15] The agencies have obligated 81 percent of these funds to partner organizations to implement program assistance; the agencies used the remainder of funding for expenses such as program administration and procurement of materials for distribution.

[16] USAID/OTI's funding supported the award and oversight of one grant and one contract. According to USAID/OTI officials, USAID/OTI does not plan to continue its Cuba program activities, which were envisioned from their inception to be temporary. The office's programs are generally intended to provide short-term assistance to take advantage of windows of opportunity to build democracy and peace.

[17] This average amount does not include additional funding that State/DRL has transferred to NED, which totaled $14.4 million from fiscal years 2005 to 2011.

[18] Cuba-specific NGOs focus only on Cuba; worldwide or regional organizations, comprising both NGOs and for-profit companies, focus on multiple countries. Although most Cuba funds have been awarded through grants or cooperative agreements to NGOs, USAID has also awarded contracts to two companies that operate worldwide. All universities that have received program funding are based in the United States.

[19] USAID and State awarded most program funding through cooperative agreements and grants, respectively. In terms of annualized funding, USAID's largest cooperative agreement was

for $6.7 million over 4 years, and State's largest grant was for $6.1 million over 3.25 years. Only two contracts were awarded, both by USAID in 2008, one for $11.6 million over 3.75 years and the other for $6.9 million over 2.5 years. Since 2006, most USAID and State awards and contracts have resulted from openly competed solicitation processes.

[20] GAO-09-165.

[21] We reviewed selected characteristics—including the use of subpartners—for all of USAID's and State's 29 awards and contracts that were active in fiscal year 2011, funded with appropriations from fiscal years 2007 through 2009. These awards and contracts were awarded to 22 partners.

[22] According to USAID and State legal officials, the Foreign Assistance Act of 1961, as amended, is the primary authority for both USAID's and State's efforts to provide foreign assistance to promote democracy in Cuba. These officials stated that because of prohibitions on assistance to certain countries and other restrictions in the act applicable to Cuba, "notwithstanding authority" is needed to provide assistance to individuals and independent groups in Cuba to support democracy-building efforts."Notwithstanding authority" allows agencies to carry out activities abroad regardless of country prohibitions or certain procurement regulations, personnel regulations, competitive process standards, or other restrictions that would otherwise prohibit or restrict programming. Both section 1705 of the Cuban Democracy Act and section 109(a) of the Helms-Burton Act authorize the provision of assistance "notwithstanding any other provision of law." See 22 U.S.C. § 6004(a) and 22 U.S.C. § 6039(a).

[23] For example, section 7034(m) of the Consolidated Appropriations Act of 2010 (Pub. L. No. 111-117) states that the term "promotion of democracy" means programs that support good governance, human rights, independent media, and the rule of law, and otherwise strengthen the capacity of democratic political parties, governments, nongovernmental organizations and institutions, and citizens to support the development of democratic states, institutions, and practices that are responsive and accountable to citizens.

[24] Under Treasury regulations at 31 C.F.R. Part 515, general licenses authorize certain types of transactions in which Cuba or a Cuban national has an interest, such as certain humanitarian assistance, without the need to apply for a specific license on a case-by- case basis. By statute, informational materials may be exported to Cuba without a license. In addition, USAID and State legal officials noted that USAID and State/DRL both have specific licenses that may extend to implementing organizations and individuals under the program.

[25] We selected a nonprobability, nongeneralizable sample of six USAID and State implementing partners and 11 of their subpartners, to assess performance and financial monitoring and oversight of their awards and contracts. These six partners were among the top 15 recipients of program funding awarded in fiscal years 2007 through 2010 and represented about 60 percent of funding for awards active in fiscal year 2011.

[26] The statutory framework for performance management of U.S. government programs is contained in the Government Performance and Results Act of 1993 (GPRA), Pub. L. No. 103-62, as amended by the GPRA Modernization Act of 2010, Pub. L. No. 111-352.

[27] USAID now uses the term "performance management plan" to refer to monitoring and evaluation plans. USAID regulations require that every country program and its partners develop and maintain a comprehensive performance management system.

[28] According to USAID regulations, these reports should compile information about activities that occurred under the award or contract during the reporting period, including any activities carried out by subpartners, if applicable.

[29] We reviewed M/E plans to determine if they had clearly defined indicators, targets set for each indicator, and data collection methods specified for each indicator, as these were some basic elements of M/E plans as defined in both USAID and State guidance. To review progress reports, we determined whether progress was clearly reported against each indicator identified in the M/E plans, as such reporting allows the agency to see the progress being

made toward the awards' and contracts' identified objectives, according to USAID and State guidance.

[30] Four of the partners did this through organizing their reporting against targets into indicator tracking tables, while another partner provided weekly quantitative updates on progress against targets to USAID/OTI through an online database.

[31] In addition, in 2007, USAID developed and began to implement a structured approach to monitoring meetings with its partners through which USAID collects information on partner performance, such as more in-depth information on activities and planning. Also, since December 2009, USAID/LAC has convened quarterly coordination meetings for all of USAID's and State's Cuba program partners to provide guidance and share experiences and program implementation strategies. According to USAID officials, since December 2010, USAID and State have also used the quarterly partner meetings to obtain information orally about which beneficiaries in Cuba receive goods and services, to identify any potential duplication of beneficiaries across partners. Since partners often consider the names of beneficiaries as sensitive, they do not generally include names or other identifying information about beneficiaries in their progress reports.

[32] During these data quality audits, the M/E contractor plans to review whether partners document lists of specific beneficiaries counted against any particular indicators, either with their specific names or, given program sensitivities, some other form of documentation to show that reported numbers are accurate and do not include any duplication.

[33] Requests for proposals (RFPs) are used in negotiated acquisitions to communicate Government requirements to prospective contractors and to solicit proposals. (See The Federal Acquisition Regulation § 15.203.)

[34] Similar to the progress reports submitted to USAID, State's partners must submit reports that compile information about all activities that occurred under their award during that quarter, including any activities carried out by subpartners, if applicable. Also, State officials may follow up with partners for more information by phone, e-mail, or, less frequently, in-person meetings.

[35] In addition, USINT officials will review the quarterly reports for these awards to ensure they are accurate.

[36] One of the awards in our sample, the third award to Partner #3 as listed in table 4, originated from the RFP State/DRL issued in fiscal year 2010, while the other three State/DRL awards in our sample were from the RFP State/DRL issued in fiscal year 2009.

[37] No State/WHA partners will be reviewed or trained by this M/E contractor.

[38] Partners are responsible for managing and monitoring each project, program, subaward, function, or activity supported by their award.

[39] These trip reports documented information such as the purpose of the trip, locations visited, number of individuals receiving assistance in Cuba, details on assistance provided, and traveler assessments of the situation on the island.

[40] We interviewed 11 subpartners under the six partners in our sample, nine of which were under USAID awards or contracts and two of which were under State awards. Some factors that we considered in our selection of subpartners included the amount of funding obligated to and the type of activity performed by the subpartner. Table 5 in appendix I describes characteristics of the six sample partners and their subpartners that we reviewed.

[41] We recommended that USAID strengthen its financial monitoring of its Cuba program in GAO-07-147 and GAO-09-165. According to GAO's *Standards for Internal Control in the Federal Government* (GAO/AIMD-00-21.3.1, Washington, D.C.: November 1999), for financial monitoring of partners, agencies' internal controls should be designed to provide for ongoing monitoring in the course of normal operations. Prior to 2008, the only ongoing financial monitoring that USAID performed during active awards involved reviewing high-level summaries of expenditures included in a standard federal financial report, known as Form 425, which partners submit quarterly.

[42] In addition to financial internal controls reviews, the external auditor has performed other reviews, as requested, such as compliance and financial systems reviews.

[43] USAID's Management Bureau performs preaward and follow-up audits. USAID's Inspector General also performs periodic audits of some ongoing awards and contracts.

[44] At each partner, the external auditor reviewed nonpayroll disbursements that occurred during a certain review period, ranging from one quarter to over a year. As of September 2012, according to USAID, 75 percent of questioned costs were adequately resolved and USAID had received $50,000 in refunded costs, while the remaining questioned costs were in the process of being resolved.

[45] According to both USAID and State Department regulations (22 C.F.R. § 226.45 and 22 C.F.R. § 145.45), some form of cost or price analysis shall be made and documented in the procurement files in connection with every procurement action. Price analysis may be accomplished in various ways, including the comparison of price quotations submitted, market prices and similar indicia, together with discounts. Cost analysis is the review and evaluation of each element of cost to determine reasonableness, allocability, and allowability.

[46] Our selected sample consisted of six partners. Out of the six selected partners, one had an award solely with State, three had awards solely with USAID, and two had awards with both State and USAID during the period from which we sampled. See table 5 in appendix I for further information on the six selected partners. During our review, we examined partners' policies and procedures, interviewed key officials knowledgeable of the partners' financial management processes, and then tested a nongeneralizable sample of transactions related to program expenditures under the partners' awards and contracts in our sample. Specifically, we verified that the transactions made were supported with adequate documentation; tested the partners' policies and procedures relating to approval of payments, segregation of duties, and subpartner approval; and reviewed the partners' processes to ensure that the payments made were allowable, allocable, and reasonable. We also verified that expenditures made were within limits approved in the contracts and agreements negotiated with the agencies and that payments made were in accordance with the scope of work established in those contracts and agreements.

[47] Specifically, we found that the USAID/LAC partner overcharged for overhead expenses, such as facility rental expenses and office supply purchases, because the partner included overhead in its labor rate multiplier (a negotiated multiplier applied to direct salary costs), while simultaneously directly charging these actual costs. We were unable to determine the total amount of overcharges caused by this double billing because the partner declined to provide us with information necessary to complete this calculation. In addition, we identified almost $1,000 in overbilled labor charges by this partner. In October 2012, we informed USAID of this issue.

[48] During the reviews, Bureau of Administration officials review financial records, documentation of expenditures, and the adequacy of financial and other administrative systems. The review allows State to trace financial statement balances through the recipient's general ledger, cash books, and other summary journals to the original detailed accounting transactions and their supporting documentation for audit trail purposes.

[49] Other than performing financial internal control reviews, State's main tool for financial oversight of its partners is to review the high-level summary of expenditures reported on their quarterly Form 425 submissions. These forms are reviewed by Bureau of Administration officials and officials in the relevant programming bureau, State/ DRL or State/WHA. State uses the information that their partners report on Form 425 to ensure that payments are commensurate with progress and that the partners do not maintain large unused cash balances.

[50] The external auditor will review approximately 9 of State/DRL's 12 current and new partners.

[51] Our selected sample consisted of six partners. Out of the six partners we selected for further review, one had an award solely with State, three had awards solely with USAID, and two

had awards with both State and USAID during the period from which we sampled (see table 5 in appendix I for further information on these partners). During our review, we examined partners' policies and procedures, interviewed key officials knowledgeable of the partners' financial management processes, and then tested a nongeneralizable sample of transactions related to program expenditures under the partners' awards and contracts in our sample. Specifically, we verified that the transactions made were supported with adequate documentation; tested the partners' policies and procedures relating to approval of payments, segregation of duties, and subpartner approval; and reviewed the partners' processes to ensure that the payments made were allowable, allocable, and reasonable. We also verified that expenditures made were within limits approved in the contracts and agreements negotiated with the agencies and that payments made were in accordance with the scope of work established in those contracts and agreements.

52 Because of overlap in partners under State's and USAID/LAC's programs, nine of State's partners active in fiscal years 2011 and 2012 have been reviewed by USAID/LAC's external auditor.

53 According to Office of Management and Budget (OMB) regulations for awardees and Federal Acquisition Regulations for contractors, partners are responsible for financial monitoring of subpartners. Partners should manage and monitor the costs incurred by subpartners to ensure all costs expended are reasonable, allocable, and allowable.

54 One partner noted that this 100 percent threshold was more than it requires of many of its subpartners, but that it maintained this requirement for its Cuba program given its high risk and monitoring challenges.

55 For more information on these subpartners and how they were selected, see appendix I.

56 The price of a fixed-price subcontract is set before the subcontract is signed and is not subject to any adjustment on the basis of actual costs.

57 We were not able to substantiate the assertions of these two partners based on the documentation provided. According to USAID officials, they are currently taking steps to address issues relating to lack of detailed information on cost/price analysis identified by USAID's external auditor for some USAID awards.

58 OMB regulations state in 22 C.F.R. § 215.25(c)(8) that a partner shall request prior approval from the federal awarding agency for the subaward, transfer, or contracting out of any work performed under an award, unless described in the application and funded in the approved award. A similar agency-specific USAID regulation is at 22 C.F.R. § 226.25(c)(8).

59 An RFA is a type of solicitation notice in which an organization announces that grant funding is available.

60 The USAID/LAC contract we reviewed required the contractor to develop USAID- approved guidelines, including criteria for selecting subpartners, and to obtain USAID approval for individual subawards and subcontracts. The technical proposal for the contract also stated that a USAID representative would participate in the contractor's subpartner proposal review process, and representatives of the contractor noted that they also submitted copies of all final subpartner agreements to USAID for approval. For the USAID/OTI contract, USAID/OTI worked closely with the contractor to select and approve all subpartners, in accordance with the contract terms.

61 These Standard Terms and Conditions apply to all overseas federal assistance awards made by State.

62 No differentiation is made in the Standard Terms and Conditions between requirements for the approval of subawards or subcontracts, including those for the use of consultants to perform work toward award objectives.

63 These requirements for preapproval of consultants reflect factors contained in OMB, *Circular A-122: Cost Principles for Non-profit Organizations*, which lays out the types of information that an agency may use to determine that consultant fees are reasonable and allowable under an award, such as the nature and scope of the service to be performed, the

qualifications of the individual or organization performing the service, and their customary fees.

[64] We reviewed the 18 awards that were State's portion of the 29 Cuba program awards and contracts funded from fiscal years 2007 through 2009 appropriations that were active in fiscal year 2011.

[65] The National Endowment for Democracy (NED), an independent, nongovernmental organization, funds programs to promote democracy in Cuba through both direct congressional appropriations and with funding that it receives through State/DRL. NED was not included in the scope of this review, because it is not a U.S. government agency and because USAID and State provide the majority of Cuba democracy assistance.

[66] A nonprobability sample is not randomly selected, but rather is selected based on the use of a set of key criteria; also, it is not a large enough sample for findings to be generalizable to the full population of partners.

[67] See GAO, *Standards for Internal Control in the Federal Government*, GAO/AIMD-00-21.3.1 (Washington, D.C.: November 1999).

INDEX

B

C

D

E

G

H

I

J

K

L

O

P

S

COMPUTER ETHICS

Editor

Zaira Segreti

scitus academics

Computer Ethics

Edited by **Zaira Segreti**

This Book is Printed on Acid-Free Paper. ♾

ISBN: 9781681178479

Library of Congress Control Number: 2017948874

SCITUS Academics LLC,

616, Corporate Way, Suite 2, 4766,

Valley Cottage, NY 10989

www.scitusacademics.com

Notice

Printed in United States of America.

Preface

Everybody talks of ethics, but computer ethics is a comparatively new term which is fast becoming widely popular because of the large-scale software piracy spree. In today's world of hackers and pirates, all of us are somehow involved in the dirty business of software piracy. Computing technologies and artifacts are increasingly integrated into most aspects of our professional, social, and private lives. One consequence of this growing ubiquity of computing is that it can have significant ethical implications that computing professionals need to be aware of. It makes us think about the "relationships among facts, conceptualizations, policies and values with regard to constantly changing computer technology. If you think that ethics are all about living a more moral and principled life, you must practice the same everywhere, even in the computer field, where sooner or later, you would face a dilemma concerning software piracy. When we talk about computer ethics, software piracy reflects in our minds first of all, followed by pornography, spamming, unauthorized access to somebody else's computer, and hacking. Computer ethics are increasingly becoming important because of the rising number of cyber crime issues, including software piracy, unauthorized access, pornography, spamming, and target marketing, and hacking. The widespread popularity and use of the Internet has given rise to a number of cybercrime issues and concerns about user privacy. Various computing applications are tampered with to invade into other's privacy. Malware, spyware, freeware, and browser cookie exploits are some of the notorious computing applications that have spurred the deliberate of importance of ethical behavior in technology.

This volume 'Computer Ethics' discusses ethical issues, technologies that they are related to, and ethical theories, as well as the methodologies that the literature employs, its academic contribution, and resulting recommendations. It discusses general trends and argues that the time has come for a transition to responsible research and innovation to ensure that ethical reflection of computing has practical and manifest consequences.

Table of Contents

CHAPTER 1

From Computer Ethics to Responsible Research and Innovation in ICT The Transition of Reference Discourses Informing Ethics-Related Research in Information Systems

Bernd Carsten Stahl [a,*], Grace Eden[b], Marina Jirotka[b], Mark Coeckelbergh[a]

[a]*De Montfort University, Centre for Computing and Social Responsibility, The Gateway, Leicester LE1 9BH, United Kingdom*

[b]*University of Oxford, Department of Computer Science, Wolfson Building, Parks Road, Oxford OX1 3QD, United Kingdom*

ABSTRACT

The discourse concerning computer ethics qualifies as a reference discourse for ethics-related IS research. Theories, topics and approaches of computer ethics are reflected in IS. The paper argues that there is currently a broader development in the area of research governance, which is referred to as 'responsible research and innovation' (RRI). RRI applied to information and communication technology (ICT) addresses some of the limitations of computer ethics and points toward a broader approach to the governance of science, technology and innovation. Taking this development into account will help IS increase its relevance and make optimal use of its established strengths.

Keywords: Ethics, Morality, Norms, Responsible research and innovation, Information systems

1. INTRODUCTION

Information systems (IS) as a field of academic research and business practice has long considered the importance of ethical considerations, including questions of what counts as right and wrong, good or bad, moral or immoral. IS also theoretically reflects on why particular acts or rules may be considered moral or immoral. Such questions touch on the design and use of computing artefacts in organizations in many different ways. A perception of moral appropriateness, or the lack thereof, can be an important component of technology acceptance. One aspect of this may be seen in the tradition of 'applied ethics' within philosophy, which focuses on more practical questions in various fields, such as medicine and biotechnology but is also relevant within computing. Organizational policies are often driven by perceptions of (morally) correct ways of acting, and these are based on broader ethical positions that tend to be reflected in organizational visions and cultures. The societal governance of technology follows public perceptions and is often driven by majority (moral) concerns, such as in the case of access to pornography and the protection of digital intellectual property.

The theoretical and practical relevance of different ethical and moral questions is reflected in a long-standing stream of research that this paper refers to as 'ethics-related research in the field of IS'. Despite considerable attention, ethics has never been a mainstream concern in IS. The reasons for this are manifold. One likely reason is that ethics, in the sense of moral philosophy, is a complex subject with a long history of scholarly work. Recent attempts to raise the visibility of ethics in IS education point toward a perceived lack of relevant competence of IS graduates [1].

This raises the question of whether traditional notions of ethics provide an appropriate discourse for discussing ethics-related issues in IS; that is, whether this type of ethics gives us a language, a vocabulary that enables us to delineate and develop a field of inquiry that is both intellectually sound (i.e., to philosophers and other academics) and practically relevant to professionals working in IS (a more specific definition of discourse will be provided below). In other words, the problems mentioned raise the question of whether we can find or create a discourse that is better integrated with the discourses that are already present in IS. What should be the 'reference discourse', the main discourse in ethics-related IS research?

This paper discusses the ways in which computer ethics (CE) and responsible research and innovation (RRI) may inform ethics-related IS research, and how these discourses, held outside of the field, may contribute to IS research and practice. We begin by arguing that research in the field of CE has provided many insights into substantive moral problems as well as the theoretical underpinnings required to address them. After demonstrating this point by summarizing the main streams of activity in computer ethics and mapping them to ethics-related research in IS, we argue that, despite these contributions, there are limitations in CE that may be addressed by adopting a novel discourse that is currently held under the title of 'responsible research and innovation in information and communication technologies (ICTs)'.

We argue that future ethics-related research in IS would benefit greatly from taking into consideration this change in reference discourse. The practical and application-oriented nature of the IS field renders it well equipped to address many of the issues arising from RRI in ICT. This is not to say that more philosophical-theoretical work is no longer welcome nor that there can be no other way of relating ethics to more practical concerns. Rather, the RRI discourse creates new opportunities for better integrating ethical concerns in IS that are directly linked to its 'home' discourse. Adopting RRI as a driving force of ethics-related research therefore constitutes the next step in build on the cumulative tradition of ethics-related work in IS. At the same time, adopting RRI means that IS work will need to look beyond its traditional boundaries and be more mindful of broader societal needs.

The argument put forward in this paper is important because it suggests new research directions for IS that fit into a broader research agenda in in Europe and elsewhere. The suggestion of considering the concept of RRI plays on the traditional strengths of the IS discipline, namely its detailed empirical understanding of the role of ICTs in organizations and society. At the same time, this study shapes new research topics by focusing on significant social challenges. The paper furthermore contributes to the self-reflection of the IS discipline by demonstrating that there are important reference discourses that inform the way that ethical questions are addressed. By explicitly reflecting on such reference discourses, this paper allows scholars involved in ethics-related research to position their work in the broader social context, thereby contributing to greater societal and scholarly relevance in the field of IS.

To develop this argument, we start by reviewing the discourse in computer ethics and discuss its relevance in past ethics-related IS research. We define the basic terms of ethics, morality, norms and values. On this basis we then discuss topics, theories and approaches to computer ethics. This background is used to demonstrate the importance of what we refer to as 'the reference discourse' in shaping IS research. Having thus demonstrated the relevance of computer ethics as a reference discourse, we then introduce the discourse on RRI in ICT. This is done by showing that the limitations of computer ethics constitute a natural starting point for engaging with questions of governance of research and innovation. This provides the basis of a discussion of the degree to which the RRI discourse is already part of the IS research agenda and which steps would follow from integrating RRI into IS. The paper concludes by reflecting on its contribution and discussing future work.

2. COMPUTER ETHICS AND INFORMATION SYSTEMS

This section begins with an explanation of the approach taken by the paper. It explains the idea of reference discourse and outlines the methodological underpinnings of the narrative. This leads to a delineation of the subject area of ethics followed by a discussion of ideas and activities that may be considered under the heading of computer ethics. The section then demonstrates how work undertaken in computer ethics has influenced or is mirrored in the field of IS.

2.1. 'Reference discourses'

At the core of this paper is the argument that ethics-related work in IS is influenced by external events related to society and computing and that the current transition from computer ethics to RRI in ICT will be, or should be, reflected by IS research. We use the term 'reference discourse' to refer to these external influences. The term was chosen in explicit recognition of the long and continuing discussion of 'reference disciplines' in IS [2–4], where scholars recognize that the field of IS draws on prior and parallel scholarly activities. However, the term 'reference discipline' is stronger than required for our purposes. It would be difficult to argue that computer ethics constitutes a reference discipline because it is likely too small to be considered a discipline. Furthermore, it has its historical roots in philosophy and computer science [5].

However, the point of this paper is not to discuss whether there is a cumulative body of knowledge that is sufficiently mature and accepted as the 'core' of the IS discipline [6]. Rather, our interest lies in the fact that there is a discourse on computer ethics that is clear enough to constitute a recognizable influence factor on IS research and practice.

The definition of discourse used here is inspired by Habermas, who states that competing claims about truth, rightness and authenticity can be clarified by relying on the power of the best argument [7]. Without having to invoke potentially problematic aspects of Habermas' position, such as the ideal speech situation, this paper argues that there is a discourse in computer ethics, (i.e., an enduring exchange of ideas that focuses on ethical questions related to computing) and that it becomes a reference discourse for IS if it influences and is influenced by research practice in IS. The following sections will broadly survey the discourse in both CE and RRI to trace the general lines of the computer ethics discourse and argue that new developments, such as RRI, should be incorporated into IS research.

2.2. Computer ethics: Theories, topics and approaches

One delimitation of computer ethics is any research activity that touches on right and wrong, good or bad, or moral or immoral in relation to computing. Philosophers may add moral intuition, character and virtue, explicit morality, ethical theory or reflection and meta-ethics insofar as it pertains to computing. However, this definition may be too broad to be practically useful because it would include a disparate body of work that is rooted in a wide range of disciplines.

The present paper therefore suggests a narrower definition that focuses on a community of practice. Willcocks [8] suggests a definition of information systems as '[....] those academics, researchers, teachers, students and indeed practitioners who gravitate around conferences such as ICIS, ECIS, HICSS, PACIS and AIS, tend to be members of the association for information systems or related/similar bodies, write research papers and books consciously within an IS discipline and publish in a self-defined group of 'IS' journals [...]'. This paper adopts an analogue definition of the field of computer ethics comprised of those individuals who would attend conferences, such as CEPE (computer ethics, philosophical inquiry), ETHICOMP or CAP (computers and philosophy), who are members of INSEIT (the international society for ethics and information technology) and who would publish in journals, such as Ethics and information technology, the *Journal of Information Communication and Ethics in Society* or the *International Review of Information Ethics*. It is important to note that computer ethics discourse journals are a less dominant publication venue than in other fields. Much of the work undertaken in computer ethics is published in single authored [9] or edited volumes, such as [10–14].

Most CE conferences and outlets were established in the 1990s. The history of some of the early members of this community and the interest in ethics and computing can be traced back further [5]. In fact, concern with the ethical aspects of computing coincides with the beginnings of digital computing and was famously expressed by Norbert Wiener [15].

A field of research spanning nearly 70 years is difficult to describe. However, the majority of research in IS has taken place since the 1980s and 1990s, and there are a number of clearly recognizable streams of activities in computer ethics that can be categorized in terms of theories, topics and approaches, each discussed below.

2.2.1.Theories of computer ethics

The discourse in computer ethics is led by a significant number of scholars with a background in philosophy. It is therefore unsurprising that philosophical ethics plays an important role in the theoretical makeup of the discourse. One can observe numerous references to philosophical ethics, most notably the big three of (Kantian) deontology [16,17], utilitarian consequentialism [18] and virtue ethics in the Aristotelian tradition [19]. Deontology and consequentialism focus on principles of rational decision making in view of dilemmatic situations, whereas virtue ethics stresses the importance of individual character in evaluating the ethical character of the agent.

Although these three ethical positions are an integral part of the theoretical composition of the field, it is important to note that they are sometimes adapted specifically to questions and applications concerning computing [20]. Additionally, there are numerous other ethical approaches that are invoked in the area of computing, such as the ethics of care [21], which may be useful for understanding particular aspects, such as the ethical side of gender issues in computing [22], or the ethics underlying the capability approach developed by Sen [23] and Nussbaum [24], which has been shown to be relevant to ethics-related research on ICT [25], [26].

In addition to these examples of ethical theory applied to computing issues, there are also attempts to develop novel ethical approaches that are specific to problems of information or technology. Notable examples of such new ethical ideas are information ethics and disclosive ethics. Information ethics, which has been promoted by, and is strongly linked to, Luciano Floridi's work [12,27,28], is an attempt to develop an ethical position on the basis of the ontological properties of information. Disclosive ethics, on the other hand, [29] is an approach that attempts to make explicit the ethical assumptions embedded in technologies. Within the Dutch philosophy of (information) technology there have also been interesting new developments focusing on the link between ethics and design, such as value sensitive design [29], the morality of artefacts [30], and thinking about responsible innovation, which has contributed to the development of RRI in general [31].

Computer ethics draws on a broad range of theoretical positions that are relevant to the development or use of computing technologies. On the more technical side, there are theoretical positions stemming from areas such as engineering ethics [32] or technology ethics more broadly [33,34] that are relevant to computing. However, there has been a long tradition of research into ethical issues in organizations and societies, much of which touches on specific questions or problems raised by computing technology [35,36]. Computer ethics therefore draws on discussions in business ethics [37–39] and corporate social responsibility [40,41].

A final set of theoretical influences on the computer ethics discourse comes from philosophical disciplines other than moral philosophy. Computer ethics benefits from insights into the relationship between normative questions and other philosophical fields, such as epistemology, philosophical anthropology or ontology. An interesting recent example of this is the 'Onlife Manifesto' [42], in which a number of leading scholars explore the ways in which computing technologies fundamentally change the world we live in and the way we relate to it.

2.2.2. Topics of computer ethics

The complexity and multiple nature of theories in computer ethics is reflected by a broad array of topics discussed in the field. Brey and Soraker [43] identified a set of central issues of the field that they list as 'privacy, security, computer crime, intellectual property, free expression, and equity and access, and issues of responsibility and professional ethics'. A recent attempt by the European Group on Ethics in Science and New Technologies [44] to provide general guidance for European ICT researchers focuses on the fields of personal identity, changes to the social sphere, political participation and citizenship, and the sphere of e-commerce.

Each of these fields is the home of a number of different sub-topics that, due to space restrictions, cannot be discussed in detail in the present paper. However, there are a number of easily identifiable core issues worth listing because they have led to political debates in various countries [45] and are often examined by IS researchers. The most visible and dominant topic of computer ethics is that of privacy. While the discourse on privacy can be traced back to Warren and Brandeis' seminal paper [46], the current relevance to computer ethics is that the availability of digital information and the possibility of storing, processing and disseminating it changes the nature of how we perceive privacy [47]. This is directly related, albeit not identical, to the question of data protection [48].

Current debates around privacy and data protection, fuelled by recent political developments and the revelation of large-scale surveillance by several government agencies, show a direct link to the tension between individual rights and national security. Security is a complex topic that has been at the center of attention in computer ethics from its inception. Attempts to ethically evaluate security principles and practices have shown the ambiguities inherent in security. On the one hand, a feeling of security appears to be a necessity for human development and can therefore be argued to be of high ethical relevance [49]. On the other hand, concerns about security can be used to enforce potentially problematic power relationships. Security can be seen here as an ethically problematic mechanism of domination [50].

A further topic of discussion for computer ethics scholars is that of intellectual property. Like privacy and security, intellectual property is not fundamentally a new problem. Questions of ethical and legal justification and protection can be traced back through the history of philosophy [51,52]. However, the use of computing technology has changed important aspects of intellectual property. Digitizing content, such as books or movies, leads to new challenges that traditional property and intellectual property governance did not consider, notably, the negligible cost of reproducing content combined with the ease of copying leads to conceptual problems when applying traditional notions such as theft [53]. A further aspect of this debate refers to specific problems of software. The ownership of software and its justification raises issues that are hotly contested on philosophical as well as legal grounds. The widely used and arguably highly misleading metaphor of "piracy" to denote intellectual property infringements [54–57] is an indication of the visceral quality of the debate.

Intellectual property has influenced further debates in computer ethics regarding justice and the equity of the availability of computing resources. The more that computing is used throughout society, the more important it is for most individuals to be able to access it to live fulfilling personal and professional lives. Computer ethics therefore has long engaged in debates about ethical issues related to access that is often framed in terms of "digital divides' [58,59]. An important aspect of this is the question of participation in public discourses and the influence that computing has on political processes and decision making [60], which links to another well-established topic of computer ethics, namely that of freedom of speech. Again, this is a well-established topic of ethical reflection that takes on a new urgency because of the capability of computers to facilitate a simple and broad dissemination of ideas while simultaneously offering new ways of censoring and prosecuting unwanted contributions [61].

The ever-broader uptake of computing and its pervasive role in many contexts has led to a more recent focus in computer ethics on issues related to the interplay between technology and humans as well as between technology and society. This covers new threats, such as identity theft, and more fundamental questions about the way we conceptualize humans and how modern technologies affect our individual and collective views of ourselves [42]. This covers topics such as the construction of gender [62] and the way we protect ourselves against crime as well as issues around what constitutes a good life and how we want to live it [20,63]. Spending our leisure time with and mediating our social contacts through technology is one example of this issue.

This list of topics in computer ethics cannot claim comprehensive coverage. It is meant to give an indication of some of the items discussed in the field. The list concludes with a reference to a central issue, namely that of professionalism. Within CE there has been a long debate about the degree to which computing can be perceived as a professional field and the way in which professionalism in computing may be used to address ethical concerns [64–66].

2.2.3. Approaches to computer ethics

The final aspect of computer ethics to be introduced here is the set of different approaches and research activities in the field. Just as there are a multitude of theories and topics, one can observe a number of research approaches. One observation of central importance to this paper is that, likely due to the strong influence of academic philosophy on computer ethics, there is a dominance of conceptual and reflective work, with empirical research playing a relatively minor role.

One likely cause of this is that philosophers realize that it is not straightforward to relate descriptions and prescriptions or, to use a different term, to move from 'is' to 'ought'. Philosophers have had long discussions of what is sometimes referred to as Hume's law or Hume's Guillotine [67], which states that description cannot lead to prescriptive or normative statements. In terms of computer ethics, this means that one cannot deduce the ethical acceptability of a particular position from an observation of actions or preferences of people with regard to this position. For example, a piece of research showing that a vast majority of respondents believe that sharing software is acceptable does not allow the conclusion that it is morally justified.

A second possible explanation for the relative paucity of empirical work in computer ethics is that neither philosophical training nor the technical training that computer scientists receive normally prepares a scholar for social inquiry. Questions of methodology that have a key place in social science-oriented publications, such as those in the field of IS, are rarely found in computer ethics.

The dominant approach to computer ethics is that of philosophical argument. Assumptions of concepts and discourses are made explicit and critically discussed to come to new insights. This is usually done using the background of philosophical ethics. The aim tends to be to improve conceptual clarity. Such improvement has recently been often linked to attempts to formulate possible consequences or policies. Examples of this are the development of a comprehensive ethics impact assessment [68] and technically addressing ethical issues in computing by, for example, developing ways of integrating values into technology through methods such as value-sensitive design [31,69,70] or privacy by design [71]. This paper argues that this practical turn is an expression of the move from computer ethics to RRI in ICT. However, before we come to this argument, the relationship between computer ethics and ethics-related research in IS must be established.

2.3. Computer ethics as a reference discourse of information systems

The argument made here relies on the recognition that computer ethics can be seen as a reference discourse for ethics-related IS research. To demonstrate this, one needs to show that theories, topics and approaches of computer ethics are reflected in IS. It is important to note that this relationship is not one way and that one can observe an inverse relationship with IS as a reference discourse for computer ethics. However, this inverse relationship is less central to the argument made in this paper.

A look at the history of ethics-related IS research (cf. [72]) shows that theories, topics and approaches of computer ethics are all reflected in IS, albeit to different degrees. In recent years there has been some interest in ethical theory and its relevance to Information Systems [73–76]. In general, one can safely say that ethical theory has not been at the center of attention for IS scholars interested in ethical questions. Where ethics is the topic of investigation, one can typically find brief dictionary-style definitions of ethics and sometimes more specific references to some of the better-known ethical positions. The majority of the more intricate ethical discussions and, in particular, meta-ethical positions that are often at the core of computer ethics play a minor role in ethics-related IS research.

The role of computer ethics as a reference discourse for ethics-related IS research becomes much more prominent when one looks at the topics being investigated. The seminal discussion of ethics in Information Systems by Mason in 1986 [77], which coined the oft-cited acronym PAPA, touched on several of the aforementioned topics of computer ethics, namely, privacy, accuracy, property and access. However, it is interesting to note that the paper was written as an opinion piece that included only two references, neither of which could be considered related to computer ethics. When tracing the history of ethics-related research in IS, one can nevertheless find numerous further examples of investigations into the topics of relevance of computer ethics.

Privacy, most likely the most dominant topic of computer ethics, plays a similarly important role in ethics-related IS research. Not only is it the first 'P' in Mason's PAPA acronym, it is also a recurring topic in numerous prominent early ethics-related IS papers, such as [78,79]. The problematic relationship between privacy and security also finds its reflection in IS research [80]. Questions of intellectual property have similarly been of interest to IS researchers, who frequently attempt to identify the attitudes of professionals or users toward the ownership of content or software. This interest in intellectual property is reflected in related investigations that touch on questions of ownership, such as open source software [81], computer fraud [82] or perceptions of plagiarism [83].

In other topic areas, there is less overlap between computer ethics and ethics-related IS research. Some of the larger problems concerning the nature of identity, the development of culture and social interaction are less prominent in IS. At the same time, there is more of a focus on the ethical aspects of the use of computing technologies in IS [84]. One can nevertheless argue that there is significant overlap in the topics of interest between computer ethics and ethics-related IS research.

However, the same would be difficult to argue for with regard to research approaches and methodologies. IS research is often perceived to be part of the social science tradition. As a result, there is a strong dominance of empirical research using the various paradigms, methods and methodologies that have been established in the field [85].

To summarize, one can say that computer ethics plays the role of a reference discourse for ethics-related IS research in the sense that much of the key areas of interest can be found in both. This does not make a strong statement about the direction of the relationship between the two. As indicated above, one can equally argue that IS research has the function of a reference discourse for at least some work in computer ethics. A further possibility would be to see both streams of research reflect the public interest in important social concerns. For the purposes of this paper, the details of the mechanisms facilitating it are of secondary importance. The paper argues that computer ethics is taking a novel turn toward RRI in ICT, which IS may wish to reflect. Although the ground for this turn was prepared by those working in computer ethics [31,86], it is worth emphasizing the new discourse and its advantages in comparison to (traditional) computer ethics. This point is made in detail in the following section, which introduces the concept of RRI in ICT as the next step in building on computer ethics.

3. RESPONSIBLE RESEARCH AND INNOVATION IN ICT

This section argues that there is a novel discourse that has the potential to build upon and integrate computer ethics but takes the theories, topics and approaches further and in a slightly different direction. The section starts by discussing the limitations of (mainstream) computer ethics, which may account for why the RRI discourse can offer novel avenues that are attractive to computer ethicists. The following subsections briefly outline the current actors, activities and underpinnings of RRI. This leads to a discussion of the implications of adoption of RRI in IS.

3.1. Limitations of the computer ethics discourse

As suggested, the discourse on computer ethics has a number of limitations whose resolution is likely to require new ideas and approaches. The first and most obvious shortcoming is that of the term 'computer'. When the discourse on computer ethics began, a computer was a clearly recognizable artefact. As computers changed their shape and role from large mainframes to distributed terminals and personal computers, they remained clearly identifiable. However, the distinction between computers and other artefacts is no longer obvious (e.g., cyber-infrastructures, mobile computing and sensor networks). Additionally, most technical artefacts, from cars to telephones, include and are often based on computing technology.

In addition to this ubiquity there is a convergence of technologies. Several types of currently developed fields of technology are intimately connected to computing. One can find reference to the convergence of nano, bio, information and cognitive technologies [87,88]. This convergence renders it difficult to see how a coherent discourse on computer ethics can be sustained. It is likely that the distinction between computing and other technologies will be even less clearly defined in the future than it is in the present [89].

The ubiquity of computers and computing technology and their convergence with other technologies may be why some authors prefer the term 'information and communication technology' (ICT). Consequently, one can observe the use of the term ICT ethics [20,90]. Another strategy is to move away from the link to the artefact, which appears to be the focus of computer ethics, to an ethical focus on the concept of information, as is done by scholars concentrating on information ethics [27,91].

This shift in focus to information from computing artefact may go some way toward defining the boundaries of computer ethics discourse. It does not necessarily address one of the other key challenges, namely, the unclear 'practical relevance' of the CE discourse to IS researchers and practitioners. Complaints about lack of relevance are not confined to computer ethics because the discussion in IS on the relationship of rigor versus relevance [92–94]demonstrates. Nor is this concern particularly novel, which was noted by Gotterbarn 20 years ago [95]. This paper does not take a position on whether the observation of the lack of relevance of computer ethics is justified. However, it is plausible to assume that the increasingly ubiquitous nature of computing technologies calls for a reflection of related ethical concerns in public discourse and policy and for a stronger presence of computer ethics in such policy discussions would be desirable.[1]

A final reason why the (mainstream) computer ethics discourse may be deemed to be in need of novel impulses is that some of the traditional ways of dealing with ethics research appear to be reaching their limits when applied to innovations in computing. A primary example is that of informed consent, the cornerstone of biomedical and research ethics, which is also used in other disciplines that conduct empirical research. It is unclear whether established procedures of informed consent are applicable or whether they are principle conceivable [96]. Traditional concepts of responsibility that focus on the clearly delineated action of the individual are increasingly difficult to apply to the mediated and networked world of computing technologies.

None of these limitations of the computer ethics discourse are insurmountable, and most have been discussed and addressed [43,97]. However, these limitations may explain to some degree why the novel discourse on responsible innovation resonates and points to activities concerning the ethics of computing.

3.2. Definition, actors, activities, normative foundations of RRI

The concept of responsible research and innovation (RRI) is a relatively new one that aims primarily at research governance. Thus, RRI it is not focused on computing or ICT. However, a number of aspects and components of RRI are directly related to the ethics of research and technology development. In many respects it is a natural extension of other discourses, such as the discourse on computer ethics, and particularly on novel approaches within that discourse.

Many of the ideas and principles underpinning RRI can be traced to the Enlightenment. However, the history of the term 'RRI' began much more recently. The term 'responsible development', which is one of the roots of RRI, first appeared in the 21st century nano-technology research and development act (Public Law 108-153 [2]) in the US, where the concept is promoted as part of an initiative to strengthen nano-technological research.

Originally aimed at preventing harm arising from research and innovation activities [98], RRI has since broadened its objectives. Its starting point is the dilemma that we may not know the future but that it seems likely that there are significant socio-economic and technical problems (often referred to as grand challenges [99]) that will require input from research and innovation systems if they are to be satisfactorily addressed [100]. These grand challenges are often global or cover large parts of humanity. Examples include demographic developments, security and sustainability. At the same time, there are no global mechanisms for identifying such challenges, much less for a unified response to them.

RRI is meant to address the gap in time between the initial phases of research strategy formulation to the point at which individuals and organizations regularly use products and services based on research output. Various definitions exist [101–103]. However, the key component of each definition is that they all express a need to develop greater democratic accountability within the innovation lifecycle. RRI is concerned with creating a new mode of research governance that can transform existing processes with a view to ensuring a greater acceptability and even desirability of novel research and innovation outcomes.

A widely quoted definition of RRI from Von Schomberg [104] follows:

Responsible Research and Innovation is a transparent, interactive process by which societal actors and innovators become mutually responsive to each other with a view to the (ethical) acceptability, sustainability and societal desirability of the innovation process and its marketable products (in order to allow a proper embedding of scientific and technological advances in our society).

This definition is critically discussed in [105]. However, for this paper, one can underline the intention of ensuring acceptability, sustainability and desirability of both the process and products of innovation. Additionally, it is important to note that this requires the mutual responsiveness of various societal actors. RRI has gained considerable prominence in European policy discussions and has been adopted by the European Commission as a cross-cutting activity that will govern all research funded under Horizon 2020, the next European research framework program that will run from 2014 to 2020 and have an overall value of more than €70 billion. This research framework has an important influence on ICT research, which will receive approximately €8 to €9 billion of this amount. Activities directly related to RRI are concentrated in a funding stream on 'Science with and for Society', which has a budget of €462 million over the lifetime of the Horizon 2020 program.[2]

We have elsewhere [106] argued that RRI can be interpreted as a higher-level responsibility or meta-responsibility that aims to shape, maintain, develop, coordinate and align existing and novel research and innovation-related processes, actors and responsibilities with a view to ensuring desirable and acceptable research outcomes. This means that RRI can and will incorporate a number of extant components (and this is likely to include numerous aspects of computer ethics), but goes beyond them by supporting actions that will allow these components to contribute to the acceptability and desirability of the outcomes of research and innovation.

Many of the activities that contribute to RRI are well established. Examples include various types of assessments, such as risk assessment and impact assessments [107,108]; technology assessment [109]; various types of research governance structures, such as research ethics reviews [110] and research integrity measures [111]; public engagement in research [112]; the education of researchers [98]; and the support for the development of professionalism [101], to name just some of the more prominent assessments. It is interesting to note that there is some overlap with the applications of computer ethics discussed above. However, RRI has a much broader goal of integrating research and innovation into social discussion and reflection.

It is worth noting here that RRI overlaps to a large degree with technology assessment, although RRI is broader and encompasses many of the ideas and principles that have been developed in technology assessment since the 1970s [113].

The focus on acceptability and desirability means that RRI must actively reflect on its normative underpinnings. In view of the global nature of research and innovation systems as well as of the global challenges that humanity faces, the search for generally binding normative underpinnings is important [114]. Normative foundations on which RRI can be based include human rights and internationally binding treaties, such as the EU treaties for EU Member States. Normative principles can furthermore be sought in philosophical ethics, with some scholars suggesting that there are generally accepted principles [86,115]that are sufficiently defined to provide the basis for collective action. Further normative principles can be deduced from established processes, such as ethics reviews, which can draw on a long history of debate in biomedical research ethics. The implementation of RRI activities through research governance measures can make use of widely shared principles of research governance, such as the integration of democratic principles into research [116], the precautionary principle [117,118] or the principle of regulatory parsimony [116].

The debate about the exact shape that RRI will take and the way in which existing and new responsibilities can be defined and realized is in its infancy. One can observe attempts to implement principles of RRI in some of the more contested areas of research and innovation, such as synthetic biology [119,120], nanotechnology [121–123] or geoengineering [124,125]. RRI must contend with a number of fundamental questions, such as the limits of foresight and the legitimacy of research governance that may be difficult to address. There is likely to be resistance to RRI from numerous actors who may see it as a threat to the academic freedom of research. The global nature of the challenges to be addressed will require novel forms of transnational research governance, which may be difficult to agree on and implement. It is thus by no means certain that RRI will be successful. At the same time, one can see that RRI has spawned a lively debate and that it offers an exciting opportunity to think about new research concepts, innovation and governance.

Returning to the discussion of the relationship between RRI and computer ethics, one can see that there is space for all of the concerns of computer ethics in RRI. At the same time, RRI is broader and specifically focused on the outcomes of research with attention to grand challenges and the desirability and acceptability of both research processes and products. The move from computer ethics to RRI in ICT is thus not a disruption, but rather can be seen as an evolutionary development that addresses some of the limitations of mainstream, traditional computer ethics and allows for the tradition of computer ethics to be embedded into a broader social context.

3.3. Consequences of adopting the RRI discourse in IS

If the thesis of this paper is correct, namely that there is a development from computer ethics to RRI in ICT and that computer ethics serves as a reference discourse to ethics-related research in IS, then an important question is how IS can react to this change in the reference discourse.

An initial response to this question is that ethics-related research in IS can examine broader topics. Traditionally, much attention has been paid to ethical issues related to the use of ICT in organizations and to questions of the legal status of activities and enforcement of organizational policies. Such research will remain relevant. However, in the spirit of RRI, a more prominent question will focus on whether and how information systems can contribute to the resolution of grand challenges. Initial developments in this direction can already be observed. There are now examples of papers in IS that look at various grand challenges [126,127]. An alternative way of framing this broader perspective has been suggested by Walsham [128] in considering whether IS research is creating a better world.

This broadening of the research agenda can be supported by paying more attention to the likely future consequences of IS. Again, this is a stream of research that already exists [129] but would need to become more central to support the RRI agenda.

RRI is explicitly normative. To reflect and translate this normative foundation, IS research, in taking RRI seriously, would need more time to reflect on ethical theory and its relevance in IS. As previously outlined, this is a recent development and should be supported.

However, a core aspect of RRI not currently strongly reflected in IS is that of public participation and engagement. Much organization-focused IS research accepts the existing socio-economic structures in current commercial organizations where participation and engagement have a place as user feedback, at best. However, it is worth noting that there is a long tradition of participative IS design [130,131], particularly strong in Scandinavian IS [132], which can serve as a bridgehead for a more general democratic and participative approach to IS governance. Additionally, IS can further learn from technology assessment, which includes approaches to organizing and public participation in decisions concerning technology.

4. CONCLUSION

This paper has argued that computer ethics constitutes a reference discourse for IS. That is to say, an ongoing body of work has inspired, and is inspired by, IS research, particularly ethics-related IS research. By discussing the underlying concepts, theories, topics and approaches to computer ethics, this paper has made the case that the correspondence between this discourse and ethics-related IS research is sufficient to justify the contention that computer ethics is a reference discourse.

The paper then argued that the computer ethics discourse is evolving to constitute part of a broader discourse on responsible research and innovation. This can at least be partly explained by some of the limitations of traditional computer ethics (and perhaps also traditional technology assessment), which can be addressed in RRI. The paper then outlined the RRI discourse, highlighting its motivation, definition, main actors, activities and normative foundations.

Returning to the idea of a reference discourse, the paper further argued that the field of IS, at least insofar as it is interested in ethics, should take this discussion of RRI seriously. The paper outlined what this may mean for the theory and practice of IS research.

One contribution of the paper is an opening to theoretical discussions that can be of relevance to IS. Much research has been undertaken in areas ranging from ethical theory [72] to future foresight [133,134], to research governance [102]. IS researchers can make use of this prior research to inform their work and integrate it into larger discussions cf. [135].

The development from computer ethics to RRI in ICT, furthermore, can lead not only to a reconsideration of the purpose of ethics-related IS work but also arguably to that of the field of IS more generally. Much IS research focuses on the organizational use of ICT, often with an explicit or implicit agenda of improving the use of technology. The majority of this work takes for granted the socio-economic context in which IS is used. Through the lens of RRI, IS researchers may obtain a better understanding of the grand challenges that humanity and most societies face and may be encouraged to consider how their research can contribute to addressing these challenges. From this perspective, it is imperative to question the way we organize and critically review our assumptions of what constitutes legitimate social practice, including ICT research and development, and the use of ICT in organizations and society. This means a fundamental reevaluation of the very nature of the technologies we use and how we use them.

This is a serious challenge for any discipline, including IS. At the same time, one can argue that in light of many of the problems we face, and the important role ICT can play in constituting as well as resolving those problems, the IS field can embrace such a redefinition.

It is important to highlight the important contribution that IS can make to grand challenges. Unlike conceptual and philosophically oriented computer ethics, the field of IS has a strong background in empirical research. Additionally, IS further has a rich repertoire of theoretical positions that aid in understanding the role of ICT and its interaction with individuals, organizations and society. ICTs have the potential to play a critical role in addressing societal challenges but doing so successfully requires a detailed understanding of both the technical capabilities and the socio-technical practices that shape the role of ICT in society. IS can thus play a key role in addressing the challenges of modern societies. Both society and IS as a discipline would benefit from a willingness and ability of IS scholars to accept this challenge.

ACKNOWLEDGEMENTS

This paper benefited from the activities undertaken within the project 'Framework for Responsible Research and Innovation in ICT' (EPSRC reference EP/J000019/1). For more information, see www.responsible-innovation.org.uk.

This study received funding from the European Community's Seventh Framework Programme (FP7/2007–2013) under grant agreements no. 230318(ETICA). Additionally, the paper benefited from work undertaken in preparation for the EU project, 'Responsible-Industry', grant agreement no. 609817.

REFERENCES

1. A.L. Harris, M. Lang, D. Yates, S.E. Kruck**Incorporating ethics and social responsibility in IS education** J. Inf. Syst. Educ., 22 (Fall (3)) (2011), pp. 183-189
2. P. Keen**MIS research: reference disciplines and a cumulative tradition** Proceedings of the First International Conference on Information Systems (1980)
3. A.S. Lee, M.I.S.R. Center, University of Minnesota Architecture as a Reference Discipline for MIS. Management Information Systems Research Center, Curtis L. Carlson School of Management, University of Minnesota (1991)
4. R.L. Baskerville, M.D. Myers**Information systems as a reference discipline** MIS Q., 26 (1) (2002), pp. 1-14
5. T. Bynum Computer and Information Ethics, Stanford Encyclopedia of Philosophy (2008)
6. I.R.W. Benbasat Zmud**The identity crisis within the IS discipline: defining and communicating the discipline's core properties** MIS Q., 27 (2) (2003), pp. 183-194
7. J. Habermas Theorie des Kommunikativen Handelns, Suhrkamp (1981)
8. L. Willcocks**Foucault, power/knowledge and information systems: reconstructing the present** J. Mingers, L. Willcocks (Eds.), Social Theory and Philosophy for Information Systems, Wiley, Chichester (2004), pp. 238-296
9. D.G. Johnson Computer Ethics (1st ed.), Prentice Hall, Upper Saddle River, New Jersey (1985)
10. K.E. Himma, H.T. Tavani The Handbook of Information and Computer Ethics, N.J.: Wiley, Hoboken (2008)
11. J. van den Hoven, J. Weckert (Eds.), Information Technology and Moral Philosophy (1st ed.), Cambridge University Press (2008)
12. L. Floridi L. Floridi (Ed.), Information ethics, Cambridge University Press (2010), pp. 77-97
13. R.A. Spinello, H.T. Tavani Readings in Cyber Ethics, Jones and Bartlett Publishers Inc. (2001)
14. D.G. Johnson, H. Nissenbaum Computers, Ethics and Social Values (1st ed.), Pearson (1995)
15. N. Wiener The Human Use of Human Beings, Doubleday, New York (1954)
16. Kant Kritik der praktischen Vernunft, Reclam, Ditzingen (1986)
17. Kant Grundlegung zur Metaphysik der Sitten, Reclam, Ditzingen (1998)
18. J.S. Mill Utilitarianism (2nd ed.), Hackett Publishing Co., Inc. (2002)
19. **Aristotle** The Nicomachean Ethics, Filiquarian Publishing, LLC (2007)
20. T.W. Bynum**Flourishing ethics** Ethics Inf. Technol., 8 (1) (2006), pp. 157-173
21. C. Gilligan In a Different Voice: Psychological Theory and Women's Development, Reissue, Harvard University Press (1990)
22. Adam**Computer ethics in a different voice** Inf. Organ., 11 (4) (2001), pp. 235-261
23. Sen The Idea of Justice, Allen Lane (2009)

24. M.C. Nussbaum Creating Capabilities: The Human Development Approach, Harvard University Press(2011)
25. J. Johnstone**Technology as empowerment: a capability approach to computer ethics** Ethics Inf. Technol., 9 (1) (2007), pp. 73-87
26. Oosterlaken, J. van den Hoven (Eds.), The Capability Approch Technology and Design(2012th ed.), Springer (2012)
27. L. Floridi**Information ethics: on the philosophical foundation of computer ethics** Ethics Inf. Technol., 1 (1) (1999), pp. 33-52
28. Ess**Luciano Floridi's philosophy of information and information ethics: critical reflections and the state of the art** Ethics Inf. Technol., 10 (2–3) (2008), pp. 89-96
29. P. Brey**Disclosive computer ethics: exposure and evaluation of embedded normativity in computer technology** CEPE Computer Ethics: Philosophical Enquiry, Dartmouth College (2000)
30. P.-P. Verbeek**The moral relevance of technological artifacts** P. Sollie, M. Düwell (Eds.), Evaluating New Technologies: Methodological Problems for the Ethical Assessment of Technology Developments, Springer (2009), pp. 63-78
31. J. van den Hoven, D. Helbing, D. Pedreschi, J. Domingo-Ferrer, F. Gianotti, M. Christen, FuturICT – The Road Towards Ethical ICT, arXiv:1210.8181, Oct. 2012.
32. Royal Academy of Engineering Engineering Ethics in Practice: A Guide for Engineers, Royal Academy of Engineering, London (2011)
33. Grunwald**Technology assessment or ethics of technology?** Ethical Perspect., 6 (2) (1999), pp. 170-182
34. L. Magnani**The mediating effect of material cultures as human hybridization** S. Hongladarom, C. Ess (Eds.), Information Technology Ethics: Cultural Perspectives(Illustrated ed.), Information Science Reference (2006), pp. 31-53
35. R.T. De George The Ethics of Information Technology and Business, Wiley-Blackwell (2003)
36. P. Duquenoy, N. Dando, I. Harris Ethics in the Provision and Use of IT for Business, Institute of Business Ethics in association with the BCS, London (2010), p. 1
37. E. Cohen, L. Cornwell**A question of ethics: developing information system ethics** J. Bus. Ethics, 8 (June (6)) (1989), pp. 431-437
38. D.G. Johnson**Corporate excellence, ethics, and the role of IT** Bus. Soc. Rev., 111 (4) (Dec. 2006), pp. 457-470
39. D.B.J.L. Payne Landry**A uniform code of ethics: business and IT professional ethics** Commun. Acm., 49 (11) (2006), pp. 80-84
40. M.E. Porter, M.R. Kramer**The link between competitive advantage and corporate social responsibility** Harv. Bus. Rev., 84 (12) (2006), pp. 78-92
41. Blue and Green Tomorrow, The Guide to Corporate Social Responsibility 2013, Apr. 2013
42. S. Broadbent, N. Dewandre, C. Ess, L. Floridi, J. G. Ganascia, M. Hildebrandt, Y. Laouris, C. Lobet Maris, S. Oates, U. Pagallo, J. Simon, M. Torseth, P.-P. Verbeek The Onlife Initiative, European Commission (2013)
43. P. Brey, J.H. Soraker**Philosophy of computing and information technology** D.M. Gabbay, A.W.M. Meijers, J. Woods, P. Thagard (Eds.), Philosophy of Technology, Engineering Sciences: 9, North Holland (2009), pp. 1341-1408
44. European Group on Ethics in Science and New Technologies, Ethics of Information and Communication Technologies, BEPA – Bureau of European Policy Advisors, Brussels, Opinion of the EGE 26, Feb. 2012.
45. European Commission Monitoring Policy and Research Activities on Science in Society in Europe (MASIS) – Final Synthesis Report, Directorate-General for Research and Innovation, Brussels (2012)
46. S.D. Warren, L.D. Brandeis**Right to privacy** Harv. Law Rev., 4 (1891 (1890), p. 193
47. H. Tavani**Informational privacy: concepts, theorie and controversies** J.V.D. Hoven, J. Weckert (Eds.), Information Technology, Moral Philosophy, Cambridge University Press, Cambridge (2008), pp. 131-164

48. L.M. Lee, L.O. Gostin**Ethical collection, storage, and use of public health data a proposal for a national privacy protection** JAMA-J. Am. Med. Assoc., 302 (1) (2009), pp. 82-84
49. W.S. Brown**Ontological security, existential anxiety and workplace privacy** J. Bus. Ethics, 23 (1) (2000), pp. 61-65
50. J. Arquilla**Conflict, security and computer ethics** L. Floridi (Ed.), The Cambridge Handbook of Formation, Computer Ethics, Cambridge University Press (2010), pp. 133-148
51. A.D. Moore**Personality-based, rule-utilitarian and Lockean justifications of intellectual property** J.V.D. Hoven, J. Weckert (Eds.), Information Technology and Moral Philosophy, Cambridge University Press, Cambridge (2008), pp. 105-130
52. G. Hull**Robert merges: justifying intellectual property** Ethics Inf. Technol., 14 (2) (2012), pp. 169-177
53. T. Thomas, M. Ahyick**Can we help information systems students improve their ethical decision making?** Interdiscip. J. Inf. Knowl. Manag., 5 (2010), pp. 209-224
54. R.D. Gopal, G.L. Sanders**International software piracy: analysis of key issues and impacts** Inf. Syst. Res., 9 (4) (1998), pp. 380-397
55. W.-B. Chiou, P.-H. Wan, C.-S. Wan**A new look at software piracy: soft lifting primes an inauthentic sense of self, prompting further unethical behaviour** Int. J. Hum. Comput. Stud., 70 (2) (2012), pp. 107-115
56. T.T. Moores, J.C. Chang**Ethical decision making in software piracy: initial development and test of a four-component model** Mis Q., 30 (1) (2006), pp. 167-180
57. T.T. Moores, A. Nill, M.A. Rothenberger**Knowledge of software piracy as an antecedent to reducing pirating behaviour** J. Comput. Inf. Syst., 50 (1) (2009), pp. 82-89
58. K. McSorley**The secular salvation story of the digital divide** Ethics Inf. Technol., 5 (2) (2003), pp. 75-87
59. G. Parayil**The digital divide and increasing returns: contradictions of informational capitalism** Inf. Soc., 21 (1) (2005), pp. 41-51
60. W. Olphert, L. Damodaran**Citizen participation and engagement in the design of e-government services: the missing link in effective ICT design and delivery** J. Assoc. Inf. Syst., 8 (9) (2007), pp. 491-507
61. Leiter**Cleaning cyber-cesspools: google and free speech** S. Levmore, M.C. Nussbaum (Eds.), Offensive ternet: Speech, Privacy, Reputation, Harvard University Press (2010), pp. 155-173
62. Adam Gender, Ethics and Information Technology, Palgrave Macmillan, Basingstoke (2005)
63. S. Vallor**Flourishing on facebook: virtue friendship and new social media** Ethics Inf. Technol., 14 (3) (2012), pp. 185-199
64. Gotterbarn, K. Miller, S. Rogerson**Software engineering code of ethics is approved** Commun. ACM, 42 (10) (1999), pp. 102-107
65. M.E. Gorman**Turning students into ethical professionals** IEEE Technol. Soc. Mag., 20 (Winter/2001 (4)) (2002), pp. 21-27
66. D.G. Johnson**Computing ethics computer experts: guns-for-hire or professionals?** Commun. ACM, 51 (10) (2008), pp. 24-26
67. Hume An Enquiry Concerning the Principles of Morals, Project Gutenberg (1777)
68. D. Wright**A framework for the ethical impact assessment of information technology** Ethics Inf. Technol., 13 (3) (2011), pp. 199-226
69. B. Friedman, D. Howe, E. Felten**Informed consent in the mozilla browser: implementing value sensitive design** Hawaii International Conference on System Sciences, Los Alamitos, CA, USA (2002), p. 247
70. B. Friedman, P. Kahn, A. Borning**Value sensitive design and information systems** K. Himma, H. Tavani (Eds.), The Handbook of formation, Computer Ethics, Wiley Blackwell(2008), pp. 69-102

71. S. Gürses, C. Troncoso, C. Diaz**Engineering privacy by design** Presented at the Conference on Computers, Privacy and Data Protection (CPDP 2011)(2011)
72. B. Stahl**Morality, ethics, and reflection: a categorization of normative IS research** J. Assoc. Inf. Syst., 13 (August (8)) (2012)
73. Bryant, F. Land, J.L. King**Editor's introduction to the special issue on ethical issues in IS research** J. Assoc. Inf. Syst., 10 (11) (2009), pp. 782-786
74. J.G. Mingers Walsham**Towards ethical information systems: the contribution of discourse ethics** MIS Q., 34 (4) (2010), pp. 833-854
75. C. Bull**A review of ethical theory in the 'upper echelons' of information systems research** Proceedings of the 17th European Conference on Information Systems, Verona, Italy, Verona, Italy (2009)
76. S. Sarker, M. Fuller, S. Chatterjee**Ethical information systems development: a baumanian postmodernist perspective** J. Assoc. Inf. Syst., 10 (November (11)) (2009), pp. 787-815
77. R.O. Mason**Four ethical issues of the information age** MIS Q., 10 (1) (1986), pp. 5-12
78. D.W. Straub, R.W. Collins**Key information liability issues facing managers: software piracy, proprietary databases, and individual rights to privacy** MIS Q., 14 (2) (1990), pp. 143-156
79. H.J. Smith, S.J. Milberg, S.J. Burke**Information privacy: measuring individuals' concerns about organizational practices** MIS Q., 20 (June (2)) (1996), pp. 167-196
80. S. Conger, K.D. Loch, B.L. Helft**Ethics and information technology use: a factor analysis of attitudes to computer use** Inf. Syst. J., 5 (3) (1995), pp. 161-183
81. K.J. Stewart, S. Gosain**The impact of ideology on effectiveness in open source software development teams** MIS Q., 30 (June (2)) (2006), pp. 291-314
82. J. Wareham, D. Robey, C. Chua**The role of online trading communities in managing internet auction fraud** Manag. Inf. Syst. Q., 31 (November (4)) (2008), pp. 759-781
83. R. Davison, N. Kock**Dealing with plagiarism in the information systems research community: a look at factors that drive plagiarism and ways to address them** Manag. Inf. Syst. Q., 27 (December (4)) (2003)
84. H.J. Smith, J. Hasnas**Ethics and information systems: the corporate domain** Mis Q., 23 (1) (1999), pp. 109-127
85. W.J. Orlikowski, J.J. Baroudi**Studying information technology in organizations: research approaches and assumptions** Inf. Syst. Res., 2 (1) (1991), pp. 1-28
86. P.A.E. Brey**Anticipating ethical issues in emerging IT** Ethics Inf. Technol., 14 (May (4)) (2012), pp. 305-317
87. Khushf**Systems theory and the ethics of human enhancement – a framework for NBIC convergence** M.C. Roco, C.D. Montemagno (Eds.), Coevolution of Human Potential Converging Technologies, 1013 (2004), pp. 124-149
88. B. Gordijn**Converging NBIC technologies for improving human performance: a critical assessment of the novelty and the prospects of the project** J. Law Med. Ethics, 34 (December (4)) (2006), pp. 726-732
89. P. Sollie, M. Düwell (Eds.), Evaluating New Technologies: Methodological Problems for the Ethical Assessment of Technology Developments, Springer (2009)
90. K. Górniak-Kocikowska**From computer ethics to the ethics of global ICT society** Libr. Hi Tech, 25 (March (1)) (2007), pp. 47-57
91. K. Mathiesen**What is information ethics?** ACM SIGCAS Comput. Soc., 34 (1) (2004), pp. 6-15
92. P. Darke, G. Shanks, M. Broadbent**Successfully completing case study research: combining rigour, relevance and pragmatism** Inf. Syst. J., 8 (October (4)) (1998), pp. 273-289
93. Benbasat, R.W. Zmud**Empirical research in information systems: the practice of relevance** MIS Q., 23 (March (1)) (1999), pp. 3-16

94. P.G.W. Keen**Relevance and rigor in information systems research: improving quality, confidence, cohesion and impact** H.-E. Nissen, H.K. Klein, R. Hirschheim (Eds.), Information Systems Research: Contemporary Approaches, Emergent Traditions, North Holland, Amsterdam (1991), pp. 27-49
95. D. Gotterbarn**Computer ethics - responsibility regained** D.G. Johnson, H. Nissenbaum (Eds.), Computers, Ethics and Social Values (1st ed.), Prentice Hall, Upper Saddle River (1995), pp. 18-24
96. E.-H.W. Kluge**Ethical and legal challenges for health telematics in a global world: telehealth and the technological imperative** Int. J. Med. Inf., 80 (2) (2011), pp. E1-E5
97. T.W. Bynum, S. Rogerson Computer Ethics and Professional Responsibility: Introductory Text and Readings, Wiley-Blackwell (2003)
98. Technopolis and Fraunhofer I.S.I Interim Evaluation and Assessment of Future Options for Science in Society Actions Executive Summary Final Report, Tchnopolis group, Brighton, UK (2012)
99. M. Boden, R. Johnston, F. Scapolo**The role of FTA in responding to grand challenges: a new approach for STI policy?** Sci. Public Policy SPP, 39 (March (2)) (2012), pp. 135-139
100. L. Levidow, C. Neubauer**Opening up societal futures through EU research and innovation agendas** EASST Rev., 31 (3) (2012), pp. 4-11
101. Sutcliffe, A report on Responsible Research and Innovation, 2011.
102. R. von Schomberg (Ed.), Towards Responsible Research and Innovation in the Information and Communication Technologies and Security Technologies Fields, Publication Office of the European Union, Luxembourg (2011)
103. R. Owen, P. Macnaghten, J. Stilgoe**Responsible research and innovation: from science in society to science for society, with society** Sci. Public Policy, 39 (6) (2012), pp. 751-760
104. R. von Schomberg**A vision of responsible research and innovation** R. Owen, M. Heintz, J. Bessant (Eds.), Responsible Innovation, Wiley (2013), pp. 51-74
105. R. Owen, M. Heintz, J. Bessant (Eds.), Responsible Innovation, Wiley (2013)
106. B.C. Stahl**Responsible research and innovation: the role of privacy in an emerging framework** Sci. Public Policy (September) (2013), p. sct067
107. R. Owen, N. Goldberg**Responsible innovation: a pilot study with the U.K. engineering and physical sciences research council** Risk Anal. Int. J., 30 (November (11)) (2010), pp. 1699-1707
108. D. Wright, R. Gellert, S. Gutwirth, M. Friedewald**Precaution and privacy impact assessment as modes towards risk governance** R. Von Schomberg (Ed.), Towards Responsible Research and Innovation in the formation, Communication Technologies , Security Technologies Fields, Publication Office of the European Union, Luxembourg (2011), pp. 83-97
109. Grunwald**Technology assessment: concept and methods** D.M. Gabbay, A.W.M. Meijers, J. Woods, P Thagard (Eds.), Philosophy of Technology, Engineering Sciences: 9, North Holland (2009), pp. 1103-1146
110. European Commission Ethics for Researchers – Facilitating Research Excellence, European Commission Directorate-General for Research, Luxembourg (2007)
111. RCUK, Integrity, Clarity and Good Management – RCUK Policy and Code of Conduct on the Governance of Good Research Conduct, Jul. 2009.
112. R. van Est**The broad challenge of public engagement in science** Sci. Eng. Ethics, 17 (December (4)) (2011), pp. 639-648
113. C. Coenen, E. Simakova**STS policy interactions, technology assessment and the governance of technovisionary sciences** Sci. Technol. Innov. Stud., 9 (2) (2013), pp. 3-20

114. European Commission Ethical and Regulatory Challenges to Science and Research Policy at the Global Level, Directorate-General for Research and Innovation, Luxembourg (2012)
115. T. Holmes, E. Blackmore, R. Hawkins The Common Cause Handbook: A Guide to Values and Frames for Campaigners, Community Organisers, Civil Servants, Fundraisers, Educators, Social... Funders, Politicians, and Everyone in Between, Public Interest Research Centre Ltd (2011)
116. Gutmann**The ethics of synthetic biology: guiding principles for emerging technologies** Hastings Cent. Rep., 41 (4) (2011), pp. 17-22
117. S. Clarke**New technologies, common sense and the paradoxical precautionary principle** P. Sollie, M. Düwell (Eds.), Evaluating New Technologies: Methodological Problems for the Ethical Assessment of Technology Developments, Springer (2009), pp. 159-173
118. S.M. Gardiner**A core precautionary principle** J. Polit. Philos., 14 (1) (2006), pp. 33-60
119. Technology Strategy Board Responsible Innovation Framework for Commercialisation of Research Findings for Use in Synthetic Biology Feasibility Studies Competition : Advancing the Industrial Application of Synthetic Biology, Technology Strategy Board (2012)
120. U.K. Synthetic Biology Roadmap Coordination Group A Synthetic Biology Roadmap for the UK, Technology Strategy Board (2012)
121. K.L. Kjolberg, R. Strand**Conversations About Responsible Nanoresearch** NanoEthics, 5 (1) (2011), pp. 99-113
122. European Commission Commission Recommendation on 07/02/2008 on a Code of Conduct for Responsible Nanosciences and Nanotechnology, European Commission, Brussels, C (2008) 424 final
123. E. Simakova**Making nano matter: An inquiry into the discourses of governable Science** Sci. Technol. Hum. Values (April) (2012)
124. P. Macnaghten, R. Owen**Good governance for geoengineering** Nature, 479 (November (7373)) (2011)
125. K. Parkhill, N. Pidgeon, A. Corner, N. Vaughan**Deliberation and responsible innovation: A geoengineering case study** R. Owen, M. Heintz, J. Bessant (Eds.), Responsible Innovation, Wiley (2013), pp. 219-240
126. D. Hovorka, J. Corbett**IS sustainability research: a trans-disciplinary framework for a 'grand challenge'** Proceedings ICIS 2012, December (2012)
127. S.J. Winter, B.S. Butler**Creating bigger problems: grand challenges as boundary objects and the legitimacy of the information systems field** J. Inf. Technol., 26 (2) (2011), pp. 99-108
128. Walsham**Are we making a better world with ICTs? Reflections on a future agenda for the IS field** J. Inf. Technol., 27 (2) (2012), pp. 87-93
129. P. Gray, A. Hovav**Methods for studying the information systems future** M. Chiasson, O. Henfridsson, H. Karsten, J.I. DeGross (Eds.), Researching the Future in Information Systems: IFIP WG 8.2 Working Conference, Future IS 2011, Turku, Finland, June 6-8, 2011, Proceedings (1st Ed.), Springer, Heidelberg (2011), pp. 299-316
130. E. Mumford, D. Henshall Participative Approach to Computer Systems Design, Halsted Press, New York (1978)
131. E. Mumford**The story of socio-technical design: reflections on its successes, failures and potential** Inf. Syst. J., 16 (4) (2006), pp. 317-342
132. M. Kyng**Designing for cooperation: cooperating in design** Commun. ACM, 4 (December (12)) (1991), pp. 65-73
133. K. Cuhls**From forecasting to foresight processes – new participative foresight activities in Germany** J. Forecast., 22 (2) (2003), pp. 93-111
134. Z. Sardar**The namesake: futures; futures studies; futurology; futuristic; foresight – what's in a name?** Futures, 42 (April (3)) (2010)

135. G. Eden, M. Jirotka, B.C. Stahl**Responsible Research and Innovation: Critical reflection into the potential social consequences of ICT** Presented at the Research Challenges in Information Science (RCIS), Paris (2013)

CHAPTER 2

AI Assisted Ethics

Amitai Etzioni[1]• Oren Etzioni[2]

[1]*The George Washington University, 1922 F Street NW, Room 413, Washington, DC 20052, USA*

[2]*The Allen Institute for Artificial Intelligence, Seattle, USA*

ABSTRACT

The growing number of 'smart' instruments, those equipped with AI, has raised concerns because these instruments make autonomous decisions; that is, they act beyond the guidelines provided them by programmers. Hence, the question the makers and users of smart instrument (e.g., driver-less cars) face is how to ensure that these instruments will not engage in uhical conduct (not to be conflated with illegal conduct). The article suggests that to proceed we need a new kind of AI program—oversight programs—that will monitor, audit, and hold operational AI programs accountable.

Keywords: Ethics bot Communiterianism Second-layer AI Driverless cars

INTRODUCTION

The question of which values should be introduced into the guidance systems of driverless cars has implications well beyond the ethical directions to be granted to these new vehicles. Namely, such guidance is needed for a great variety of robots, machines, and instruments (instruments, from here on) that are already equipped with artificial intelligence (AI)—and many more in the near future (The Economist 2015). These instruments are often referred to as "smart." As

Ed Lazowska of the University of Washington put it, "During the next decade we're going to see smarts put into everything. Smart homes, smart cars, smart health, smart robots, smart ence, smart crowds and smart computer–human interactions" (Markoff 2013). According to Francesca Rossi, a computer entist at the University of Padova, "Until now, the emphasis has been on making machines faster and more precise—better able to reach a specific goal set by humans. Today, the aim should be to design intelligent machines capable of making their own good decisions according to a human-aligned value system" (Rossi 2015). Gary Marcus of New York University holds that in the near future a moment will arrive that will herald an "era in which it will no longer be optional for machines to have ethical systems" (Marcus 2012).

One should note, a note essential for all that follows, that these smart instruments are able not only to collect and process information in seconds much more efficiently than human beings can do in decades or even in centuries—but also to form decisions on their own. That is, AI provides these instruments with a considerable measure of autonomy in the sense that they often will not inquire of their human users how to proceed and instead will render numerous decisions on their own (Mayer-Schönberger and Cukier 2014: 16–17). Stuart Russell discusses the development of algorithms that closely "approximate" autonomous human behavior and values (Wolchover 2015). Autonomy in computer ence thus refers to the ability of a computer to follow a complex algorithm in response to environmental inputs, independently of real-time human input. That is, autonomous robots are "robots that can figure things out for themselves" (2015). For instance, self-driven cars decide when to speed up or slow down, when to hit the brake, how much distance to keep from other cars and so on.

It follows that if these smart instruments are not to act like amoral machines, their AI guidance programs will need to include substantial moral components.[1] To put it differently, given that decisions tend to have a moral dimension (Etzioni 1988)—the programs that guide all these smart instruments need moral programming (Rossi 2015; Tegmark 2015). For instance, several scholars have asked under what conditions driverless cars would be instructed to swerve into a parallel lane to avoid hitting a kitten—even if such a move could cause several human fatalities (Marcus 2012; Bonnefon et al. 2015). The same question has numerous permutations, such as whether a car on a busy road should swerve to avoid hitting a child or adult if doing so would risk causing a pileup that could kill several people, or whether it should swerve to avoid hitting a non-living but solid obstacle to protect its own occupants even if such a move will lead to hitting a car in another lane. Moreover, driverless cars will need to be instructed whether they should slow down in order to stop when they see a hitchhiker or a car accident down the road, how to react to the road rage of a driver of an old fashioned car, whether to join a long queue or try to cut in, and many other such value-laden questions.

These questions may remind readers of the moral dilemma involving a trolley coming down a track and a person at a switch who must choose whether to let the trolley follow its course and kill five people or to redirect it to another track and kill just one. (This rather popular mental experiment has been used in several permutations.)[2] However, the subject at hand is rather different because the trolley decisions are made by a person; when AI is used many decisions are made often by the instruments themselves. These programs are correctly referred to as "black boxes" and as lacking accountability and even "traceability" (Mayer-Schönberger and Cukier 2014: 141, 178). Hence, they need to be given a priori and continuous moral guidance if they are to heed the values of their users and of the community.

The article next turns to examining several suggestions that have been made about the way moral guidance is to be provided to smart instruments, those equipped with AI, and then adds a distinct approach.

Social moral values ensconced in law

One major answer to the question of which values should be embodied in AI guidance systems is that the values shared by a particular community should be used. For many issues, this community would be the nation augmented by local communities (in the United States, these would be states and municipalities). Obvious examples are values such as thou shalt not kill, steal, rape, harm others, or harm the environment.

Thus, the guidance systems of driverless cars will be expected to ensure that these cars observe various speed limits, keep a safe distance from other cars, and so on as the drivers of old-fashioned cars are required to do. In short, one part—the easy part—of the answer to the question of which values are to be implemented in the guidance systems of smart instruments is: the values ensconced in the law of the community or communities in which the smart instruments are employed.

Several questions, though, remain even about the values embodied in law. First, who should be charged if instruments violate the law? The owner, the user, the designer, the manufacturer—or the computer that in effect operates the instruments and renders many decisions on its own (Kaplan 2015)?[3] The law clearly treats cases in which there was intent to cause harm much more harshly than cases in which there was no such intent. Compare the ways the law treats murder and involuntary manslaughter. (The question of intent also figures in assigning liability.)

But how is one to determine whose intent (if any) was the cause when one cannot trace the process by which the decisions were made?

Second, what level of law enforcement does society seek, given that smart instruments make a high level of surveillance very easy to achieve? (For instance, one could determine from a central location whether truck drivers are driving too long without taking a break, the speed of any car on the road, and the location of all who use cell phones and much more.) Third, what guidance is to be given to instruments when the law and ethics diverge? (For instance, should driverless cars be programmed to refuse to violate the law even when the driver takes over, or to allow speeding in an emergency?) Wrestling with these questions is left for a separate discussion in order to focus here on the social and moral values not ensconced in law (Etzioni and Etzioni 2016).

A communitarian approach

Numerous social and moral values are not ensconced in law. These include the extent and scope of one's commitments to one's children, spouses, friends, neighbors, the various communities to which one belongs, the nation, and even the international community (Wrong 1995). These values include taking risks for others (such as fighting overseas and donating organs), volunteering, giving to charity, resolving differences with others civilly, and many others. Indeed, many values ensconced in law are paralleled by considerable additional moral commitments above and beyond those required by law. For instance, most of what the moral culture of communities expects parents to do for their children greatly exceeds what the law commands. A communitarian position holds that it is essential to include these values in the guidance systems of smart instruments because these values make for a good, civil society well beyond a stable and even liberal state.

How should one determine which communal values to incorporate into the AI guidance systems of smart instruments? Those communal values ensconced in law can be identified relatively readily—they are values that legislatures and courts draw upon when they enact laws and interpret them. But how is one to determine which additional social moral values the community seeks to foster?

Some suggest that these additional values should be those shared by the community (Walzer 1984). This position runs into several difficulties. First, people belong to different communities and to the encompassing society, which subscribes to different values. For instance, famous attempts to use community standards to determine obscenity failed because of disagreement about what behavior qualifies (Jacobellis v.

Ohio 1964). Even when there is considerable consensus about values at a high level of abstraction, there is often much less consensus about what specifically these values require. Thus many people agree that the environment ought to be protected, but disagree about which mileage standards cars should have to abide in by 2030, what level of emissions they should be allowed to produce, whether one should idle at stop signs, and whether they should drive slower than the law requires. This poses great difficulties for programmers.

A related question is how to determine what the values are of whatever community the instruments are to heed. Several scholars have suggested using focus groups or public opinion polls to determine what the relevant values are.[4] One notes, however, that the results of public opinion polls vary significantly depending on who is surveyed, question wordings, the sequence in which questions are asked, the context in which questions are asked (e.g., at home versus at work), and the attributes of those who ask the questions (e.g., are they the same race as the person queried). Even when the same question is asked of the same people by the same people twice, rather different answers follow (Institute for Statistics Education). Hence such polls cannot be used as a reliable base.

The suggestion that the time has come to engage in what might be called "teaching machine ethics," that is, teaching instruments to render moral decisions on their own (Lin 2013), runs into different difficulties. Driverless cars and other such instruments are unable to form moral guidelines out of thin air. They will need, at least to begin with, first principles and some guiding philosophy, say a utilitarian or a deontological one. After all, even humans do not start with a tabula rasa. They gain ethical foundations from those that raise them and from their communities and then modify or replace these foundations over time. Which principles and methodologies should be given to smart instruments? Is there a reason they should all be given the same foundations? Could cars come with ethical options—some equipped with utilitarian principles, or deontological ones, or Buddhist philosophies? (One may wonder whether buyers would understand the implications of their choice of car unless they took some philosophy classes.) Some kind of polling approach may be unavoidable when deciding whether or not to use instruments that are used by communities rather than individuals. For instance, such an approach may be necessary when determining whether to post in public spaces cameras equipped with AI systems that scan the footage to identify people who act in an uncivil manner. However, there seem to be serious difficulties associated with relying on a communitarian approach to determining which values the instruments should embody for most instruments that are owned and operated by individuals. Is there no way to more closely align the values that instruments are expected to heed and the ethics of their intended users?

A libertarian approach

Libertarians and some liberals hold that each person should define the good and the values they are to heed, and that the state should remain neutral (Boaz 1999). It would be compatible with this position if smart instruments came with a rich menu, which would allow each individual to choose which options are in line with their values, as well as an opened ended category which would allow them to include attention to moral preferences not included in the menu.

The difficulties this libertarian approach raises are illustrated by the development of privacy options. Many websites initially had merely a statement of the privacy policy they follow, which put the users in a "take it or leave it" position—assuming they understood the legal statements. Next, an increasing number of websites offered users a small menu of choices regarding the level of privacy they preferred. Facebook, for instance, offers five main privacy settings. Even at this level, people complained about the complexities of these settings. A Google representative recently stated that Google would provide up to a hundred such options (Fleischer 2015). Strong evidence from psychological studies and experience suggests that most users will find the requirements to make that many choices on their own overwhelming (Kahneman 2011).

These difficulties are much more challenging if one takes into account that individuals would need to personally provide individual moral guidance to all the growing number of smart instruments one uses as the world is moving into the "inter of things." In short, this approach seems highly impractical.

AI assisted ethics (ethics bots)

A paradigmatic agenda

An ethics bot is *an AI program that analyzes many thousands of items of information—not only information publicly available on the Inter but also information gleaned from a person's own computers—about what the acts of a particular individual that reveal that person's moral preferences are.* Basically what ethics bots do for moral choices is rather similar to what many AI programs do for ferreting out consumer preferences and targeting advertising to them accordingly. 5 Only in this case, the bots are used to guide instruments that are owned and operated by the person, in line with their values, rather than by some marketing company or political campaign seeking to advance their goals. For instance, an ethics bot may conclude that a person places high value on environmental protection if the ethics bot finds that the person purchases recycled paper, drives a Prius, contributes to the Sierra Club, prefers local food, and never buys Styrofoam cups. It would then instruct that person's driverless car to purchase only environmentally friendly fuels, to turn on the air conditioning only if the temperature is high, and to idle the engine at stops.

Much of what follows about other attributes of ethics bots is *paradigmatic*, in the sense that the article outlines the qualities of these as-yet to be developed ethics bots. Some readers may well consider the following pages somewhat visionary in terms of what they assume AI will be able to accomplish in the not-too-distant future. However, given the inability to implement communitarian and libertarian approaches, it seems better to employ even weak ethics bots, at least initially, than to continue to leave driverless cars and the large and growing number of other smart instruments without ethical guidance. Moreover, we shall see that some, albeit rather simple, ethics bots have already been constructed, road tested, and used by a considerable number of people.

To illustrate: nest built a smart thermostat. It first "observed" the behavior of the people in various households for merely a week and drew conclusions about their preferences. It then used a motion-detecting sensor to determine whether anyone was at home. When the house was empty, the smart thermostat entered a high energy-saving mode; when people were at home, the thermostat adjusted the temperature to fit their preferences. This thermostat clearly meets the two requirements of an ethics bot, albeit a very simple one. It assesses people's preferences and imposes them on the controls of the heating and cooling system. One may ask what this has to do with social moral values. This thermostat enables people with differing values to have the temperature settings they prefer. The residents of the home do not need to reset the thermostat every day when coming and going. This simple ethics bot also reduces the total energy footprint of the community (Lohr 2015: 147).

The ethics bot so far depicted is the most basic version. Additional features and more sophisticated ethics bots that might be developed, are outlined below. They all share, though, two major features. First, they enable the people who *employ AI to guide AI*. This is, instead of treating the AI world as if it were one unitary field AI should be restructured along the same lines as the rest of the world is. That is, some AI programs should serve as the first line "worker bees" that provide directions to an ever growing number of instruments—from robot soldiers to Barbie dolls, from voter mobilization drives to refrigerators. Second line AI programs will act as supervisors, auditors, accountants, and as ethics bots of the first line AI programs.

Second, because people bring their same basic values to different pursuits, once an ethics bot is able to carry out an analysis of the moral preferences of a person, the same findings will help guide a variety of smart instruments the person uses. Thus, a person does not need one ethics bot for shopping, another for driving, and still another for volunteering. For example, if an ethics bot determines that a given person's moral preferences are to maximize their self-interest, that bot would instruct the person's instruments to shop at places they find the lowest costs and best quality but disregard whether the sellers have been charged with employing workers at unsafe locations overseas; paying less then minimum wages; and polluting. It would also instruct the person's financial AI system to make donations to charity only if those donations generate enough deductions to make up for the "loss" or if they engender for the donor a great deal of goodwill. And so on.

Of course, the ethical preferences of most people are more complicated than the simple examples here used the purposes of exposition. Hence, in the longer run, ethics bots would need to be similarly internally diverse. Moreover, ethics bots of the future should be able to self-update at regular intervals in order to take changes to people's moral preferences into account. Last but not least, ethics bots should have an override feature, discussed below.

Basically what ethics bots do for moral choices is rather similar to what many AI programs do for ferreting out consumer preferences and targeting advertising to them accordingly.[5] Only in this case, the bots are used to guide instruments that are owned and operated by the person, in line with their values, rather than by some marketing company or political campaign seeking to advance their goals.

Ethics bots, once developed, should be able to provide a superior interface between a person and smart instruments compared to unmediated interaction. There are several reasons for this. First, if most people will be required to render a large number of ethical choices, they will quickly give up because a sort of psychological fatigue sets in akin to the one felt by people who are trying to consider their best chess move and after a while just give up and move. (By contrast, AI is very patient, which is one reason it now beats even the world chess masters.)

Second, people accommodate their inability to make a great number of choices by making lexicographic choices.[6] That is, they ignore the information about most facets of the object of their choice, and focus on a few that they hold to be most important. Thus, when they buy a car, they may examine its relative price, miles per gallon, and color, or some other such mix of features—but ignore scores of other attributes. The same holds true of moral choices. People, when making a major donation, may take into account the goals the given charity serves, whether it services people in their own community or overseas, and whether it has a reputation as an honest agency, or some other such mix. They will ignore, in the process, many other features of the given charity such as its long-term record, recent changes in leadership, its ratio of expenses to payouts, and so on. In contrast, ethics bots, given their computing power, have no such limitations. They hence will help people to ensure that their choices about how and where to donate, shop, vote and more much more closely reflect all their moral preferences than a few selected ones.

Third, ethics bots are likely to compare favorably to other means—such as interviews, self-administered forms, and mental exercises such as those used in lifeboat ethics[7]—that seek to ferret out a person's moral preferences. This is largely the case because these subjective means draw their conclusions mainly on the basis of *expressed attitudes*, while ethics bots mainly note the moral choices revealed in *actual behavior*. For instance, one may say that one attends church regularly, but an ethics bot would note that the person played golf often at the time religious services are carried out and parked at the place of worship only a few times a year. This attribute is of special importance because attitudes, a great deal of data shows, correlate poorly with behavior.[8] Also, people have difficulties in articulating their preferences.

The introduction of ethics bots would raise serious privacy concerns. Many people may well seek to encrypt them and call for laws that would treat "reading" another person's ethics bot without written prior permission as akin to reading their medical record or other sensitive information.[9] One should note, though, that ethics bots would not be mandated, and hence would be used only by those who see their merits and benefits as exceeding the bots' privacy risks.

Moreover, those who adopt ethics bots are sure to note that most of the information bots use to ascertain their preferences is already publicly available. Several corporations maintain very detailed dossiers on most people and sell these dossiers to all comers. For instance, Axciom maintains dossiers on most Americans. These dossiers include "age, behavior, buying activity, financial, household, interest, real property, life events," and more, up to 1500 items per person. SeisInt dossiers include individuals' "criminal histories, photographs, property ownership, SSNs, addresses, bankruptcies, family members, and credit information."[10] These dossiers can even include sensitive medical information. As Eli Pariser reports, "Search for a word like 'depression' on Dictionary.com, and the site installs up to 223 tracking cookies and beacons on your computer so that other websites can target you with antidepressants" (Pariser 2011). It seems that if others are free to use a very great deal of personal information to promote products and politicians and to seek to shape people's preferences, there is little to be lost and much to be gained if that person uses the same information to ensure that the instruments *they* employ will comport with *their* values.

Individuals can at any point override the guidance an ethics bot provides to their instruments. Thus, when a smart thermostat programmed by Nest did not follow individuals' preferences, but rather set the thermostat within two degrees of their preferences at a setting more favorable to the environment, many people rejected this setting (Lohr 2015). In short, ethics bots (or AI assisted ethics) are a marriage between libertarianism (because the particular person provides the moral preferences—the definition of the good) and AI programs (which provide moral guidance to smart instruments).

Individuals can use their ethics bots for self-assessment

Adolescents, people in psychotherapy, and many others often engage in self-examination, including asking themselves whether they are good people, whether they do enough to serve others, and more. Studies of such self-assessment suggest that people often greatly over- or under-evaluate themselves (Dunning et al. 2003). In the future, these individuals will be able to draw on their ethics bots to provide them a more objective—and candid—evaluations of themselves. These evaluations may well include how they compare to others in their community and whether they improve or not over time. Here ethics bots serve as tools of moral self-improvement.

Ethics bots can be programmed in ways that allow their human users to modify the program the bots chose for them, based on their preferences. They can augment, distract, and override—all moves that are much less taxing than forming the self's ethics profile de novo. For instance, a person may note that his ethics bot reveals that over the last 10 years he donated rather little to various charities then he thought he did—and instruct his financial app to increase these allocations.

Ethics bots can also help people implement pre-commitment strategies. The term refers to taking steps before a situation in which one expected to be tempted to act uhically—to fend off the temptation. Odysseus employed this strategy by instructing his sailors to tie him to the mast of the ship and to plug their ears before they entered the sea of Sirens so that they would not be tempted by the Sirens' calls (Homer 1978). Thus, an ethics bot could be set to instruct a driverless car not to yield to attempts to override the car's system to engage in road rage, or to mute the car's horn if the person blows it long, hard, and often.

Pragmatic and operational considerations

Some critics may well argue that the agenda charted so far is well beyond what AI can achieve. They may point to programs that try to divine consumer preferences much simpler than their ethical preferences and did not fare well, such as, programs that recommend books and movies. In general, arguments about what AI can and cannot accomplish swing between overblown hype and overblown despair. Some point to AI programs that play winning chess, Jeopardy, and Go, while others bemoan the difficulties computers have in accomplishing tasks that are simple to humans, such as reading graphic designs.

In response, one must reiterate that humans *cannot* on their own provide more than very elementary ethical guidance to smart instruments without AI assistance. Hence, even weak ethics bots seem preferable to none, unless one can come up with a still different way to provide moral guidance to smart instruments. The best test of whether ethics bots can be created is to invest more in trying to build some. There seem to no obvious, a priori, logical reasons to hold that such bots cannot be constructed—and some, albeit very simple ones, have been developed.

Determining the moral preferences of a person in some areas may well be less daunting than in others. For instance, determining what a person considers fair might be indeed very difficult. By contrast, people's privacy preferences seem easier to grasp. Most people are either privacy fundamentalists (as their use of multiple personal email addresses, frequently changing their passwords, and so on reveals); privacy pragmatists (who will share personal information if the price is right); or privacy unconcerned (Westin 2003). In short, ethics bots are badly needed. Whether more and more accomplished ones can be constructed in the near future remains to be seen. Meanwhile, driverless cars are roaming the streets and so far have not been granted any moral guidance. Even if we cannot be able to construct a high-fidelity model of people's ethical preferences, we will be able to approximate and the approximation will improve as the technology gets better.

IN CONCLUSION

The incorporation of AI into more and more instruments makes them much smarter—more efficient, and more effective. In the process, these instruments are acquiring a measure of autonomy in the sense that they render many decisions on their own, well beyond the guidelines that the programmers introduced and sometimes even counter to these guidelines. There is hence growing concern about how society and millions of individuals can rest assured that instruments they use, which are equipped with AI, will not render uhical decisions (Dellinger 2015).

The answer to these challenges cannot be found in bare human controls, because human beings cannot determine on their own whether uhical (and even illegal) acts carried out by smart instruments were the results of the human programming—or AI processes "under the hood." Hence, a major conclusion of this article is that *to ensure proper conduct by AI instruments—people will need to employ other AI systems. Implementing ethical preferences, when dealing with smart instruments, will need to be AI assisted.*

These second order AI programs would have to vary a great deal from one another. For instance, those second order programs that would ensure that instruments do not violate laws, the dictates of which are relatively clear, are likely to differ greatly from those second order programs that would direct instruments to heed moral values, which are often fuzzy. Whether a particular instrument is used by individuals or by a community is also be a factor. This article focuses on those AI programs dealing with compliance with social and moral values for instruments used by a person, such as driverless cars.

The article finds that relying on guidance in these matters on values shared by this or that community raises many difficulties. The same holds if one seeks individuals to directly instruct their smart instruments on their own. A preferred method, here outlined, is to develop AI programs to be used to determine the moral preferences of people, ethics bots, and for these ethics bots to guide the smart instruments.

FOOTNOTES

[1]Stuart Russell of University of California Berkeley stated, "You would want [a robot that does everyday activities] preloaded with a pretty good set of values" (Goldhill 2015).

[2]Philosophers Philippa Foot and Judith Jarvis Thomson described a situation called "the Trolley Problem," which raises the question whether a runaway train is about to run over a group of five people on the tracks, but their deaths could be averted by flipping a switch that would divert the train onto another track, where it would kill one person (Lin 2013).

Joshua D. Greene describes a number of ethical dilemmas that generally fit into the category of the "trolley problem." These include "switch" cases, in which throwing a switch will turn the trolley away from some number of people toward a single person, or "footbridge" cases in which one must push a person into the path of the trolley to save others' lives. He also discusses similar famous ethical questions studied by researchers, such as the question whether it is immoral to allow a child to drown in a shallow pond to avoid muddying one's clothes, whether one is morally obligated to donate money to save others' lives, and more (Greene 2014).

Jean-Francois Bonnefon, Azim Shariff, and Iyad Rahwan apply this question to the issue of driverless cars; as driverless cars ("autonomous vehicles," or AVs) and other forms of use of artificial intelligence become more widespread. They examined whether individuals would be comfortable with AVs programmed to be utilitarian and found that the answer was generally yes (Bonnefon et al. 2015).

[3] One scholar at the National ence Foundation points out that technology currently outstrips knowledge of how to assign liability for robots' ethical and legal failures (The Economist 2014).

Professor Patrick Lin points out that algorithms cannot make "an instinctive but noheless bad split-second decision" the way humans can, and thus the threshold for liability may be higher (Lin 2013).

[4] Slobogin and Schumacher (1993: 757) recommend that the Supreme Court draw on public opinion polls to determine that about societal expectations of privacy. Similar suggestions were made by Fradella, Morrow, Fischer, and Ireland. They conducted a survey of 589 individuals (Fradella et al. 2010–2011: 293–94).

Bonnefon et al. (2015) applied this idea to finding which values ought to guide self driving cars.

[5] For example, Nielsen has developed a marketing system for targeting very specific demographics with financial and investment products based on age, affluence, the presence of children in the home, and certain purchasing habits. These include such specific target consumer groups as "Y2-54: City Strivers" and "F4-56: Economizers" (Nielsen 2015); Ted Cruz' campaign in Iowa relied on psychological profiles to determine the best ways to canvass individual voters in the state (Hamburger 2015).

[6] Lexicographic preferences are those in which "respondents have a ranking of the attributes [consider important], but their choice of an alternative is based solely on the level of their most important attribute(s)" (Campbell et al. 2006).

[7]"Lifeboat ethics" refers to an ethical dilemma outlined by Garrett Hardin in 1974, which describes a situation in which a lifeboat nearly full to capacity must consider whether or not to bring aboard ten additional passengers out of 100 people in the water. The purpose of lifeboat ethics-style philosophical discussions is not to tell anyone what is ethically correct in any given situation, but rather to help individuals to clarify their own values.

This is one component of a larger school of ethics, "moral reasoning." Moral reasoning encourages "individual or collective practical reasoning about what, morally, one ought to do" (Richardson 2014).

[8]Across many different situations, it is well-established that "attitudes are poor predictors of behavior" (Ajzen and Fishbein 2005). See also: Azjen et al. 2004.

[9]For an in-depth discussion of the different treatment afforded to less and more sensitive information, see: [redacted].

[10]*Seeid.*

REFERENCES

1. Ajzen, I., & Fishbein, M. (2005). The influence of attitudes on behavior. In D. Albarracín, B. T. Johnson, & M. P. Zanna (Eds.), *The handbook of attitudes.* Mahwah, NJ: Lawrence Erlbaum Associates Publishers.
2. Azjen, I., Brown, T. C., & Carvajal, F. (2004). Explaining the discrepancy between intentions and actions: The case of hypothetical bias in contingent valuation. *Personality and Social Psychology Bulletin, 30*(9), 1108–1121.
3. Boaz, D. (1999). Key concepts of libertarianism. Cato Institute. January 1. http://www.cato.org/publications/commentary/key-concepts-libertarianism.
4. Bonnefon, J., Shariff, A., & Rahwan, I. (2015). Autonomous vehicles need experimental ethics: Are we ready for utilitarian cars? *Computers and Society.* October 12. http://arxiv.org/abs/1510.03346.
5. Campbell, D., Hutchinson, W. G., & Scarpa, R. (2006). Lexicographic preferences in discrete choice experiments: Consequences on individual-specific willingness to pay estimates. Working Paper, Fondazione Eni Enrico Mattei. http:// ageconsearch. umn.edu/bitstream/12224/1/wp060128.pdf.
6. Dellinger, A. J. (2015) Tim Wu says Google is degrading the Web to favor its own products. *The Daily Dot.* June 29. http://www.dailydot.com/technology/google-search-tim-wu-yelp/.
7. Dunning, D., Johnson, K., Ehrlinger, J., & Kruger, J. (2003). Why people fail to recognize their own incompetence. *Current Directions in Psychological ence, 12*(3), 83–87.
8. Etzioni, A. (1988) *The moral dimension.* New York: The Free Press.
9. Etzioni, A., & Etzioni, O. (2016). Keeping AI Legal. *Vanderbilt Journal of Entertainment & Technology Law* (Forthcoming).
10. Fleischer, P. (2015). Privacy and future challenges. Speech, Amsterdam Privacy Conference. Amsterdam, The herlands, October 23–26.

11. Fradella, H. F, et al. (2010–2011). Quantifying Katz: Empirically measuring 'Reasonable Expectations of Privacy' in the fourth amendment context. *American Journal of Criminal Law 38*, 289–373.
12. Goldhill, O. (2015). Human values should be programmed into robots, argues a computer entist. *Quartz*. October 31. http://qz.com/538260/human-values-should-be-programmed-into-robots-argues-a-computer-entist/.
13. Greene, J. D. (2014). Beyond point-and-shoot morality: Why cognitive (neuro) ence matters for ethics. *Ethics, 124*(4), 695–726.
14. Hamburger, T. (2015). Cruz campaign credits psychological data and analytics for its rising success. The Washington Post. December 13, https://www.washingtonpost.com/politics/cruz-campaign-credits-psychological-data-and-analytics-for-its-rising-success/2015/12/13/4cb0baf8-9dc5-11e5-bce4-708fe33e3288_story.html.
15. Hardin, G. (1974). *Lifeboat ethics: The case against helping the poor*. Psychology Today.
16. Homer. (1978) *Odyssey* (J. H. Finley, Jr. Trans.). Boston: Harvard University Press.
17. Institute for Statistics Education. Glossary of statistical terms test–retest reliability. http://www.statistics.com.
18. Jacobellis v. Ohio. (1964). 378 U.S. 184.
19. Kahneman, D. (2011). *Thinking, fast and slow*. New York: Firrar, Straus, and Giroux.
20. Kaplan, J. (2015). Who put the robot in charge? *Medium*, September 22. https://medium. com/the-wtf-economy/who-put-the-robot-in-charge-408a47335176#.mb8mqqs9p.
21. Lin, P. (2013). The ethics of autonomous cars. *The Atlantic*. October 8. http://www.theatlantic.com/technology/archive/2013/10/the-ethics-of-autonomous-cars/280360/.
22. Lohr, S. (2015). *Data-ism: The revolution transforming decision making, consumer behavior, and almost everything else*. London: OneWorld Publications.
23. Marcus, G. (2012). *Moral machines*. New York: The New Yorker.
24. Markoff, J. (2013). The rapid advance of artificial intelligence. The New York Times. October 14. http://www.nytimes.com/2013/10/15/technology/the-rapid-advance-of-artificial-intelligence.html?pagewanted=all&_r=0.
25. Mayer-Schönberger, V., & Cukier, K. (2014). *Big data*. New York: Houghton Mifflin Harcourt.
26. Nielsen. (2015). Nielsen P$YCLE Lifestage Groups. https://segmentationsolutions.nielsen.com/mybestsegments/Default.jsp?ID=8010&pageName=Learn%2BMore&menuOption=learnmore. Accessed 17 Dec.
27. Pariser, E. (2011). What the Inter knows about you. CNN. May 22. http://articles.cnn.com/2011-05-22/opinion/pariser.filter.bubble.
28. Richardson, H. S. (2014). Moral reasoning. *The Stanford Encyclopedia of Philosophy* (Winter Edition), Ed. Edward N. Zalta. http://plato.stanford.edu/entries/reasoning-moral/.
29. Rossi, F. (2015). How do you teach a machine to be moral? The Washington Post. November 5. https://www.washingtonpost.com/news/in-theory/wp/2015/11/05/how-do-you-teach-a-machine-to-be-moral/.
30. ence Daily. (2015). New algorithm lets autonomous robots divvy up assembly tasks on the fly. May 27. http://www.encedaily.com/releases /2015/05/150527142100.htm.
31. Slobogin, C., & Schumacher, J. E. (1993). Reasonable expectations of privacy and autonomy in fourth amendment cases: An empirical look at understandings recognized and permitted by society. *Duke Law Journal, 42*, 727–775.
32. Tegmark, M, et al. (2015). An open letter: Research priorities for robust and beneficial artificial intelligence. *Future of Life Institute*. http://futureoflife.org/ai-open-letter/.
33. The Economist. (2014). That thou art mindful of him. March 29.

34. The Economist. (2015). Rise of the machines. http://www.economist.com/news/briefing/21650526-artificial-intelligence-scares-peopleexcessively-so-rise-machines.
35. Walzer, M. (1984). *Spheres of Justice: A defense of pluralism and Equality*. New York: Basic books.
36. Westin, A. (2003) Social and political dimensions of privacy. *Journal of Social Issues 59*(2). http://onlinelibrary.wiley.com/doi/10.1111/1540-4560.00072/epdf.
37. Wolchover, N. (2015). Concerns of an artificial intelligence pioneer. Quanta. April 21. https://www.quantamagazine.org/20150421-concerns-of-an-artificial-intelligence-pioneer/.
38. Wrong, D. (1995). *The problem of order: What unites and divides society*. Boston: Harvard University Press.

CHAPTER 3

Doxing: A Conceptual Analysis

David M. Douglas[1]

[1] *Department of Philosophy, University of Twente, Enschede, The Netherlands*

ABSTRACT

Doxing is the intentional public release onto the Internet of personal information about an individual by a third party, often with the intent to humiliate, threaten, intimidate, or punish the identified individual. In this paper I present a conceptual analysis of the practice of doxing and how it differs from other forms of privacy violation. I distinguish between three types of doxing: deanonymizing doxing, where personal information establishing the identity of a formerly anonymous individual is released; targeting doxing, that discloses personal information that reveals specific details of an individual's circumstances that are usually private, obscure, or obfuscated; and delegitimizing doxing, which reveals intimate personal information that damages the credibility of that individual. I also describe how doxing differs from blackmail and defamation. I argue that doxing may be justified in cases where it reveals wrongdoing (such as deception), but only if the information released is necessary to reveal that such wrongdoing has occurred and if it is in the public interest to reveal such wrongdoing. Revealing additional information, such as that which allows an individual to be targeted for harassment and intimidation, is unjustified. I illustrate my discussion with the examples of the alleged identification of the creator of Bitcoin, Satoshi Nakamoto, by Newsweek magazine, the identification of the notorious Reddit user Violentacrez by the blog Gawker, and the harassment of game developer Zoe Quinn in the 'GamerGate' Internet campaign.

Keywords: Doxing Internet harassment Anonymity Journalism Hate speech Privacy

A spectre is haunting the Internet—the spectre of doxing.[1] Doxing, sometimes spelt 'doxxing' or 'd0xing', involves releasing someone's personal details onto the Internet in an easily accessible form. These details may include full legal names, residential addresses, unique identifiers for governmental records and services (such as social security numbers in the US), business records and documents, and personal photographs of one's self and loved ones. These details may already be publicly available, but in difficult to access forms or distributed across various sources that obscure them from casual discovery. These details might also be government, company, or organization records obtained via a security breach. In some cases, they might even have been obtained directly from the person herself, either willingly or unknowingly. Doxing can occur to anyone, from high-profile public figures to obscure everyday people. All that is necessary to become a victim of doxing, it seems, is to be of interest to someone else on the Internet.

There are various motives for doxing someone. It may be motivated by a desire to expose wrongdoing and to hold the wrongdoer to account. It may be used to humiliate, intimidate, threaten, or punish the identified individual. It is often a tool for 'cyber stalking', as the information may be released in a context that would cause a reasonable person to fear for her life (Citron 2014). It can also serve as a tool for Internet vigilantism, where those opposed to someone's actions retaliate by revealing her identity and personal information, leaving the victim open to public ridicule, harassment, and vilification (Solove 2007). And information released onto the Internet is easy to access and difficult to remove: entering a doxing victim's name into a search engine may reveal her personal details and the abuse associated with the doxing attack for years. The potential for harm and disruption are obvious when a person's professional life and reputation depends on her visibility on the Internet (Citron 2014).

As the above suggests, doxing may have a devastating impact for its victims. It assists in harassing and stalking individuals, both physically and on the Internet (Citron 2014). Such stalking creates significant distress and increases the risk of physical harm, especially if the personal information is used to encourage others to abuse the victim. A parallel can be drawn with sexual harassment on the Internet. Mary Anne Franks (2012) lists three factors that contribute to the harm online sexual harassment causes: the harassers' anonymity, the amplification of the harassment caused by the accessibility of the harassing content which may encourage further harassment, and the permanence that results from the difficulty of removing harassing content from the Internet. Even if doxing is not used as a tool for sexual harassment, these factors also contribute to the harms of having personal information revealed on the Internet.

Despite these harms, doxing is sometimes presented as a tool of protest and for exposing wrongdoing. Corruption by Chinese government officials is often the target of the so-called 'Human Flesh Search Engine', composed of Chinese Internet users who search for and release evidence of private and public transgressions and wrongdoing (Gao and Stanyer 2014). For example, an investigation into two Chinese local government officials was launched after documents listing travel expenses for research trips to the US and Canada were anonymously released onto the Internet. These documents provided evidence that public funding had been used to pay for trips to tourist attractions (Gao and Stanyer 2014).

This paper is an attempt to untangle the intertwined concepts and issues raised by doxing. I present and justify the claim that significant differences exist between various cases of doxing that justify placing them into different categories. I call these categories deanonymization, targeting, and delegitimization. Deanonymizing doxing releases personal information establishing the identity of a formerly anonymous or pseudonymous individual. Targeting doxing discloses personal information that reveals specific details of an individual's circumstances that are usually private, obscure, or obfuscated. Finally, delegitimizing doxing reveals intimate personal information that damages the credibility of that individual. I use this classification to highlight the significant differences between three cases of doxing: the alleged identification of Bitcoin creator Satoshi Nakamoto, the identification of the notorious Reddit Internet forum user Violentacrez, and the harassment of several female game developers in the 'GamerGate' incident. I conclude that in cases where exposing wrongdoing is in the public interest, deanonymizing and delegitimizing doxing is permissible only to the extent necessary to reveal that wrongdoing has occurred. Using any form of doxing to humiliate or threaten the subject, and revealing more information than necessary to establish wrongdoing, is unjustified.

DEFINING DOXING

The term 'doxing' comes from the phrase 'dropping documents' or 'dropping dox' on someone, which was a form of revenge in 1990s outlaw hacker culture that involved uncovering and revealing the identity of people who fostered anonymity (Honan 2014). The term is already prominent enough to be included in formal dictionaries. For instance, the Oxford British and World English Dictionary defines doxing as to "[s]earch for and publish private or identifying information about (a particular individual) on the Internet, typically with malicious intent" (Oxford Dictionaries 2015). As the Oxford definition suggests, doxing does not necessarily have to be motivated by malice. Several high-profile incidents of so-called 'doxing' involved journalists revealing the identities of formerly pseudonymous Internet identities (Chen 2012a; Goodman 2014). Despite this, doxing is a term with negative connotations: labeling these accounts as 'doxing' suggests that the journalists involved have acted wrongly in revealing personal information about a pseudonymous individual (Beaujon 2014). Examining the concept in more detail by considering the different kinds of personal information that may be released will help to determine whether doxing is necessarily or primarily a malicious act.

A more nuanced account of doxing can begin by considering what it actually establishes: it removes some degree of anonymity from a specific person. Marx's (1999) concept of identity knowledge offers a useful tool for this task. The seven broad types of identity knowledge Marx describes are listed in Table 1. Perfect anonymity, according to Marx (1999), is the inability to be identified according to any of these seven types of identity knowledge.

Table 1 Types and examples of identity knowledge [based on Marx (1999)]

Type	Description
Legal name	The name under which someone is known for official and legal purposes
Locatability	Information that reveals where someone lives or where she can be contacted personally, such as an address or a telephone number
Pseudonyms linked to name or location	A name or code that represents a single individual (such as a bank account number) in a system that is related to their legal name or some other potentially unique characteristic (such as an address)
Pseudonyms not linked to name or location	
(a) For policy reasons (b) Audience is	(a) A name or code representing someone in a system that is not related to her legal name, such as an anonymized medical record

unaware it is a pseudonym	(b) A name someone uses instead of her legal name as a disguise or for deception, such as an alias
Pattern knowledge	Someone who can be recognized by her regular public actions or habits, such catching the same bus every morning at the same time
Social categorization	Information that can be used to place someone into social categories (or stereotypes), such as physical appearance, accent, style of dress, and so on
Symbols of eligibility/non-eligibility	Possessing artifacts or knowledge that identifies someone as being entitled to particular privileges and treatment, such as a uniform, a password, or a train ticket

Being identified by some of these types will be greater threats to anonymity than others. For example, being identified as an adult male in a large European city does little to reduce my anonymity, as it does not easily allow someone to gain other types of identity knowledge about me. However, being identified by name and address makes maintaining my anonymity more difficult as this information can be easily used to establish other types of identity knowledge. Knowing my name allows someone to search public records and databases (to say nothing of the Internet) for further information about me. Knowing my address allows others to encounter me in person and observe my movements, habits, physical appearance and characteristics. In Marx's classification, these observations establish *pattern knowledge* and *social categorization*identity information about me.

Using Marx's categorization, I suggest that doxing should be understood as releasing publically a type of identity knowledge about an individual (the *subject* of doxing) that establishes a verifiable connection between it and another type (or types) of identity knowledge about that person. The verifiability of doxing distinguishes it from other forms of exposure and publicity. As the origins of the term 'doxing' ('dropping documents' or 'dropping dox') suggest, it utilizes documentary evidence of identity knowledge.

Different types of identity knowledge are documented in different forms. Identity knowledge relating to personal details used for administrative purposes may be recorded in official records or documents, such as birth certificates, tax returns, employment records, and so on. Such documents may reveal legal name, locatability, and pseudonyms that are connected to an individual's name or location. Documents that describe unique characteristics possessed by an individual in a pseudonymous record that is unrelated to her name or location may reveal further identity knowledge if it can be cross-referenced with other information. This possibility exists where medical records are not sufficiently anonymized. Symbols of eligibility may document themselves (such as railway tickets) or may be documented through records of such symbols being granted to an individual, such as graduating from a university. Similarly, official documentation will exist for symbols of eligibility being withheld or taken from an individual.

Other types of identity knowledge, such as pattern knowledge and social characterization, are documented in other ways. Frequently updated location information, such as stored by mobile devices that record their location, may reveal an individual's daily routine, and so establish pattern knowledge about that individual.[2] Social characterization may be established through photographs and imagery recorded about a person and her behaviour. Such characterization will often be up to the interpretation of the observer, and may be misleading if the images are taken out of context or presented in a biased manner. This is especially the case with activities that are invested with social or symbolic significance, or which challenge entrenched beliefs and expectations. For example, images of a woman wearing revealing clothing or expressing her sexuality may be used to mock or humiliate her for not conforming to traditional notions of female behavior and gender roles (Poole 2013).

Doxing should be distinguished from related concepts such as blackmail, defamation, and gossip. Unlike blackmail, doxing does not involve making a demand to the subject to prevent information being released. A blackmailer only releases information if the victim does not comply with the blackmailer's demands. While the threat of doxing can serve as blackmail, doxing itself is not blackmail.

Defamation also involves the public release of information with the intention to humiliate, threaten, intimidate, or punish the subject. However, for information to be defamatory it must reveal something damaging to the reputation of the person (or people) described. Doxing does not necessarily have to reveal something questionable or embarrassing about the person involved. As I will describe later, while one form of doxing aims to harm the subject's reputation, doxing itself does not necessarily involve releasing such information.

Finally, doxing differs from gossip (even malicious gossip) in that it relies on releasing actual (or believed to be actual) identity knowledge rather than suggestion, hearsay and innuendo. Bok (1989:93) defines gossip as "informal personal communication about other people who are absent or treated as absent" (numbering of features omitted). While doxing can be formal or informal (i.e. consist of official documents or records, or accurate informal accounts), it is the difference between communicating information *about* someone and communicating information *of* someone. To illustrate this with a benign example, consider the difference between claiming 'X wore a pink tutu at a funeral' and releasing a photo of X wearing a pink tutu at a funeral. The first is an instance of gossip, while the second is a form of doxing.[3] The photograph serves as documentation of the claim being made about X, and is evidence that can be verified. Under Marx's classification of identity knowledge, it is social characterization knowledge as it documents X's apparent disregard for social norms. Merely telling a friend about X's poor taste in funeral attire does not provide this documentary evidence.

THE VALUE OF ANONYMITY AND OBSCURITY

Before examining the different forms of doxing in detail, I will establish the value of what doxing endangers: the subject's obscurity and anonymity. Doxing undermines what Ruth Gavison calls "our concern over our accessibility to others: the extent to which we are known to others, the extent to which others have physical access to us, and the extent to which we are the subject of others' attention" (1980:423). The subject no longer controls some aspect of identity knowledge about her, which reduces her ability to decide what she reveals about herself and to whom she reveals it. This control is an important aspect of a person's identity. We reveal some aspects of ourselves to some people but not to others. Our relationships with others are shaped by what we choose reveal to them and what they decide to reveal to us. Our identities and the social value attached to them (i.e. reputation and public persona) are difficult to build and easy to lose. Even forfeiting some degree of such control is a way of establishing one's own identity. Choosing to publicly document one's experiences and movements are decisions individuals make about how they wish to present themselves to others. Influencing how others perceive you is a vital part of establishing who you are (and crucially, who you are *not*) as a person.

To further illustrate the value of anonymity, I again turn to the work of Marx (1999), this time for his list of the rationales for anonymity. These are listed in Table 2. There is much to say about the significance of each of these rationales and whether they should be accepted in all cases. For reasons of space and scope, however, I will only make a few general comments here.

Table 2 Rationales for anonymity [based on Marx (1999)]

Rationale for anonymity	Explanation
1. Facilitating the flow of information	Encourages information to be disclosed where there may be risks and penalties associated with doing so (such as whistleblowing)
2. Obtaining personal information for research	Allows individuals to be honest in their responses without fear of being punished or stigmatized if the information became public
3. Encouraging attention to the message content instead of the messenger	The identity of the informer may prejudice the reception of the information
4. Encouraging reporting, seeking information, and self-help	Individuals can report activities or to seek out information without fear of being stigmatized or victimized if others knew they were seeking certain information or were reporting certain activities
5. Obtaining resources or encouraging actions that involve illegality	Encourages individuals to seek help for problems that are linked with illegal actions (such as illicit drug addiction) or hand in illegal items (such as amnesties for contraband goods)
6. Protecting donors or those taking controversial but socially useful actions	Encourages individuals to contribute goods or actions without fear of intimidation, harassment, or creating future obligations
7. Protecting strategic economic interests	Allows someone to interact in the marketplace without their identity affecting their transactions (such as being charged more if someone is known to be wealthy)
8. Protecting one's time, space, and person	Allows someone to maintain their isolation from unwanted attention or interruption by others
9. Aiding judgments based on specified criteria	Promotes the unbiased assessment of something without being influenced by the identity of those involved
10. Protecting reputation and assets	Prevents an individual's personal information being used by someone else to deceive and

	defraud others
11. Avoiding persecution	Allows individuals to avoid harms that may result from belonging to a particular group (such as belonging to a persecuted minority)
12. Enhancing rituals, games, play, and celebrations	Allows individuals to interact in particular contexts without affecting their status and relationships in other contexts
13. Encouraging experimentation and risk-taking	Allows individuals to experiment with different behaviours and actions so that they may explore different ways of living without affecting their current relationships, commitments, and reputation
14. Protecting personhood and autonomy in sharing information	Allows individuals to control who has access to information about themselves and what information they choose to share and when
15. Traditional expectations of anonymity	Individuals expect that certain interactions do not involve revealing personal information about themselves to others, such as paying for items with cash

As Marx's list suggests, anonymity and obscurity are both forms of protection.[4] It can disguise attributes that may prejudice how others receive someone's work and ideas, such as gender, race, ethnicity, or class. It is anonymity's protective value that makes doxing particularly harmful in Internet communication, as it removes the subject's anonymity without an equivalent loss of anonymity for the attacker.

Collecting different types of identity knowledge about an individual can be regarded as building a 'dossier' on that person.[5] Dossier building involves the "compilation, maintenance, use, and dissemination of personal information on individuals" (Reichel 1977: 265). The opportunities created by the Internet for gathering personal information have effectively democratized dossier building. Now almost anyone with the desire and the time to search for another's personal information has the tools and information sources available for her to do so.

TYPES OF DOXING

I propose categorizing doxing into three types: deanonymization, targeting, and delegitimization. Each attempts to remove or damage something different from the subject: anonymity, obscurity, and credibility, respectively. Each type of doxing also creates new possibilities to further interfere with the life of the person involved. Deanonymization makes it easier to obtain other types of identity knowledge about the subject, and so creates greater opportunities for the other types of doxing to occur. Whatever advantages or protection the subject sought to gain by seeking anonymity or adopting a pseudonym will be lost. Targeting doxing creates the possibility that future harassment may take a physical form, with the uncertainty and risks of harm that it brings. The subject may be harassed and inconvenienced by others using her personal information to impersonate her. Finally, delegitimization presents a motivation for carrying out harassment and potentially further doxing by detailing how the subject is somehow unworthy of respect. These categories are listed and summarized in Table 3.

Table 3 Types of Doxing

Type of doxing	Description	Loss to the subject	Examples
Deanonymization	Reveals any kind of identity knowledge about a person	Anonymity	Revealing the legal identity of someone using a pseudonym
Targeting	Reveals information that allows an individual to be physically located	Obscurity	Revealing someone's home address
Delegitimization	Reveals information intended to damage an individual's credibility, reputation, or character	Credibility	Evidence of supposed immoral activity, hypocrisy, or willful deception

I now describe each category of doxing in further detail.

Deanonymizing doxing

Deanonymizing doxing releases information that reveals the identity of the person (or persons) who has previously been anonymous or known by a pseudonym. It also covers instances where someone's identity is revealed publically regardless of whether she has deliberately sought to conceal her identity or not.

Deanonymization is the broadest of the three categories of doxing as it can affect every type of identity knowledge and negates every rationale for anonymity. Depending on the subject's rationale for anonymity and the type of identity knowledge released, it may not cause significant harm to the subject, and there may be plausible public interest justifications for disclosing it. For example, there is at least a *prima facie* public interest justification for revealing someone's identity when anonymity or a pseudonym is being used to deceive others for personal gain (a con artist impersonating someone else to gain money or prestige, for example). Literary hoaxes are an example that I will return to later in discussing the potential justifications for doxing.

A famous instance of deanonymizing doxing is the reveal of the supposed identity of the person behind the pseudonym 'Satoshi Nakamoto'. Satoshi Nakamoto is the name adopted by the creator (or creators) of the Bitcoin crypto-currency (Nakamoto n.d.). The true identity of Bitcoin's creator is still uncertain. An article in *Newsweek* identified Dorian Satoshi Nakamoto as Bitcoin's creator, a claim he has repeatedly and consistently denied (Goodman 2014). The creator of Bitcoin has a number of clear rationales for anonymity: avoiding interference, protection, and a desire not to draw attention away from the creation itself.

Another example is the deanonymization of the notorious Reddit moderator 'Violentacrez', who was revealed to be Michael Brutsch by the blog *Gawker*.[6] Brutsch was a volunteer moderator who contributed to and oversaw various forums (or 'sub-reddits') on the Reddit website. Violentacrez was heavily involved in deliberately provocative sub-reddits such as 'creepshots' (which featured voyeuristic photographs of unsuspecting women) and 'jailbait' (which featured photographs of girls under the age of consent) (Chen 2012a). Brutsch claimed in a television interview that he treated his activities on Reddit as a game (Chen 2012c). This is the rationale of play from Marx's list of rationales for anonymity. Violentacrez was an example of an Internet 'troll': someone who deliberately flouts social norms and provokes others for her own amusement, often under some form of anonymity (Phillips 2012).[7] The pseudonym allowed Brutsch to entertain himself and others by deliberately offending people and breaking social taboos with the material he posted on the website. Protecting his reputation (and employability) is another important justification (and another of Marx's rationales for anonymity), and an accurate one given that Brutsch lost his job as a result of his legal identity being connected with that of Violentacrez (Chen 2012b).

Targeting doxing

Targeting doxing reveals specific information about an individual that allows her to be physically located. It reveals physical locatability (rather than communicative locatability, like a telephone number or email address) identity knowledge about the subject.[8] Targeting doxing increases the subject's physical accessibility by removing the obscurity surrounding where a person lives or works. Losing this obscurity makes someone more vulnerable to physical harassment because of whom specifically she is.

Targeting doxing often follows from deanonymization. As Marx's rationales for anonymity suggest, seeking anonymity is frequently adopted to reduce the risk of being targeted. The identity knowledge revealed through deanonymizing doxing makes it easier to uncover further identity knowledge, such as the subject's physical location and workplace.

The forms of harassment made possible by targeting doxing range from irritating pranks to physical assault (or worse). Relatively harmless but annoying pranks can range from calls from car dealers responding to supposed interest in a car to having to cancel unwanted deliveries ordered in the subject's name (Matisse 2015). Even these seemingly minor annoyances can serve as a form of intimidation. Mantilla (2015) mentions a case where a subject received an unordered pizza that had been ordered under the name of an accused murderer known to that individual. Another possible form of harassment is 'swatting', where an attacker makes a hoax call to the police claiming that there is a violent disturbance at the subject's address, prompting an armed police response (Mantilla 2015).[9] The identity knowledge gained through targeting doxing can be used to impersonate the subject, and in extreme cases, has been used to make it appear that the subject herself is encouraging others to attack or sexually assault her (Jouvenal 2013; Citron 2014).

'The Nuremberg Files' website is a notorious example of targeting doxing. This web site began in 1997 and listed the names and personal details of doctors who performed abortions in the US. The site also listed the personal details of the doctors' families (Solove 2007). The Nuremberg Files example illustrates the importance of the context within which identity knowledge is presented.[10] At least some of the information presented there (particularly, the addresses of abortion clinics) would already be publicly accessible. What makes it targeting doxing (beyond the additional identity knowledge about the doctors and their families) is presenting this information in a manner that promotes harassing the subjects.

Delegitimizing doxing

Delegitimizing doxing releases private information with the intention of undermining the subject's credibility, reputation, and/or character. It attempts to shame and humiliate the subject, often by portraying her as a transgressor of an established (or supposed)[11] social norm. Whether the subject herself accepts or promotes the social norm is irrelevant. Revealing the subject to be a hypocrite (by publicly supporting a social norm while privately breaking it) is certainly an attempt at delegitimizing her, but delegitimization goes beyond revealing actual or supposed hypocrisy. It can serve as a tool for maintaining the 'tyranny of the majority' that concerned John Stuart Mill. By drawing attention to how the subject differs from "prevailing opinion and feeling", delegitimizing doxing serves to "fetter the development, and [...] prevent the formation of, any individuality not in harmony with its ways" (Mill 1989[1859]: 8).

Reporting information and seeking advice, information, or assistance (rationale 4 in Marx's list) presents the possibility for delegitimization depending on the information sought or reported, and the help requested. The traditional confidentiality of medical records and library borrowings is also motivated by a desire to keep potentially embarrassing or easily misunderstood information secret, so that people can seek medical help or read controversial books without fear of being ostracized for doing so (Rindfleisch 1997; Bowers 2006). A straightforward example is a teenage girl anonymously seeking a pregnancy test or an abortion. If her identity was revealed, she risks being stigmatized and shamed for being sexually active at a young age, especially if unmarried motherhood and/or abortion are unacceptable in her society.

Sexuality is frequently used to delegitimize others. The violent and misogynist language surrounding many instances of delegitimizing doxing implies the objectification of the subject, portraying her as a thing to be used and discarded rather than an autonomous person worthy of respect (Nussbaum 2010). An example is involuntary pornography or so-called 'revenge porn', where intimate or explicit photographs and videos of individuals are posted online without their consent, either by former lovers or by third parties who have somehow acquired them (Mantilla 2015). These images are sometimes accompanied by personal information identifying the person (Citron 2014). As Citron writes, "Harassers post women's nude images because they know it will make them unemployable, undateable, and at risk for sexual assault" (2014:17). While men are also victims of involuntary pornography, the overwhelming majority of victims are women. For example, Reynolds (2016) reports that images of women were involved in 80 % of the 139 cases of involuntary pornography reported in the UK between January and April 2015. Involuntary pornography and other forms of delegitimizing doxing of women based on their sexuality are only the latest instances of the long-lived and surprisingly resilient activity of 'slut-shaming', where women and girls are ridiculed and harassed for their real or imagined sexual activity. Such harassment has a history going back to at least Roman times (Webb 2015). It also reveals a double standard in the social norms associated with sexuality, as male heterosexual activity does not share the same social disapproval (Poole 2013; Citron 2014).

Part of the harm delegitimizing doxing causes is what Franks (2012) calls 'virtual captivity': the abuse directed at someone on the Internet is potentially available to everyone who interacts with her, and so might affect every social relation she has. The possibility that everyone the subject interacts with (personally or professionally) has been exposed to the delegitimizing material is enough to cause significant emotional distress and social withdrawal. Martha Nussbaum describes something similar with her concept of subjectivity-violation, where for an abuser, "pleasure is taken in invading and colonizing the person's inner world" (2010: 72).

The context or framing within which delegitimizing doxing occurs is significant, and much of the harm it can cause is a result of taking documentary evidence out of context. An incident described by Boyd (2011) presents a good illustration of this problem. A college admissions officer asked boyd about an apparent contradiction in a prospective student's application: the student claimed to want to leave the 'gang-ridden' community he lived in, but the admissions officer found the student's MySpace page included gang insignia. The officer questioned why the student would lie in his application; Boyd's (2011) response is that adopting gang insignia is a necessity for survival in such a community, and that there is no contradiction in adopting the social norms of a community and secretly desiring to be free of their influence.

Delegitimizing doxing is often accompanied by targeting doxing, and so it might be questioned whether there is a significant difference between them. The difference between is that delegitimizing doxing supplies 'evidence' for targeting the person involved. If targeting doxing supplies the means for harassing the subject, delegitimizing doxing supplies the supposed 'motive' for doing so.

The combination of targeting and delegitimizing doxing is demonstrated by the 'GamerGate' incident, where several high-profile female computer game developers were subjected to prolonged harassment, intimidation, and vilification. The catalyst of this incident was an account posted on the Internet by Eron Gjoni, a former boyfriend of the independent game developer Zoe Quinn, of their failed relationship (Mantilla 2015). Quinn's personal details were released on the Internet and she became the target of prolonged and sustained harassment, intimidation, and vilification (Mantilla 2015). While attempts were made to justify these attacks as attempts to expose wrongdoing in computer games journalism, as one of the men Quinn was alleged to have had a relationship with was a video games journalist (who had not even written about Quinn's game), misogynism is a more convincing explanation (Mantilla 2015). Following the attacks on Quinn, other prominent women associated with computer games, including developer Brianna Wu and critic Anita Sarkessian, were also targets of sustained harassment and intimidation that included targeting and delegitimizing doxing (Mantilla 2015).

CAN DOXING BE JUSTIFIED?

I now discuss whether any instances of doxing are justifiable. I will argue that deanonymizing doxing may be acceptable depending on the rationale for anonymity and if there is a compelling public interest justification for revealing someone's identity. I also claim that delegitimizing doxing may be permissible if it exposes evidence of actual wrongdoing of public interest, and that the information revealed must only be sufficient to establish that such wrongdoing has occurred. I will argue that targeting doxing is unjustifiable, as it deliberately increases the risk of physical harm to the subject. In all cases, however, the burden of proof is on whoever wishes to disclose identity knowledge about the subject to justify why her anonymity or obscurity should be removed.

The motivation behind doxing is significant for deciding whether it is defensible or not. Doxing as a form of intimidation is unacceptable as it attempts to silence the subject and prevent her from participating in social, political, and public activity. All three types of doxing may be used for intimidation. Deanonymization intimidates those who adopt a pseudonym or seek anonymity to express unpopular or controversial views that they are otherwise uncomfortable in expressing. Targeting doxing increases the ease with which someone may be physically harassed or harmed. Delegitimization vilifies the subject, inspiring further harassment and reducing the likelihood that her opinions will be given the public respect that they might otherwise receive.

My arguments place considerable weight on the concept of 'public interest'. My interpretation of the concept is based on two claims by Bok (1989): "[t]he public has a legitimate interest [...] in all information about matters that might affect its welfare" (1989:258) and that information reported to the public that only satisfies their curiosity rather than affects their welfare must take into account the privacy of those affected. These two claims reflect the public/private distinction common to liberal political philosophy. Deanonymizing and delegitimizing doxing are acceptable only if they concern matters that affect the welfare of the public. If we accept that individuals should have control over who has access to identity knowledge about herself, the burden of proof should be on whoever attempts to reveal such information to justify why revealing it is in the public interest. If they are to be justified, deanonymizing and delegitimizing doxing cannot be indiscriminate: it must reveal *only* the identity knowledge that is relevant to establish wrongdoing by a specific individual. Such doxing can be considered analogous to whistle blowing that reveals wrongdoing by or within organizations.

Particular instances of doxing could be justified if there are allegations of legal wrongdoing, or if there is a legitimate public interest reason for establishing someone's identity. This would seem to rule out most (if not all) cases of targeting doxing, as these cases are intended to intimidate and promote further harassment of the subject. The legitimacy of doxing depends on the motivations behind publically releasing some of the subject's identity knowledge and the foreseeable risks of harm to the subject from doing so. An instance of deanonymizing doxing, therefore, might be justified on a consequentialist basis if the benefits to the public of exposing wrongdoing or deception outweigh the foreseeable harms to the subject due to her loss of obscurity.

An author using a pen name, for instance, does not seem to be a compelling target for deanonymization on public interest grounds if the pen name is not used to deceive readers and is merely a way for a writer to adopt different personas for different styles and genres of writing. Revealing that Charles Dodgson was Lewis Carroll does not seem to be particularly compelling from a public interest standpoint as there is nothing inherent deceptive in the claims made by the two personas. The rationales of play and promoting experimentation are frequently the motivations for adopting these personas (rationales 12 and 13 from Marx's list), rather than any attempt to deceive. Pen names for authors and stage names for performing artists are often little more than designations of public personas, and offer little in the way of anonymity or obscurity. Revealing that the musician Bono was born Paul Hewson does little (if anything) to affect the meaning of his music or his political and social activism. While deanonymization does not seem particularly troubling in these cases, there also seems to be little reason for doing so other than satisfying curiosity.

So-called 'cross-penning', where an author adopts a pen name of a different gender, is slightly more problematic as it may be intended to mislead the reader. However, here the pseudonym is often adopted to lend the work credibility and allow it to be judged on its own merits rather than unfairly influenced by gender bias. This is the rationale of wishing to keep attention to the work itself rather than to its creator (rationale 3 from Marx's list). George Sand, the pen name of Armandine Dudevant, is just one example (Levmore 1996). Given the disproportionate chances of women receiving misogynistic hate speech for their writings on the Internet, female authors sometimes adopt masculine names as a means of avoiding becoming targets for online abuse (including doxing) (Citron 2014). Cross-penning in these cases is both an attempt to have their work judged fairly and as a means of protection against harassment. However, this does not extend to literary hoaxes where the author falsely claims to have personal experiences or attributes that lend unjustified credibility to her work. Autobiographies that feature elaborate false accounts of the author's circumstances and experiences are one example (Manning 2012). It is more plausible to argue in such cases that the author is being deceptive in these cases, as they are not motivated by a desire to focus attention on the work itself but on how the falsely claimed characteristics of the author lend credibility to the work.

Ghostwriting is an interesting case where the problem is reversed: the actual author is not the attributed author. Ghostwriters are often an 'open secret': it is assumed that many public figures use ghostwriters to produce works published in their name (Goldacre 2012). At best, employing a ghostwriter allows for the attributed 'author' to better express her own ideas. At worst, the 'author' is misrepresenting her abilities to the readers of 'her' work. If ghostwriting is used to obscure the source or interests of the actual author, then there is a public interest justification for revealing this. For example, such justifications exist in the case of medical research, where the information published may be used to decide on medical treatment and to direct future research (Ngai et al. 2005). The names of seemingly independent researchers are sometimes attached (with their permission) to pharmaceutical studies to obscure the fact that were primarily designed and conducted by pharmaceutical company researchers (Goldacre 2012). Revealing the use of ghostwriters to gain unwarranted credibility for published works would be delegitimizing doxing with a public interest justification.

I now return to the two specific instances of deanonymizing doxing described earlier: Satoshi Nakamoto and Violentacrez. Is there a public interest justification for deanonymizing Satoshi Nakamoto? It depends on whether identifying Nakamoto is merely satisfying public curiosity or establishing information that benefits the public. There is certainly historical interest in establishing the identity of the creator (or creators) of such an influential technology. The anonymity of Bitcoin's creator may also raise suspicions about the intent behind creating it. However, given that both the theory behind Bitcoin and the source code of the software implementing it are open to public review and revision, it seems unlikely that there is anything malicious within the design and implementation of Bitcoin itself. The pseudonym is unlikely to have been adopted to deceive others for Nakamoto's benefit, and the adopting the pseudonym offers Nakamoto protection against unwanted interference and outside interest. Discovering Satoshi Nakamoto's identity would certainly be interesting given Bitcoin's influence and technical merit, but there seems little public benefit (in the sense that it would better inform the public in matters that affect it) in revealing this information beyond satisfying this curiosity. There does not appear to be a strong reason for removing the actual creator (or creators) of Bitcoin from self-imposed obscurity.

The case of Satoshi Nakamoto raises another important point: does doxing in any form have to be accurate to be harmful? I suggest that the credibility that releasing documentary evidence has is important for what makes doxing particularly harmful, as it cannot be simply dismissed as gossip or hearsay. Through no fault of his own, Dorian Satoshi Nakamoto lost his obscurity after the allegations that he was 'Satoshi Nakamoto' were published. Inaccurate or out-of-date personal information released as targeting doxing could lead to unconnected individuals being harassed. Inaccurate or false doxing may not be as harmful to the subject as accurate doxing, but it is still an attempt to remove the subject's anonymity or obscurity, and may harm others who are wrongly identified as the subject individual.

Is there a public interest in revealing the identity of Violentacrez? Again, it is not straightforward that disclosing Violentacrez's identity is in the public's interest or just something to satisfy the public's curiosity. Unlike Dorian Satoshi Nakamoto, Michael Brutsch acknowledged that he was 'Violentacrez', and so his own actions under that pseudonym led to his loss of obscurity. Violentacrez's actions were certainly (and deliberately) offensive to many people, and the deanonymizing doxing forced him to stop. Distinguishing between offensive speech and hate speech offers a potential justification for this deanonymization. Hate speech expresses claims that those with certain characteristics (such as gender, race, or sexual preference) are inferior in moral worth and little more than objects to be used and exploited (Citron 2014). There is a public interest in resisting such expression as it promotes harmful divisions within society. Hate speech damages the perception (and if left unaddressed, eventually the treatment) of such people as moral agents equal to ourselves that we have duties toward and with rights of their own (Waldron 2010). In the case of Violentacrez, many of the sub-reddits that he created or moderated (such as 'chokeabitch' and 'rapebait') may be classified as 'hate speech' or objectification. Deanonymization might be justified as a means of limiting or stopping such hate speech by increasing the speaker's accountability.

A strong objection to this conclusion is that such deanonymizing doxing risks of turning into the private enforcement of public laws and moral standards, and has the potential to further develop into vigilantism. Trottier suggests the term 'digital vigilantism' for "a process where citizens are collectively offended by other citizen activity, and respond through coordinated retaliation on digital media" (2016:2). The identity knowledge revealed through deanonymization makes it considerably easier to perform targeting and delegitimizing doxing of the subject. While it may appear to be 'just desserts' for a hate speaker to be harassed, it should be rejected on the same grounds that the intimidation promoted by targeting doxing is rejected.

Another objection is the concern that the costs and harms of deanonymization to the individual concerned outweigh the social benefits of making her accountability for offensive behavior. Consider a situation where someone uses a pseudonym to express controversial views that could be portrayed as harmful to the public interest, such as seditious comments or questioning strongly held religious or social beliefs in ways that are not hate speech. Those who object to such views might justify deanonymizing this person on public interest grounds. This would return deanonymizing doxing to being a tool for intimidating those with unpopular views rather than as a means of making those who cause harm through their anonymous or pseudonymous actions accountable.

The Violentacrez case is a good illustration of this second objection. The deanonymization was used to shame Michael Brutsch and harm him materially, as he lost his job through being deanonymized.[12] Had he not been protected by his pseudonym, it is likely he would have acted differently. If we accept that at least some of Violentacrez's postings were hate speech and objectification, that there is a public interest in controlling expressions of such speech. However, given the material harms that he would foreseeably suffer as a result of anonymization, it is worth considering what alternatives were available for stopping the hate speech from taking place. In this instance, there is a clear alternative: the site operators should have removed his deliberately offensive postings and sub-forums from Reddit. This alternative may be challenged by claims that it limits freedom of expression, but it may be defended if freedom of expression is not considered to be an absolute right that cannot be limited by other rights (Waldron 2010).

Another response is to emphasize that accountability should go both ways in deanonymizing someone: whoever performs the deanonymization should not be anonymous or pseudonymous herself. In both the 'Nakamoto' and Violentacrez examples is that the persons revealing identity knowledge about the subjects were not themselves anonymous. The journalists involved did not use anonymity as a means of avoiding responsibility for their actions. Journalists also have editors and their own professional judgment about what is in the public interest to reveal in news stories. Revealing information under a legal name makes it easier to be held accountable for doing so. This accountability does not in itself legitimize doxing, as there are other protections that those revealing information may enjoy that are unavailable to the subject. Private individuals are unlikely to be able to afford costly legal disputes with media enterprises that reveal their personal information. Nonetheless, this offers a potential solution for difficult cases where it is not clear whether there is a public interest in revealing someone's identity or not.

Finally, I will briefly consider the possibility of targeting and delegitimizing doxing that is claimed to be the public interest. Exposing corruption is often used to justify doxing or Internet campaigns that feature doxing. An example is the so-called 'Gamergate' controversy, which its defenders claim is an attempt to expose corruption in computer games journalism.[13] Quinn's relationship with the journalist Nathan Grayson had no effect on how her game was reviewed, as Grayson's published work only mentioned her game in passing and did not actually review it (Mantilla 2015). The 'exposing corruption' justification is further weakened by the forms of doxing used on Quinn (particularly targeting doxing) and Gjoni's apparent motivation behind releasing delegitimizing information, which appears to be a desire to punish his former girlfriend (Mantilla 2015). The public interest should not justify any instance of doxing that objectifies the subject, as objectification portrays the target as unworthy of personhood, without legitimate interests of her own that should also be recognized (Nussbaum 2010).

CONCLUSION

Anonymity and obscurity protect us from the unwanted intrusion of others into our lives, and allow us to express our ideas and ourselves in circumstances where we otherwise could not. They are also useful tools for deception and hiding wrongdoing. Doxing hinders all of these purposes by deliberately removing some of the subject's obscurity. In the three forms I have described here, doxing can be a tool for establishing accountability for wrongdoing, a means of intimidation and incitement to cause harm, and a way of silencing minority or dissenting views. I have argued that only revealing personal information that is in the public interest to disclose and only to the extent necessary to establish wrongdoing is justified. There is much more to say about doxing and the role it plays in public discourse, and I hope this brief account of how acts of doxing can be classified will be useful to those who examine these issues in the future.

FOOTNOTES

1. With apologies to Karl Marx and Frederich Engels.

2. The possibility of revealing pattern knowledge is why mobile device metadata (information about its usage) is so sensitive. For an example of how metadata analysis can reveal pattern knowledge about an individual, see Ockenden and Leslie (2015).

3. Specifically, this is a form of what I call delegitimizing doxing, as I will describe later.

4. This protection can of course be abused, as ‘poison pen’ letters and anonymous inflammatory comments on the Internet demonstrate.

5. I thank Michael Nagelborg for this point.

6The Violentacrez example may also be interpreted as an instance of delegitimization, given his reputation for deliberately and publicly breaking social norms.

7. Trolling is a complex phenomenon with nuances and its own cultural norms that I cannot explore here. Whitney Phillips (2015) presents a detailed account that places trolling into a broader cultural context.

8. Revealing someone’s telephone number is deanonymizing doxing, since it reveals a connection between a pseudonym (the phone number) and the subject’s legal identity.

9. The term ‘swatting’ is derived from the name of police SWAT (Special Weapons And Training) squads who respond to potentially violent situations involving armed suspects.

10. Bowman-Grieve (2009) discusses the Nuremberg Files example in more detail and places it into the broader context of violent anti-abortion activism in the US.

11. A supposed social norm is one held by a minority in society that they believed should hold sway over the majority. The lack of widespread recognition means it is not an established norm (even if it had been

historically), but for the group who hold it, they believe that everyone in society should accept it and judge others accordingly.

12. It should be noted that there is a difference between this shaming and that covered by delegitimizing doxing. Unlike delegitimizing doxing, the material used to delegitimize Brutsch was already revealed: his offensive contributions to various sections of Reddit.

13. By (willfully?) misinterpreting academic studies into video games, some have even made the outlandish claim that it has revealed a 'Communist' and 'feminist' conspiracy to undermine video games (Chess and Shaw 2015).

ACKNOWLEDGMENTS

I wish to thank the participants of the 2015 Amsterdam Privacy Conference and my colleagues in the philosophy department of the University of Twente for their encouragement and suggestions on earlier drafts on this paper. I also like to thank Kevin Macnish for a stimulating discussion of privacy, Nolen Gertz and the students in the University of Twente's Philolab class for their comments and questions, and the anonymous reviewers for their helpful comments, suggestions, and encouragement.

REFERENCES

1. Beaujon, A. (2014). Redditors furious Newsweek 'doxxed' Bitcoin Founder. *Poynter.* http://www.poynter.org/news/mediawire/242348/redditors-furious-newsweek-outed-bitcoin-founder/
2. Bok, S. (1989). *Secrets: On the ethics of concealment and revelation.* New York: Vintage Books
3. Bowers, S. L. (2006). Privacy and library records. *The Journal of Academic Librarianship, 32*(4), 377–383
4. Bowman-Grieve, L. (2009). Anti-abortion extremism online. *First Monday* 14(11). http://firstmonday.org/ojs/index.php/fm/article/view/2679
5. Boyd, D. (2011). 'Real Names' policies are an abuse of power. *Apophenia.* http://www.zephoria.org/thoughts/archives/2011/08/04/real-names.html
6. Chen, A. (2012a). Unmasking Reddit's Violentacrez, the biggest troll on the web. *Gawker.* http://gawker.com/5950981/unmasking-reddits-violentacrez-the-biggest-troll-on-the-web
7. Chen, A. (2012b). Reddit's biggest troll fired from his real-world job; Reddit continues to censor Gawker articles. *Gawker.* http://gawker.com/5951987/reddits-biggest-troll-fired-from-his-real-world-job-reddit-continues-to-censor-gawker-articles

8. Chen, A. (2012c). Reddit troll Michael Brutsch defends himself on CNN: 'I treated Reddit as a game.' *Gawker.* http://gawker.com/5953097/reddit-troll-michael-brutsch-defends-himself-on-cnn-i-treated-reddit-as-a-game
9. Chess, S., & Shaw, A. (2015). A conspiracy of fishes, or, how we learned to stop worrying about #GamerGate and embrace hegemonic masculinity. *Journal of Broadcasting & Electronic Media, 59*(1), 208–22
10. Citron, D. K. (2014). *Hate crimes in cyberspace.* Cambridge, MA: Harvard University Press
11. Franks, M. A. (2012). sexual harassment 2.0. *Maryland Law Review, 71,* 655
12. Gao, L., & Stanyer, J. (2014). Hunting corrupt officials online: The human flesh search engine and the search for justice in China. *Information, Communication & Society, 17*(7), 814–829
13. Gavison, R. (1980). Privacy and the limits of law. *The Yale Law Journal, 89*(3), 421–471
14. Goldacre, Ben. (2012). *Bad pharma: How drug companies mislead doctors and harm patients.* London: Fourth Estate
15. Goodman, L. M. (2014). The face behind Bitcoin. *Newsweek.* http://www.newsweek.com/2014/03/14/face-behind-bitcoin-247957.html
16. Honan, M. (2014). What is doxing? *Wired.* http://www. wired.com /2014/03/doxing/
17. Jouvenal, J. (2013). Stalkers use online ads as weapon against victims. *The Washington Post.* http://www.washingtonpost.com/local/i-live-in-fear-of-anyone-coming-to-my-door/2013/07/14/26c11442-e359-11e2-aef3-339619eab080_story.html
18. Levmore, S. (1996). The anonymity tool. *University of Pennsylvania Law Review, 144*(5), 2191–2236
19. Manning, M. G. (2012). A tale of three hoaxes: When literature offends the law. *Columbia Journal of Law & the Arts, 36,* 127–156
20. Mantilla, Karla. (2015). *Gendertrolling: How misogyny went viral.* Santa Barbara, CA: Praeger
21. Marx, G. T. (1999). What's in a name? Some reflections on the sociology of anonymity. *The Information Society, 15*(2), 99–112
22. Mattise, N. (2015). Anti-doxing strategy—Or, How to Avoid 50 Qurans and $287 of Chick-Fil-A. *Ars Technica.* http://arstechnica.com/security/2015/03/anti-doxing-strategy-or-how-to-avoid-50-qurans-and-287-of-chick-fil-a/
23. Mill, J. S. (1989). *On liberty and other writings.* Edited by Stefan Collini. Cambridge: Cambridge University Press [1859]
24. Nakamoto, S. (n.d.). Bitcoin: A peer-to-peer electronic cash system. *Bitcoin.* https://bitcoin.org/bitcoin.pdf
25. Ngai, S., Gold, J. L., Gill, S. S., & Rochon, P. A. (2005). Haunted manuscripts: Ghost authorship in the medical literature. *Accountability in Research, 12*(2), 103–114
26. Nussbaum, M. C. (2010). Objectification and internet misogyny. In S. Levmore & M. C. Nussbaum (Eds.), *The offensive internet: Speech, privacy, and reputation* (pp. 68–87). Cambridge, MA: Harvard University Press
27. Ockenden, W., & Leslie, T. (2015). 'Scarily accurate': What you found in our reporter's metadata. *ABC News.* http://www.abc.net.au/news/2015-08-24/metadata- what-you-found-will-ockenden/6703626.
28. Phillips, W. (2012). What an academic who wrote her dissertation on trolls thinks of Violentacrez. *The Atlantic.* http://www.theatlantic.com/technology/archive/2012/10/what-an-academic-who-wrote-her-dissertation-on-trolls-thinks-of-violentacrez/263631/.
29. Phillips, Whitney. (2015). *This is why we can't have nice things: Mapping the relationship between online trolling and mainstream culture.* Cambridge, MA: The MIT Press
30. Poole, E. (2013). Hey girls, did you know: Slut-shaming on the internet needs to stop. *University of San Francisco Law Review, 48,* 221–260

31. Reichel, P. L. (1977). Dossier building as a social problem topic. *Teaching Sociology, 4*(3), 293–306
32. Reynolds, E. (2016) Revenge porn: There's no 'silver bullet' for ending non-consensual pornography. *Wired UK.* http://www.wired.co.uk/news/archive/2016-03/08/revenge-porn-facebook-social-media
33. Rindfleisch, T. C. (1997). Privacy, information technology, and health care. *Communications of the ACM, 40*(8), 92–100
34. Solove, D. J. (2007). *The future of reputation: Gossip, rumor, and privacy on the internet.* London: Yale University Press
35. Trottier, D. (2016). Digital vigilantism as weaponisation of visibility. *Philosophy & Technology,*1–18. doi:10.1007/s13347-016-0216-4
36. Waldron, J. (2010). Dignity and Defamation: The Visibility of Hate. *Harvard Law Review,123*(7), 1596–1657
37. Webb, L. M. (2015). Shame transfigured: Slut-shaming from Rome to cyberspace. *First Monday, 20*(4). http://firstmonday.org/ojs/index.php /fm/ article/view/5464

CHAPTER 4

Ethics of Virtual Reality Applications in Computer Game Production

Daniel Kade

Innovation, Design and Technology, Mälardalen University, Högskoleplan 1, Västerås, Sweden

ABSTRACT

A current trend in the gaming industry is to provide more realistic and believable looking animation. To support this, motion capture has been used to create such animation. Lately, immersive virtual environments have been further developed and can be used to support motion capture actors with their work. Using immersive virtual environments as work environments has already been explored, but no ethical analysis or applied ethical code has been provided for such situations. In this paper, we investigate the ethical implications of introducing a highly immersive virtual environment for motion capture acting and discuss under which circumstances it is ethically justified to place an actor in such an environment. Moreover, we provide an overview of research in computer games ethics, ethics for virtual realities and acting, as well as an investigation of potential moral and ethical issues in motion capture. Finally, a discussion shall help in finding an ethical consensus within the field of motion capture and for related situations.

Keywords: ethics; virtual realities; computer games; motion capture; acting

1. INTRODUCTION

Computer games are becoming more and more realistic nowadays [1]. This is not only because of hardware and graphical improvements but also because human motions are captured and mapped to computer game avatars. This makes animation and motion in games as well as in other entertainment applications, such as movies, advertisements and video clips increasingly realistic. Through the help of motion capture actors and their performance, virtual characters are brought to life. Thereby, actors use their own motions and emotions to shape the virtual character and give it a more human-like touch.

For an immersive entertainment or gaming experience, not only the aesthetic appearance of a game or application is of interest but also the level of immersion, the level of realism, as well as the way of playing or using the application. Therefore, applications have already been developed to combine gaming with virtual reality (VR) environments [2,3]. Currently VR technology and computer games, created for such technology, experience a hype. This is especially because technology becomes more affordable and usable for home entertainment.

Even for theatre plays, the combination of computer game technology, motion capture and traditional theatre has already been explored [4]. One could think easily of multiple application scenarios where the combination of VR technology with computer games or graphics could be applied.

In this analysis, we use motion capture acting as an example of how VR technology in combination with computer game environments could be used to help actors with their work. A motion capture actor's job is to contribute to the task of letting motions and expressions of emotions of an avatar appear in-game as realistic as possible. Nonetheless, the current acting environment for motion capture actors is very simplistic and the actors' work is not supported by technology in general motion capture shoots, in particular for computer games. Placing an actor "inside the game" for which an actor is acting can be assumed to support the actor's work.

When applying technological advancements, such as a highly immersive virtual reality environments to the field, it is worth investigating potential ethical issues that could occur when actors are supposed to work with it.

This paper is organized into four main sections. The state-of-the-art overview covers the ethical discussions addressed in this paper, namely: the ethics of computer games, virtual reality and acting. Then, we describe and discuss the possible ethical implications of motion capture actors when being placed into a highly immersive virtual work environment. This section gives an overview of the issues identified and describes the scenario that we use as a basis for discussions. The following section provides an ethical analysis on three questions extracted from the previous section to look at the raised questions in more detail. On these findings, Section 5 builds up a discussion towards an ethical consensus that can be applied in motion capture. The term "ethical consensus" is seen in this paper as a more defined ethical code or agreement between the stakeholders involved in motion capture. This includes, for example, a consensus between the gaming industry, the actor and the gamer as well as actors placed in a highly immersive work environment.

2. STATE-OF-THE-ART

To understand ethics that could be applicable for motion capture, we need to look at professional ethics that have been discussed in related areas. Professional ethics have been discussed in many fields; one of those is computer science where discussions about known and potential issues are ongoing and constantly evolving into new areas [5]. As computer technology and their applications are evolving, ethics for virtual environments [6,7,8] and computer games [8,9,10,11] have moved into focus, and discussions in the fields are already established. On the other hand, a brief literature search has shown that fewer ethical studies have been written about the ethical implications of computer games in combination with virtual reality as a work environment.

To provide an overview of the related topics, we address the three areas of interest: ethics in computer games, virtual realities and acting in the following sub-sections.

2.1. Ethics in Computer Games

Ethical aspects of computer games have been discussed within different contexts already [9,10,12]—especially the contents shown in computer games and their ethical implications have been discussed [12]. In previous research, it was suggested that the actual effect of computer games needs to be determined and that one must be cautious about the "widespread indiscriminate use of video games" [12]. In addition, the researchers' concern was that computer games have increasing graphic capabilities that could provide gamers with open-ended possibilities of immoral actions that might even be rewarded. In another paper, it was additionally pointed out that the "ethical decision-making" in video games and video game curricula should be changed and adopted to a more realistic approach [9]. A general point of discussion around the ethics of computer games is about the content and how computer games are designed.

To improve this, suggestions have been made that "game developers should follow an ethical consensus because they reach a large population of young people" [10]. Others have a slightly deviated understanding and have argued that: "players are actually reflective, and responsible for the choices they take while playing games. Games are objects designed with affordances that suggest a certain experience that is evaluated by its players' moral sense" [13]. Thus, basically, it was pointed out that players of computer games are "moral beings" and construct an "implicit code of values" that can be used by game designers [13].

This leads us to the issue that "game designers are still struggling with what kind of ethics code they should adopt" especially "because video games are a newer medium" [14]. In the same research, it was mentioned that the distinction if a game is unethical or not depends on the "eye of the beholder". It is furthermore stated that consumers steer the market and, in some cases, even demand e.g., violent games. Through this demand, the gaming industry feels the pressure to deliver the requested material to not loose market share to other competitors in the field. One of the interviewed industry professionals in their research even mentioned that violent computer games can allow for the exploration of behaviors that would not be possible in the real world and allow to "clear negative emotions" [14]. On the other hand, it has been mentioned in their article that there should be a clear distinction between what happens in the game and what happens in the real world. Thus, the player should be aware that it is just a game.

Another piece of research suggested to follow a model from Bertolt Brecht which makes the audience aware that they are being played and teaches them "not only to act, but emotionally and ethically respond in a specific manner" instead of implementing an ethical code [15].

The discussions on which kind of ethical code should be used for computer games and entertainment applications is ongoing and should, in our view, constantly be adapted to new situations, technologies or applications. Combining virtual realities with computer games might be such a situation.

2.2. Ethics in Virtual Realities

Research within ethics for virtual realities seems to mainly consider the ethical implications of the technology and their application areas [16,17,18,19], as well as the representation, interactions and behavior in virtual realties [6,8].

Within the field of ethics for virtual realities, the issue has been raised that there is a need to regulate the actions of a user through an ethical consensus [7]. The call expressed was that designers should implement barriers that do not allow immoral actions or that even virtual or real punishments could be a result for immoral actions. The concern of immoral actions in virtual realities has also been raised by other researchers who mentioned that VR could imaginably allow the freedom to commit e.g., rape and murder [6,7]. It was pointed out that actions might be performed which would be morally questionable in the real world. Especially because technology could allow the creation of VR applications that can graphically depict and create "immoral acts including murder, mutilation, torture, rape, robbery, and grand theft" [6], ethics in virtual environments become even more of interest. In the event that such environments become available to the public, an ethical code for VR applications must be of importance. When thinking about actors working in such a virtual environment (VE) for several hours, it is questionable how it affects the actors mind and if there are psychological issues that need to be considered.

Ethics in virtual realities have also been discussed in the field of psychological sciences. VE-related side effects have especially been pointed out [11,20]. Cybersickness, aftereffects, altered sense of reality, limited self-awareness and the loss of choice have been identified as possible occurrences in virtual realities. A "general discomfort, fatigue, headache, eyestrain, difficulty in focusing eyes, increased sweating, nausea, difficulty in concentrating, stomach awareness and blurred vision" was also observed as more present when showing a fast moving simulation in comparison to a slower moving simulation [11]. Another research project tested the perception of crime by using a virtual environment. It was also compared how the perception of crime is perceived in the real world and in comparison to the virtual world. Within this research, it was found that participants appeared more observant in the virtual environment (VE) and sometimes even more fearful than in the real environment [21]. This also adds to the question on how moral it is to expose a motion capture actor to such conditions and how this could be avoided by technology solutions and ethical guidelines.

It has been mentioned before that, when IT is introduced into our lives, it will change our experiences and the way of living or working with it. Therefore, it was stated that it needs to be ethically analyzed how to prevent harm to human beings and to look at the effect of the introduced IT on the users [22]. In our scenario, it is not IT but another form of technology that shall be introduced and ethical analyses need to be conducted. It has even been mentioned that actions in virtual environments can be harmful to others and raise moral issues within all major traditions in ethics [23]. This is why we need to investigate how this could affect actors in a VR environment.

As mentioned by others, "morality has always been in a state of flux" where the discussion about moral questions might not lead to an ethical consensus [7]. This might be the case, but, nonetheless, guidelines that are moral at this present time can be given to steer and control the morality of the current society and secure a VR experience that is not harmful and secures the general acceptance of the technology to be used in work and entertainment environments.

2.3. Ethics in Acting

Even in acting education, ethics have already been discussed. The use of power and the problem of inter-corporeal actions like in extreme cases like seduction, rape, nakedness and sexual intercourse that actors were forced to act out by acting teachers have been ethically questioned. In this specific study, the approach of the acting teachers and their use of power to enforce such scenes was addressed [24].

The job of a motion capture actor is to act out motions and emotions in a believable and naturally looking way, for example for a computer game. To trigger more natural emotions, actors might be placed in a highly immersive virtual environment to support their act. Directors and animators could have the power to create immoral scenes and situations enforcing actors indirectly to act out scenes that are ethically questionable. This implies two ethical issues; first, it comes back to the question of content shown to the actors, and, second, it addresses the problem of the effects on the actors' psychology when placed in a highly immersive environment for a longer time.

3. POSSIBLE ETHICAL ISSUES WITHIN MOTION CAPTURE

From the state-of-the-art, we can see that a futuristic motion capture environment deals with similar ethical issues than other fields. Nonetheless, we see that there are specific differences to motion capture and, therefore, we need to look at the possible ethical issues that could occur within motion capture acting.

Before discussing possible ethical issues and the implications when placing an actor in a highly immersive virtual environment we need to describe the imaginative environment that we are using for explanations. The environment is fictive but yet not far from reality and already technologically possible to implement. Figure 1 shows an actor in a motion capture environment wearing virtual reality glasses to see and "feel" the virtual environment.

Figure 1. Motion capture actor in a futuristic virtual environment supporting the actor's work.

For the depicted environment, we have to know that different stakeholders and also different ethical fields come together. Actors, directors, game developers and game publishers are interested in the results of a motion capture shoot. While acting, the fields of computer games and virtual environments come together to support the actors with their work. The goal of all stakeholders is to deliver realistic and natural-looking animation. Game developing companies get their demand from the market, generally speaking from computer game consumers. In this market, it seems to be a trend that gamers seek to play games that are increasingly realistic in terms of movements and emotions. Thus, it is in the interest of the gaming industry to support the needs and wants of realism that gamers seek and pay for.

This implies that game designers and developers work on creating storylines, characters and environments that are fairly realistic. Actors and directors then support the need of realistic motions and emotions through motion capture performances.

To create a certain amount of realism in a game, the actor is responsible for bringing the game character to life and providing realistic and believable motions. This alone does not necessarily mean that any moral or ethical issues might occur. It depends on context and content of the performance an actor is asked to play and later shown in-game. Games and animation can be perceived as very violent or disturbing and sometimes even immoral. It is then questionable who should be held responsible for immoral scenes and which norms or ethical code should be applied.

Actors enrich the perception of realism through their skills, the games industry designs the games and story and the gamers' clientele buy games and seek for realism in unethical games. One can see that there are multiple stakeholders, from the consumer to the game publisher, involved before a game gets on the market, and it might not be easy to answer the question of who is responsible—especially, when we consider that content can be seen as immoral by one person and seen as acceptable by another [23]. However, when we talk about professional ethics, responsibility lies with different professionals involved, each of them responsible for their own conduct. Regardless, one point needs to be mentioned, animation can appear quite realistic with modern technology and techniques like motion capture. Therefore, it needs to be considered if ethical issues are involved within the application or the content of that animation. Ethical guidelines that result from discussions within computer games ethics might help to apply an ethical consensus on this matter.

More ethical issues within motion capture can occur when thinking about an actor working in a highly immersive virtual environment. Here, we imply that an actor experiences a strong feeling of being immersed and "present" in the virtual world. Presence is a term that has been introduced by other research to describe the degree of a persons feeling of "being there" in a virtual environment [25]. For such a scenario, we think of using an immersive environment where an actor can feel present to a large degree and immoral scenes are displayed that the actor needs to act out. As mentioned before, research has shown that virtual environments can increase the subjective experience such as the feeling of fear [21] and possibly other negative feelings as well. However, a professional actor is trained to distinguish between the real world and the world on stage [26]. In other words, an actor must be aware of the actions of the character and the actions of the actor as a person. Still, it has not yet been tested what effects a highly immersive virtual environment would have on an actor under extreme conditions and even on a sub-conscious somatic level, especially when extremely violent and touching scenarios are shown to the actor.

One could argue that this might trigger more natural reactions and actions, but the general question remains: what happens when virtual reality becomes more real and the border between reality and play becomes harder to distinguish? In other research, it has even been argued that "virtuality threatens to alter our perspective on reality, causing us to see it as yet another sign or simulation" [27]. Psychological issues that actors experience because of the work in a virtual environment (VE) could occur. In such a case, one would have to make very clear to the actor, and the actor to himself or herself, that the performance is not real and only the character to be played experiences the reality, not the actor. Professional actors have the experience and training to deal with such circumstances better but for motion capture acting, the term "actor" is quite vague. Here, an actor might also be an inexperienced actor, a stunt man or a professional athlete. The effects of a highly immersive virtual work environment when using it over a long time and in an inappropriate way have not yet been studied. Therefore, we must find guidelines or techniques to prevent issues due to acting in highly immersive virtual environments. "Greater than life" experiences can even challenge experienced professional actors. A scenario where such an ethical issue is possible is when actors need to perform cases of torture, extreme interrogations or training of special forces under severe conditions.

Furthermore, it needs to be tested if and how the end users of a motion capture performance are affected when these performances become even more realistic and believable. This relates back to content and context of computer games but also includes the gamers' actions and their effect on the gamers' real life. As mentioned before, professional actors are trained to distinguish between what is real and what is performance. It is questionable if gamers can make this distinction too. When computer games are becoming more realistic through better graphics and hardware but also the contribution of motion capture performance that enrich 3D characters through motions and emotions that allow to appear human-like, we should also consider the effects that might occur when gamers are exposed to games that appear close to realism and depict unethical content.

4. ETHICAL ANALYSIS

The moral issues pointed out in the previous section resulted in three questions that we think summarizes their essence and therefore are used in the following to analyze the ethical issues in more depth. Stakeholders that we address in this analysis are actors, directors, game developers, game publishers and gamers.

The resulting three questions to be analyzed are the following:

(1) What are the responsibilities of stakeholders of a performance and how can unethical performances be averted?

(2) Could an actor be psychologically affected when working with disturbing content within a highly immersive virtual environment (such as e.g., violent scenes)?

(3) How are gamers affected when being exposed to games with more human-like 3D characters that are behaving unethically?

Analysis of Question 1:

To analyze the first moral issue mentioned-above, we need to investigate who is responsible for the content of a performance. This question is not as simple to answer as it might seem at first. An actor is performing and gives the 3D avatar a human touch, helps shaping the personality of the virtual avatar and makes the animation more realistic and believable. After that, the directors add their thoughts and guide the performance of an actor. As another instance, game developers create the idea, the storyboard and therefore create content and context of a game. The question of who is responsible could also be redirected to the consumers that are buying computer games with unethical content. It seems like a viscous cycle where demand and offer are closely related. Figure 2 shows how the demand and offer of unethical computer games content is related between the stakeholders of the scenario chosen in this paper. The cycle also reflects, in a very abstract way, the processes and connections between the game development industry, motion capture and the consumer. One needs to keep in mind that this figure is meant to illustrate the relations of stakeholders when producing a game with unethical content.

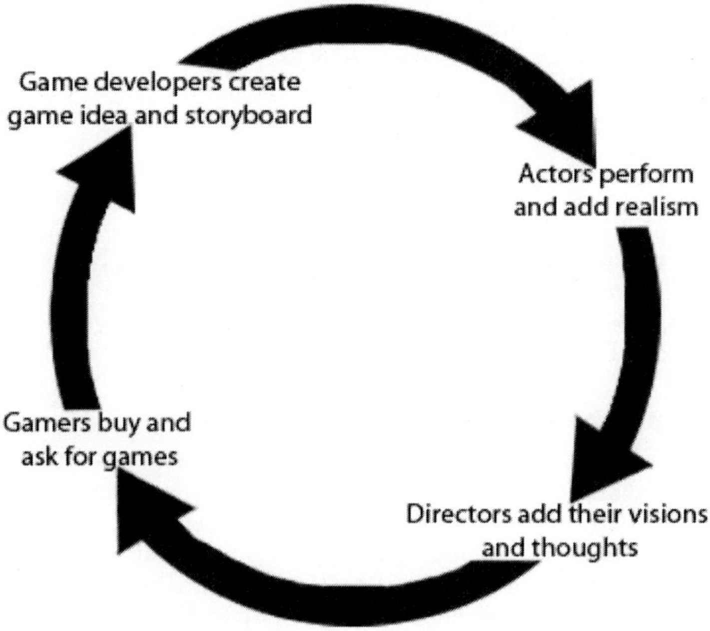

Figure 2. Cycle of demand and offer within computer games development, including motion capture, that can become viscous in the case of unethical games.

It might not be the right way to find a solution to who is responsible for the development of computer games with unethical content but rather to apply an ethical consensus to steer the use of content in computer games. As others have already addressed, a more realistic ethical consensus for computer games needs to be found [9]. One solution could be similar to what has been done for the film and movie industry by applying an ethical code to the creation of content for computer games [14]. This could help to steer the development of computer games to reduce highly unethical content and the demand for it. Other research suggests to "extend the age-based censorship for media, ...including VR" and to encourage designers to establish high ethical standards with the help of professionals [7]. This might be an additional option towards an ethical consensus but it cannot be the complete solution, we need to discuss a minimal ethical consensus that is applicable to the gaming industry first.

Analysis of Question 2:

For the question of if an actor could be psychologically affected when working with unethical content within a highly immersive virtual environment, the stakeholders are actors being exposed to a highly immersive environment and game developers providing unethical content that is shown in a virtual reality environment.

Previously, (highly violent, *etc.*) games have been produced for computers, game consoles or mobile devices. There, gamers control the games from outside the game world. This creates a physical distance between the game world and the real world, even though games can pull the gamers attention towards the game and allow for the feeling of presence. When bringing real humans into the virtual world, immersed and exposed to the game content shown to them, the physical distance is taken away. The player, actor or user of such a system perceives the happenings and the virtual environment now directly.

For motion capture acting in a highly immersive environment, there are even more distinguishing characteristics. Normal users of a virtual reality do not necessarily act, react or move in a realistic way. Movements and interactions can be symbolic and do not resemble everyday reactions and movements. Motion capture actors need to live up to the experience so that they can get to the state of "being the character". A normal VR user does not internalize the VR environment to such an extent. VR users only explore and experience the environment where the feeling of presence and "being the character" might be reached. An actor in a VR on the other hand, needs to internalize the scene to act and deliver a believable performance. Therefore, we can say that the actor is working towards reaching the feeling of "being the character" and "living the scene". In comparison to traditional acting, we introduce a highly immersive VR and increase the "greater than live" feeling which is common for VR environments. This leads to a much higher cognitive load than in ordinary acting and just exploring a VR.

One problem to consider is, as research has shown, that highly immersive virtual realities can increase the feeling of fear [21]. This could support the assumption that highly immersive environments can place actors in situations with extreme unethical content, and this might psychologically affect motion capture actors working in it.

To avoid such events from happening, it is of importance that such a scenario might not even be needed. Ethical guidelines and the non-existing need to perform unethical performances could be a way of preventing the need to capture unethical scenes. By unethical performances, we mean scenes that depict extreme acts of violence, torture, extreme interrogations, drug abuse or even extreme depiction of sexual acts. In case actors are obliged to work in such an environment, what is moral and what is not must be defined. As this is not yet answered, the exact implications on users of a highly immersive system with unethical content needs to be researched and applicable ethical codes need to be discussed.

Analysis of Question 3:

The question on how gamers are affected when being exposed to games with more human-like 3D characters that are performing unethical actions can be addressed through the comparison of ethical concerns that could occur in our scenario with the ethics of virtual reality, computer games and acting.

The fact of having a 3D character performing realistic human-like motions and actions might not be an issue in itself. It could even be helpful for educational and sports games. This is clearly up to the context and the actions that the 3D character performs or the gamer is allowed to perform. We can also assume that a game character performing unethical actions in more realistic, human-like appearance has a larger, negative impact on the gamers' reactions and perceptions as current lower fidelity characters. Earlier research supports this statement by saying that more realistic virtual environments are more powerful [6]. On the other hand, research in ethics for VR claims that sometimes "well-constructed action, plot, and play-ability" can have a greater effect than "gorgeous renderings that mirror the everyday world" [8]. As this issue has not yet been fully discovered, it can only be the goal to prevent gamers from affecting them in a negative manner. Especially, protecting underaged gamers from being exposed to unethical content must be avoided as they could be negatively affected by e.g., 3D avatars performing unethical actions with believable motions and emotions. This issue might even become more imminent when we introduce a highly immersive virtual reality application to gamers, similar to what could be used for actors. It must then be of strong interest to look into ways of preventing gamers from harm.

Again, a refined ethical consensus, as well as the social perception of demanding such games must be addressed to help limit negative effects on consumers.

5. DISCUSSION TOWARDS AN ETHICAL GUIDELINE

We see the need for a discussion towards more defined ethical guidelines for motion capture acting in a highly immersive environment, as no guidelines for our scenario exist yet. From the analysis above, we can see that the main ethical issues within a futuristic acting environment are fairly similar to what is being discussed within ethics in virtual environments and computer games. Therefore, existing ethical codes from these areas could be applied, but we could not find an ethical code that can be applied out of the box.

There are many interesting concepts that could lead towards an ethical consensus; some of them have been mentioned in the state-of-the-art section of this paper. On the other hand, it might be hard to implement ethical codes when discussions within ethics for gaming and virtual realities face different opinions and contradict each other. It can even be said that in some of the discussed areas, contradictions or the struggle to apply the "right" code exist. One largely discussed example is the content of computer games or the discussion on health issues and their acceptability within virtual realities. To get to an ethical code, it might not be the right way to only restrict unethical content in computer games. This would probably lead to scenarios where unethical games will be designed and distributed through non-profit or private persona. It must certainly be the case that harm should be prevented and unethical actions should not be rewarded.

A punishment system to steer what is seen as moral or immoral was suggested to implement in VR systems [7]. This could be one way of achieving this but what remains questionable is in which situations this applies—for computer games, VEs and in which use cases (work or entertainment environment). The discussion on what is immoral and what is not, is still ongoing and it might not be simple to answer this. Other research talks about indecent representations that depict shocking and offensive scenes which still might be seen differently by people, some might find it unethical and others might accept it [23].

Changing the industry habits but also the thinking of the community that uses unethical digital content seems to be of more value to achieve an ethical consensus or even to come to an understanding on how to deal with unethical content. This could e.g., mean to make designers aware of the ethical issues and how to avoid them. Unethical actions should not be rewarded, the choices of behavior can be up to the users themselves but the understanding of what is real or virtual must be supported through the design of the system. Such thoughts are not new to ethical discussions but what needs to be pointed out is that the society that uses or designs virtual reality systems must come to an understanding of what can or should be depicted. It should be made clear that if unethical actions are shown, the difference of what is real and what cannot happen in the same way in real life is understood.

Such guidelines could help towards an ethical code in motion capture as well. We think that an ethical guideline for a futuristic motion capture environment needs to consider and cover the most imminent issues within motion capture acting. This implies the physical and mental health of actors, the content generated and released for computer games, unethical performances and acting conditions that actors are dealing with. As these moral issues are strongly related to ethics from computer games, virtual realities and acting, we see a larger potential to use a subset of ethical codes from these areas to create an ethical guideline for motion capture actors. We need to prevent the "greater than life" experiences that actors might have from including extreme unethical content. It also needs to be made sure that discomfort and cyber sickness, when using a VR environment, does not occur.

As the ethical codes in VR, computer games and acting are not yet evolved enough to deal with these issues, it might be a good way to make actors aware of the circumstances they might be placed in and teach them how to handle them. This could be similar to what other research suggested for visitors of a theatre that where taught to "emotionally and ethically respond in a specific manner" and made aware that the play is not real [15]. Motion capture actors could even learn to distinguish between the real world and the world on stage from professional acting training. At this point, we see that more discussions are needed to define a more concise ethical code that can be applied for motion capture acting, especially when introducing VR technology as acting support.

6. CONCLUSIONS

Discussions to reach an ethical consensus for motion capture acting are still rather basic. This is mainly because of a lack of an existing ethical code in motion capture. We see an ethical code, applicable for motion capture, resulting from a subset of ethical codes within the areas of virtual reality, computer games and acting. These areas have a larger impact, are widely discussed and are closely related to a futuristic motion capture environment.

In this paper, we also discussed possible ethical implications of technology, applied to a new environment, involving different areas of ethics. The specific case of virtual reality technology applied to a motion capture environment as acting support has been chosen for analysis. To approach this specific domain of applied ethics, an overview of ethics in computer games and virtual environments has been given.

The ethical implications that were discussed and considered as important when placing actors in a virtual environment as a support for their work can be summarized as follows:

Game content as well as the performance to act out could already be ethically questionable. Especially in such a case, new ethically questionable situations can occur when the scenery of the act is depicted in a highly immersive virtual environment. Actors might experience extreme emotions ("greater than life") that they have to overcome to not be affected psychologically. It also needs to be understood that working in a VR environment can cause physical implications for actors. This needs to be considered, further researched and possibly prevented.

For the scenario, used in this paper, the ethics within the areas of computer games, virtual environments as well as acting were considered. When introducing technology into a new area where ethics of multiple areas need to be taken into account, such as game ethics, theatre ethics and occupational ethics, it must be considered if those ethical approaches still apply and if there are no new ethical issues that have not yet been studied.

The ethical analysis that we performed has shown that VR technology used in a work environment could be covered by using ethical codes from computer games ethics and ethics of virtual environments, even though there are some questions that we specifically raise in the context of motion capture acting. More research is needed when new technologies are introduced to special working environments that connect multiple ethical fields. In this respect, the work in this paper shall serve as a starting point for discussions and for further research on ethics for motion capture acting within virtual worlds.

REFERENCES

1. Wages, R.; Grünvogel, S.M.; Grützmacher, B. How realistic is realism? Considerations on the aesthetics of computer games. In Proceedings of the 3rd International Conference on Entertainment Computing (ICEC 2004), Eindhoven, The Netherlands, 1–3 September 2004; Springer Berlin Heidelberg: Heidelberg, Germany, 2004; pp. 216–225.
2. Yoon, J.W.; Jang, S.H.; Cho, S.B. Enhanced user immersive experience with a virtual reality based FPS game interface. In Proceedings of the 2010 IEEE Symposium on Computational Intelligence and Games (CIG 2010), Copenhagen, Denmark, 18–21 August 2010; pp. 69–74.

3. Liarokapis, F. An exploration from virtual to augmented reality gaming. *Simul. Gaming* **2006**, *37*, 507–533.
4. Andreadis, A.; Hemery, A.; Antonakakis, A.; Gourdoglou, G.; Mauridis, P.; Christopoulos, D.; Karigiannis, J.N. Real-time motion capture technology on a live theatrical performance with computer generated scenery. In Proceedings of the 2010 14th Panhellenic Conference on Informatics (PCI 2010), Tripoli, Greece, 10–12 September 2010; pp. 148–152.
5. Moor, J.H. What is Computer Ethics? *Metaphilosophy* **1985**, *16*, 266–275.
6. Brey, P. The ethics of representation and action in virtual reality. *Ethics Inf. Technol.* **1999**, *1*, 5–14.
7. Whitby, B. The virtual sky is not the limit: Ethics in virtual reality. *Digit. Creativity (Intell. Tutoring Media)* **1993**, *4*, 23–28.
8. Ford, P.J. A further analysis of the ethics of representation in virtual reality: Multi-user environments. *Ethics Inf. Technol.* **2001**, *3*, 113–121.
9. Gotterbarn, D. The ethics of video games: Mayhem, death, and the training of the next generation. *Inf. Syst. Front.***2010**, *12*, 369–377.
10. Dodig-Crnkovic, G.; Larsson, T. Game ethics-Homo Ludens as a computer game designer and consumer. *Int. Rev. Inf. Ethics* **2005**, *4*, 19–23.
11. Bruck, S.; Watters, P.A. Estimating cybersickness of simulated motion using the simulator sickness questionnaire (SSQ): A controlled study. In Proceedings of the IEEE 2009 Sixth International Conference on Computer Graphics, Imaging and Visualization (CGIV 2009), Tianjin, China, 11–14 August 2009; pp. 486–488.
12. Gotterbarn, D.; Moor, J. Virtual decisions: Video game ethics, Just Consequentialism, and ethics on the fly. *ACM SIGCAS Comput. Soc.* **2009**, *39*, 27–42.
13. Sicart, M. Game, player, ethics: A virtue ethics approach to computer games. *Int. Rev. Inf. Ethics* **2005**, *4*, 13–18.
14. Takahashi, D. Ethics of game design. 2005. Available online: http://webdocs.cs. .ca/games/299/Gamasutra.pdf(accessed on 25 March 2014).
15. Pinchbeck, D. A theatre of ethics and interaction? Bertolt brecht and learning to behave in first-person shooter environments. In *Technologies for E-Learning and Digital Entertainment*; Springer Berlin Heidelberg: Heidelberg, Germany, 2006; pp. 399–408.
16. Cranford, M. The social trajectory of virtual reality: Substantive ethics in a world without constraints. *Technol. Soc.***1996**, *18*, 79–92.
17. Whalley, L. Ethical issues in the application of virtual reality to medicine. *Comput. Biol. Med.* **1995**, *25*, 107–114.
18. Thurmel, S. Ethische Aspekte der Virtuellen Realität. In Analyomen 1: Proceedings of the 1st Conference on Perspectives in Analytical Philosophy; De Gruyter: Boston, MA, USA, 1994.
19. Beardon, C. The ethics of virtual realitySussex. *Digit. Creativity (Intell. Tutoring Media)* **1992**, *3*, 23–28.
20. Rizzo, A.; Schultheis, M.T.; Rothbaum, B.O. Ethical issues for the use of virtual reality in the psychological sciences. In *Ethical Issues in Clinical Neuropsychology*; Bush, S.S., Drexler, M., Eds.; Swets & Zeitlinger Publishers: Lisse, The Netherlands, 2002; pp. 243–280.
21. Park, A.J.; Spicer, V.; Guterres, M.; Brantingham, P.L.; Jenion, G. Testing perception of crime in a virtual environment. In Proceedings of the 2010 IEEE International Conference on Intelligence and Security Informatics (ISI 2010), Vancouver, BC, Canada, 23–26 May 2010; pp. 7–12.
22. Van den Hoven, J. Moral methodology and information technology. In *The Handbook of Information and Computer Ethics*; Himma, K.E., Tavani, H.T., Eds.; Cambridge University Press: Cambridge, UK, 2008; p. 49.
23. Brey, P. Virtual reality and computer simulation. In *The Handbook of Information and Computer Ethics*; Himma, K.E., Tavani, H.T., Eds.; Cambridge University Press: Cambridge, UK, 2008; p. 361.

24. Isaac, T.H. Film and Ethics. Available online: http://isaacth.com/ ethics. html (accessed on 23 May 2013).
25. Slater, M.; Usoh, M. Body centred interaction in immersive virtual environments. *Artif. Life Virtual Reality* **1994**, *1*, 125–148.
26. Sawoski, P. *The Stanislavski System Growth and Methodology*; Santa Monica College: Santa Monica, USA, 2010.
27. Borgmann, A. *Holding on to Reality: The Nature of Information at the Turn of the Millennium*; University of Chicago Press: Chicago, IL, USA; London, UK, 1999.

CHAPTER 5

Towards More Inclusive IP Analysis by Frontier Tools

Yoshiyuki Osabe[1]* and Mari Jibu[2]

[1] *Japan Patent Office (JPO), Tokyo, Japan*

[2] *Japan Science and Technology Agency (JST), Tokyo, Japan*

ABSTRACT

This chapter introduces multilateral analysis on IP rights: (1) a new indicator "Innovation Front" and its use, (2) analysis of patent quality, and (3) future prospect in pharmaceutical field. Through these items, more inclusive IP analyses have been conducted. We introduce the origin, trajectory, and destination of knowledge spillovers in the science and technology system, especially in pharmaceutical field. "Innovation Front" is also covered, where it is possible to find major hotspots in basic research, which give a great influence to technologies. Readers of this chapter will find (1) the major hotspots in basic research, (2) patent quality analysis ranging from basic research to application research, and (3) an overview of drug R&D and future competitiveness in the pharmaceutical field.

Keywords: patents, non-patent literature, knowledge spillovers, knowledge flows, patent quality, pharmaceutical, drug pipelines

1. INTRODUCTION

There are no doubts about the importance of knowledge spillovers for creation of intellectual properties (IPs) and furthermore economic growth.

Because the spillovers allow a better penetration and diffusion of innovation and stimulate cooperation in R&D for innovators to try to internalize knowledge flows, the role of knowledge exchange and dissemination is often as important as the role of direct investment in a knowledge- and technology-driven economy. Various reports have been published to study the origin, pass way, and end products of knowledge spillovers in the science and technology ecosystem. Patents and citations between patents and non-patent literature (NPL) are analyzed to make comprehensive grasp of knowledge spillovers [1] or to estimate patent quality [2].

Recently, we have developed a new indicator, named "Innovation Front" by calculating papers cited in patents by co-citation analysis [3]. Since papers cited in patents are close to technology, it is possible to find major hotspots in basic research, which give a great influence to technologies.

Apart from our analysis, a new indicator "Patent-Science Link" is developed by the Organisation for Economic Co-operation and Development (OECD). This is able to understand how the technical knowledge has been flowing from the science-based study to the innovation activity [4]. The Patent-Science Link indicates that pharmaceutical patents reckoned the large part of citations made from patents to scientific papers. That is, in the fields of pharmaceutical, the science-based study is much closer to the innovation activity comparing with other fields of technology.

Therefore, we analyzed the intellectual properties in pharmaceutical science using Innovation Front and OECD's Patent-Science Link. Based on these indicators, we clarified how the knowledge flows on each pharmaceutical R&D stage and how a drug has been created as a final end product. In this study, papers, patents, and drug pipelines represent each pharmaceutical R&D stage. Especially, the indicators of the drug pipelines also showed an overview and future prospects of pharmaceutical industries.

This chapter introduces multilateral analysis on IP rights: (1) a new indicator "Innovation Front" and its use, (2) analysis of patent quality, and (3) future prospect in the pharmaceutical field.

Note that the opinions expressed in this chapter are those of the authors and do not represent those of the institutions that the authors belong to.

2. INNOVATION FRONT

The citations between scientific papers and patents have been analyzed since Narin started the study of science linkage, which is seen in the administrative process as patent examiners refer scientific papers in examining a patent [5, 6]. Japan Science and Technology Agency (hereafter JST) is one of the funding agencies in Japan with aims of promoting technology transfers and technology-based innovation. As such, it is important to investigate the JST's contribution as a result of its funding. JST has a point of view that non-patent literature (NPL) in patents is seemed to provide a hint of knowledge flows between science-based study and innovation ecosystem. JST has developed as such an indicator, the "Innovation Front" in order to show specific research areas where science significantly influences technology by calculating co-citation between patents and research papers [7].

2.1. Innovation Process

For the purpose of stimulating innovation, the innovation process needs to be carefully analyzed. **Figure 1** shows that the innovation develops on the bases of the two internal actions, "knowledge embodiment (in other words, development part in R&D)" and "knowledge creation (in other words, research part in R&D)" [8]. Yamaguchi developed his idea, so-called innovation diagram that shows the way of visualizing an innovation process using two intellectual elements. They represent the element of the "knowledge embodiment (development)" and the element of the "knowledge creation (research)" (see **Figure 1**). Once the technological exploitation come to a total deadlock (A" in **Figure 1**), the R&D need going back to the basic scientific knowledge ("Induction" in **Figure 1**) and force it to move with scientific exploitation ("Abduction" in **Figure 1**) for the purpose of creating another innovation by overcoming the deadlock above (A* in **Figure 1**). Making a breakthrough and creating a new innovation will be possible by the actions above. As you see, by linking science-based knowledge to technological information, we can see a whole way of the innovation process, as shown in **Figure 1**.

Figure 1 indicates an overview of how to understand and visualize the innovation process by connecting the results such as patent analysis, bibliometric analysis, and clustering analysis by the following innovation process. In this attempt, **the linkage between patents and papers** is a key to obtain an understanding of the innovation process, which is evidence-based, such as the review of science-based knowledge from technology stage ("induction" in **Figure 1**). Because patent examiners

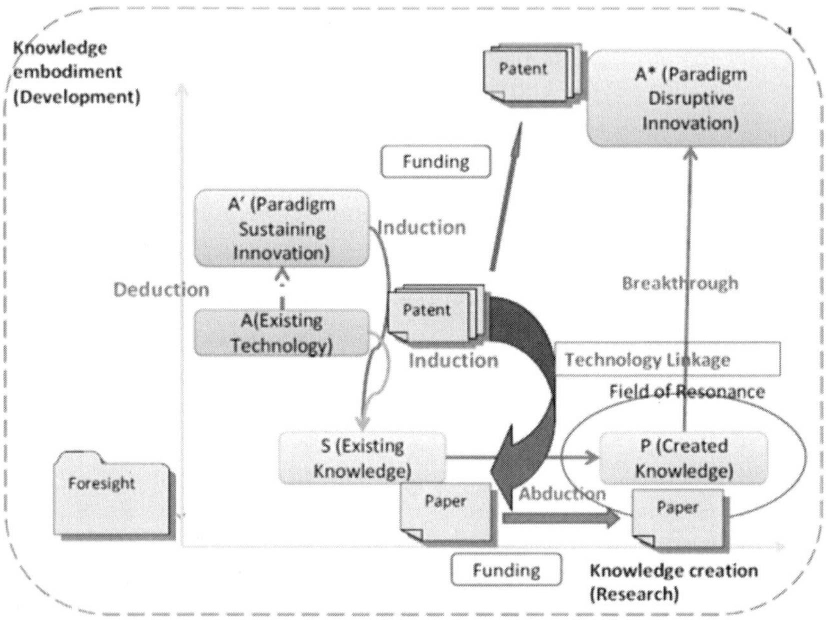

FIGURE 1. Innovation process and scientometrics.

generally cite patent and paper information as a reference, we can develop an index that is called "the science linkage index," that is, the number of forward citations and the NPL share (share of non-patent literature) citations, and the index may show whether a patent is technology-oriented or science-oriented. Many studies show that backward citations to the non-patent literature (NPL) relate the closeness between a patented invention and science-based knowledge. On the other hand, the importance of a patent for the R&D is related to forward patent citations.

In order to create a new indicator "Innovation Front," we utilized the above-mentioned relationship between patents and papers. As shown in **Figure 2**, pairs of papers are co-cited by different patents. For example, Paper A and Paper B are co-cited by both Patent X and PatentY. So, we can recognize that Paper A and Paper B are close to each other **in the point of technical (patent) view**. On the other hand, the relationship between Paper A and Paper C or Paper B and Paper C is not technically close because they"do not share a group of patents which co-cite them. In this way, we can create a set of clusters consisted of papers which are technically close to each other.

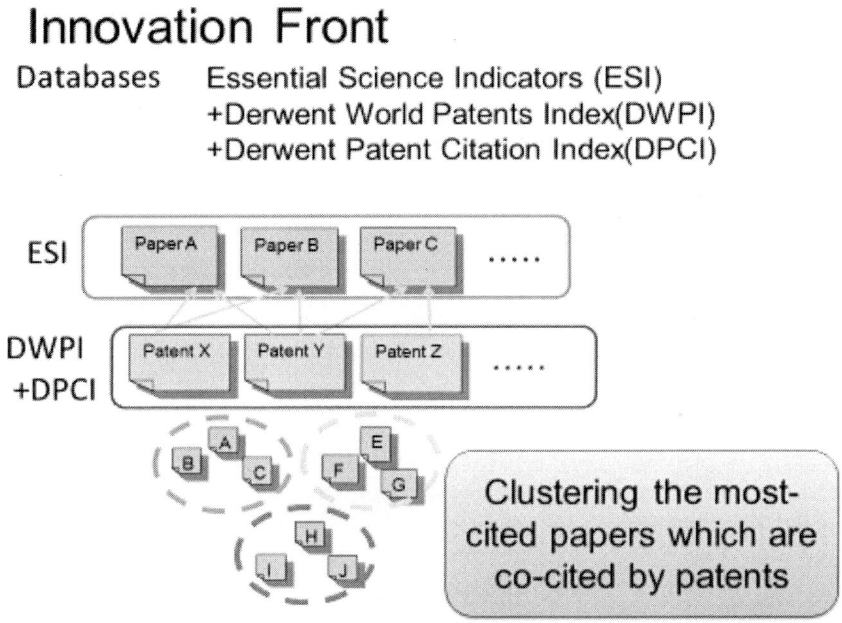

FIGURE 2. Structure of Innovation Front.

"Technical closeness" is the keyword in this research. Innovation front is the first indicator who can show the technical closeness between papers by calculating patent co-citations.

2.2. Methodology

2.2.1. Series of database

The series of databases used for the Innovation Front are as follows:

Papers: "Essential Science Indicators," "Web of Science" by Thomson Reuters.

Patents: "Derwent World Patents Citation Index," "Derwent World Patents Index" by Thomson Reuters.

2.2.2. Classification

Papers:

Note that 22 category codes from Thomson Reuters "Essential Science Indicators" for Innovation Front.

Patents:

1. International Patent Classification (IPC) codes for Information and Communication Technology (ICT), biotechnology, environment-related technologies, nuclear energy, and fuel cells [1].
2. ECLA codes for nanotechnology.

2.2.3. Calculation

1. 1. Extract two papers arbitrarily (the papers were published between 2006 and 2010. These papers were extracted from "Essential Science Indicators."). 2. Calculate the frequency of forward co-citation by patents. This calculation is done with "Derwent World Patents Citation Index." 3. Then, calculate how frequent forward citation patents are in the two arbitrary papers.

2. Derive cosine coefficient N from Eq. (1).

$$N = \frac{F(A,B)}{\sqrt{F(A) \times F(B)}}$$

$$F(A,B) \geq 2$$

3. where A and B of formula (1) show arbitrary papers, $F(A)$ and $F(B)$ show the cited frequency of the arbitrary papers, and $F(A, B)$ also shows the co-citation frequency of the arbitrary papers. Note that definition of $F(A, B)$ is larger than 2 or equal to 2.

4. Then define $N \geq 0.3$ (note that N is larger than 0.3 or equal to 0.3).

5. 1. Compile the papers that are extracted under the condition (3) above as nodes. 2. Connect the linkage among papers as the edge function. 3. Visualize the network of nodes and linkages, by using Cytoscape Web.

2.3. Analytical Results

"Innovation Front" shows scientific hotspots by means of making clusters by co-citation analysis between scientific papers and patents. Since scientific papers cited by patent examiners are approximate to technology in the patent, we can find who are the major researchers and important scientific specialties in terms of papers having an influence on technology. Thomson Reuters publishes a "research front," that is the papers which have a strong impact on Science. It is based on the calculation of co-citation frequency of papers. JST has a similar approach in terms of patent. JST clusters the top 1% in terms of the most frequently cited papers that patent co-cite and show them as a cluster indicating an impact in terms of innovation. It has named as "Innovation Front." The clustering by the Innovation Front is shown in **Figure 3**.

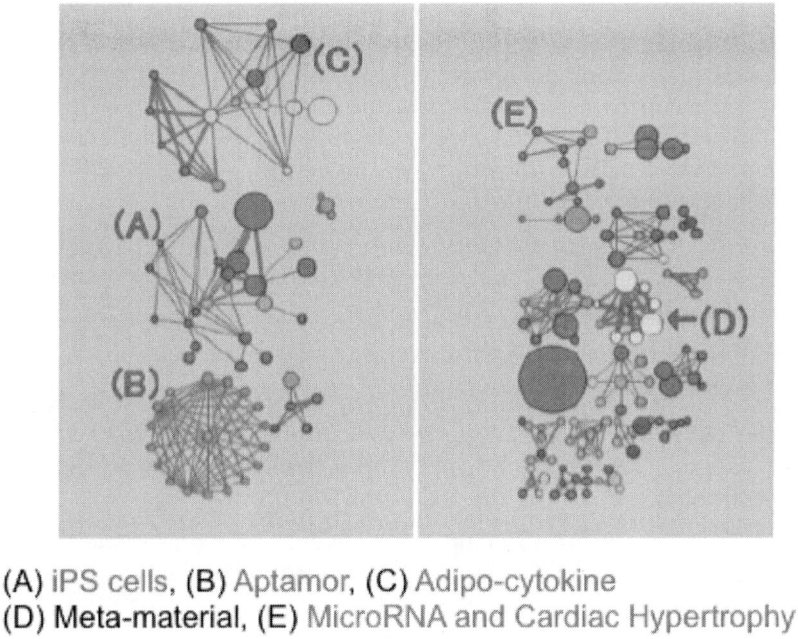

FIGURE 3. Result of the Innovation Front.

Each node represents a paper and the nodes' colors express 22 kinds of fields. The node's size represents how many citations are there in papers. The line width between nodes represents the number of co-citations by patent families. As a result, generated 24 clusters composed of 183 papers in total. The total citations of patent families and those of papers are 1095 and 46,038, respectively. The most frequently published journal is *Science*, with 18 papers and the second top journal is *Nature* with 17 papers. As to subjects of papers in clusters of Innovation Front: the largest subject is clinical medicine (45.4%) and the second top is chemistry (14.2%).

Cluster (A) is the field of induced pluripotent stem (iPS) cells composed of 22 papers. The iPS cells represent "induced pluripotent stem cells" which was discovered by Shinya Yamanaka in Kyoto University in 2006. Among the 22 papers, the title of the core paper is "Induction of Pluripotent Stem Cells from Adult Human Fibroblasts by Defined Factors," written by Shinya Yamanaka et al. This paper was published in *Cell* in 2007.

Cluster (B) is the field of aptamor composed of 18 papers. Aptamor is a peptide or oligonucleotide that binds to a specific target molecule. The main fields of the 18 papers are engineering and chemistry. This is slightly different from the main field of the definition of aptamar above. In terms of 18 papers composing the cluster (B), the title of the core paper is done by Lee et al. in Northwestern University and the title is "Colorimetric Detection of Mercuric Ion (Hg^{2+}) in Aqueous Media using DNA-Functionalized Gold Nanoparticles." This paper was published in *Angewandte Chemie International Edition* in 2007.

Cluster (C) is the field of adipocytokine (cytokine secreted by adipose cells) composed of 15 papers. The main field is clinical medicine, immunology, and biology/biochemistry.

Cluster (D) is the field of meta-material, composed of nine papers. The main field is physics.

Cluster (E) is the field of microRNA and cardiac hypertrophy composed of nine papers. The main fields are clinical medicine and neuroscience and behavior.

Figure 4 shows the detail mapping of cluster A that represents the field of iPS cells composed of 22 papers and the number of citation papers are 7517. Squares indicate papers written by Yamanaka group and the largest square is titled "Induction of Pluripotent Stem Cells from Adult Human Fibroblasts by Defined Factors," written by Shinya Yamanaka, published in *Cell* in 2007. The aim of the core paper by our indicator is the iPS cells derived from **human fibroblast**. On the other hand, the Nobel Prize in Physiology or Medicine 2012 was awarded jointly to Sir John B. Gurdon and Shinya Yamanaka, and the Press Release by the Nobel Assembly announced" Induction of pluripotent stem cells from mouse embryonic and adult fibroblast cultures by defined factors" by Shinya Yamanaka as one of the key publications. This paper's aim is to develop iPS cells derived from **mouse embryonic and adult fibroblast cultures**. As the study advances from basic to practical research, subjects for study are changing from *in vitro*, *in vivo*, mice, apes, and to humans. Our indicator is able to find the papers and fields (clusters) that affect the **technology, which is closer to applied stage than science study**. This represents the result of iPS cell cluster above. In other words, Innovation Front showed the application study papers that target humans instead of *in vitro* study or mouse in the analysis of iPS cell field. Other papers of iPS cells also target humans in cluster (A). The results of Innovation Front composed of roughly six groups are presented below.

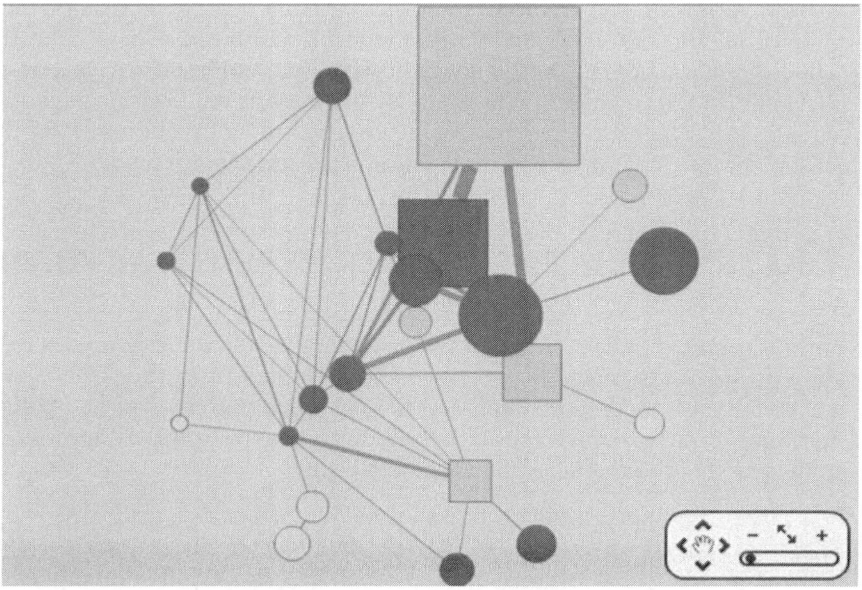

FIGURE 4. Innovation Front: a cluster of iPS cells.

As to papers of Node 1, this is by a group of the International Consortium of Stem Cell Networks, which is a consortium of human embryonic stem (ES) cell researchers from around the world. As to Node 2, this is by Yamanaka group, which made a success in converting mice and human skin cells into iPS cells. As to Node 3, this is written by Thomson J. A. (University of Wisconsin) group. Thomson group succeeded in isolating the human embryonic stem (ES) cells in 1998 and also known as a study of human iPS cells. They published the paper related to iPS cells at almost the same time as Yamanaka group, 2007. As to Node 4, this is the group of Schoeler H. R. (Max Planck Institute) group. Schoeler groups succeeded in reprogramming of adult mouse neural stem cells by introducing on 4 October. As to Node 5, this is the group of Jaenisch R. (MIT) group. Jaenisch group had a success in terms of reprogramming from mature, differentiated mouse B cells. Finding alternatives instead of the cancer-causing retroviruses for making iPS cells is also their study. As to Node 6, this is the group of Hochedlinger, K. (Harvard University) group. Hochedlinger group used an adenovirus in order to transport the transcription factors into the DNA.

As shown above, Innovation Front shows "new geography of innovation hotspots." According to analysis by Innovation Front, the hotspots are the science fields of induced pluripotent stem cells (iPS cells), aptamor, microRNA, cardiac hypertrophy and other fields of pharmaceutical science. Apart from Innovation Front, the OECD also introduced a new indicator called "Patent-Science Link" and indicated that patented inventions in pharmaceuticals account for the majority of citations to scientific papers in patents [4]. Therefore, we have moved forward with our aims to the analysis of intellectual properties in pharmaceutical science fields.

3. PATENT QUALITY ON PHARMACEUTICAL FIELD

Various reports have been published to show the study the origin, pass way, and end products of knowledge flows and the delays in the science and technology system. Patents and citations between patents and non-patent literature (NPL) are analyzed to make comprehensive grasp of knowledge spillovers [1] or to measure patent quality [2].

Pharmaceutical innovation is particularly important for drug discovery. There is a steady decrease in R&D productivity of drugs over the last number of years [10]. According to Scannell and Bosley, inflation-adjusted industrial R&D costs per novel drug increased nearly 100 fold between 1950 and 2010 [11]. R&D efficiency per billion US dollars of R&D spending has declined fairly steady, measured simply in terms of the number of new drugs brought to market by the global biotechnology and pharmaceutical industries. They call this trend "Eroom's Law in Pharmaceutical R&D" [12]. On the other hand, several cases of success have been found recently. For example, the reports present that drugs sourced via open innovation have a higher chance of later-phase clinical success, among 281 biopharma companies, between 1988 and 2012 [13].

We present an analysis of knowledge flows in the pharmaceutical innovation process. Backward citations, citations to NPL and forward citations that link patents, scientific papers and pharmaceutical pipelines data are analyzed and visualized to provide a more holistic understanding in hotspots of R&D. Because new drug discovery is the global issue and based on science knowledge, like biotechnology and chemistry, the analysis related to this field is eligible for science and innovation for global challenges.

3.1. Methodology

3.1.1. Dataset and its preparation

The datasets below by Thomson Reuters, which is covered from 1981 to 2011, are prepared. (1) For patent data, the Derwent World Patents Index (DWPI). (2) For papers' data, the Web of Science (WoS) database. (3) For drug pipeline data, the Thomson Reuters Cortellis for Competitive Intelligence database (hereinafter "Cortellis") including detailed information of drugs. (3) For citations data, the Derwent Patents Citation Index (DPCI). (4) For linkages between patents and papers, the WoS-DPCI Linktable computed by Thomson Reuters and JST that prepares backward citation data from patents to the NPL (non-patent literature) from the DPCI. Data were prepared on December 11, 2013.

Because of interest to find patents and their relationship to the non-patent literature in pharmaceutical fields, we prepared all 833,376 patents having one of the IPC (International Patent Classification) codes A61P that represents "specific therapeutic activity" from the Derwent Patents Citation Index and also we extracted their citations from DPCI, and we named them as "Pharma_Patents." Then, we prepared 57,800 patents that are linked to drug pipeline data from the Cortellis, named as "Drug_Patents (DP)." Then, the DP were subtracted from the patents having A61P resulting in a dataset of 325,576 "Non-Drug Pharma Patents (NDPP)." In other words, NDPP is a patent that has the code "A61P" but is not linked to drug pipelines. **Figure 5** shows the relationship between DP and NDPP.

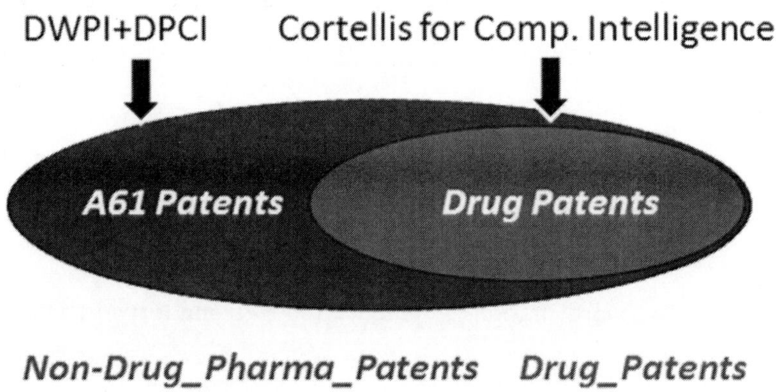

FIGURE 5. Relationship between Drug_Patents (DP) and Non-Drug_Pharma_Patents (NDPP).

Finally, all 115,252 NPL for Drug_Patents (DP) and 718,269 Non-Drug_Pharma_Patents (NDPP) were retrieved using the WoS-DPCI Linktable.

3.1.2. Calculation

As to patent quality analysis, patent family size, IPC counts, forward citations, backward citations, and citations to NPL were compared between 701 DP of random sampling and 701 NDPP for logistic regression analysis.

The citation lag: it can be calculated for forward citations to identify the speed by which patents are cited by future patens. It can also be used for backward citations to identify how prompt the existing works are cited by patents, see **Figure 6**. The citation lag is calculated as the average time gap of the years when the focus patent published (see **Figure 6**, patent A) minus the publication years of all cited works (see **Figure 6**, patents B, C, and NLP D). Similarly, the citation lag of all forward citations is defined as the average time gap of the publication years from all citing works (patents E–G in **Figure 6**) to the publication year of the patent A.

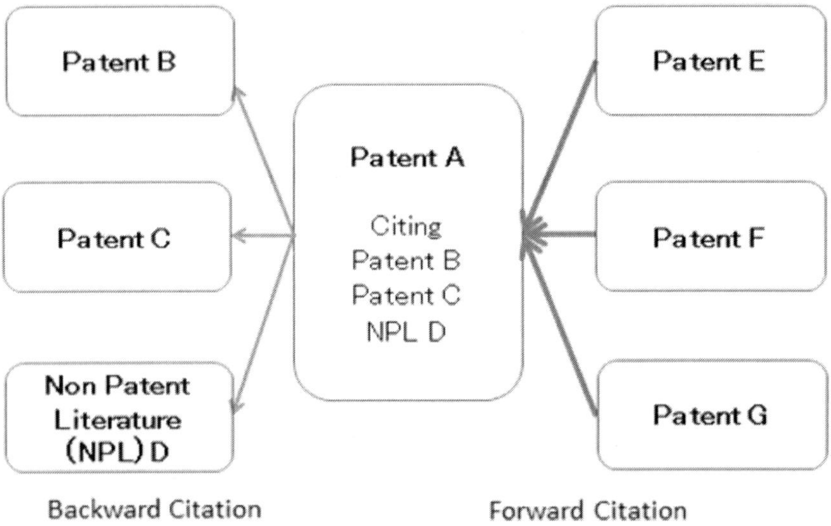

FIGURE 6. Backward citations to patents and NPL and forward citations to patents.

The generality index G_X: This is a quantitative index that represents the technical diversity of patents that are cited by a given focal patent (patent A in **Figure 6**). It also represents the technical diversity in terms of a group of patents that cite one focal patent. The count of different IPCs, which are associated with citing and cited patents, is used for calculation of the diversity. The generality index will be high if citing and cited patents occupy a wide spread of technology fields. For example, when x is the focal patent and y_i patents are citing the focal patent x, with i = 1, ..., N, then G_X can be calculated by following formula:

$$G_x = 1 - \sum_{j=1}^{M_i} \left(\frac{1}{N} \sum_{i=1}^{N} \frac{T_{ji}^n}{T_i^n} \right)^2 \tag{2}$$

Where T_i^n is the total number of IPC n-digit classes in y_i, T_{ji}^n is the total number of IPC n-digit classes in the jth IPC 4-digit class in y_i, and j = 1... M_i is the cardinal of all IPC 4-digit classes in y_i.

The index was calculated using all different 6-digit IPC subclasses for all patents in Drug_Patents (DP) and Non-Drug_Pharma_Patents (NDPP).

The subject index S_X: this is a new indicator we proposed. It is based on the generality index above, but the difference is that it is computed for NPL. For example, when x is the focal patent which cites y_i, i = 1, ..., N scientific papers (NLP), then S_X can be calculated as following formula:

$$S_x = 1 - \sum_{j=1}^{j} \left(\frac{N_{ij}}{N_i} \right)^2 \tag{3}$$

Where N_i is the total number of subject code in y_i and N_{ij} is the total number of subject code in jth the subject code in y_i.

The subject codes attached in each scientific paper are counted using basic element 1. The subject index was calculated for all non-patent literatures cited by patents in DP and NDPP.

The patent scope "SCOPEp": this is often associated with the technological and economic values of patents. It is said that broad scope patents tend to have a higher value (Lerner, 1994). For each patent P, the patent scope is defined as:

$$SCOPE_p = n_p; n \in \left\{ IPC_1^4; \ldots IPC_i^4, \ldots IPC_j^4; \ldots; IPC_n^4 \right\} \text{ and } IPC_i^4 \neq IPC_j^4$$

where n_p denotes the number of distinct 4-digit IPC subclasses listed in the patent P and is normalized. The patent scope was calculated for all distinct 6-digit IPC subclasses for all patents in Drug_Patents and Non-Drug_Pharma_Patents.

3.2. Analytical Results

3.2.1. Patent Quality

Table1 shows the results of logistic regression analysis. As a result, forward citations ($P < 0.001$), IPC count ($P < 0.001$) and also citations to NPL ($P < 0.05$) are significantly associated with patent quality. Therefore, forward citations, IPC count, and citations to NPL are emerged as new indicators for distinguish genuine patents that have strong linkage with R&D processes from other patents related to drug.

TABLE 1. Logistic regression analysis.

	Coefficient	Std. error	*z*-value	*P*-value	Sig.
Intercept	−0.05474	0.0824	−0.664	0.50638	
IPC count	−0.009	0.0018	−4.89	1.01E−06	0.1%
Forward cites by patent	0.020355	0.0031	6.645	3.03E−11	0.1%
Backward cites to patent	0.003728	0.0025	1.516	0.12958	
Backward cites to NPL	0.003718	0.0014	2.626	0.00864	1%
Patent family size	−0.00062	0.0041	−0.153	0.87859	

3.2.2. Citation Lag: Technology Delays

Comparison of citation lags for Drug_Patents and Non-Drug_Pharma_Patents shows the dynamics of knowledge spillovers. **Table2** represents that the forward citation lag of Non-Drug_Pharma_Patents is 2.17 years later on average while Drug_Patents are cited faster—after 1.89 years on average. Here, we can see the high-quality patent (Drug_Patents) tend to be referenced faster than other patent (Non-Drug_Pharma_Patents) The backward citations lag of Non-Drug_Pharma_Patents to patents is 3.4 years earlier on average and they go to much more recent non-patent literature (NPL)—published on average only 1.69 years earlier. Surprisingly, Drug_Patents cite older works comparing with Non-Drug_Pharma_Patents : cited NPL are 2.5 years old on average and cited patents are 5.64 years old. All values are statistically significant at the 1% level. In resume, we can see that Drug_Patents cover wider ranges and are cited more quickly comparing with Non-Drug_Pharma_Patents.

TABLE 2. Forward and backward citation lags.

	NDPP	DP
Forward cites by patents	2.17	1.89
Backward cites to patents	3.40	5.64
Backward cites to NPL	1.69	2.50

3.2.3. Generality Index, Subject Index: Knowledge Diversity

The generality index was calculated for 4- and 6-digit IPCs. It was analyzed for backward and forward citations, also for Non-Drug_Pharma_Patents and Drug_Patents, see **Table3**. Drug_Patents have higher generality index and subject index than Non-Drug_Pharma_Patents. That is, in general, Drug_Patents is based on wider "technologically based knowledge" and is cited by a wider range of set of patents that have more diversified IPCs. All values are statistically significant at the 1% level.

TABLE 3. Generality index.

		NDPP	DP
Generality index (4-digits)	Forward citations	0.36	0.37
	Backward citations	0.40	0.54
Generality index (6-digits)	Forward citations	0.46	0.50
	Backward citations	0.52	0.73
Subject Index	Backward citations to NPL	0.22	0.28

3.2.4. Scope: Technology Value

The patent scope was calculated for NDPP and DP, see **Table4**. Contrary to expectation, in the pharmaceutical fields, the scope of DP (Drug_Patents) tends to be narrower than that of NDPP (Non-Drug_Pharma_Patents). This is unexpected as, in general, patents linked to drugs are seemed more valuable than those not linked to drugs.

TABLE 4. Scope for Non-Drug_Pharma_Patents and Drug_Patents.

Scope	NDPP	DP
4-digit	0.13	0.11
6-digit	0.16	0.15

Some possibilities are conceivable below: (1) NDPPs are used for protect peripheral technologies surrounding one core DP and broader patents are useful to protect an inconsequential broad area, (2) because it takes long time to get an approval of drug, a patent owner has to obtain the other patent which is narrower than original patent, like second use patent or formulation patent, in order to extend a patent term, and (3) pharmaceutical companies might hide their core patent, therefore, Thomson Reuters cannot link drug pipelines to an appropriate patent and so on.

3.3. Knowledge Flows in Pharmaceutical Innovation

Our study compared and contrasted patents that are linked or not linked to drugs to understand knowledge flows and delays in pharmaceutical innovation. Results are summarized in **Figure 7**. As can be seen, Drug_Patents ("patents linked with drugs" in **Figure 7**) draw from a more diverse set of technologies and are cited more widely across the technology landscape. However, they tend to be more technically specialized (lower scope) than Non-Drug_Pharma_Patents ("patents linked without drugs" in **Figure 7**). Concerning citation lag, Drug_Patents tend to refer to older patents and scientific papers and are cited faster than Non-Drug_Pharma_Patents.

4. AN OVERVIEW AND FUTURE PROSPECTS OF PHARMACEUTICAL INDUSTRY

For the sake of providing evidence that contribute to policy making or strategy planning in national governments and pharmaceutical companies, we tried to show an overview and future prospects of the pharmaceutical industry based on data related to science papers, patents, drug pipelines and other various data of enterprises.

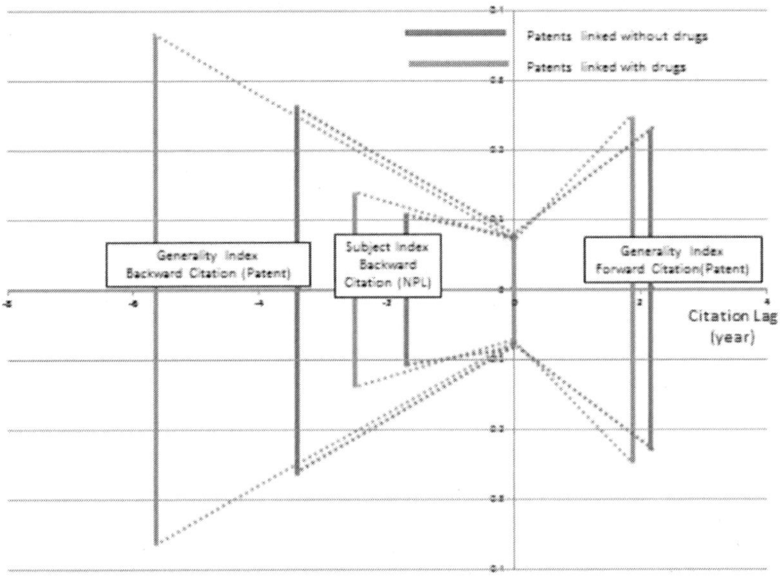

FIGURE 7. Comparison of Non-Drug_Pharma_Patents and Drug_Patents.

As to the evidence for policy making, there have been many analyses on current status, based on drug sales and R&D expense. For example, OECD has published the report overviewing the pharmaceutical industry in a global market based on the indicators of R&D expense, trade balance and term of drug approval in September 2009 [9, 14]. In addition, many reports on drugs of top sales by research companies and publications of in-house products by pharmaceutical companies have been published. What is more, the global competitiveness ranking of pharma companies is announced by collecting and arranging the information. Although these indicators represent the R&D capacity or competitiveness of the past and the present, they cannot foresee the future status.

We aimed to focus on the R&D pipelines that pharmaceutical company in each country have. The results indicated the R&D potential of each country in the current stage and in the future, by compiling the pipeline data in each R&D step, comparing the number of marketed pharmaceuticals and so on.

4.1. Drug Development and Pipelines

Drug development is the process of bringing a new pharmaceutical lead compound to the market. There are several stages in drug development. It starts from "Basic research" stage, then moves to "Pre-clinical," "Clinical

Phase," "Filed and Approved," and finally it goes to "Marketed" stage. Sometimes, it starts from intermediate stage like "Clinical Phase" when a company installs a lead compound from other company by licensing or M&A. Lead compounds in these stages are called as "R&D pipelines" or simply as "pipelines."

"Pre-clinical" is a stage of research that begins before clinical trials. Typically, animal testing and *in vitro* testing (test with microorganisms or cells) will be performed.

"Clinical Phase" is a stage of research on human participants, tested for safety and effectiveness in humans in order to be validated for a therapeutic use by a ruling authority of a government. "Clinical Phase" is composed of three stages. "Phase I" trials determine safety and dosing, usually in healthy volunteers. "Phase II" trials are the test in order to obtain an initial reading of efficacy and presumable safety in small groups of patients suffered by the disease targeted by the lead compound. "Phase III" trials are large and pivotal tests to decide its safety and also efficacy in sufficiently large groups of patients suffered by the targeted disease.

When these trials prove its adequate safety and efficacy, drug development goes to "Filed" stage, where a new drug application is filed to the ruling authority. After safety and efficacy are adequately confirmed by the authority, the application is approved. It means that drug development goes to "Approved" stage. When sales of the drug starts, the stage goes to "Marketed." Because the several stages above are need to be achieved in order to bring a lead compound to the market, it needs immense of R&D expense and dozen years by selecting an appropriate one from several tens of thousands of compounds and bring it to upper stages.

4.2. Methodology

4.2.1. Data acquisition and preparation

Two datasets of Evaluate and Thomson Reuters are used in this analysis. The dataset of Evaluate we used is "EvaluatePharma," which includes the pipelines data (about 45,000), licensing data, and M&A data from big pharma and biotech companies (about 7560 companies) in the world. It includes the data from 1986 to 2012. Also, the dataset of Thomson Reuters we used is "Cortellis for Competitive Intelligence," which includes timely global information on over 61,000 drugs, 6,000,000 patents, and 44,000 deals from big pharma to biotech companies.

4.2.2. Pipelines

Pipelines datasets from the EvaluatePharma of Evaluate are used in this analysis. The EvaluatePharma is the database containing R&D pipelines data, licensing data and M&A data, etc. in about 7560 pharma or biotechnology companies. We extracted 43,057 products, which are drug products and pipelines, related to small molecule drug or biomedicine covering 1986–2012 on May 2013. As to licensing data, we extracted in-licensed and out-licensed data covering January 2006–May 2013 from the EvaluatePharma. The Cortellis for Competitive Intelligence database by Thomson Reuters is also used in the pipelines analysis. Note that 21,086 pipelines are extracted and compiled on 11 December 2013.

4.2.3. Categories of business entities

Focusing on the sizes and categories of the business entities possessing pipelines, we classified 43,057 products data from the EvaluatePharma into eight categories, which are SMEs&VBs (Ventures), Majors, Generics, Specialties, Universities, Government, NGOs, and Others.

4.3. Analytical Results

Figure 8 represents the R&D pipelines and marketed drugs covering from 1986 to 2012 in terms of small molecule drugs. **Figure 8** reveals that the number of US "Marketed" is immense, Japan is the second, followed by Germany, Switzerland, the United Kingdom, and France. The same goes for bio-medicines that include recombinant product, bioengineered vaccine, monoclonal antibodies, cell therapy, gene therapy, and DNA&RNA therapeutics. In terms of bio-medicines, the USA is the most competitive among the countries [15]. Later analysis revealed that the R&D pipelines in Korea are also high, however, they include many generic drugs [16].

Then we focused on the types of business entities that have R&D pipelines.

The conventional R&D process was "**closed innovation**" which happened in closed environment, like inside of laboratories in major pharmaceutical companies, where they discovered a lead compound and brought it into the market by themselves. This was because they had strength in chemical synthesis and took advantage of the strength to R&D of small molecular drugs. By bringing a blockbuster to the market, they got huge income to

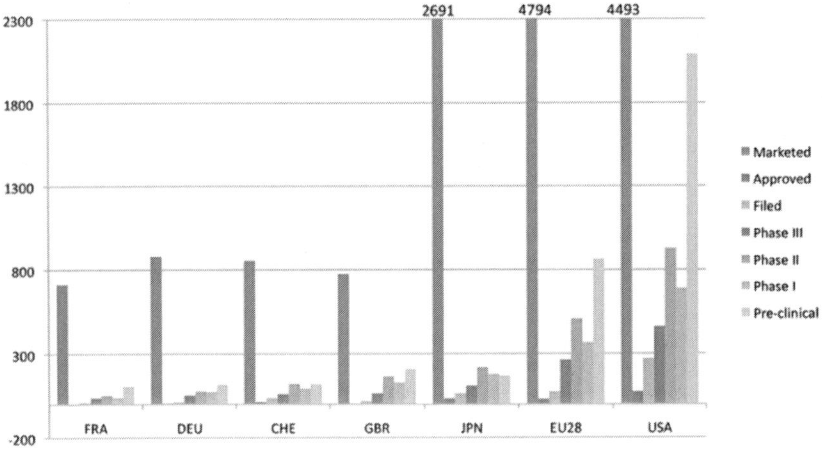

FIGURE 8. R&D pipelines and marketed drugs by country.

make an investment to the R&D of next drug. However, a new business model has recently been necessary for the pharmaceutical industry in order to cope with higher risk of drug development caused by longer term of R&D, increasing R&D expense. Examples are the capital expansion by a merger of major pharma companies and the role-sharing between major pharma companies and ventures. Recently "**open innovation**" has been attention-getting, which looks for the R&D pipelines outside of companies in order to reduce the higher risk. A report said that the success rate of drug R&D with open innovation would be three times as one with conventional R&D [13].

Therefore, we tried to find what kind of business entities have an important role in the drug R&D process, by analyzing the types, scales, and licensing activities of business entities having R&D pipelines. The business entities having R&D pipelines are divided into eight categories such as "SMEs, VBs," "Majors," "Generics," "Specialties," "Universities," "NGOs," "Governments," and "Others" by using the database EvaluatePharma. **Figure 9** shows licensing activity by country. The number of licensing in the USA is predominantly large. This is because the USA has the largest number of pipelines and its market is the biggest in the world. Following to the USA, the number of licensing activities in Japan is second large, but about one-third of the USA.

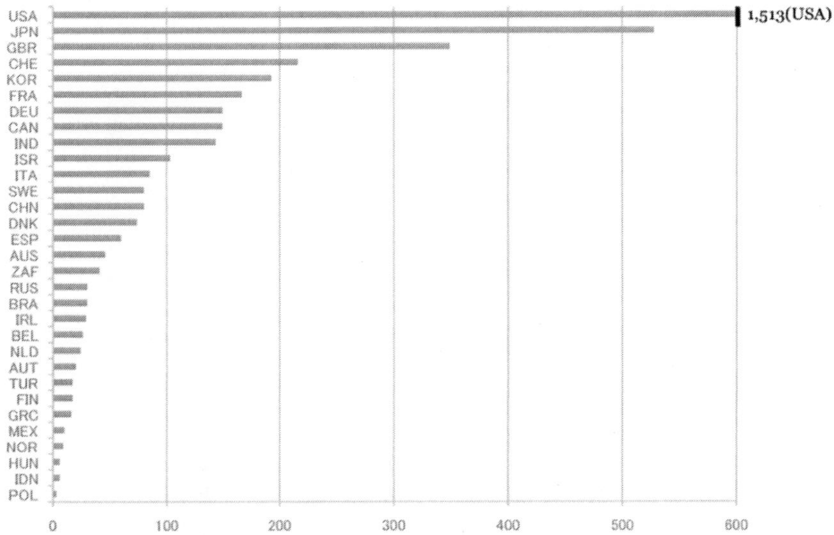

FIGURE 9. Licensing activities by country.

Then, **Figure 10** shows licensing activities by categories of business entities by country. Although the largest number of licensing in the USA is "SMEs, VBs," the largest number in Japan and EU countries are "Majors." As we show the predominance of the USA in **Figure 1**, it seems that a key role of the predominance is strong activity of SMEs and ventures. Later analysis revealed that the pipelines in the USA flowed not only from a university to a major via SME&VB, but also they flowed from a major to a SME&VB and from an SME&VB to other SME&VB. The multilateral flows of pipelines constitute, so to speak, "the roundabout of drug R&D" and that is the strength of the USA [16].

4.4. Drug R&D Status By IPC

The International Patent Classification (IPC) has a subclass "A61P" which represents "specific therapeutic activity of chemical compounds or medicinal preparations." We analyzed the R&D status by providing IPC subclass to the therapy field that each pipeline has.

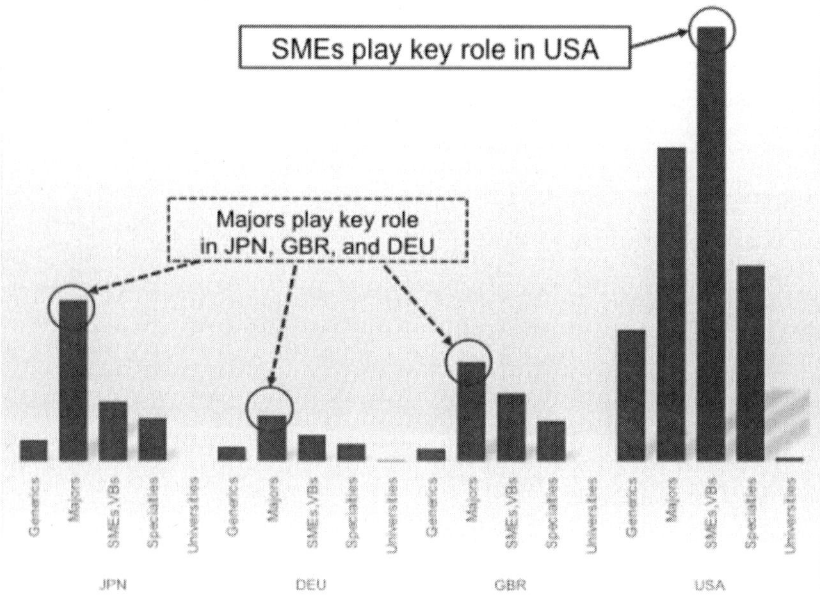

FIGURE 10. Licensing activity by category of business entities.

Figure 11 shows the origin countries of marketed drugs by IPC. We prepared the drug data of the countries (the USA, Japan, the UK, Switzerland, Germany, France, and Korea) that have many pipelines. As to the marketed drugs, the fields of infections and cancers are hotspots. The number of Japan's original drugs is also relatively large and the ratio to total is about 20%, although the number of the US origins is the largest (about 44%). **Figure 12** shows the origin countries of R&D pipelines by IPC. In order to overview the present R&D status, **Figure 12**does not include the marketed drugs. It reveals that, as to drug candidates that are in process of R&D, the fields of cancers, infections, and mental disorders are largest. It also reveals that the pipelines of the US origin are largest.

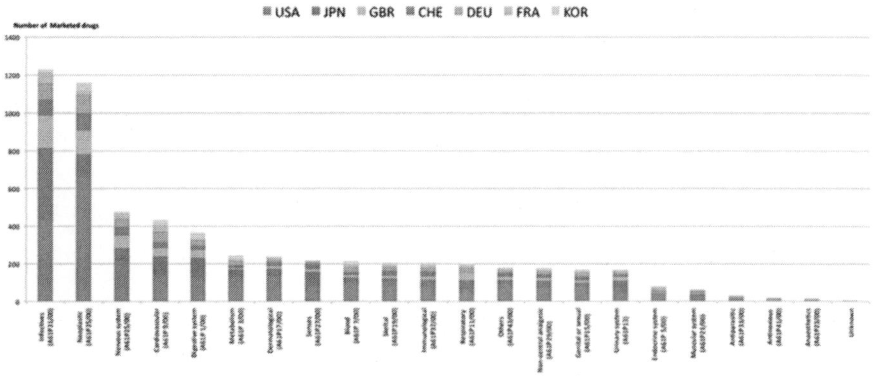

FIGURE 11. Origin countries of marketed drugs by IPC.

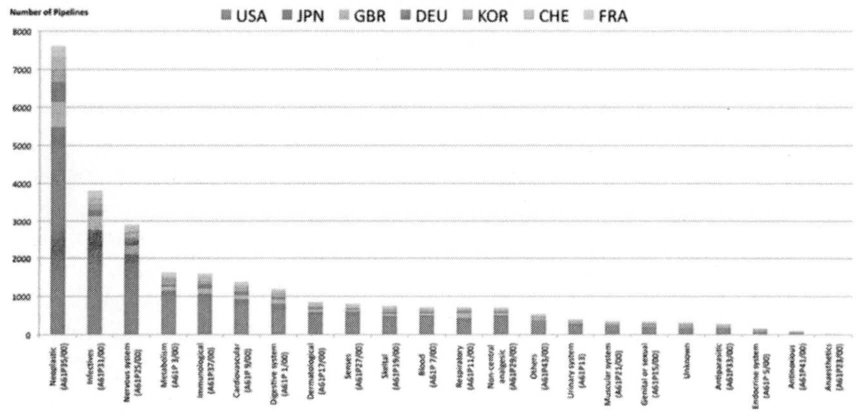

FIGURE 12. Origin countries of pipelines by IPC.

Cancer is the top rank in cause of mortality among developed countries. It is also the top rank in the DALY (disability-adjusted life year; a new metric based on the sum of years of potential life lost due to premature mortality and the years of productive life lost due to disability) and in the YLD (years lived with disability) [17, 18]. We can see that anti-cancer drugs have been under development by many countries.

As to infection, IPC code A61P31/00 includes antivirus for HIV, influenza, antimicrobials, and antifungal. Infections are a major cause of high mortality and low DALY among developing countries. Disorders of the nervous system (A61P 25/00) include neurodegenerative disorders like

Alzheimer's disease or Parkinson's disease, antidepressants, and antipsychotics like schizophrenia. Comparing to "Marketed," the number of pipelines is larger, especially in "discovery" stage that represents basic research and pre-clinical stage. It is noted that this is the field with high future growth potential.

5. FINAL DISCUSSION

This chapter discussed multilateral analysis on IP rights: (1) a new indicator "Innovation Front" and its use, (2) analysis of patent quality, and (3) future prospect in the pharmaceutical field.

In the part (1), a new indicator "Innovation Front" showed scientific hotspots by focusing on "Technical closeness," that is to say, by means of making clusters by co-citation analysis between scientific papers and patents. It shows that the science fields of induced pluripotent stem cells (iPS cells), aptamor, adipocytokine, meta-material, and microRNA are the hotspots. In the part (2), we prepared patent datasets that are linked or not linked to drug pipelines in order to understand knowledge flows in the drug R&D field. The results indicate that DPs (Drug_Patents) are based on a wider range of technologies and are cited wider technology landscape. Contrary to expectation, they have a narrower scope. It means that they tend to be more technically specialized than NDPP (Non-Drug_Pharma_Patents). Concerning citation lag, DPs seem to refer to older documents and are cited faster than NDPPs. In the part (3), we showed an overview and future prospects of pharmaceutical industries, focusing on drug pipelines, size of business entities, and the International Patent Classification (IPC). The intricacies of patenting of pharma products by country-based and business entity-based were discussed.

Since many years, diverse studies have been conducted to study the origin, trajectory, and destination of knowledge spillovers in the science and technology system. The work presented here also contributes to a study for innovation analysis by showing a newly developed indicator (Innovation Front) and a new way of analysis, like comparison of Drug_Patents to Non-Drug_Pharma_Patents. Our next approach will be a linkage of "trademark" database to other database like patents and drug pipelines and also "financial analysis" linked to IP rights.

REFERENCES

1. Lukach R, Plasmans J. Measuring knowledge spillovers using patent citations: Evidence from the Belgian firm's data. CESifo Working Paper NO.754 Category 9: Industrial organization; July 2002
2. Squicciarni M, Dernis H, Criscuolo C. Measuring patent quality: Indicators of Technological and Economic Value. Vol. 3. OECD/DSTI/DOC;2013.
3. Jibu M, Osabe Y. Knowledge flows and delays in the pharmaceutical innovation. Business Management Studies. 2016;2(1):88-94
4. OECD. OECD Science, Technology and Industry Scoreboard 2013: Innovation for Growth. Paris, France: OECD Publishing; 2013
5. Narin F, Pozek RP. Bibliometric analysis of US pharmaceutical industry research performance. Research Policy. 1988;17:317-330http://dx.doi.org/ 10.1016 /0048-7333(88)90039-X
6. Narin F, Hamilton K, Olibastro D. The increasing linkage between US technology and public science. Research Policy. 1997;26:317-330. http://dx.doi .org/ 10.1016/S0048-7333(97)00013-9
7. Jibu M. Analysis achievements of JST operations through scientific patenting: Linkage between patents and scientific papers. In: Atlanta Conference on Science and Innovation Policy, IEEE; 15-17 Sept. 2011, Atlanta, GA, USA. pp. 214-220. http://dx.doi.org/10.1109/acsip.2011.6064473
8. Yamaguchi E. Innovation theory leading to breakthrough—How will the memory device industry face the "post-Moore" future? Oyo Buturi. 2010;79(12):1077-1083
9. OECD. Compendium of Patent Statistics, 2008. Organization for Economic Co-operation and Development; 2008.
10. Booth B, Zemmel R. Prospects for productivity. Nature Reviews Drug Discovery. 2004;3:451-456
11. Scannell JW, Bosley J. When quality beats quantity: Decision theory, drug discovery, and the reproducibility crisis. PLoS ONE. 2016;11(2):e0147215
12. Scannell JW, Blanckley A, Boldon H, Warrington B. Diagnosing the decline in pharmaceutical R&D efficiency. Nature Reviews Drug Discovery. 2012;11:191-200
13. Deloitte Development LLC. Executing an Open Innovation Model: Cooperation is Key to Competition for Biopharmaceutical Companies [Internet]. 2015. Available from: https://www2.deloitte.com/content/dam/Deloitte/us/Documents/life-scienc es-health-care/us-lshc-open-innovation.pdf(Accessed: January 19, 2017)
14. Lerner, J. (1994). The Importance of Patent Scope: An Empirical Analysis. The RAND Journal of Economics. 25(2):319-333.
15. OECD. OECD Health Policy Studies: Pharmaceutical Pricing Policies in a Global Market. OECD [Internet]. 2008-09-24. Available from: http://www .oecd. org/els/pharmaceutical-pricing-policies-in-a-global-market.htm (Accessed: January 19, 2017)
16. Osabe Y, Jibu M. Development of new indicators for the launch of a Japanese version of the NIH (1): An overview and future prospects of pharmaceutical industry based on new indicators, Journal of Information Processing and Management. 2013;56(7):448-458
17. Osabe Y, Jibu M. Development of new indicators for the launch of a Japanese version of the NIH (3) : Analysis of entities conducting pharmaceutical R&D activities, Journal of Information Processing and Management. 2013;56(10):685-696
18. World Health Organization (WHO). About the Global Burden of Disease (GBD) Project [Internet]. Available from: http://www.who.int/healthinfo /global _burden_disease/about/en/index.html, (Accessed: January 19, 2017)

CHAPTER 6

Game Mechanics and Technological Mediation: An Ethical Perspective on The Effects of MMORPG's

Christian Klemm[1] Wolter Pieters[2]

[1]*TH Köln, Steinmüllerallee 1, 51643 Gummersbach, Germany*

[2]*Delft University of Technology, University of Twente and TH Köln, TU Delft, TBM -3S, P.O. Box 5015, 2600 GA Delft, The Netherlands*

ABSTRACT

In the past decades, video games have grown from a niche market to one of the major entertainment media, enticing millions of players worldwide. When ethical aspects of video games are being debated, the discussion oftentimes revolves around effects of their content, such as violence. This paper argues that effects of game mechanics, such as reward mechanisms, should be considered as well, as these are at the core of the appeal of games. We analyze the ethical dimension of behavioral game design present in Massively Multiplayer Online Role Playing Games (MMORPG's). Using the framework of technological mediation, we show how opaque and seductive game mechanics can invite problematic usage patterns, such as excessive use with negative effects on well-being, and how designers can take responsibility for morally acceptable impact of their games. Having a practical focus, the paper concludes with several proposals for better design.

Keywords: Behavioral design Choice architecture Design ethics Game mechanics MMORPG Technological mediation

INTRODUCTION

Problem statement

Currently, a major part of the ethical discussion about video games revolves around their content and its impact on users. Just like with books or music, video games are often being evaluated by the level of violence or sexually explicit character design (e.g., Anderson and Bushman 2001; Anderson and Dill 2000; Copeland 2004; Scott 1995; Sherry 2001). "Morally problematic" games are then the ones with substantial violent or sexual content, and the associated risk is that such content will adversely affect the real-life behavior of players.

However, such a discourse leaves out an important aspect at the core of the appeal of video games. Many effects of video games on users can be attributed to game mechanics rather than content. For example, the design of reward mechanisms may affect the way in which users interact with the game, and the duration of play. This is particularly visible in Massively Multiplayer Online Role Playing Games (MMORPG'S). While all games revolve around the player and share the general "psychology underlying how players learn and react to the game" (Hopson 2001), the oftentimes subscription-based business model of MMORPG's makes it necessary for game designers to keep players enticed for a longer period of time. While this could be achieved by better visuals, story and diverse gameplay, it can also be achieved by limiting the speed of progression and carefully working with reward mechanisms that influence the playing habits without the players being aware of it.

At the same time, MMORPG *content* is very diverse, ranging from hardcore science fiction with ego shooter elements to a persistent Lego Universe for a younger audience (game service shut down in 2012). You could play some MMORPG's and spend most of the time healing other player's characters or becoming a rich space-merchant without ever harming any virtual characters. While other types of games such as ego-shooters almost always include some form of violence, MMORPG's can hardly be lumped together and evaluated only based on their content.

For these two reasons, the game mechanics are a central theme in an ethical evaluation of MMORPG's. This paper will look at the problem space outlined above from the perspective of computer ethics, in particular focusing on the element of *behavioral design* inherent in such games.

Behavioral design is user-centered design beyond simple usability of a product (or game), and encompasses design for behavioral-level responses (Norman and Ortony 2003, p. 4). Video game design is generally quite prescriptive when it comes to the options on how users interact with the game and what choices are available, which may have negative effects on users. Addressing game design ethics from this game mechanics perspective complements the discussion on game contents such as violence. In addition, this perspective is relevant in relation to the broader philosophical question on how technologies co-design behavior. Whereas Sicart (2009) discusses mechanics mostly in relation to specific morally sensitive behaviors in games (again for example violence), we focus on the attention issue, which is essential for MMORPG's. Compared to Yee (2006a), our focus is more on design ethics than on the psychology of adverse effects as such. Our main questions are how this ethical dimension of game mechanics design can be understood, which exemplary problems exist in current designs, and what possible solutions might look like.

To comprehend the interconnected set of issues and possible problem-fields, i.e. influencing behavior and the underlying game mechanisms, this paper will engage with related literature regarding ethics, social sciences, psychology and game design. Given the topic and practical focus, we also use Internet sources as a basis for identifying concerns. Our target audience includes game designers as well as scientists from various disciplines, so the depth of discussion of ethical theories may be less than in papers targeted at a philosophical audience.

After a short introduction to the genre, potentially harmful effects of MMORPG's and usage patterns are introduced. Section "Materials and methods" then provides the theoretical framework and methods applied from the perspective of ethics of technology. The concepts of technological mediation, choice architecture and behavioral design are presented as explanations of the influence of design on user experience and behavior.

In the ensuing evaluation ("Ethical evaluation" section), we will focus on ethics of the good life as a means to evaluate such influences. Mechanics receive strong attention because of the behavior they evoke in a user, but also because of the oftentimes opaque way that they can exert the influence. This provides the backdrop to assess the responsibilities of game designers, who are responsible for the design of mechanics and the way they are implemented. Focusing at design choices with potential for causing negative effects such as addiction, the "Better design" section will investigate how those effects could be engineered out or at least be suppressed. To come up with innovative proposals for better design, we will apply the concepts of technological mediation and choice architecture to the design process of an MMORPG.

Massively multiplayer role playing games (MMORPG's)

MMORPG's are online games in which a large number of players interact in the role of a character in a virtual world. In the past decades, MMORPG's have developed from a niche in the gaming market to a highly successful genre, enticing millions of players around the world. As the business model of many MMORPG's relies on monthly subscriptions, it is paramount for their design to keep a user's attention for a long period of time. In some cases, this can lead to users excessively investing time and money for their hobby. MMORPG players with problematic usage patterns have not been associated with school shootouts, but are confronted with more inward-directed adverse effects. The genre itself and some of its more prominent games are the subject of extensive studies that cover nearly all aspects from demographics and social behavior or even in-game economics (e.g. Taylor 2006; Feng et al. 2007; Papagiannidis et al. 2008).

MMORPG's evolved from so-called "muds" (massively multiplayer text-based online RPG's), being run on first-generation computers around the 1980s (Koster 2001). With the evolution and popularization of the Internet and video games, the genre ultimately grew above its original community of "nerds" and sported graphical online environments with thousands of simultaneous players (Yee 2006a, p. 3). The general core mechanics that can also be found in classic table-top role playing games are still in place to date, though: "a combination of interactive story-telling and logistical optimizations under the guise of slaying monsters and attaining higher levels and skills" (ibid.). Users can create a virtual character with specific attributes to interact with other players and "achieve structured goals" (ibid., p. 2). Modern MMORPG's include state-of-the-art graphics and include the latest features of game design and (social) interactivity with the vast game world.

The classical business model of MMORPG's is subscription-based (Koster 2001). Users usually have to buy the game or game client and then pay a monthly fee to play the game. To recover the substantial development and infrastructure costs, it is important for the game publishers that users stay attracted to the game for a longer period of time measured in months rather than days or weeks. Newer business models try to attract a larger player base by offering the game for free, but only sporting a limited experience or slowing progress to a slog, unless the player buys certain boosts or items from a shop. This system is called "free to play" and is financed through so-called micro-transactions.

Effects of MMORPG's on users

Addiction

Probably the most obvious and visible effect that playing an MMORPG can have on a user is the consumption of time. Yee states that MMORPG players spend an average of 22.72 h a week with their game and 70% of his respondents spent, at least once, 10 h straight playing the game (Yee 2006a, p. 10, 22). While 22 h spent on a single activity each week is already quite a lot, 8% of the users even invest 40 h or more each week. More than half of the respondents considered themselves addicted (ibid., p. 23). In the worst cases, they may withdraw from their social contacts, neglect hygiene, diet and ultimately risk their health. In several cases, extended play of MMORPG's has even led to the death of people (Blackburn 2013). Motivation for and importance of playing may vary considerably, making the time spent on gaming not the major factor for addiction (Griffiths 2010).

The complex of addiction, also in regard to technology and the Internet and gaming as well as its comparison and classification within the topic area is the subject of wider research (Shaffer 1996; Ng and Wiemer-Hastings 2005; Ko et al. 2009). Already in 1995, it was estimated that the (offline) gaming behavior of 7.5% of 11–16 year olds might be considered addictive (Phillips et al. 1995). Later, others addressed the relation of gaming addiction with physical and mental health (Mentzoni et al. 2011), similarity to other addictive behaviors (Gentile et al. 2011), and systematic overview studies (Kuss and Griffiths 2012). A major difference between game addiction and some other addictions is the fact that the causes of addiction are due to design activities rather than natural properties. Whereas certain natural phenomena have *inherent*addictive properties, behavioral game design involves *imposed* addictive properties (Blakley 1996).

Addiction is defined by Yee as "a recurring behavior that is unhealthy or self-destructive which the individual has difficulty ending" (2002, p. 2). When comparing Yee's definition of MMORPG addiction to Hyman's definition of drug addiction "as compulsive drug use despite negative consequences" (2005, p. 1414), both definitions share that the addiction is not controllable by the subject and has negative consequences. While not encompassing the component of possible physical dependences like drug use or smoking, psychological dependence and the negative consequences are palpable. Looking at the possible consequences, a paper by Dr. Kimberly Young, clinical director of the Center for Internet Addiction Recovery lists negative visible effects that come with addiction to online games, including hiding or lying about the gaming, withdrawal from family and friends as well as other activities (2009). In extreme cases, players might even neglect their diet and hygiene, putting their health in danger (ibid.). When asked, nearly 20% of players answered that their "playing habits have caused [...] academic, health, financial or relationship problems" (Yee 2006a, p. 22) due to the large amount of time spent playing. This paper will use the terms of *problematic usage patterns or behavior* when talking about playing behavior that results in negative consequences for the player, regardless of the actual time spent with the game.

Motivation and attraction

Yee stresses that game mechanics and their design are only partly responsible for problematic usage. The users themselves and their motivations for playing also need to be taken into account. Yee distinguishes between the spheres of attraction addiction and motivation addiction, also listing the resonating attraction factors. Attraction addiction encompasses the sphere of the game itself, its content and mechanics as well as social factions, whereas motivation addiction focuses on the internal human condition that may cause addiction. It is noteworthy that attraction addiction does not need a matching motivation factor, whereas motivation addiction only comes into existence when the game offers a matching attraction factor for compensation. This view is further confirmed by a qualitative analysis by Hussain and Griffiths (2009), showing that addicted gamers play to relieve negative moods and emotions. In Germany, a study found out that gaming addiction is connected to higher anxiety and depression rates, present in up to 3.5% of the younger internet users (Peukert et al. 2010). An interesting study by Ko et al. (2005) looked closer at demographics as well as other factors related to online gaming addiction in Taiwan. One of the key findings was that addiction correlated with lower self-esteem, a latent dissatisfaction with life and mostly with older age. With this kind of motivation addiction, it is rather likely that the user would focus on another thing when games are not available as the main factor is intrinsic to the player. Nevertheless, game designers should be aware of potential misuse or potential in their product.

A user might play an MMORPG not only for entertainment, but to "empower them in ways specific to their anxieties" (Yee 2006a, p. 23). A shy and introvert person with few social ties for example "can become needed and valued members of groups and guilds" (ibid.). In this example, the motivation factor of missing social interaction finds a match in the attraction factor of possible relationships the game offers. It would be a point for further research to examine the correlation between motivation and attraction addiction further and look closer at users suffering from game addiction to find out to what extent factors for addiction are already present in players and what factor the video game and mechanics play. Ng and Wiemer-Hastings (2005, p. 112) argue that players who "could be labeled as anti-social or introverted by most [...] spend their social time and energy in-game rather than socializing in the real world".

In the context of this paper, we will mainly focus on the domain of the game itself and its design ethics, and therefore on the aspect of attraction addiction and evaluate its consequences from an ethical perspective. We will also discuss ways of how game design could possibly help avoid problematic usage that is being caused by attraction addiction. In order to enable such an evaluation, we first discuss our method.

MATERIALS AND METHODS

Before an ethical evaluation of the problem space outlined in the previous section is conducted in the "Ethical evaluation" section, this part of the paper first provides an overview over concepts and ideas dealing with ethical design of technologies, and assesses the way they apply to the realm of MMORPG's. This serves as the backdrop for the upcoming evaluation.

Technological mediation and choice architecture

Two main theories for discussing effects of technology design on human experience and behavior are technological mediation and choice architecture. In the paper "Materializing Morality", Peter-Paul Verbeek presents the concept of technological mediation (Verbeek 2006). From an ethical perspective, an engineered artifact is not completely neutral of values. The engineer puts a "script" to the artifact, encouraging a certain way of behavior in the user or society itself (a speed bump, for example, forces a driver to drive slower). Such scripts may be the result of conscious design as well as unintentional side effects. Although technology and its design can be used for a variety of purposes that cannot be completely anticipated once it has been released to the public use, a part of responsibility for its use rests with the engineer. He can design a technology to serve as a mediator; encouraging a certain desired behavior and discouraging misuse (see Verbeek 2006, p. 3, 10). As the reward mechanisms constitute a script, this theory applies directly to behavioral game design.

Excluding any specifications that might be set by the publishing company, a developer of a MMORPG is able to modify and script the virtual world he is creating as desired, thereby imposing properties that mediate the behavior of users. The context of use (gaming for entertainment purposes) is very narrow. Taking ethics into account, the game design could incorporate the concept of technological mediation to design its core mechanics, by evaluating the mediating role of the mechanics in the experience and behavior of the gamer.

Acknowledging the mediating role a technology or video game can have, it is necessary to look at the way that "scripts" outlined in the paragraph above can be consciously integrated during the design process. One way of thinking about this is in terms of how choices of users (e.g., in terms of time spent) are influenced by the design. This can happen via deliberate choice architecture, setting a certain environment for the user of an artifact. The person creating this environment is a "choice architect" (Thaler et al. 2010, p. 1), *imposing* the choice architecture on the game. For example, better accessible options are more likely to be chosen, both in supermarkets as well as in online environments. Clever and ethical choice architecture tries to help people to "do the right thing" and avoid any harm for the user without applying force. This could be accomplished by providing sound feedback mechanisms or engineering in incentives that reward desired behavior, such as the pursuit of responsible and moderate playing times. Thus, the problematic effects of the technological scripts could be alleviated by providing better choice architectures in the game design.

Behavioral game design

The effect of design on behavior may occur via emotions. Norman and Ortony point out that design can influence user emotions on three levels: visceral, behavioral and reflective. The first mostly covers style and surface and triggers rudimentary, unconscious emotional reactions, whereas reflective level design affects self-reflection as well as aspects of "pride of ownership" (Norman and Ortony 2003, p. 3–5). In our context, behavioral level design and responses are most crucial, as it focuses on "function and use of a product" (ibid., p. 4) and affects the area where learned sub-conscious "skills and routine behavior reside and are controlled" (ibid.).

With a few exceptions and varying characteristics, video games are designed to be attractive and entertaining for their users, which is reflected in their choice architectures and mediating effects. There is a continuing evolution in technology, but the underlying psychology remains constant (Hopson 2001). To better grasp the mechanics at work that evoke problematic usage, this subsection takes a closer look at behavioral game design and psychology. The latter can easily be applied to a variety of fields, including the design of MMORPG's.

Rewards

Part of the cause for problematic playing patterns can certainly be understood as mediating effects of the game mechanics themselves:

Goals and rewards in MMORPGs typically use a random-ratio reinforcement schedule based on operant conditioning. Early achievements are quick, almost instantaneous, and gradually take more and more time and effort until progression becomes almost imperceptible. (Yee 2006a, p. 4)

The operant conditioning is at the heart of behavioral game design we see applied in MMORPG's. The psychologist B. F. Skinner conducted experiments on rats that were rewarded when performing certain tasks in their boxes, for example pressing a lever. After going through the reward circle for some time, the rat would still press the lever, even if not being rewarded every time (Skinner 1938; Yee 2001).

In an article on behavioral game design that builds upon cognitive psychology and applies Skinners research, John Hopson (2001) exhaustively explains the concept of behavioral game design and the various factors that designers can use to influence the "pattern of activity" that they want from the players. While the rat is rewarded with food pellets, the game offers human players other kinds of rewards such as "power-ups" or a level gain for their character. To gain the reward, the player usually has to conduct a set of learned actions. Reward mechanisms can thus be very powerful choice architectures.

In most games, rewards are being provided at a fixed or variable rate and interval. MMORPG's extend these parameters by random elements, for example in random item drops after slaying a monster and elements of chance in crafting success. Aforementioned operant conditioning is most successful when using a random ratio schedule for rewards. MMORPG's actively exploit these mechanics to keep players attracted and adjust ratios and intervals for rewards in a way that maximizes the player's activity in the game and keep him or her attracted over a longer period of time. Working with these mechanisms needs careful balancing, as the application of rewards plays a major role when it comes to addiction, as it is a chronic disease of "brain reward, motivation, memory and related circuitry. [...] This is reflected in an individual pathologically pursuing reward and/or relief by substance use and other behaviors." (ASAM 2011).

Naturally, reward mechanisms make sure we survive as individuals and species, by reaching "survival-relevant natural goals [...] pursued with the anticipation that their consumption (or consummation) will produce desired outcomes (i.e., will "make things better")" (Hyman 2005, p. 1414). A human being would even take some pains like setting out for a hunt or going to work every day to obtain the desired rewards, oftentimes resulting in complex "behavioral sequences" that become automated to achieve the goal and rewards. The same sequences are exploited in games. In order to accomplish new goals, the players need more experience, more money, etc., requiring them to chase all of these at the same time: "You can't keep up with mobs if you level but don't buy new gear. You can't continue blacksmithing if you run out of money. What this means is that you're always close to a goal—a reward. You are seldom far away from all possible rewards." (Yee 2001, p. 71).

When it comes to addiction, the behavioral patterns of (e.g., drug) addicted beings become so strong as "to supplant almost all other goals" (ibid., p. 1415). Gamers who are showing problematic user patterns also show core features of addiction, including symptoms of withdrawal, craving and relapse as well as modifications of mood and the development of tolerance to the subject (Chappell et al. 2006). In a sense, game developers are creating a "Skinner Box [that] is tailored to its host's needs and reinforcement schedule, and where individuals can interact with each other without sacrificing the integrity of their own construct." (Yee 2001, p. 72).

Social interaction

In an MMORPG, social interactions between players make up a huge part of the user experience, and often cooperation with other players is necessary to advance (Yee 2006a, p. 4). The most basic interactions are talking to other players via chat, trading items, fighting each other or forming groups to accomplish quests. Designers further encourage the latter by setting up specific quests that can only be accomplished with a (sometimes large) group of other players. A lot of effort can be put into the design of the quests, making them a rewarding experience, also in terms of experience points and virtual items for the player's character. Oftentimes, this results in the formation of online friends (Yee 2002; Søraker 2012a). More or less fixed playing scheduled together with the need to keep pace with or outpace the peers leads to a higher time investment.

Many modern MMORPG's include another layer of social interaction, binding players more strongly to the game: guilds (Yee 2006b). Guilds are comparable to an association or club, sporting general members and people in hierarchical functions, e.g., a guild leader. The scale of a guild can vary from several hundred members from all around the world to a small group of friends that know each other personally. In either case, the guild contributes meaning and a sense of belonging to something bigger. Game design can further enhance that feeling by offering features such as special guild emblems, the possibility to acquire virtual real estate or offering special rewards for guilds, which also create choice architectures for players. Oftentimes, guilds would meet at specific times to play the game together. In-game social structures may also emerge in game culture without being specifically designed, but with similar implications (Taylor 2006).

Being engaged in a guild or other in-game social structure does also come with certain responsibilities and could evoke subliminal peer pressure: A quote from a guild leader in Nicholas Yee's Daedalus project, exploring the psychology of MMORPG's is a good example: "[...] after a while you just feel that you have to log on so that you don't let down the other people in the guild, sometimes even if you are not really in the mood to play the game" (Yee 2005). The guild itself can transcend the game and become "a motivation for playing" (ibid.). While the first MMORPG's offered none at all or only very basic mechanics for representing virtual groups, game designers now explicitly engineer in features that make their games appealing to social gamers and guilds.

SUMMARY

The effect of a user playing MMORPG's for a long time, not considering buying or playing other games is a direct consequence of the game mechanics, deliberately engineered into the game. At the same time, the mechanics that evoke the desired behavior also encourage a problematic level of usage in a significant part of the players. Up to this point, we have discussed possible adverse effects of MMORPG design, and the psychological (design) principles underlying such effects, notably conditioning, reward mechanisms and peer pressure. Our focus is on the relation between design ethics and such adverse effects. This relation can be understood in terms of technological mediation and choice architecture, which describe how user behavior is influenced by design decisions. Before discussing the role of the designer, the abovementioned effects require ethical evaluation, to enable discussion of possible alternatives, which we will take up next.

Ethical evaluation

Following the presentation of existing concepts for ethical design of technologies in the previous section, this section provides an ethical analysis of the mechanics leading to potentially harmful usage patterns of

MMORPG players. Although technological mediation and choice architecture provide a framework to discuss effects of designs, they do not necessarily provide means to discuss the desirability of such effects. In this section, we assess several approaches that would enable such an evaluation, and then choose ethics of the good life as the most promising one. This approach encompasses not only the results of the behavior the mechanics evoke, but also takes the way they are implemented and transparent to the user into account. The last section will focus on the creators of said mechanics. There, we will discuss the responsibilities of the MMORPG designer and evaluate whether he or she can at least partly be held accountable for the effects that emerge from a user's interaction with the game.

Classical theories

To sketch the ethical issues at hand, we will first look at classical ethical theories. Although classical ethical concepts have been developed without modern IT in mind, they can actually be applied to this paper's problem space, but may be somewhat rigid in their focus.

From a deontological perspective, the theories of German philosopher Samuel Pufendorf (1672) deal with the "foundational principles of obligation", regardless of effects. In the context of this paper, the main question is whether MMORPG's have a negative impact on the fulfillment of duties classified by Pufendorf. Problematic usage patterns and addiction to playing an MMORPG would mean the user is no longer able to completely fulfill the duties of soul (including skill development) and the duties of the body, in particular not harming oneself. Focusing on the duties of soul, it can be argued that the game does actually promote skills and talents in some cases. Someone leading a guild might become more responsible and learn about organization and leadership. Someone who's usually shy might become more outgoing and find it easier to communicate with others, even in the real world.

Also on the positive side, playing an MMORPG brings millions of people pleasure and fun or help people to socialize with others. Long-lasting friendships or even relationships can be sparked. From a utilitarian perspective, these effects have to be balanced against the significant and severe possible negative effects that MMORPG's can have on their users, as they may even threaten their physical well-being (Dodig-Crnkovic and Larsson 2005).

Finally looking at the development of virtues, a player may excel in exploring the vast world of an MMORPG or improving the character he controls. Nevertheless, those achievements can be seen as void of true value as they lack the "physical and mental capabilities of [the] practitioner" (Briggle n.d., p. 12). Although "achieving excellence in the game world", this achievement does not carry over to the real world. Briggle terms this "artificial arête" (ibid.). The time spent achieving shallow excellence in the MMORPG is ultimately wasted, as it could have also been used to achieve excellence in the real world. The underlying virtues are being devalued, as they are so easy to achieve and people become oblivious to the difference between real and simulated virtue (ibid.)

Impact on the good life

Still, an integrative perspective seems to missing from these theories in terms of the psychological effects. Part of modern ethics looks at technologies and their effects on people from a more applied perspective, namely how they contribute to or hamper a good life. Oftentimes, a good life is measured in a person's happiness and longevity—the first being described as the ultimate goal in human life in Aristotle's Nicomachean ethics. Problematic usage patterns of MMORPG's seem to fit this perspective very well, integrating the separate issues outlined by the classical theories.

With users that show problematic levels of usage, devaluation of life itself in the real world has major impact on the good life (Briggle n.d.). Requiring less effort and providing faster rewards, the MMORPG can seduce users to spend their time in the virtual realm and abandon reality. Resulting problems for example at school, oftentimes lead to people withdrawing even more into the MMORPG. The person gets to value success and rewards in the MMORPG as equal as or even higher than those in the real world, devaluating the latter.

In the paper "Quality of Life in Technological Society", Ruut Veenhoven presents a model to assess quality of life, which "serves as a catchword for different notions of the good life" (Veenhoven 2012, p. 58). Veenhoven's model looks at (a) outer and inner qualities of life and further distinguishes both by (b) the chances they provide for a good life and the ultimate results of life. The combination of the two dimensions leads to four quadrants of qualities. In the following paragraphs, we will examine the four quadrants in regard to a person spending a problematic amount of time playing MMORPG's.

Although *livability of the external environment* focuses mainly on society as a whole as well as external factors such as housing and habitat, the quadrant also encompasses "close networks, strong norms and active voluntary associations" (ibid.: 58). Spending most of the time with the game could lead to the decay of close personal networks outside the virtual world and would keep the player from being an active part of society.

The quadrant *life-ability of the person* compromises a person's physical and mental health. The negative impacts of problematic usage patterns affect both areas. In extreme cases, hygiene and nutrition can be ignored to the point of death. Additionally, there might be severe impacts on "autonomy, reality control, creativity and inner synergy of traits and strivings" (ibid.), when losing oneself in the game world and neglecting the outside world. Addicted users would heavily suffer by addiction through the loss of autonomy, as their playing habits are no longer under their control and even when negative consequences come to the conscious mind, it is hard for them to stop.

Now looking at the results of leading such a life in regard to *utility of life*, the time spent playing a video game would be void of "functionality for the environment" (ibid.), as there would be no contribution to culture or society.

MMORPG's are being played for a person's enjoyment, so one could expect that they would at least contribute to a good life in the *satisfaction* quadrant. For people with conscious and controlled playing behavior, this is certainly the case, as they experience fun, relieve stress or combat boredom without negatively impacting their life. Unfortunately, people (most noticeably those spending a lot of time with the game) also might continue to play an MMORPG, even if they are frustrated with it. Last but not least, the question whether enjoyment from playing or virtual friendships to be experienced in MMORPG's are comparable or inferior to the real world experiences is the topic of some debate and cannot be answered universally, as many individual factors play into the equation.

In his article "Virtually Good?", Johnny Søraker examines virtually versus actually good in the context of six presuppositions ranging from the value of something for the good life to the opportunity costs necessary to achieve it in the real or virtual world. Søraker ultimately recommends choosing the actual thing over the virtual, but makes clear that each case has to be carefully examined and evaluated individually—the claim is ultimately only veritable when all presuppositions hold true (Søraker 2012b). Ultimately, problematic usage patterns, blending into addiction can have noticeable negative impacts on leading a good life, encompassing all four qualities of life, with "life-ability of the person" and "utility of life" being the quadrants most noticeably affected by addiction to a MMORPG. Because not only inner qualities are being affected, but also outer qualities, addressing the issue might not only benefit the affected individuals, but society as a whole.

Evaluation of mechanics

After having evaluated the impact of problematic usage patterns on users, this section will investigate the role of the design. As choice architects, video game developers choose certain mechanics for their games, and choose how these mechanics are implemented. While a seamless and immersive experience is desirable for the gamer, the mechanics can foster both positive and adverse effects by the way they are apparent to the user and the way they exert their influence through user choices. Using a refined model of technological mediation by Tromp et al. (2011), we will assess the game mechanics from the way they exert their influence on the users.

Tromp et al. state that "a design can exert influence that can vary from weak to strong (force), and a design can exert influence that can vary from an implicit to a more explicit manner (salience)" (ibid., p. 11). Again, the two dimensions can be combined into four quadrants. The four types of influence of the model are coercive (apparent and strong), persuasive (apparent and weak), seductive (hidden and weak) and decisive (hidden and strong). Although the types of influence are subjective and might differ from person to person (someone being aware of behavioral game design might classify the mechanics as persuasive), MMORPG's mechanics are mostly *seductive*. That means that players would be "not aware of the influence and most probably regard the behavior as internally motivated" (Tromp et al. 2011, p. 12). For the sake of immersion, underlying algorithms for item drops or crafting success are not available to the user. The way MMORPG's are designed, developers chose not to make the game mechanics apparent and instead chose to use hidden seductive mechanics.

Seductive mechanics are not bad in themselves, as they can also help to encourage positive behavior through choice architecture. For example there is a minor incentive built into the newer MMORPG's to encourage shorter playing sessions: resting from play for several hours provides a boost to the earned experience points when logging in next time. The boost has a cap, so that it does not last very long. The other side of the coin is that the mechanic encourages players to log in daily to exploit the experience bonus—the cap is not extended when resting for a longer period of time (a week will provide the same bonus as a day of rest).

Ethical responsibilities of game designers

Given that (a) games contain imposed scripts and choice architectures, (b) these may have undesirable effects on the good life, and (c) there are clear ethical problems with specific aspects of the current scripts and architectures, the question of the responsibility of designers comes to mind. At least part of the responsibility for the way a certain technology is used by a person rests with the designer; "when technologies fulfill their functions, they also help to shape the actions of their users" (Verbeek 2006, p. 4). "Engineers design a new technology with specific functionalities in mind, without explicitly aiming to influence the actions and behavior of users" (Verbeek 2006, p. 14). The latter is not always the case, as technologies such as speed bumps actively aim at influencing the behavior of people. Likewise, the design of video games aims straight at the experience and behavior of a user, creating a whole environment with specific mechanisms and rules with which the user can interact (also with other users). It is the designer who acts as a conscious choice architect and ultimately sets the script for the artifact, evoking a certain way of behavior and discouraging misuse.

The game developer takes a mediating role between the publishers and the players who ultimately interact with the game. While the game publishers like Blizzard Entertainment CEO Mike Morhaime say that they just offer "the best entertainment values available today" (Reilley 2010), there are business objectives that the company aims at. "The fact that most games are for-profit endeavors opens the door to accusers who say that games profit at the expense of others' misery" (Takahashi 2004). It is not just important to entertain the user for some time, it is important to keep a user's attention for as long as possible to ensure the revenue flow. For the game to be commercially successful, the developer has to create a game that is appealing to a large audience and sets itself apart from the competition. Taking ethical responsibilities into account that were discussed in the "Behavioral game design" section, the designer also needs to take the impact on his product on a user into account.

MMORPG's core mechanics revolve around making the player play over a long period of time and keeping him or her busy so that the player won't consider other games or forms of entertainment. This intention is ethically questionable, taking Putendorf's duty ethics "promote the good of others" and Kant's categorical imperative into account:

By merely focusing on sales and profit, game developers fall into the trap of treating the players as mere means (to an end), thereby violating the categorical imperative to treat humanity "as an end in itself, and never simply as a means." (Dodig-Crnkovic and Larsson 2005, p. 4)

Following an ethical code of conduct, game designers are thus challenged "to assess whether the product they are designing has undesirable mediating capacities" (Verbeek 2006, p. 14). Ethical choice architecture in a game can go beyond the presentation of ethical concepts via the game's content. It can encompass the game mechanics, namely feedback mechanisms, the provision of incentives or making the usually opaque and seductive playing mechanics more transparent and thus empower the player to pursue the "options that will make them better off" (Thaler et al. 2010, p. 10). For an MMORPG, that could mean that it encourages moderate playing times and keeping mechanics transparent. In the following section, we will outline in detail how design could change for the better.

Better design

In this section, we will outline a process for coming up with a better design, before looking at possible ways to engineer out undesirable effects and facilitate positive ones through choice architecture and technological mediation. In the last subsection, we discuss the possible implications and check the feasibility of changing existing games. The proposed solutions are not meant to be comprehensive, and should be understood as initial ideas and proposals based on the framework outlined above. The main focus is to bring the subject to attention with involved parties and hopefully making it the focus of additional research or a consideration within the game development process.

To assess possible changes in MMORPG design that could also be applied to running games it is important to discern to what extent the games can be classified as product or service. While the latter would result in a more static game that would only see minor adjustments after launch, games with characteristics of a service could also be easily changed after launch. As it is, MMORPG's are a hybrid between the two: after finishing the game's development, it sports characteristics of a product: its consumption is separated from the production and the game is usually sold to the customers with a fixed price—regardless if it is bought on a physical medium or from a download shop. The service part consists of the hosting of the game, regular content updates as well as fixing bugs and improving the game balance. These services are produced and consumed (near) simultaneously and covered by the monthly subscription fee of an online roleplaying game (Niessink and Van Vliet 2000; Pieters 2013). When a game is delivered as a service, this makes it easier for designers to change features on the fly.

Design process

To come up with ethically better design, developers of seductive media such as MMORPG's should, in a first step, become aware of possible implications of their design decisions. This encompasses both the game's content as well as its mechanics. A better understanding of the issue and discussing mechanics with the team, not only based on the expected attractiveness to a user, but also from an ethical perspective could be a good starting point (Fleuriot 2012, p. 3).

For assessing the possible impact of design choices, developers can apply a couple of techniques. Comprehending an “informed prediction”, they could assess the impact of game mechanics from an ethical point of view and analyze their influence on human behavior (Verbeek 2006, p. 22). MMORPG’s have been on the market for nearly two decades and developers can draw on a wealth of research and case studies covering MMORPG’s from technological to social aspects, enabling the developers to come up with an accurate prediction (e.g., Taylor 2006; Feng et al. 2007).

The second approach could consist in using the augmented constructive technology assessment (CTA), involving “all relevant stakeholders in the design” (Verbeek 2006, p. 22). Currently, closed and public beta testing mainly serves to optimize the netcode of a game, fine-tune the balancing or minimize bugs. It could also be used to actively monitor how people play the MMORPG, how much time they spend on performing certain tasks and how they make progress around the game. Following the philosophy of CTA, feedback of the various player groups should also be incorporated to assess the game in the various use contexts, e.g. hardcore gamers vs. more social gamers that mainly focus on their guild. This would also make the game deployment more responsible as a social experiment (Van de Poel 2011). Coming back to the hybrid characteristics of product and service outlined in this section’s introduction, there is ample opportunity for changing existing MMORPG’s. While core mechanics are unlikely to be drastically altered in the course of continued support, service-based games could also be adjusted after launch when it becomes clear that certain mechanics lead to players spending a problematic amount of time in the game.

Of course, it is not only the game developers making choices, but also the publishing company that has its own targets in terms of copies sold, monthly subscriber base and revenue. It’s a balancing act the developers have to accomplish—developing a game that is commercially successful, while minimizing possible problematic usage by its players. The game must attract players without trapping them. This should become an integral part of the industry code of conduct, not only for the developers, but also for the management of developers and publishing companies. The last two groups can be said to have the meta-task responsibility to create an environment in which the developers have the freedom to design their games in a responsible way, not only maximizing for sales (Ahmed and van den Hoven 2010, p. 6).

Engineering out undesirable effects

Reverse engineering the choice architecture of MMORPG's, one could look into the design choices leading to undesirable effects on choices made by users, and find alternative void of or with less adverse effects. Limiting the effect of the random reward ratio in operant conditioning described under "Effects of MMORPG's on users" would be a major factor in designing a more ethical MMORPG. The oftentimes random reward ratio could change towards a more transparent, fixed ratio. The player would be able to estimate exactly how long it would take to collect a certain amount of gold or gain the next level, changing the effect of the choice architecture from seductive to persuasive.

Hopson argues that fixed ratios are producing quite a problematic pattern for games, though: replacing randomness with a fixed ratio reward scheme can have direct influence on a player's behavior. With a random scheme, each click could bear the next reward, enticing the player to keep up a high level of activity. With a fixed ratio, this would be different: After having received a reward by logging wood (e.g., by clicking 10 times on a tree trunk), the player would know that the next nine clicks will bear no direct reward and usually pause for a longer period of time. Combined with the non-linear progress of MMORPG's, players might no longer find the reward worth their time. To forego this effect, progress could be engineered in a way as to be more linear. It would then take around the same amount of time to gain the character levels 1–10 as the levels 40–50. Of course, that would take away some of the prestige and reward of reaching higher levels, but the "grind" could be replaced with more sophisticated and challenging gameplay for example. In further research, it would be interesting to empirically examine the impacts of changing reward schemes for gamers and their influence on playing times and enjoyment of the game.

A second mechanism that could effectively help to encourage more modest playing times, while enticing the player to stay attracted over a longer period of time, would be the implementation of fixed interval schedules for rewards. Hopson notes that the usual reaction to fixed intervals is "pausing for a while after a reward" before picking up more frequent activities before the next reward. Carefully scheduling those rewards, a designer could influence the style of play in a persuasive fashion. Quests with high rewards could be designed in a way that they have to be completed over a period of several days or even weeks with fixed breaks in between the single tasks.

To reduce the need for players to spend a lot of time on simple mindless tasks like gaining experience points through killing monsters at the same spot over and over again or collecting crafting materials, the game could offer to perform one of the tasks while the player is actually offline. (Some MMORPG's like "Star Wars: The Old Republic" already have features that go into that direction.) This could help minimize the player's feeling of falling behind when not being online and add a possibly interesting layer of management to the game. This would make the mechanics less coercive, and leave more choice to the player.

Facilitating desirable effects

Technologies can exert their influence and mediating role in a variety of ways, outlined under "Impact on the good life". Next to engineering out undesirable effects, desirable effects could therefore be engineered in. An MMORPG client could be programmed to not allow a player to play more than 20 h a week, forcefully shutting itself down after the time has been filled up. This would most likely be viewed as very coercive by most players and not necessarily result in positive feedback. A clever design, on the other hand, could seduce the player not to play longer than 20 h per week by rewarding him, with experience points or special items and virtual badges. A mixture between seduction and coercion could be a change in design that lessens a character's effectiveness during combat or reduces the amount of experience points received when playing more than 8 h straight. Already, there are boosts in some games that apply when a character is "rested". A certain amount of experience points is earned faster when the game was not played for several hours, thereby changing the choice architecture of resting choices.

With motivation addiction in the gamer being a factor not controllable by the designer, he or she could possibly work on creating an environment that makes the player aware of his or her playing habits. As a choice architect, the game designer could implement feedback mechanisms for gamers that pull their attention towards potential problematic usage. This could happen by making the time spent in the game more transparent via displaying it on the graphical user-interface of the game (the figure changing color or consuming more space on the GUI over time) or subtle, by showing signs of fatigue on the played character. Those signs would also be visible to other players, who in turn, could encourage their peer to stop playing for a while. Again, this delegation of responsibility to other users is part of the new possibilities that online services offer in terms of mediation.

The designer could further draw upon the strong social aspects that an MMORPG has. One possible example would be to (voluntarily) provide the statistics of weekly playing time to guild leaders, who in turn, could talk to members with problematic usage patterns and encourage them to reduce the pace a little. Such designs make use of the gamers' own responsibility and social mechanism as part of the choice architecture. Not only the game mechanics themselves provide seduction and persuasion here, but part of the behavioral incentives are looped via the user's own values (by viewing time statistics oneself) or via others (when other players or guild leaders view them).

Putting even more responsibility on the user, the designer could provide the user with settings to set off warnings after a certain amount of time or when a certain threshold is reached per week. Making that a mandatory part of the installation process or the creation of a new character would force the player to deal with the general issue and bring it to attention. In this case, the values to be protected are (partly) chosen by the player rather than the designer, although the general mechanism and choice architecture are part of the design.

Last, but not least, the design could help users to become more aware of the underlying mechanics and make the game overall more transparent and persuasive instead of seductive. Fleuriot stresses that designers of pervasive media "have a responsibility to make the implications of their actions clear to the user" (2012, p. 3). This could be supported by bringing transparency to the hidden mechanics and algorithms that govern the game through a database that is directly accessible from within the game. A user could look at the time that others on average spent on certain quests or to gain the next level. The database could also list the explicit drop rates of rare items and the various algorithms that govern the game. There is little threat regarding theft of intellectual property by other studios, as drop rates are finely balanced with the overall game and cannot simply be applied to other titles.

Possible implications for users and publishers

Acknowledging their mediating role and developing games that deliberately tackle problematic user behavior with a combination of new mechanics or adjusting current games without appalling users is something that can be done, but can also pose a risk for a game's (continued) commercial success: Hopson warns that any changes in that area of reward mechanisms would have to be done "carefully and gradually" as to avoid that the player is no longer attracted at all to the game or would even quit playing altogether. Also, changing the mechanics must not lead to a ruination of the overall business models of MMORPG publishers.

Still, making the mechanics more transparent and providing strong feedback mechanisms together with certain measures to address already existing problematic usage patterns as discussed in the previous sections is something that could be included without requiring a drastic change of the games. Better design is not limited to the range of MMORPG's currently in development, as the genre incorporates aspects of a service, for example continued support through regular enhancements to the game's content and design via patches. It is not feasible to completely change the core mechanics, because that could alienate existing users, but small and more subtle features like displaying fatigue on a player's character or the inclusion of rewards could be introduced over time to already running games.

DISCUSSION

In the course of this paper, we have shown that the ethical dimension of games goes beyond their content. Problematic investment of time into an MMORPG that might otherwise be void of explicit sexual or violent content can have consequences for the good life of a person. Seductive game mechanics and exploiting behavioral game design claim a user's attention well past the initial interest in a game's story or visual presentation.

MMORPG's differ in a way from most other games. They do not really have a true ending. There is always a new quest to fulfill or a dungeon to explore. Even if someone would run out of content, he or she could start anew with a totally different class. The virtual worlds are continuously expanding and social components further bind people to the game. "Normal" single player video games usually offer entertainment for a finite time only and have a limited "replayability". Once the game's story has been completed, the game usually ends with the credits, just like after watching a movie. Players might play through the game for multiple times, but there is little incentive to do so.

It is rather unlikely that developers of video games are consciously set to make players addicted to their games and possibly create negative consequences for them. Their main goal is to create an enjoyable and enduring experience for players so that they would keep up playing the game and pay a subscription fee over a longer period of time. Most developers would be proud when their MMORPG excels at maximizing the attention it gets from a user for a long period of time and seduces players to stick with the game. Still, looking at the effects holistically, developers also need to consider how their games influence users in a negative way: some users play the games even past the point of enjoyment. The consequences can be dire, including death in the most extreme case. Although some governments in Asia restrict playing times (Blackburn 2013), developers are not yet fully addressing the issues stemming from their games' mechanics, possibly because they are not looking at the impacts of their design choices from all angles.

Contributing to personal enjoyment for many people, MMORPG's are not inherently "bad", nor are the mechanics—they just have to be designed and used responsibly. The possible negative impacts resulting from problematic usage patterns cannot solely be attributed to MMORPG game mechanics, and the scope of this work did not include comprehensive research on motivational factors or the player's environment. It is likely that a number of people would not lead a more satisfactory life were they playing no MMORPG's at all. For them, the game is just an outlet for problems in other areas of life. While said underlying causes of motivation addiction cannot be addressed by a game developer, game mechanics could still play a mediating role to reduce problematic usage adversely affecting the good life. This could either happen by mitigating negative effects, such as uncontrolled playing times through hard contingencies or make the players themselves aware of their problematic usage patterns. By ethically evaluating the impact of game mechanics on a user and taking the necessary steps to influence behavior for the better, developers have the possibility and responsibility to help contribute to a good life for millions of users. Future research could further explore the ethical implications of the business models of MMORPG's, especially the controversially discussed microtransactions (Hamari and Lehdonvirta 2010), and assess their impact on user behavior.

More generally, the discussion of ethical MMORPG design links up with broader questions on technology and autonomy. Autonomy is "[...] commonly viewed as a key component of human well-being" and "[...] referenced as a fundamental principle of ethics in Kantian deontology" (Piper n.d.). The autonomy issue is especially prominent in the privacy domain (Van den Hoven et al. 2014), where one of the main questions is how leakage of personal data affects our ability to make choices, for example because we are refused certain services based on what is known about us. This autonomy dimension is very prevalent in game design as well, albeit through a different mechanism. In this case, autonomy is not lost because technology enables others to make decisions for us based on what they know about us, but rather because our ability to make our own choices is affected by the mechanisms in the technology, impacting self-control. Gamers may be seduced into spending (excessive) time on the game, presumably without conscious choice, and with negative effects on the good life. In this sense, MMORPG's can be considered persuasive technologies (Lockton et al. 2008; Versteeg 2013).

One can discuss whether the effects of (excessive) gaming on autonomy may be considered morally problematic even if the gamers are happy with it themselves. A positive psychology approach (Søraker 2012a) would focus on subjective well-being and therefore evaluate the effect of the technology based on self-reported happiness. However, in a democratic society, we will typically not accept systems limiting our autonomy, such as dictatorship, even if people would be happier under such a regime. Similarly, the fact that people are happier or have more fun when playing a game does not imply that the design is therefore good. In particular, game design may inhibit the ability of users to behave responsibly themselves, therefore failing to meet the criterion of meta-task responsibility (Van den Hoven 1998). In this sense, MMORPG's are just an example of a wider technological development, and the broader application of the suggested solutions may provide inspiration for further study.

REFERENCES

1. Ahmed, M. A., & van den Hoven, J. (2010). Agents of responsibility—Freelance web developers in web applications development. *Information Systems Frontiers, 12*(4), 415–424.
2. American Society of Addiction Medicine (ASAM). (2011). Definition of Addiction. Resource document. asam.org. http://www.asam.org/quality-practice/definition-of-addiction. Retrieved May 25, 2014.
3. Anderson, C. A., & Bushman, B. J. (2001). Effects of violent video games on aggressive behavior, aggressive cognition, aggressive affect, psychological arousal, and prosocial behavior: A meta-analytic review of the scientific literature. *Psychological Science, 12*, 353–359.
4. Anderson, C. A., & Dill, K. E. (2000). Video games and aggressive thoughts, feelings, and behavior in the laboratory and in life. *Journal of Personality and Social Psychology, 78*, 772–790.
5. Blackburn, T. (2013). Chinese Gamer Dies After 40 Hour MMO Session. Resource document. gamebreaker.tv. http://www.gamebreaker.tv/game-industry-news/chinese -gamer-dies-after-40-hour-mmo-session/. Retrieved April 20, 2014.
6. Blakley, B. (1996). The Emperor's old armor. In *Proceedings of the 1996 workshop on New security paradigms* (pp. 2–16). ACM.
7. Briggle, A. (n.d.). *Draft: Gamers and the Good Life*. Resource document. Game Philosophy Research Initiative. http://game.unimore.it/Papers/Briggle_Paper.pdf. Retrieved April 25, 2014.
8. Chappell, D., Eatough, V., Davies, M. N., & Griffiths, M. (2006). EverQuest—It's just a computer game right? An interpretative phenomenological analysis of online gaming addiction. *International Journal of Mental Health and Addiction, 4*(3), 205–216.
9. Copeland, L. (2004). Battle over violent video games heating up. *USA Today, 3 A*, January 29, 2004.
10. Dodig-Crnkovic, G., & Larsson, T. (2005). Game ethics-Homo Ludens as a computer game designer and consumer. *International Review of Information Ethics, 4*(12), 19–23.
11. Feng, W. C., Brandt, D., & Saha, D. (2007). A long-term study of a popular MMORPG. In *Proceedings of the 6th ACM SIGCOMM Workshop on Network and System Support for Games*(pp. 19–24). ACM.
12. Fleuriot, C. (2012). Avoiding vapour trails in the virtual cloud: developing ethical design questions for pervasive media producers. *Culture Machine, 13*, 1–9.

13. Gentile, D. A., Choo, H., Liau, A., Sim, T., Li, D., Fung, D., & Khoo, A. (2011). Pathological video game use among youths: A two-year longitudinal study. *Pediatrics, 127*(2), e319–e329.
14. Griffiths, M. D. (2010). The role of context in online gaming excess and addiction: Some case study evidence. *International Journal of Mental Health and Addiction, 8*(1), 119–125.MathSciNet
15. Hamari, J., & Lehdonvirta, V. (2010). Game design as marketing: How game mechanics create demand for virtual goods. *International Journal of Business Science and Applied Management, 5*(1), 14–29.
16. Hopson, J. (2001). Behavioral Game Design. Resource document. Gamasutra. com. http://www.gamasutra.com/view/feature/3085/behavioral_game_design.php . Retrieved April 04, 2014.
17. Hussain, Z., & Griffiths, M. D. (2009). The attitudes, feelings, and experiences of online gamers: A qualitative analysis. *CyberPsychology & Behavior, 12*(6), 747–753.
18. Hyman, S. E. (2005). Addiction: A disease of learning and memory. *American Journal of Psychiatry, 162*(8), 1414–1422.
19. Ko, C. H., Liu, G. C., Hsiao, S., Yen, J. Y., Yang, M. J., Lin, W. C., ... & Chen, C. S., et al. (2009). Brain activities associated with gaming urge of online gaming addiction. *Journal of Psychiatric Research, 43*(7), 739–747.
20. Ko, C. H., Yen, J. Y., Chen, C. C., Chen, S. H., & Yen, C. F. (2005). Gender differences and related factors affecting online gaming addiction among Taiwanese adolescents. *The Journal of Nervous and Mental Disease, 193*(4), 273–277.
21. Koster, R. (2001). On "Pay To Play" Or, MMORPG Business Models 101. Resource document. Raphkoster.com. http://www.raphkoster.com/gaming/busmodels.shtml.
22. Kuss, D. J., & Griffiths, M. D. (2012). Internet gaming addiction: A systematic review of empirical research. *International Journal of Mental Health and Addiction, 10*(2), 278–296.
23. Lockton, D., Harrison, D., & Stanton, N. (2008). Design with intent: Persuasive technology in a wider context. In *Persuasive Technology* (pp. 274–278). Berlin: Springer.
24. Mentzoni, R. A., Brunborg, G. S., Molde, H., Myrseth, H., Skouverøe, K. J. M., Hetland, J., & Pallesen, S. (2011). Problematic video game use: estimated prevalence and associations with mental and physical health. *Cyberpsychology, Behavior, and Social Networking, 14*(10), 591–596.
25. Ng, B. D., & Wiemer-Hastings, P. (2005). Addiction to the internet and online gaming. *CyberPsychology & Behavior, 8*(2), 110–113.
26. Niessink, F., & Van Vliet, H. (2000). Software maintenance from a service perspective. *Journal of Software Maintenance, 12*(2), 103–120.
27. Norman, D. A., & Ortony, A. (2003). Designers and users: Two perspectives on emotion and design. In *Proc. of the Symposium on Foundations of Interaction Design at the Interaction Design Institute*, Ivrea, Italy.
28. Papagiannidis, S., Bourlakis, M., & Li, F. (2008). Making real money in virtual worlds: MMORPGs and emerging business opportunities, challenges and ethical implications in metaverses. *Technological Forecasting and Social Change, 75*(5), 610–622.
29. Peukert, P., Sieslack, S., Barth, G., & Batra, A. (2010). Internet-and computer game addiction: phenomenology, comorbidity, etiology, diagnostics and therapeutic implications for the addictives and their relatives. *Psychiatrische Praxis, 37*(5), 219–224.
30. Phillips, C. A., Rolls, S., Rouse, A., & Griffiths, M. D. (1995). Home video game playing in schoolchildren: A study of incidence and patterns of play. *Journal of Adolescence, 18*(6), 687–691.
31. Pieters, W. (2013). On thinging things and serving services: technological mediation and inseparable goods. *Ethics and Information Technology, 15*(3), 195–208.
32. Piper, M. (n.d.). Autonomy: Normative. Resource Document. Internet Encyclopedia of Philosophy. http://www.iep.utm.edu/aut-norm/. Retrieved April 12, 2014.

33. Pufendorf, S. (1927). *On the Duty of Man and Citizen According to the Natural Law* (W. A. Oldfather, Trans.). Oxford: Clarendon Press. (Original work published 1672).
34. Reilley, J. (2010). World of warcraft reaches 12 million subscribers. Resource document. ign.com. http://www.ign.com/articles/2010/10/07/world-of-warcraft-reaches-12-million-subscribers. Retrieved April 12, 2010.
35. Scott, D. (1995). The effect of violent video games on feelings of aggression. *Journal of Psychology, 129*, 121–132.
36. Shaffer, H. J. (1996). Understanding the means and objects of addiction: Technology, the internet, and gambling. *Journal of Gambling Studies, 12*(4), 461–469.
37. Sherry, J. L. (2001). The effects of violent video games on aggression; A meta-analysis. *Human Communication Research, 27*, 409–431.
38. Sicart, M. (2009). The banality of simulated evil: designing ethical gameplay. *Ethics and Information Technology, 11*(3), 191–202.
39. Skinner, B. F. (1938). *The behavior of organisms: An experimental analysis*. New York: Appleton-Century-Crofts.
40. Søraker, J. H. (2012a). How shall I compare thee? Comparing the prudential value of actual virtual friendship. *Ethics and Information Technology, 14*(3), 209–219.
41. Søraker, J. H. (2012b). Virtually Good? In: Philip Brey, Adam Briggle and Edward Spence (Eds.) *'The Good Life in a Technological Age'*, Routledge, New York, Series Studies in Science Technology and Society, Chap. 15, 225–237.
42. Takahashi, D. (2004). *Ethics of Game Design*. Resource Document. Gamasutra. com. http://www.gamasutra.com/view/feature/2181/ethics_of_game_design.php. Retrieved April 14, 2014.
43. Taylor, T. L. (2006). Does WoW change everything? How a PvP server, multinational player base, and surveillance mod scene caused me pause. *Games and Culture, 1*(4), 318–337.
44. Thaler, R. H., Sunstein, C. R., & Balz, J. P. (2010). Choice Architecture. http://papers.ssrn.com/sol3/papers.cfm?abstract_id=1583509.
45. Tromp, N., Hekkert, P., & Verbeek, P. P. (2011). Design for socially responsible behavior: A classification of influence based on intended user experience. *Design Issues, 27*(3), 3–19.
46. Van de Poel, I. (2011). Nuclear energy as a social experiment. *Ethics, Policy & Environment, 14*(3), 285–290.
47. Van den Hoven, J., Blaauw, M., Pieters, W. and Warnier, M. (2014). Privacy and Information Technology, *The Stanford Encyclopedia of Philosophy* (Winter 2014 Edition), Edward N. Zalta (ed.), URL = < http://plato. stanford. edu/ archives/ win2014/entries/it-privacy/&gt.
48. Van den Hoven, M. J. (1998). Moral responsibility, public office and information technology. *Public administration in an information age: a handbook*, 97–112.
49. Veenhoven, R. (2012). Quality-of-life in technological society. In: Philip Brey, Adam Briggle and Edward Spence (Eds.) *'The Good Life in a Technological Age'*, Routledge, New York, Series Studies in Science Technology and Society, Chap. 3, pp 55–76.
50. Verbeek, P. P. (2006). Materializing morality design ethics and technological mediation. *Science, Technology & Human Values, 31*(3), 361–380.
51. Versteeg, M. J. J. M. (2013). *Ethics & Gamification design: a moral framework for taking responsibility*. Master thesis, New Media & Digital Culture, University of Utrecht. http://dspace.library.uu.nl/handle/1874/281831.
52. Yee, N. (2001). *The Norrathian Scrolls: A Study of EverQuest*. Resource document. nickyee.com. Retrieved April 08, 2014, from .http://www.nickyee.com/report.pdf.
53. Yee, N. (2002). *Ariadne: Understanding MMORPG Addiction*. Resource document. nickyee.com. http://www.nickyee.com/hub/addiction/addiction.pdf. Retrieved April 16, 2014.
54. Yee, N. (2005). *In Their Own Words: The Social Component*. Resource document. nickyee.com. http://www.nickyee.com/daedalus/archives/001301.php. Retrieved April 18, 2014.

55. Yee, N. (2006a). The psychology of massively multi-user online role-playing games: Motivations, emotional investment, relationships and problematic usage. In *Avatars at work and play* (pp. 187–207). Netherlands: Springer.
56. Yee, N. (2006b). Life as a guild leader. *The Daedalus Project, 4*(2), 4–2.
57. Young, K. (2009). Understanding online gaming addiction and treatment issues for adolescents. *The American Journal of Family Therapy, 37*(5), 355–372.

CHAPTER 7

Refining The Ethics of Computer-Made Decisions: A Classification of Moral Mediation by Ubiquitous Machines

Marlies Van de Voort, Wolter Pieters

University of Twente EEMCS-PS, Zilverling, P.O. Box 217, 7500 AE Enschede, The Netherlands

TU Delft and University of Twente TBM-ICT, P.O. Box 5015, 2600 GA Delft, The Netherlands

Faculty of Science, Radboud University Nijmegen Philosophy and Science Studies, P.O. Box 9010 (internal code 77), 6500 GL Nijmegen, The Netherlands

ABSTRACT

In the past decades, computers have become more and more involved in society by the rise of ubiquitous systems, increasing the number of interactions between humans and IT systems. At the same time, the technology itself is getting more complex, enabling devices to act in a way that previously only humans could, based on developments in the fields of both robotics and artificial intelligence. This results in a situation in which many autonomous, intelligent and context-aware systems are involved in decisions that affect their environment. These relations between people, machines, and decisions can take many different forms, but thus far, a systematic account of machine-assisted moral decisions is lacking. This paper investigates the concept of machine-assisted moral decisions from the perspective of technological mediation. It is argued that modern machines do not only have morality in the sense of mediating the actions of humans, but that, by making their own decisions within their relations with humans, mediate morality itself. A classification is proposed to differentiate between four different types of moral relations. The moral aspects within the decisions these systems make are combined into three dimensions that describe the distinct characteristics of different types of moral mediation by machines. Based on this classification, specific guidelines for moral behavior can be provided for these systems.

Keywords: Ubiquitous computing, Moral reasoning, Technological mediation, Moral decisions, Human computer relations.

INTRODUCTION

The role of computers in society has changed profoundly over the last decades. While the computer began as a specialized tool, used for specific calculations, it has now conquered the world, and is present in every corner of modern civilization. Not only is it being used for a wide array of applications, it is also at the same time embedded in our environment, present without the user being aware of it, and assisting us with tasks that are not even recognized as tasks that we are assisted with. Thereby, computers also have become more and more involved in decisions that have moral impact.

While the role of the computer is changing, two fields of technology are rising. The first field is that of robotics. Examples of recent developments leading to computers being more involved in daily life can be found in socially interactive robots, including examples like robotic pets with artificially modeled behavior, robots that do housekeeping, and health care robots (Fong et al. 2003). Developments in robotics increase the potential for computers to interfere in their environment.

The second field, which is related to both robotics and ubiquitous computing[1], is the field of artificial intelligence. In this field, developments lead to more advanced reasoning methods for computers, enabling them to use information acquired from their environment in increasingly complex tasks (Ramos et al. 2008). Decisions made may then lead to output in the form of information for humans, or in the form of actions through (robotic) actuators.

Using advancements in robotics and artificial intelligence, machines acquire more advanced reasoning, and have increased awareness about their environment. They do not only have more action capabilities, but are also often functioning without any human intervention. This way, these systems become more human-like. In particular, the role of computers has become more active, in the sense of making explicit decisions. The issue at stake is what should be done when such computers become involved in decisions that are now made by humans, and thereby impact moral patients.

Several people have already identified the need for the computer to have ethical behavior (Allen et al. 2006). Wallach and Allen (2009) argue that computers taking more responsibilities implies the need for learning ethics. This need emerges from the fields of both robotics and computing science. Engineers as well as ethicists already attempted to model ethical or moral decision making in algorithms, mainly using techniques originating from the field of artificial intelligence (McLaren 2006). Also, the question has been asked to what extent we should transfer (moral) decision-making authority to machines (Kuflik 1999), and, from the moral patient perspective, whether new entities such as robots have rights (Coeckelbergh 2010; Hildebrandt et al. 2010).

Relational perspective

In this paper, we take a relational perspective, and analyze decisions with moral characteristics within the various types of relations between humans and ubiquitous computing devices. This will augment existing approaches to technology assessment, by not focusing on direct effects of the technology, but rather on the effect of the technology on the distribution of morality. In this sense, answering the question of changes in moral relations will enable technology designers to deal with their *meta-task responsibility*, by enabling technology users to act responsibly in the context of decision-making machines (Hoven 1998). For example, a designer could choose to leave more responsibility with the user, or instead let the machine decide to avoid moral overload for the user (Van den Hoven et al. 2012).

We focus on the moral aspects of computer contributions to complex decisions in human-machine constellations. This means that we consider the machines as mediating entities in human-world relations, in particular contributing to outcomes that have moral implications. From the perspective of relations between humans and machines, most existing work focuses on computers as moral mediators in a rather limited sense, namely as mediating entities in decision-making processes in which the human is still the central agent (Magnani and Bardone 2008; Verbeek 2006). Within the existing frameworks, examples relate to ways in which machines influence the processes in which humans acquire information, make decisions and act (im)morally. These theories thus consider the computers as rather passive mediators, indeed changing human moral decisions, but not so much as mediators that make their own decisions.

In ubiquitous computing, we have exactly the aforementioned situation where machines make explicit decisions about acquiring and selecting information themselves, and provide suggestions for action, or even act themselves. In the end, like in a traditional technologically mediated context, the outcome can be partly ascribed to humans and partly to technology, if distinguishable at all. For example, an expert system may acquire information and use this to suggest a solution to a problem, or a care robot may decide to take a certain action affecting a patient. In both cases, the machines change human-world relations, but by explicitly making decisions. Existing frameworks are not tailored to situations where explicit decisions are made within the machines themselves. Simply accounting for the machines as moral mediators of human decision-making does not suffice for understanding human-machine relations in such settings.

This paper aims at filling this gap. We analyze moral relations between humans and systems that make explicit decisions, with the purpose of enabling more refined judgments on the acceptability and more refined possibilities for steering such developments. Like in existing approaches, machines are considered as moral mediators, but in an active rather than a passive sense. In particular, the machines, when making explicit decisions, can be said to have increased control over the outcome of events, knowledge about such outcomes, and the choice between different possible actions (Noorman 2012).

The objective of this research is to provide a classification for the kind of contributions to moral decisions that these systems make, and what this means for the human-machine relation. In this sense, the classification is similar to those for passive forms of mediation, such as those by Ihde (1990) and Verbeek (2006). Our claim is not that machines necessarily have characteristics of moral agents. Rather, we are interested in the (moral) contributions of machine decisions to more complex decisions made by human-machine combinations. We do so by relating our work to the work on technological mediation by Ihde and Verbeek. We argue that the existing types of mediation distinguished are inadequate when machines are explicitly involved in decision-making.

Contributions

This paper has two distinct contributions. Firstly, we provide a classification of human-computer relations of ubiquitous computing devices capable of making decisions that have moral aspects. Secondly, we derive a classification of moral decisions for these machines.

At some point in the (near) future, rules or guidelines will be needed to deal with the increased capacities of computer systems and their increased involvement in our society. By defining types, names and categories in this context, we intend to create tools (e.g. a classification and vocabulary) to facilitate the discussion about the role and the advisability of and room for these systems in our society. Using these categories, we can discuss boundaries for what we would allow these systems to do, we can use the vocabulary for (discussion about) future legislation to prevent undesirable situations with these machines or for describing design criteria for the engineers who develop these systems. In terms of design criteria, the classifications can for example be used to guide the design of algorithms for moral decision making, complying with the rules applicable to the relevant classes of relations. This also enables reuse of basic design concepts across different systems with the same classification. Finally, designers can explicitly design a trade-off between delegating decisions to machines and delegating decisions to humans, by considering the changes in moral relations induced by the technology.

This paper is organized as follows. An overview of related work is provided in Sect. 2, followed by the methodology of our study in Sect. 3. The definitions given in Sect. 4 are used to propose a taxonomy to classify different relations between humans and computers in Sect. 5. This model is combined with a four-stage function model for automation and used to analyze the moral aspects within the different types of relations between humans and computers in Sect. 6. The results of these analyses are then integrated, and result in a list of different types of moral decisions found within different relations between humans and computers in Sect. 7. The resulting list can be used as a guideline for modeling moral behavior in these systems. The paper is concluded in Sect. 8.

Related work

Within existing literature, there are various views on the (moral) relationship between humans and technological artifacts in general and computer-like devices in particular.

In the first approach, technological artefacts are said to have properties that influence human actions and decision-making. This approach includes notions such as scripts (Akrich 1992) and technology affordances (Gaver 1991). These approaches view the relation between humans and technology from an external perspective, in which there is mutual influence.

The second approach studies the relations from an internal, phenomenological perspective. In such an approach, the role of technology in the "directedness" of humans to the world is analyzed. From this point of view, Don Ihde (1990) describes four different relationships between humans and technology. The first relation is the embodiment relationship, in which the technology becomes part of the human: the human establishes a relationship with the world through the technology. A magnifying glass forms this kind of relation with a person. The second relationship is the hermeneutic relationship, in which the technology interprets the world and provides a representation of it to the user of the technology. An example of this relationship is a thermometer: the thermometer provides interpretation of a specific property of the environment, namely the temperature. The third relation is the alterity relation, in which the technology is experienced as a "quasi-other", which is the case with, for example, a robot. The fourth relation is the background relationship, in which the human is not necessarily conscious of the presence of the technology, although the technology influences the experiences the human has within its environment. An example of technology that has this background relationship with the human are the lights in a room: they influence how the human experiences the room, but the human is not always conscious of the absence of dark.

Also from the phenomenological perspective, an overview of views on relations between humans and technology is provided by Verbeek (2008a). The first approach to human-technology relationships according to Verbeek is externalism. Within this view, which he regards as inadequate, the interaction between humans and technology is described using the terms means and goals. Within this view, the technology is used by the human as means to achieve its goals (instrumentalism). Another view is that that of substantivism, which includes the view that technology determines our culture (determinism) and develops a certain autonomy in its own development, creating an unstoppable development of technology (technological imperative).

The second approach to the relation between technology and humans Verbeek describes, is that of transhumanism. This view includes the impossibility to separate human and technology, and also to distinguish between human and technology. Some transhumanists believe that the homo sapiens will soon be extinct because it has been superseded by technology, or question the value of human life in the transhumane world.

The third approach to the relationship between humans and technology according to Verbeek, is that of technological mediation. Within this relationship, the technology mediates in the relation between the human and its environment. Technology is neither a neutral artifact, nor a transhuman replacement for the human. Verbeek (2006) describes the embodiment relationship and the hermeneutic relationship of Ihde as technological mediation. In addition, he includes the notion of mediation of action, in which machines change the action possibilities of humans, analogous to the notion of script.

Also using the concept of mediation, Magnani and Bardone (2008) speak about technological artefacts as moral mediators. They argue that by externalizing part of our moral tools into moral mediations, we can improve our moral decision making, particularly under conditions of uncertainty. As an example, he discusses the Internet, which influences human moral decisions by changing (and improving) the possibilities for acquiring the relevant information.

In all these approaches, the human is still the central decision-making agent. The moral decisions made are influenced by technological artefacts such as computers or the Internet, thereby justifying the classification of machines as moral mediators. Both in the classification Ihde provides and in the description of Verbeek of mediation, the influence of technology on the human is recognized. However, this influence only exists as part of the relationship between the human and the environment, via the technology. The technology is not conceived as an agent making decisions itself.

Attempts to generalize these approaches have already been made. In particular, developments that extend the notion of agency in relation to intelligent machines are important here, such as extended agency Akrich (1992), surrogate agency Johnson and Powers (2008), and mind-less morality Floridi and Sanders (2004). All these conceptualizations seem to suggest that technological mediation may require a broader notion of agency as well. This would imply a re-interpretation of the central concept in technological mediation, intentionality: the "directedness" of people towards their environment, in which technology can play a role.

Verbeek (2008b) takes an important step towards a broader conception of mediation. In particular, he introduces the notion of "cyborg intentionality", in which a complex composition of human and technology is directed towards the world, rather than a clearly distinguishable human. However, Verbeek limits the application of this approach to cases where "pieces of technology are actually merged with the human body". Therefore, this approach is not directly applicable to the situation where the *decision making process*, consisting of both experience and action, is shared within a cyborg construct, or, more modestly, at least in particular types of human-technology-world relations.

The technology that is considered in the present study has some sort of intentionality that is independent of this relationship with the human or the presence of the human. The technology can function autonomously and is dedicated to a task. This comes close to what Verbeek (2008b) calls "composite intentionality": the technology itself has a particular kind of "directedness" towards the world, which may be different from the human one, thereby generating its own representations of the world, which may then augment or construct a particular representation for humans. Again, Verbeek does not explicitly consider decisions made by such technologies.

The technology we consider also explicitly makes decisions, rather than the "affordances" of analogue technology, which merely makes certain actions easier or harder, or amplifies or reduces aspects of experience. We do not claim that this constitutes full moral agency, but rather that this calls for different mediation relations than the approaches outlined above. To describe the differences between the types of human-technology relationships for the kind of semi-autonomous technology prevalent in ubiquitous computing, a different classification is therefore provided in this paper.

This paper fits in a larger class of analyses that aim at extending work in the ethics of technology to new types of technology enabled by the information revolution. Other examples include reconfigurable technology (Dechesne et al. 2013) and services as opposed to products (Pieters 2013). With reconfigurable technology, the central issue lies in the possibilities to change the effects of the technology after it has been deployed. This means that mediating effects cannot always be identified upfront, as the behavior of the technology can be changed later on. However, it is still assumed that the reconfiguration is a human decision. With services as opposed to products, the central issue lies in the direct relation between production and consumption, making it possible for the service provider to monitor service use and intervene if necessary. Again, the focus is on the responsibilities of the provider rather than the service itself. Thus, this paper provides a complementary perspective, focused on mediation of moral relations through explicit decision making by (computing) technology.

METHODOLOGY

The research objective of the present study is a categorization of contributions to moral decisions by ubiquitous computing environments.

To find these categories, we use a structured approach to analyse these systems.

Our approach consists of three steps. First, we determine a categorization of different *relations*these ubiquitous systems have with humans. Secondly, for each of these relations we determine the different *decisions* with moral implications that the system can make using a model for information processing. Thirdly, we analyze the resulting decisions from the previous step and we construct based on their properties different *categories* of moral decisions from the perspective of the information processing done by the machines.

For the relations, our approach is similar to that of Ihde and Verbeek, in the sense that we focus on the role of the computer in the human-world relation. From this perspective, we identify the different possible "positions" the machine can have as a mediator. For the decisions, we rely on the sequential model of information processing proposed by Parasuraman et al. (2000). For each identified relation, and for each step in the information processing, the moral aspects of the decision are mapped systematically. For the final categories, we align and compare the moral aspects within the relations in order to identify cross-cutting concerns.

Definitions

In the following sections, we derive a classification of moral mediation by machines that make explicit distinctions. This complements existing classifications of moral mediation by passive artefacts, such as those of Ihde and Magnani. To be able to do this, we first provide definitions of computer and moral behavior within the scope of this research.

Computer

In this research, the term *computer*, *system*, or *machine* is used to merely refer to a category of systems that calculates output based on a given input, following a predefined script of instructions. In this definition, *system* is meant in the abstract sense, meaning that for example a cluster of computers—a distributed system—can also be seen as a "computer" within the scope of this research.

The kind of systems that are considered in this research, are computers that fit (partly) into the definition of Poslad (2009) of ubiquitous computing systems. In Poslad's definition, five properties of ubiquitous systems are identified:

1. *Distributed*—Computers need to be networked, distributed and transparently accessible;
2. *Implicit HCI*—Human-computer interaction is hidden, the users do not necessarily know that they are using a computer;
3. *Context-aware*—Computers are context-aware, information about the environment is used to optimize their operation and make informed decisions;
4. *Autonomous*—Computers can operate autonomously, without human intervention. Human agents in its environment may not be able to manipulate or configure the steps the computer takes in its calculations resulting in its behavior;
5. *Intelligent*—Computers can handle complex problems by using intelligent algorithms for decision making. Making decisions that depend on several factors, like context, is considered as intelligence within this research.

The systems considered must be context-aware, intelligent and autonomous to a certain extent. Being distributed or having implicit HCI is not necessary for the systems in this article.

Systems that fit within this definition are for example a household robot, which is context-aware (it can see dust, dirty dishes, and is able to navigate within its environment without bumping into objects), it functions autonomously (when there are dishes to be done, it finds them and cleans them, when there is visible dust, it starts cleaning without having to wait for the owner of the house), and it is intelligent (it can anticipate to moving objects, find solutions to deal with moving objects, finds hidden dust in a room in which objects are sometimes being moved). Another example of such a system is an UAV (Unmanned Aerial Vehicle) or a drone. Some of these drones are used for surveillance and fly around autonomously in a specified range while observing the area. Many of the current drones send data to a control center where the data is being analyzed, but it would be imaginable that in time these machines are responsible for the analysis of the data as well. For the gathering of data and for the autonomous flying it is necessary that the drone is context-aware. For the data analysis it needs intelligence. In this case, if there is interaction between a user and the drone, this might be explicit interaction by an operating, using a computer interface or a remote control to operate the drone. To control a large area, in some cases the observation system might consist of multiple drones that work together, together forming a distributed system.

Although the systems that are considered in this research are all intelligent, context-aware and function autonomously, the extent to which they have these properties might vary. Intelligence, context-awareness and the level of autonomy is of course variable. A regular thermostat that can commonly be found in a house can be described as being intelligent (it makes decisions based on incoming information), it is context-aware (it measures the temperature in the room) and it functions autonomously (it works without any human intervention). However, the decisions it takes are based on simple rules that are programmed in the system (if the temperature is lower than x, put on the heater) and this would be a very limited level of intelligence, if it were considered intelligence at all. It is also questionable whether such a system is functioning autonomously: most of these systems just react to a timer, check the temperature, and wait again for the timer to send a signal. No human intervention is needed, but the system is also not making complicated decisions by itself. The system measures one environmental parameter: temperature. It is in that sense context-aware, but its awareness is very limited.

Determining whether a system fits within the definition is hard: for each of the properties a scale on which the level of ubiquitousness of a system could be measured would be very useful. However, developing such a scale would be beyond the scope of this research, because it introduces a number of issues that need their own space to be addressed. For example: one might argue that a scale for properties like autonomy, context-awareness and intelligence, might be defined using a range from simple one-dimensional calculations on one side to human properties on the other side of the scale. In this research, we do not address this question further, and use the definitions of the properties as defined above, leaving room for discussion and interpretation about which systems do fit and do not fit within the definition (a family resemblance).

Moral behavior

In this research, the assumption is that now that computers have become complex systems that are able to run autonomously and automate large tasks, it is no longer feasible nor fair to place the full responsibility for the actions of the system with developers of the system or with the owner of the system. Many systems, and especially the autonomous, intelligent and context-aware systems we consider in this research, work with self adapting techniques. This means that the developers deliver a machine with guidelines about how to process information and how to learn from this information, but they no longer deliver a system that does exactly what they programmed. The outcome of the calculations of the computer is in this case the result of the experiences the computer has, and these experiences and the resulting outcome of the computer's calculations are often beyond the control of either developers or owners of the systems. This leads to the need for the computer to have ethical behavior (Allen et al. 2006), or as argued by Wallach and Allen (2009), that since computers take more responsibilities, the need for computers learning ethics has risen.

Within this research, the computer's moral behavior is described. A definition of moral responsibility is used, which is translated into a definition of moral behavior of a computer. According to Noorman (2012), most analyses of moral responsibility share at least the following three conditions.

- The actor that is held responsible should have had control over the outcome of events;
- The actor that is held responsible should should have knowledge about and insight into the outcome of its actions;
- The actor should have the choice to act in a specific way.

Within the definition of Noorman, the actor should have control over the outcome of events and he should have the choice to act in a specific way. These conditions might imply intentionality of the machine and will lead to the discussion about whether the computer can actually make moral decisions, whether it has a human-like consciousness which it uses during the decision making, or whether the computer has free will. Whether computers have a consciousness or free will is not important for this research: in this research moral decision making of computers is defined as the observation that the computer makes decisions of moral content. This means that a decision is a moral decision when the decision were a moral decision if it were a human agent that would be making the decision.

Within this research, we consider moral behavior of a computer an act (or a sequence of multiple acts), or lack thereof, if (Noorman 2012):

- the computer influences its environment for better or for worse with this act (Causal contribution);
- the computer decides for this act based on moral arguments (Considering the consequences);
- the computer chooses between two or more possible courses of action (Freedom to act).

Types of relations between humans and computers

Between individuals and computers, different relational constructs are possible, in which different kinds of moral responsibilities and moral behavior play a role. For example, a medical expert system might be used by a third party (a doctor) to ask for advice about an individual (the patient), or a (very futuristic) care robot takes care of elderly people. In this section, a classification is provided that identifies different types of relationships between individuals and computers, following the definition of computers in Sect. 4.1. The classification can be considered analogous to the one by Ihde for passive mediation.

The classification of the relations between individuals and computers is based on the kind of interaction between the machine and the individual. Two kinds of partial interaction are used in this classification: interaction consists of either the machine providing output or processing input, or a combination of both. In this context, input means that the machine records and analyzes data. With output is meant that the machine interferes in the situation of the individual using movement, speech, gestures or any other kind of communication or action. This input and output, in turn, forms part of the human experience with the machine in their relation.

In this classification, four different types of relations are identified, based on whether the relation contains observation, interference or both.

1. Observation relation: the system observes individuals and reports information to a third party;
2. Interference relation: the system acts based on a task that is given to it. The system influences an individual, but not having meaningful interaction with the individual, meaning that the goal of the interaction is finishing its task, and the reaction of the individual is only used for finishing the task;
3. Interaction relation: the system interacts with individuals, using both observation and interference;
4. Advice relation, or observation and interference by proxy: the system gives advice to the third party about the action it should take towards the individual.

We will discuss these relations in more detail in the corresponding subsections below. Three actors play a role in the classification:

1. The computer: The computer is a ubiquitous computing system as defined in the previous section. The computer has a purpose that is determined by its designers or by a third party. The computer can observe its environment with sensors or can interfere in its environment using its actuators. Some systems only have sensors or actuators, other systems have both;
2. The individual: The individual is the one that interacts with the computer. The individual can be a person, but it can also be an animal or another computer. The individual has behavior and can react to changes in its environment or to actions of others;

3. The third party: A computer runs a program, which has a specific task. The third party is the actor that starts the program, that assigned the computer to this task and that benefits from the work the computer carries out. The third party can be an individual, a company, a computer system or a government agency, etc.

The actors and their relations can be seen in Fig 1. In the remainder of this section the four relationships are described in more detail and some examples are given for each type of relationship.

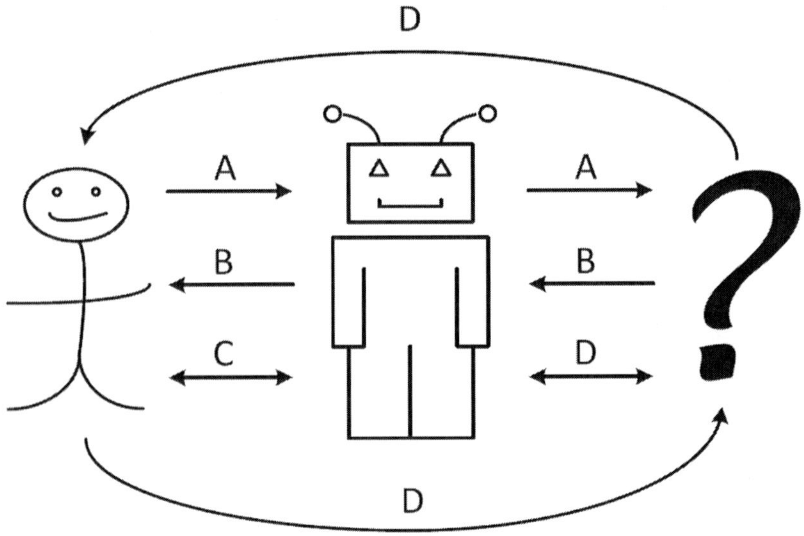

Figure. 1 Four different relations between individual and machine. From *left* to *right* the individual, the computer, the third party. The relations are labeled as follows: *A*Observation, *B* Interference, *C* Interaction, *D* Advice

Observation

In the observation relationship, the system observes the individual. It does not act towards the individual, and it also does not react to the individual based on actions of this individual.

Characteristics of this observation relationship are that the system can only observe and pass the information to a third party. The decisions this system can make are to pass specific information, or to withhold this information from the third party. The importance of this decision depends on the kind of system and on the kind of information the system observed.

An example of a system that has an observation relationship with the individual is a surveillance system watching parts of the entertainment district in a large city, which is prone to violence, for suspect behavior. Still, also systems that observe people without surveillance as a primary objective might run into the same decisions as surveillance systems. For example a smart house, keeping track of the inhabitants for the purpose of making their life more comfortable, also may observe changes in patterns of behavior, or might recognize specific behavior. Depending on the options the developer gave the system, the system can store information, label information, or even send the information to a third party.

In Ihde's classification, the observation relation would constitute a hermeneutic relation *third party - (system - world)*, in which the third party interprets and acts upon information provided by the system. However, from the perspective of the current study, the observation and selection of the information also provides a contribution to the moral decisions made within the relation, as the observation contains information about a moral patient, not just an abstract world. In addition, rather than providing a fixed perspective on the world such as a thermometer, explicit decisions on the selection of information are involved. Such decisions taken by the system may directly affect decisions being made about the moral patient.

Interference

In the interference relationship, the system interferes with the life of the individual, mostly to execute a task, or to fulfill a mission. These systems can take over full functions or tasks that have to do with human interaction, such as carrying out an arrest, or finding and assassinating a terrorist. The system acts according to its task, and it only reacts to the actions of the individual when this is necessary to finish its tasks. This means that all the feedback from the user is only used to fulfill its task.

The characteristics of the interference relationship are that the system interferes with a human, based on an assignment the third party gives. The purpose of the system is not to judge the situation it finds while carrying out its assignment, but to finish the assignment. The third party decides on the assignment, but the system is still able to choose how to execute its tasks and adapt its strategy to changing conditions. The issues that are involved in making decisions while carrying out the assignment are related to the amount of freedom the third party gives the system.

Examples of systems that are involved in an interference relationship are a care robot that makes sure that all the patients in a hospital ward receive their medication in time, a drone that drops a bomb on enemy soil, a doctor robot that replaces an organ, a police robot that arrests a suspect, etc.

As Ihde's classification focuses on perception only, this is not directly relevant for the interference relation. Compared with Verbeek's phenomenology of action, in which systems mediate human behavior, the focus is on decisions to act and actions performed by the system itself. Where a speed hump may change human behavior in the sense of inviting slower driving, thereby contributing to the moral value of safety, ubiquitous computing systems may contribute to outcomes and values by explicitly chosen actions. Therefore, the technology does not merely make certain moral actions by humans more likely or unlikely, thereby mediating moral choices of humans, but is actively involved in the manipulation of the environment, potentially affecting moral patients.

Interaction

In an interaction relationship, the system has full interaction with the individual. It acts towards the individual and it reacts to the actions of the individual. The third party does not have a strong influence on the actions of the system, or has no influence at all.

The interaction relation includes both interference and observation. The system chooses its actions based on its (self-)assigned role in the situation, and its actions are either its own initiative, or a reaction to the individual it is interacting with. Depending on its capabilities, its behavior can approach human behavior, and therefore its relation with the individual can approach a well-matched human-like relation.

In this relation, the system has to make decisions that are comparable with decisions humans make in their relationships with each other. Examples of systems that fall into this category are robot-pets and (future) autonomous humanoids. The category may also include humanoid robots that have dedicated tasks, but are also able to interact on an equal level, such as care robots and household robots with which one can also socialize.

The interaction relation appears to have a strong similarity to Ihde's alterity relation, in which the technology appears as quasi-other. In this relation, the system can make decisions both about the selection of information and about actions based on this information. However, the explicit decisions being made by the system make the moral status of the relation more complicated, especially when physical actions are involved. Whereas a computer game can easily be understood as quasi-other, care robots are meant to act in the real world. Whereas part of the interaction can certainly be explained in terms of quasi-others (Coeckelbergh 2011), the real-world consequences of the system's decisions should be part of the moral considerations as well.

Advice

In the advice relationship, the system gives advice to a third party about how to act towards the individual based on the information that is provided by the third party. This system embodies both observation and interference, but the machine only implements these forms of relation by proxy: the observation and the interference are executed by the third party or are overseen by the third party.

The main characteristic of the advice relation is that the system does not actually do anything itself. A third party provides the system with data, and asks what the best course of action should be. The system only provides advice to the third party. This means that the computer cannot judge the validity of the information or the way in which its advice is executed and the impact it has on the environment.

Examples of systems that give advice are expert systems, medical diagnosis systems that provide a diagnosis based on photo's or lab results, or forecasting systems that predict the stock market.

Compared to the observation relation, the system does not merely select and distribute information, but proposes an action. Therefore, the advice relation contains elements of both Ihde's hermeneutic relation (providing information about the world to the third party) and Verbeek's phenomenology of action (influencing the choice of action of the third party). This combined role makes it particularly powerful in contributing to the decisions made within the human-machine relation, even though the *direct* effects are more limited than in an interaction relation, in which the machine may act itself.

Using the classification

This classification identifies the type of relation a computer has with a person. However, there are many systems in which the relationship with the a person includes aspects of different types of relationships. This is not a problem: one system can have multiple types of relations with its environment. The objective of the classification is not to put one system in one category, but to identify different types of relationships within one system, which reveals different moral aspects of the behavior of the system that are bound to these relationships.

An example of a system in which more than one type of relationship can be identified is the care robot. Firstly it socializes with the elderly people it takes care of. In this role the relationship that the system has with the individual is equal, and the relationship would be classified as an interaction relationship. Secondly, the robot is responsible for giving the patients their medication. To complete its task, the system sometimes needs to force patients to take their medication. In this role, the relationship would fit within the description of an interference relation. Thirdly, the system also has the task to observe the ward and report any incidents that require medical attention. Within this role, the system has an observation relationship with the people involved. Within this system, three relations exist next to each other. This classification is used to identify these different relations.

Moral aspects in the computer's behavior

In the previous section, four different relations between humans and computers are described. The behavior of the computer or machine within each type of relationship can be seen as a sequence of steps that are repeated as long as the relationship or the interaction between the computer and the individual exists. To find the moral aspects of the behavior of the machine, we will first analyze these relations in more detail, to determine the separate steps that occur within these relations. We use a four stage model of information processing in automation provided by Parasuraman et al. (2000). This model is an adaptation of a model that describes how humans process information, and consists of the following steps, which can also be seen in Fig. 2:

- information acquisition;
- information analysis;
- decision and action selection; and
- action implementation.

In this section, we use this model to analyze each relationship. Based on the outcomes, moral aspects within the relation between human and machine are identified. The moral aspects are then further analyzed using the three properties of moral behavior (freedom to act, causal contribution, considering the consequences, see also Sect. 4.2). It is assumed that for each moral aspect the computer has freedom to choose, and therefore *freedom to act*, within the limits of the specific type of relation. For each moral decision the computer makes within these relationships, the *causal contribution*, the moral aspects and the *consequence* the computer should *consider* in its decision are described.

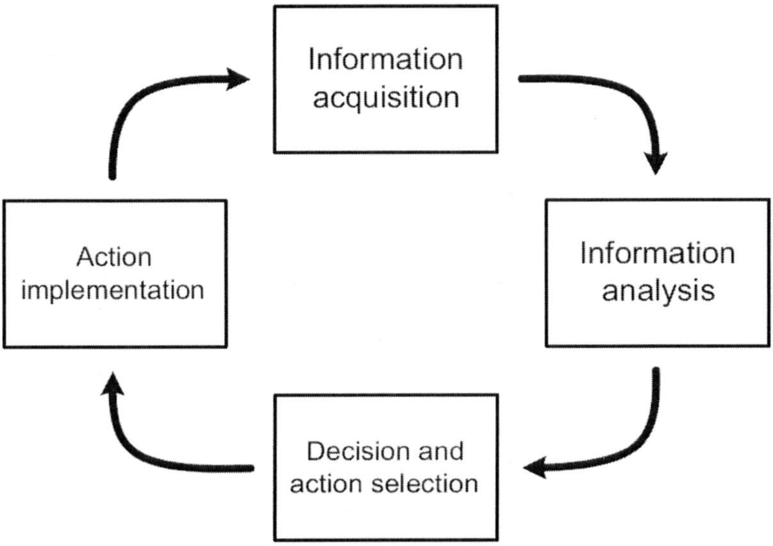

Figure. 2 Steps for information processing in automation by Parasuraman et al. (2000)

Observation

The process of observation can be found in Table 1. The decision and action selection phase within the cycle of information processing for the observation relationship includes moral aspects of the behavior of the computer. The computer compares the information it gathers by observing the individual with the goal of the observation. If the information fits within the goal of the observation, it chooses between three options: it can decide to forward the information to the third party (a), to store the information (b) or to delete the information(c). It might also decide to forward, store or delete information that does not fit within the goals of the observation(d).

In for example a smart house, each of these four decisions can be recognized. To make the life of the inhabitants more comfortable, the house keeps track of what the inhabitants do during the day. When one day the fourteen-year old living in the house starts using drugs, the computer at first only observes that this inhabitant uses drugs once. The computer can pass this information on to a third party, causing this information to end up in a database where it might influence his future options for health insurance or for getting a job. The computer can also decide to store this information. Observing the inhabitant daily and storing the new observations, after a couple of days the computer can conclude that the inhabitant is now a regular user. Since this conclusion is stronger than the observation of only using drugs once, this conclusion can be even more harmful when the computer decides to share this information. The computer might also decide to delete the information: in this case, the drug usage history is "erased" and does not end up in any systems where it might harm the inhabitant— at the same time this might be useful medical information which can be used by doctors to help the patient, or maybe even to inform the parents about their child's behavior, to give them a chance to interfere. While observing the activities of the inhabitants, the house might also observe the activities of visiting friends of the inhabitants. These activities are outside the scope of the assignment. However, the system might decide that it is the responsible thing to do to report the drugs dealer that provides the child with drugs, to store this information or to delete this information.

The decisions to pass, store or delete information, in- or outside the scope of the assignment, include a number of moral aspects:

a) By passing information to a third party, the computer can influence the life of the individual and its environment in many ways. The influence that the passing of information has on the individual depends on what the third party will do with it, and on what kind of information it is. Passing the information might result in actions from the third party or from others towards the individual, which can either be harmful or beneficial for the individual, but this might also result in a change of the impression that people have of the individual in the future.

In the case of the drug-using inhabitant of the smart house, if the usage of this individual becomes known, the third party or people informed by this third party might want to intervene in the life of this person. When the drug usage is known by insurance companies, this might influence the possibilities of getting insurance. When this information is passed on to future employers, the chances of getting a job in the future might decrease. But, when this information is passed on to experts in treatment of drug addiction, the result of the act of passing the information might also be beneficial for this inhabitant, also when this person would not choose this solution at first.

By passing information, the computer should thus determine what the interests of the involved parties are, and what effect the forwarding of information might be on all parties.

b) When the computer decides to store the gathered information, this can be either beneficial or harmful for the individual, depending on information the computer has about the individual. By storing information, the computer might be "prejudiced" towards the individual in the future, depending on how exactly the data is stored, processed and used. The computer may also be able to predict future behavior or recognize patterns in behavior. Storing this information when the information includes something the individual wants to hide, can be either bad or good for the individual.

For example: drug users might want to hide their addiction for various reasons, but it might be better for their health if their addiction is discovered. This addiction can only be detected by storing multiple incidents of drug usage, adding these events and concluding that there is a pattern usage indicating an addiction. When the system observes an individual, it might also be able to recognize specific patterns that indicate illness, psychiatric disorders, specific character traits. This information might be helpful for the individual itself, for example for detecting a dangerous disease early.

Storing information can also be detrimental to the individual: by storing information, it is like the computer never forgives a bad act (e.g., the inhabitant of the house in the example will always be a (former) drug user). Also, saving information always creates a risk: as long as the information is stored somewhere, it might become available to someone with unknown intentions some time in the future. The impact of this will be unknown. Saving information therefore always contains a risk.

c) By deleting the information, the computer might lose valuable information. This might either protect or harm the individual, depending on the kind of information and the interests of the third party. When the system determines that the individual has had several symptoms of a disease over the last days, but it deleted the data about each symptom, it will not be able to combine the information into a full diagnosis. This might also be a problem when certain private information relevant to diagnosis is deleted, such as drug use. When a computer observes personal or sensitive information about the individual which it does not have to report (directly) to the third party and its stores it, the information might still be retrievable at a later moment by, for example, the police or the third party. To make sure that the privacy of the individual is guaranteed, it would be safest to remove the data immediately.

By deleting data, also another issue is raised: the computer might consider that the gathered information is currently not useful, but the information might become useful in the future because of advancing scientific knowledge or future changes in laws or regulations.

d) Even when the assignment of the computer does not include observing specific events, the computer might decide to forward, store or delete the information. For storing, forwarding or deleting this information, the same issues play a role as described in a, b and c.

However, the issue here is that this is not part of its assignment. For example: the system might be able to detect drug usage of friends of the inhabitant, or even behavior that indicates that some friends are providing the inhabitant with drugs or otherwise have a bad influence on health of the inhabitant. This information might be used by a third party in such a way that it might be beneficial or harmful for either the inhabitant or the friends of the inhabitant.

Table 1 Stages of information processing in the observation process

Information acquisition	1. The computer observes an individual and gathers data about the individual
Information analysis	2. The computer evaluates and interprets the gathered information
Decision and action selection	3. The computer compares the information with the goal of the observations and decides to
	- forward the information to the third party
	- store the information
	- delete the information
	forward, store or delete information that does not fit within the goals of the observation.
Action implementation	4. The computer forwards, deletes or stores the information and goes back to step 1

Interference

In the process of interference, the decision and action selection phase includes two types of decisions: the computer decides on whether to proceed with or abort its mission (a), and it decides on how to proceed (b). This process can be found in Table 2.

An example of an interference system is an autonomously flying drone, which might be used by a third party to drop a bomb on enemy territory with the intent of eliminating a terrorist. This system is sent to a remote area, where it has to locate its target and destroy it. Its goal is fixed, but its exact strategy is not determined yet. The drone can first try to locate the target. When the target is clear, it can drop a bomb. Afterwards, it should evaluate whether the target is destroyed. If this is the case, it can go home. Otherwise, it needs to retry: it finds a new strategy, relocates the target and drops another bomb. It can also be problematic to destroy the target because there are innocent people standing close to the target, or because the target cannot be found. In this case, the drone might have to abort the mission and fly back back because it is impossible to finish the goal.

Within the two decisions of the interference relationship, the computer's decisions can have large impact on the individual and its environment. The moral aspects of these two decisions are the following:

a) The computer determines to what extent its goal is achieved, and decides on how to proceed. The computer might decide that the goal is (partly) achieved and quit trying, or the system might decide to continue trying to achieve its goal. It might also decides to quit trying because it realizes that its strategy is (becoming) too harmful for the environment. The computer should weigh the importance of carrying out the assignment against the harm or benefit its strategy causes. In the case of the drone attack: it might have the objective to destroy a target, and try to reach this goal several times, but while trying it might conclude that the chance of successfully finishing the mission is low, while with every attempt innocent bystanders get hurt. When at some point the environmental parameters change, e.g. the weather increases the chances of hurting these bystanders, the best solution might be to decide that this goal cannot be received without doing too much damage to the environment.
b) The computer has to find a way to achieve its goal. The choice for a specific strategy influences the individual: the strategy might be comfortable for the individual or it might be unpleasant. A strategy might even cause harm or benefit to the individual. The strategy is chosen within the scope of a fixed goal. This means that in certain cases, the computer might be forced to choose an option that is considered bad or immoral, because no alternative strategies are available within the scope of the goal. When the drone has the assignment to destroy some target, it might have several ways to achieve this goal. In choosing the strategy wisely, the system might optimize the number of casualties. For example, by waiting with carrying out the assignment until midnight, the number of bystanders is minimal compared to destroying the target at the middle of the day.

When the system considers the consequences of the strategy, not only the experience of the affected individual, but also the anticipated actions of the ones affected play a role. When the computer chooses for a specific strategy, the individual or the environment will react in a specific way. This reaction of the individual might have consequences for the individual, for the robot or for the environment.

The goal of the computer is determined by the third party. The computer can decide to execute a strategy targeted at achieving the goal, or decide to give up, but it cannot alter the goal. This means that the moral responsibility for the goal itself is limited within this relationship.

Table 2 Stages of information processing in the interference process

Information acquisition	1. The computer receives feedback about its action from the individual he is interacting with, or from the environment.
Information analysis	2. The computer evaluates the feedback, evaluates to what extent the goal is achieved
Decision and action selection	3. The computer: - decides that it did not achieve its goal yet, and retries - decides that the goal is achieved, or that the goal could or should not be achieved, and finishes the interference procedure - keeps or changes its strategy
Action implementation	4. The computer acts, according to its strategy.

Interaction

For the interaction relation, the decision and action selection phase includes moral aspects of the behavior of the computer, as can be seen in Table 3. During this phase, the computer evaluates the feedback it received from the environment and the individual. The computer can decide whether it already achieved its goal, but it can also determine how the environment and the individual experience behavior of the system. When the behavior of the system is not appreciated, or when the system does not get the results it expected, it can either change its strategy and hold on to its goal (b), or reformulate or renew its goal and find a new strategy (a).

As an example, we look at the situation where robots work in our houses to do our housekeeping. Within the relationship between a household robot and the inhabitant of the house, an interaction relationship exists. Within this relationship, the function of the system is clear: it is responsible for cleaning the house, assisting with the dishes and doing the laundry. The main goal of the system is to keep the house clean, but this results in changing sub-goals: cleaning the dust next to the cupboard, cleaning a pile with dirty laundry from the bedroom, collecting dirty cups and glasses in the house and bringing these to the kitchen. For each of these goals, the system has to find a strategy. In each of these strategies or goals, conflict between the goals might arise, of which some of them might also include moral dilemmas. The conflicts might range from simple scheduling problems to moral decisions: is it more important to do the dishes or clean the bathroom in the time available, is the robot allowed to for example destroy an object to allow itself to clean behind it, or is the robot allowed to kill vermin like roaches, or even small mammals like mice and rats?

The decisions within the interaction relationship, both changing a strategy and changing the goal, include a number of moral aspects:

a) The system has a goal or multiple goals which determine the behavior of the system within the interaction with its environment. Goals can usually be explained or substantiated by a certain moral reasoning. A toy-robot might have "entertaining the individual" as a goal. This goal displays positive intentions towards the individual. A robot that is designed to build cars has as goal "to assist humans with heavy tasks within the process of building cars". This goal can be read as a positive goal towards the individual. However, the goal can also be explained as a bad goal towards the environment: a robot which acts "good" should not help build a machine that poisons the environment with air pollution. Determining the goal itself can be seen as a moral decision. Within the interaction relation, the goal of the system is partly determined by the system itself. The household robot is of course dedicated to keeping the house clean, but within that scope, it decides its tasks: cleaning the dishes, vacuuming the house, organizing loose objects in the living room.
b) Within the scope of a specific goal, the computer can use a specific strategy to achieve this goal. This issue is the same as issue (b) for the interference relationship, but the difference is that the goal of the computer is decided by a third party for the interference relationship, while the goals within the interaction relationship can be adjusted by the system itself (within the scope of its general assignment). This means that the moral responsibility of the computer within the interference relationship for accepting a strategy has a different weight than the moral responsibility within the interaction relationship: within the interaction relationship the computer has control over the goal, which means that the choice for a strategy is always a positive choice and not a negative choice, while the choice for a strategy within a fixed goal might be a negative choice (the choice is the best option of the available options).

The interaction relationship can be seen as an equal relationship. The communication between the individual and the computer is both ways. The influence of the third party is absent. This means that the goals and strategies of the computer influence the impression the individual or the environment have of the computer. This, in turn, implies that not only the morality of current goals and strategies play a role, but also the effect that the current choices have on future decisions and the effects it might have on the options to choose strategies and goals in the future.

Table 3 Stages of information processing in the interaction process

Information acquisition	1. The computer observes an individual or receives feedback from the individual about its own action
Information analysis	2. The computer evaluates and interprets the gathered information, and evaluates to what extent its goals are achieved
Decision and action selection	3. The computer evaluates the feedback, evaluates to what extent the goal is achieved The computer either: - decides that it did not achieve its goal yet, and tries to achieve its goal again - decides to change its strategy and hold on to its goal, - adjusts or renew its goal and find a new strategy.
Action implementation	4. The computer acts or stays still, in accordance to its goals

Advice

The moral aspects of the behavior of the computer for the advice process can be found in the decision and action selection phase in Table 4.

Within the advice relationship, the impact the computer has on the individual completely depends on the question the third party asks the computer. Within this relation, there are three moral aspects: two have to do with the fact that the gathering of information (a) and the action that results from the advice (b) are not performed by the system that gives the advice. The third moral aspect is about the advice itself (c).

An example of an advice system is a medical expert system. This system receives information from a doctor. It first needs to judge the reliability of this information before it can use it. The doctor might unknowingly have provided incorrect or incomplete information, which might influence the diagnosis. The system then provides the doctor with an advice about a medical strategy. By providing this advice, the system should consider how the doctor's behavior is influenced by this advice. The doctor might get the advice to do something that he is not sufficiently trained in, or not skilled enough for. In that case, the system should maybe not give this advice. The last moral aspect might be in the decision itself, for example when the doctor asks for advice about a situation involving controversial treatments like life extending treatment for terminally ill patients, abortion or euthanasia.

a) The computer receives information from the third party which it has to base its decision on. This information might be unreliable, so balancing the possible impact of the advice on the individual with the perceived reliability of the information is a moral aspect within this relation, e.g. in a diagnosis.
b) The computer gives advice about actions towards an individual to a third party. This means that the computer has no control over how the action is understood or how the action is executed. If the action is executed and the expected result fails to materialize, the computer cannot intervene and correct its advised action. In the medical example, the system cannot observe or interfere with the treatment and its effects.
c) The computer gives advice, and depending on the kind of advice, the advice might also include moral aspects. Here, the computer can assume that the third party might adopt and execute its advice without any consideration. It is like the computer acts itself; the moral responsibility for the outcome of the advice lies therefore with the computer. This would be the case in advice about life extending treatment.

Dimensions in moral decisions

In the previous section, the for each type of relationship between human and computer in the context of this research, the decisions with moral implications are listed. Within these decisions, three distinct dimensions can be identified: scope, impact and involvement. In these three dimensions the action that is being carried out, the actors that carry out the action, and the situation in which the action is being carried out are being considered. These dimensions can be used to distinguish between different types of moral mediation.

Table 4 Stages of information processing in the advice process

Information acquisition	1. The computer receives a case description about an individual from a third party
Information analysis	2. The computer evaluates and interprets the gathered information
Decision and action selection	3. The computer: - decides on the reliability of the information - decides on the expected execution of the advice - decides on the advice
Action implementation	4. The computer advises the third party about the case

The dimension "scope", is about how actions fit within the responsibilities of the system. A distinction can be made between decisions that are inside or outside of the scope of the assignment of the system. This decision is about the borders of the responsibility that is assigned to someone or something in an implicit or explicit way. For example, a smart house has the task of making the life of its inhabitant more comfortable by keeping track of what its inhabitants do during the day. While observing its inhabitants, it may also observe behaviors of others than the inhabitants. In some situations, it might be helpful (or harmful) for the inhabitants when the house also takes action based on behaviors of others than these inhabitants, for example when visitors exhibit harmful behavior towards its inhabitants. However, this is outside the scope of the assignment of the system, and these actions might impact the life of both the inhabitants and the visiors, so there clearly is a scoping dilemma here.

The dimension "impact", is about the situation in which the decision has impact. A distinction can be made between decisions that make a direct impact or that have a more indirect impact on the environment of the system. When making a decision with direct impact, the situation is known or can be known: all the factors that influence the immediate outcome of the decision are already present. An example is a decision about dropping a bomb from a drone. The damage is immediate: people die or get hurt. The information about this decision is already present: the drone has a location, the weight of the bomb, the speed of the drone and the wind velocity is known, the location where it will land can be calculated and the people that will be on that location of the drop are already there or very close. When the decision is about whether to store or delete information, the decision might have impact in the future, depending on how the rules about using this information are at that point, depending on which party uses the information and a number of other factors. This information is not known yet when the decision is made to store the information.

The dimension "involvement" is about the actors that are involved in the decision: the decision can be an independent decision in which the system is the only one who's judging, deciding and implementing the decision into action. The decision might also be a dependent action in which multiple actions are involved. This might be in delivering the input to the system, or in carrying out the action that results from the decision. A medical advice system, for example, uses information that is given to the system by doctors. The doctor has already made a prejudgement to give this specific information, and possibly to hold back other information that is being perceived as being redundant, irrelevant or otherwise not of any value to the system. The system gives advice based on this information, and then the doctor will (or won't) carry out the intended action. The system might give an advice that results in a medical procedure being performed by the doctor in an incorrect way because of lacking skills, or for any other reason. Even though the system gives the right advice based on the assumption that the information that it used is correct and that the advice is being carried out correctly, the outcome of the advice might still be a disaster.

Using the dimensions above, decisions identified in the previous section can be classified. For example, adjusting a strategy, which is a decision which the system has to make in the interference and the interaction relationship, can usually be classified as a decision inside the scope of the assignment, the impact is direct, and it is an independent decision. The decision to store information is a decision inside its scope, also an independent decision, but the impact is indirect. The decision about changing goals or choosing an empty strategy for the interference and interaction relationships are decisions that are direct, independent and outside the scope. Forwarding information to a third party has indirect impact, is a dependent decision and is inside the scope.

CONCLUSION

In phenomenology of technology, analyses of the relation human-technology-world have mostly been directed to situations where the technology forms a more or less constant "lens" through which experience and action can be changed. In that sense, the technology contributes to experience of and action by humans, and as such may affect moral patients.

Several approaches have proposed extensions to this framework, as discussed in the Related Work section, including cyborg intentionality and composite intentionality, reconfigurable technology, and service technology. These approaches indicate that in emerging technologies clear boundaries between humans and technology may no longer be observable or relevant, and that these technologies allow for a dynamically changing type of mediation, rather than a fixed one. However, none of these approaches pay attention to technologies that make explicit decisions, and are thereby able to *reconfigure themselves*. Their moral impact is therefore different than in the traditional approaches to mediation, and new types of mediation relations need to be considered.

Within this research, we have investigated moral aspects of computer-made decisions, in order to identify ways in which new technologies change moral relations. We considered computer systems that are autonomous, context-aware and intelligent. These systems can have four types of relationships with a human and, in some cases, a third party. These relationships are the observation relation, the interference relation, the interaction relation and the advice relation. Each of these relations can be modeled as a loop in which the same stages are always repeated. Within the decision and action selection stage, each system has a number of decisions it can make. When all these decisions are combined, it can be concluded that for these kind of systems, there are three dimensions of the decisions of the system in these four relations that require moral insights or moral behavior:

- Scope: Operation inside/outside the scope of its assignment
- Impact: Considering the short-term/long-term consequences
- Involvement: Dependency of the result on self only/on others

Based on these distinctions, it becomes possible to derive guidelines for the design of machines that are involved in these aspects of decision-making. Specific guidelines for the three moral aspects of decision-making can be provided, for example providing requirements for the considerations that have to be taken into account in the decision-making processes. Using the classification for the types of relations, a distinction in advisability and desirability can be made in different relations and for the different moral aspects of the decision-making. For example, in terms of involvement of others, it can be required that the machines take into account the possible actions taken by others based on the decision made by the machine. Specific details for suitable reasoning approaches may be provided as well. Such guidelines can form the basis for future regulatory measures on different levels of abstraction, as well as accompanying design approaches and certification standards. In addition, the changes induced in moral relations can form the basis for grounded design choices on which decisions should be delegated to machines and which to humans.

As mentioned, several approaches have discussed the particular characteristics of emerging technologies that would require extensions to ethics of technology: reconfigurability, service-orientation, and, in this paper, explicit decision-making. In future work, these different approaches could be combined in an overarching analysis of responsibilities in such systems, and associated requirements for regulation and design. To this end, existing approaches such as value-sensitive design (Friedman et al. 2006; Van den Hoven 2007) can be used, but with specific attention to aspects of reconfigurability, inseparability of production and consumption, and—following the analysis in this paper—explicit decision making by the computers themselves. We assume here that assigning responsibilities and proposing regulation is still a task to be performed by humans, but it might be justified to question even that assumption.

FOOTNOTES

1 For a definition of ubiquitous computing, see Sect.4.

ACKNOWLEDGMENTS

We would like to thank N. Moseley for his help with the graphics.

REFERENCES

1. Akrich, M. (1992). The de-scription of technical objects. Shaping technology/ building society pp. 205–224.
2. Allen, C., Wallach, W., & Smit, I. (2006). Why machine ethics? *Intelligent Systems, IEEE*, *21*(4), 12–17. doi:10.1109/MIS.2006.83.
3. Coeckelbergh, M. (2010). Robot rights? Towards a social-relational justification of moral consideration. *Ethics and Information Technology*, *12*(3), 209–221.
4. Coeckelbergh, M. (2011). Talking to robots: On the linguistic construction of personal human-robot relations. In M. Lamers & F. Verbeek (Eds.), *Human-Robot Personal Relationships* (Vol. 59, pp. 126–129)., Lecture notes of the institute for computer sciences Springer, Berlin, Heidelberg: Social Informatics and Telecommunications Engineering. doi:10.1007/978-3-642-19385-9_16.
5. Dechesne, F., Warnier, M. & Hoven, J. (2013). Ethical requirements for reconfigurable sensor technology: A challenge for value sensitive design. Ethics and Information Technology pp. 1–9, doi:10.1007/s10676-013-9326-1.
6. Floridi, L., & Sanders, J. (2004). On the morality of artificial agents. *Minds and Machines*, *14*(3), 349–379. doi:10.1023/B:MIND.0000035461.63578.9d.

7. Fong, T.W., Nourbakhsh, I. & Dautenhahn. K. (2003). A survey of socially interactive robots. Robotics and Autonomous Systems.
8. Friedman, B., Kahn, P. H, Jr, & Borning, A. (2006). Value sensitive design and information systems. *Human-Computer Interaction In Management Information Systems: Foundations, 5*, 348–372.
9. Gaver, W.W. (1991). Technology affordances. In: Proceedings of the SIGCHI conference on human factors in computing systems, ACM, pp. 79–84.
10. Hildebrandt, M., Koops, B. J., & Jaquet-Chiffelle, D. O. (2010). Bridging the accountability gap: Rights for new entities in the information society? *Minnesota Journal of Law, Science & Technology, 11*(2), 497–561.
11. Ihde, D. (1990). *Technology and the lifeworld*. Bloomington, IN: Indiana University Press.
12. Johnson, D. G., & Powers, T. M. (2008). Computers as surrogate agents. In J. Van den Hoven & J. Weckert (Eds.), *Information technology and moral philosophy* (pp. 251–269). Cambridge: Cambridge University Press.
13. Kuflik, A. (1999). Computers in control: Rational transfer of authority or irresponsible abdication of autonomy? *Ethics and Information Technology, 1*(3), 173–184. doi:10.1023/A:1010087500508.
14. Magnani, L., & Bardone, E. (2008). Distributed morality: Externalizing ethical knowledge in technological artifacts. *Foundations of Science, 13*(1), 99–108. doi:10.1007/s10699-007-9116-5.
15. McLaren, B. M. (2006). Computational models of ethical reasoning: Challenges, initial steps, and future directions. *Intelligent Systems, IEEE, 21*(4), 29–37.
16. Noorman, M. (2012). Computing and moral responsibility. In: Zalta, E.N. (ed). The stanford encyclopedia of philosophy, fall 2012 edn.
17. Parasuraman, R., Sheridan, T., & Wickens, C. (2000). A model for types and levels of human interaction with automation. *Systems, Man and Cybernetics, Part A: Systems and Humans, IEEE Transactions on, 30*(3), 286–297. doi:10.1109/3468.844354.
18. Pieters, W. (2013). On thinging things and serving services: Technological mediation and inseparable goods. *Ethics and Information Technology, 15*(3), 195–208. doi:10.1007/s10676-013-9317-2.
19. Poslad, P. S. (2009). *Ubiquitous computing: Smart devices*. Wiley, Chichester: Environments and Interactions.
20. Ramos, C., Augusto, J. C., & Shapiro, D. (2008). Ambient intelligence—The next step for artificial intelligence. *IEEE Intelligent Systems, 23*(2), 15–18. doi:10.1109/MIS.2008.19.
21. Van den Hoven, M. J. (1998). Moral responsibility, public office and information technology. In I. T. M. Snellen & W. B. H. J. Van de Donk (Eds.), *Public administration in an information age: a handbook, IOS* (pp. 97–112). DC: Amsterdam; Washington.
22. Van den Hoven, J. (2007). Ict and value sensitive design. *The Information Society: Innovation*(pp. 67–72). Ethics and Democracy In honor of Professor Jacques Berleur sj, Springer: Legitimacy.
23. Van den Hoven, J., Lokhorst, G. J., & Van de Poel, I. (2012). Engineering and the problem of moral overload. *Science and Engineering Ethics, 18*(1), 143–155.
24. Verbeek, P. P. (2006). Materializing morality : Design ethics and technological mediation. *Science, Technology & Human Values, 31*(3), 361–380.
25. Verbeek, P. P. (2008b). Cyborg intentionality: Rethinking the phenomenology of human-technology relations. *Phenomenology and the Cognitive Sciences, 7*(3), 387–395. doi:10.1007/s11097-008-9099-x.
26. Verbeek, P. (2008a). De grens van de mens. over de relatie tussen mens en techniek. In I. L. Consoli & R. Hoekstra (Eds.), *Annalen van het Thijmgenootschap*. Nijmegen: Valkhof Pers.
27. Wallach, W., Allen, C. (2009). Moral machines: Teaching robots right from wrong. Oxford University Press, http://books.google.com/books?id=tMENFHG4CXcC

CHAPTER 8

Ethical Requirements for Reconfigurable Sensor Technology: A Challenge for Value Sensitive Design

Francien Dechesne • Martijn Warnier

F. Dechesne Energy and Industry Section, Department of Technology, Policy and Management, TU Delft, Delft, The Netherlands

F. Dechesne (&) _ J. van den Hoven Philosophy section, Department of Technology, Policy and Management, TU Delft, Delft, The Netherlands

M. Warnier Systems Engineering section, Department of Technology, Policy and Management, TU Delft, Delft, The Netherlands

ABSTRACT

Information technology is widely used to fulfill societal goals such as safety and security. These application areas put ever changing demands on the functionality of the technology. Designing technological appliances to be reconfigurable, thereby keeping them open to functionalities yet to be determined, will possibly allow the technology to fulfill these changing demands in an efficient way. In this paper we present a first exploration of potential societal and moral issues of reconfigurable sensors developed for application in the safety and security domain, in the context of a large scale R&D-project in the Netherlands. We discuss the subtle distinction between the relevant notions of reconfigurability, function creep, and unrestricted or unforeseen technological affordances. We argue that the feature of reconfigurability makes *context of use* the central issue in the assessment of the societal and moral impact of the technology. It follows that the design of good policies for new application contexts has to be central in a value sensitive design approach to reconfigurable technology.

Keywords: Reconfigurability Sensors Security Contextual integrity Policy Value sensitive design

INTRODUCTION: RECONFIGURABLE SENSORS

Sensors are devices that measure physical properties and convert them into signals interpretable to an observer, for the purpose of recording or responding with action. Sensors such as cameras and motion detectors have proven to be practical sources of information that can be used in the effort to provide safety and security for a society in general, and to help resolve crisis situations. Sensors are connected in networks, thereby facilitating collecting the information, analyzing it, and making it accessible to human decision makers.

The application areas of safety and security have given a strong impulse to the development of sensor technology, because it provides efficient ways of monitoring various kinds of situations, both involving technological and human behavior. This impulse has been clearly visible, for example, in the post 9/11 aviation sector [with mixed results: Johnson (2006), King (2011)]. Applications to safety and security also motivated a large scale, 4.5 year research project in The Netherlands: Sensor Technology Applied in Reconfigurable systems for Sustainable security (STARS 2010). The STARS project involves both academic and private research partners. The goal of the project is the development of "necessary knowledge and technology to be able to build reconfigurable sensors and sensor networks." By making sensors reconfigurable, the project aims to deliver a continuous and affordable infrastructure for societal security. The reconfigurability contributes to the affordability by keeping other application areas as open as possible: this should reduce the need to develop new technology for every new desired functionality. Reconfigurable parts of sensor networks that will be looked at are antennas, receivers, transmitters, on-chip and off-chip communication. As an example, reconfiguration of the sensors should make it possible to transform a sensor network installed in a harbor for security purposes, e.g. to prevent theft or sabotage, into an information system for rescue workers during a fire in the same harbor.

In this paper we intend to show how the reconfigurability of sensor technology, as envisaged in STARS, adds an extra challenge for efforts towards including moral and societal considerations in the technology design. The extra challenge of reconfigurable technology arises from pushing the specification of the intended functionality and intended use forward, outside of the technology design phase. We present a first conceptual exploration of the societal and moral implications of reconfigurable sensor technology such as the technology developed within the STARS-project. We present the questions we think will be the relevant ones, both for the conceptual, and for the empirical and technical aspects of the design process.

Reconfigurable sensor technology for the security domain

Among the characteristic aspects of the security domain are the diversity of threats, and the absence of warning time. The opponent is unpredictable: he cannot be expected to comply with any rules, and will be creative. Hence, maintaining security means to be able to anticipate and respond adequately to new situations. The societal problem is that it takes too long, and that it is too expensive, to invest over and over again in the development of new systems to protect against changing threats. Successful security technologies should therefore satisfy a number of requirements: to be reliable and affordable, sustainable and effective, multi-domain and multi-service. In the STARS-project, reconfigurable sensors are developed to have these characteristics. Their design is intended to allow for flexible application, where the possible functionalities of the system should be relatively easy and quickly to adapt.

In the initial architecture definition of the STARS-sensors, there are three different types of base stations, to which sensor modules of different types (e.g. radio frequency, infrared, sonar) can be attached. For flexible functionality, the base stations can be (re)positioned strategically, the sensor modules can be replaced by other types, and the settings of the modules can be adapted (e.g. one can switch to other frequencies).

With this specific setting in mind, we distinguish three reconfiguration levels: reconfiguration by tuning *parameters* (almost instantaneous, possibly remote), reconfiguration of the *arrangement* of sensors at a certain sensor node (which may require those sensors to be offline for a short time), and substantial structural reconfiguration, where components are removed or added (which may require the sensor node to be relocated and out of order for a longer period of time). In STARS, all three types of reconfiguration are envisaged, with different levels of functional flexibility associated to them. For example, instantaneous reconfiguration should be available to optimize for *foreseen* situations and the corresponding tasks. In future scenarios that are yet unforeseen, the reconfiguration may have to be more structural: Functionality that is radically different from the one envisaged in the current configuration, could also require more radical reconfiguration. (Note however that this is not necessarily the case: we will come back to the relationship between configuration and functionality in "Reconfigurable technology" section.)

With the flexibility and adaptability of the functionality as motivation, the feature of reconfigurability is leading in the design and development of the architecture and technologies in the STARS-project. The use cases that are initially defined within project, primarily restrict the envisaged users to police, fire brigade, security- and information services. However, it is expected that the technology, if successful, will cover a broader application area by a broader range of users. During the project, system concepts and application potential are to be defined and explored.

Restricted perspective on reconfigurability

To avoid misunderstanding about the scope of this paper, we should point out that we take a restricted perspective on reconfigurable sensor technology, one that stays close to the reconfigurability expected in STARS. It is worthwhile to make some remarks on the relationship between this restricted account, and the topics of ambient intelligence, and pervasive/ubiquitous computing. In short, we could say the latter topics are essentially about the presence of computing power in (interconnected) devices and how this can be used for a broad range of applications: they are about 'smart' objects (mostly equipped with sensors), which are reconfigurable by virtue of their computing power. Such systems do share with STARS a broad (and potentially flexible) range of functionality. An ambient intelligence/pervasive computing system as a whole may be considered to be reconfigurable. However, it is reconfigurable in a more general sense than the sensor technology in STARS, where the reconfigurability lies in the sensors rather than in (adding) computing power.

To illustrate this, we mention some work from the literature on ubiquitous computing and ambient intelligence. For example, Bellotti and Sellen (1993) describes a case study ("the RAVE network") which features a sensor network where the users are at the same time objects of observation—whereas the users in STARS are enforcement agencies who observe an external environment. It uses a framework [of Gaver et al. (1992)] focusing on control and feedback: control encompasses "empowering people to stipulate what information they project, and who can get hold of it", and feedback "informing people when and what information about them is being captured and to whom the information is being made available." These are general principles that can be guiding for ensuring a reasonable degree of privacy in surveillance systems. In particular the feedback aspect could be useful for STARS sensors. The fact that the issue of control will be hard or impossible for the objects of observation, points to a fundamental problem with STARS sensors (and surveillance for security in general). The ubiquitous computing platform in Ortmann et al. (2007) also differs from STARS in the fact that the sensors are intended for a different type of users: users who both observe and are being observed.

We will come back to our restricted perspective on reconfigurability in "Reconfigurable technology" section.

Anticipating societal and moral impact of reconfigurable sensor technology

The development of reconfigurable sensor technology is aimed at impacting the societal goals of safety and security. But the overall societal impact is not just determined by the technical features and their intended functionality. We approach this technology as a socio-technical system, which means that we take the embedding of the technology in social and societal structures to be of essential importance to its effect: What data will be gathered and by whom? Who will handle the data? How will the data be used? Who determines the priority of functionalities when the system is intended to serve different goals? We believe the aspect of reconfigurability makes these questions both more complex, and into the crucial part of the societal success of the technology.

The open functionality of reconfigurable technology and its intended wide applicability within society ['logical malleability' in the terminology of Jim Moor's seminal 1985 article Moor (1992)], require that societal and moral values are considered in the application phase. Ideally this is anticipated already in the design phase. This is the main principle behind the programme of *value sensitive design* (VSD), a "theoretically grounded approach to the design of technology that accounts for human values in a principled and comprehensive manner throughout the design process" (Friedman et al. 2006). VSD distinguishes three types of activities: conceptual, empirical and technical, which are integrative and iterative.

This paper aims to show how the reconfigurability of sensor technology, as envisaged in STARS, adds an extra challenge to VSD. The flexibility of reconfigurable technology necessitates new tools, for instance to keep track of different and evolving contexts of use. In particular, the design of good usage policies—including: what are boundary conditions for the use of the technology in a certain configuration in a particular context?—becomes crucial.

The paper is structured as follows. In the next section, we describe a use case from the STARS project to illustrate the reconfigurable sensor technology and a possible context of use for which it is being developed. In "Reconfigurable technology" section, we reflect from a more theoretical perspective on the feature and concept of reconfigurability. In "Ethical impact of reconfigurability?" section, we connect the issues that come with reconfigurability to VSD. In "Applicability of the framework of contextual integrity" section, we argue that *contextual integrity*is a useful notion for reconfigurable technology, indicating that the specification of contexts of use will add importance to policy design. We end the paper with a discussion of our findings, some concluding remarks and issues for further work.

USE CASE: SAFETY AND SECURITY DOMAIN

The intended application of the reconfigurable sensors and sensor networks in STARS is the safety and security domain. A use case for the sensor networks is the situation at a large port area (for example, the port of Rotterdam or Shanghai). Radar systems are used in large ports to monitor the movement of ships. Ship sizes can also be determined by these systems. Such radar systems consist of a number of radar devices, which send their collected data to a central control center. Here the data is processed to provide a full overview of the whole area. Other sensor data, for example from camera surveillance systems [closed circuit television (CCTV)] or motion detectors (around security gates) are also sent here, providing even more information in case of an incident.

Numerous issues around safety and security can arise in a port environment, including fire hazards, drug or people trafficking, terrorism or transport of hazardous chemicals. During an incident all sensor data can be combined to coordinate emergency services. Reconfigurable sensors could be especially useful in such environments, since they are intended to be usable for different tasks as the need arises, whereas previously multiple sensor systems would have been required. Consider, for example, the case where a small plane crashes into the port area. The police might be worried that this is part of an organized terrorist attack, in which case (part of) the radar system can be reconfigured to look for other (low flying) planes. Information provided by the reconfigured radar system can be crucial for the police (and other services) to gain control of the situation.

The reconfigurability of the sensors provides more flexible functionality, but it also introduces a number of potential problems. First of all, by reconfiguring the radar system, the 'normal' radar view of the ships in the harbor is compromised: the spatial resolution will go down, making it harder to distinguish different ship sizes. Part of the harbor may not be visible at all. This might be acceptable in a crisis situation, but it does lead to another issue: Who decides if the radar system may be reconfigured, and under which circumstances? Is the fire brigade in charge or the police? Or perhaps the port authorities or the government? Clear policies need to be defined for this, policies that can become more complex as the sensor systems' reconfigurable functionality increases. Even if the aim is to make the technology almost instantaneously reconfigurable, it is still likely that there will be some processing time needed for each reconfiguration. This can be crucial in crisis situations: during reconfiguration sensors cannot be used, leaving the control center in essence blind to the current situation. This may be acceptable if reconfiguration time is in the range of fractions of seconds, but longer delays may compromise the usability of the technology. So, only certain types of reconfiguration are realistic for certain situations.

The potential problems with reconfigurability mentioned above, stem from the same core problem: reconfigurable systems may provide multiple functionality and be flexible in extending functionality, but they cannot necessarily provide the functionalities concurrently. One can either search for ships or for low flying planes, not both (at the same time). An important part of the activity of VSD (Friedman et al. 2006, Section 6) consists of identifying the values associated with the benefits and harms of the technology, analyzing them, and identifying potential value conflicts. Reconfigurability of the technology significantly adds complexity to this already difficult task of identifying, balancing and prioritizing values, especially if different values are supported by different functionalities. Who gets to decide which value should be given priority in such situations?

A policy is a plan of action or procedure put in place by a governing body to determine actions and decisions with the aim to achieve a certain goal. A simple example of a policy for STARS-technology in the port case, could be the identification for three general situations in the context of which the sensor network will be used, for example Normal Operation of the Port (code Green), Hazard or Accident (code Orange) and Malicious Threat or Attack (code Red). The policy then prescribes for each of these situations how the (initial) configuration of the sensors should be, and who will be in control and responsible for the operation of the sensor network in this situation (e.g. the Fire Brigade for code Orange, and the counter-terrorism coordinator of the Ministry of Interior Affairs for code Red). This implies policies on different levels: policies concerning control over the sensors on the lower level, and higher level policies on who is responsible for declaring the general state (Green, Orange or Red).

Although policy design may not in itself be part of technology development, the VSD perspective requires that in the design of the technological artifact attention is paid to the aspect of governance (e.g. to fulfill values such as trust). With each policy comes an information flow of a certain granularity and with a certain focus. Such aspects of the policies will most probably and efficiently be implemented into the technology. We also believe that with the reconfigurability in STARS, the flexible functionality asks for the design of default policies, at least for foreseen usage, for example by defining default configurations for default situations. This should be anticipated when designing the technology, because controlling and evaluating mechanisms may be hard to implement as add-ons.

The framework of VSD suggests to start in practice from either a value, a technology or a context of use (Friedman et al. 2006). Whereas reconfigurable technology deliberately aims to remain as flexible as possible to serve different goals (values), it becomes central to take a context of use as starting point. Indeed, the questions raised above are practically unanswerable without delineating a specific context of use first. It is important to realize that the ethical evaluation of the technology highly depends on the context of use, making the specification of the context a *sine qua non* for the integration of societal and ethical aspects into the design and application of the technology.

RECONFIGURABLE TECHNOLOGY

The VSD approach distinguishes *conceptual activities* as one of the three types of activities of the approach. When addressing the ethical and societal impact of reconfigurable sensor technology, it is useful to clarify the notion of 'reconfigurable technology.' In particular, we deem it to be useful to analyze the relationship between (re)configuration and (flexible) functionality and usability of the technology. We will try to clarify the relationship between reconfigurability on the one hand, and concepts referring to the use of technology on the other, such as *function creep* and the flexibility of technological *affordances*.

Literally, 'reconfiguration' means: modification of the configuration, i.e. rearrangement of the parts (of a system). For example: the modification of a radar system so that it can scan for low flying planes instead of boats. Implicitly, at least within the STARS-project, it is taken that new configurations will enable new functionalities and usages, in particular: functionalities that may not yet have been specified when the technology was developed. Hence, reconfigurability should serve to provide a flexibility in functionality beyond the design phase. Because the term 'functionality' often bears the connotation of being the particular use *for which something is designed*, it helps to also bring in Gibson's terminology of *affordance* (Gibson 1986) in order to also talk about more general usage.

Affordances of a technology can be defined as the action possibilities latent in the technology, and need not be designed-in intentionally. This is demonstrated in the *dual use*-problem: technologies designed with a peaceful functionality, such nuclear radiation technology for cancer treatment, or aviation technology for the transportation of people, also bring the affordance of harmful usages.

We would like to point out that configuration on the one hand, and functionality and affordance on the other, by no means have a one-to-one relationship. A piece of technology can have different functionalities (or affordances) without being reconfigurable. A simple stone can be a missile, but also a "paper weight, a bookend, a hammer, or a pendulum bob" (Gibson 1986, p. 134). Also, a car can be both a means of transportation and a deadly weapon if intentionally used to drive into a group of people. Conversely, not every rearrangement of parts will necessarily lead to new affordances of the technology. So, while the STARS project focuses on making the sensors reconfigurable in order to achieve flexible functionality, part of the research should also address the question to which extent this ultimate goal is achieved in the developed technology.

We described "reconfiguration" very generally as a modification of the arrangement of parts, but this does not draw clear borders as to what counts as reconfiguration and what type of modification goes beyond reconfiguration. To which extent can we speak of reconfiguration when we don't just rearrange *existing* parts but (also) add *new* parts, or even technologies? For example, it is current practice to extend the affordances of sensor systems by processing the signals using computers. Think for example of the enhancement of CCTV systems with software that processes faces and compares these to a database with known subjects (Zhao et al. 2003) in order to identify them. In a sense this extension could be described as a reconfiguration of the CCTV system, since the original configuration of the system is changed for a specific purpose. In this paper however, we take it that not every alteration or extension of a technological system necessarily counts as a *reconfiguration*. In particular we will not include situations where new technologies are brought into the system, such as the face recognition layer to the camera observation (we would call this *synthesizing* technologies rather than *reconfiguring* the sensor system).

We can see the STARS-project as a specific case of a development process for reconfigurable technology. In this concrete case, what kind of reconfigurability can we expect? The ultimate goal of the project is to develop sensors and sensor networks with as much (potential) functionality as possible. The project proposes to achieve this by making the hardware reconfigurable, which involves mainly analogous front-ends (infrared, radar, etc.) and digital signal processing. We think the resulting range of possible reconfigurations will be limited to the three types of reconfiguration we have identified in the introduction: adaptation of parameters, switching modules and more structural reconfiguration of the base stations.

Although this is a rather limited range of reconfiguration, we believe it provides an interesting starting point to our reflection on reconfigurability and the applicability of VSD. Methodological questions are already raised by making parts of the architecture reconfigurable, such as those concerning testing procedures, software-hardware partitioning and composability [as pointed out for reconfigurability in the context of software architecture in Guo (2006)].

On a higher order level, one could state that STARS aims to create the affordance to address future, yet unknown, applications by making the technology reconfigurable. In our involvement in the STARS-project, we aimed to identify specific ethical challenges related to the reconfigurability of technology, although this also touched upon more general issues of multiple and flexible functionality. The goal of this endeavor, in line with the VSD approach, is to create awareness and anticipate these challenges in the research and development phase of the technology.

ETHICAL IMPACT OF RECONFIGURABILITY?

When looking for ethical challenges raised by the feature of reconfigurability, it is natural to turn to ethical theories for what seems the ultimate reconfigurable technology: the 'universal machine', i.e. the computer. In his seminal paper "What is Computer Ethics?" (Moor 1992), James Moor refers to the *logical malleability* of computers as the essence of the revolutionary character of computer technology, from which the need for a separate attention for computer ethics follows:

"The essence of the Computer Revolution is found in the nature of a computer itself. What is revolutionary about computers is logical malleability. Computers are logically malleable in that they can be shaped and molded to do any activity that can be characterized in terms of inputs, outputs, and connecting logical operations. [...] This is all I need to support my argument for the practical importance of computer ethics. In brief, the argument is as follows The revolutionary feature of computers is their logical malleability. Logical malleability assures the enormous application of computer technology. This will bring about the Computer Revolution. During the Computer Revolution many of our human activities and social institutions will be transformed. These transformations will leave us with policy and conceptual vacuums about how to use computer technology.

Such policy and conceptual vacuums are the marks of basic problems within computer ethics. Therefore, computer ethics is a field of substantial practical importance." (Moor 1992)

Here the logical malleability of computers is taken as the central cause of several effects computers will have on society, and from these effects, the need for computer ethics follows. We explore what ethical issues follow from the aspect of reconfigurability in itself (hence, not just from the instantiated effects) in reconfigurable technology. Does reconfigurable technology ask for different types of functional and non-functional requirements? Do we need to specify meta-requirements to capture requirements on the level of the reconfiguration process?

An important aspect of reconfigurability is that it challenges the type of stable, knowable, unambiguous function ascriptions to artifacts and systems. It raises epistemological issues related to agency. The STARS-project, one could say, takes it as a goal to defer the specification of functionalities for the technology past the design phase, even past the implementation phase, to remain flexible during the use phase. In that sense, the central feature of reconfigurability may ask for an extension of existing theories of technical functions (Houkes and Vermaas 2010). An implication of this observation, is that the developers of the technology can only to a lesser extent be expected to anticipate values and social consequences of the use of the technology. Because of the reduced knowability of the usage of the technology, agency with respect to social and moral consequences shifts from those involved in the development to those involved in the actual use (users, or policy makers). This puts limits on a VSD approach to the actual technology development, and shifts importance to value sensitive policy design. On a meta-level however, one could say that the developers should know this, hence bear responsibility for technology design that enables good policy design for as wide a range of applications as possible. Hence, guidelines for the use of reconfigurable technology can (and must) be given by its designers. These can include meta-rules that state how to deploy the system for a specific context, such as the example of the large port area discussed above, and make the reconfigurable usage of the system explicit in policies.

The epistemological issue connected to the open functionality clearly bears on the principle of informed consent. A prerequisite of that principle is a knowable impression of what the system will do under which circumstances. One can argue that this prerequisite is hard to fulfill for many of today's (socio-technical) systems, as they are developed for a certain goal, but once in place, easily used for or combined with other functionalities. This is called function creep; a well known example is the use of cameras that are put in place to implement a road pricing system, also for the detection of stolen cars, or tax evaders. This issue is even more prominent if the system is intended to be reconfigurable to changing circumstances, or even designed to fit yet unthought of functionalities and affordances. This could be called: "function-creep-by-design", open-ended design intended for open-ended use. At what level of abstraction can the system's behavior be specified for people subject to it, and is that enough of a basis for them to be able to consent or as a basis to justifiably assume their consent? A clear specification of the context of use becomes crucial for the users to be able to truly decide on consent. In the example of the port area: if the goal and potential use of the reconfigurable sensor system are not know and the main actors (police, fire brigade, counter-terrorism coordinator) are not identified, it becomes impossible to write policies on how the reconfigurable system should be used.

The specification of the behavior of the system requires a sophisticated and complex balancing of the different values the different functionalities of the technology serve. Combining technology for flexible and multiple functionality into one sensor, adds the restriction that only one functionality at a time can be actually used: concurrent use of different functionalities may not be possible. This means that more crucially than usual, priorities of the different functions must be assigned. This adds an extra dimension to the design process, namely the necessity of designing policies to specify priorities. But these policies should also be flexible to deal with the flexible functionality of the technology. Thus, in the port example, it is important to specify who (fire brigade, counter-terrorism coordinator) should have control about the reconfigurable sensors in which situation (chemical fire, terrorist attack). But the responsible actors should be able to decide for themselves how the sensor are to be configured and used; the policy should not be overly restrictive.

Reconfigurable technology comes with two sources of complexity: the technical complexity raised by the reconfigurability (which is deterministic), and the complexity associated with the fact that the application of the technology is deliberately left open (which is even non-monotonic). The latter will be the biggest challenge to address. Indeed, the observations above show that the reconfigurability leads to an increased range of choices that need to be made to put the technology to use. These choices address not only practical aspects, but more essentially higher order choices: who will be in control of such (practical) choices? Who will bear responsibility for the different functionalities, or for the system as a whole? This indicates that the development of policies around reconfigurable systems will bring in new complexities. Such complexity may compromise the expected efficiency of reconfigurability.

APPLICABILITY OF THE FRAMEWORK OF CONTEXTUAL INTEGRITY

The reconfigurable sensor networks primarily envisaged in STARS may be intended as *closed*systems, in the sense that the network will only be open to explicitly authorized users, and this group of users will be more or less stable and uniform (in particular: order preserving authorities, such as fire fighters, police, port authorities). The problem with the reconfigurability is that the contexts in which the technology may be used is deliberately left as open as possible in the design of the technology.

While the initial use case for the reconfigurable sensor networks is not primarily related to the observation of persons and their behavior, we deem it useful to look at the ethical issues related to sensor networks like camera surveillance and RFID access control systems. There is extensive literature discussing how sensor networks for observation of individuals and their environment bring up issues concerning privacy and the protection of personal data, such as Chan and Perrig (2003), Shi and Perrig (2004), and the legally oriented account in Solove (2008). Also, we expect that the technology may in the future be applied in privacy sensitive ways. This not just because the functionality is left open to future use and might include observation of individuals, but also because with increasing data collection surrounding all transactions in society, and linking of databases, objects and transaction traces can be more and more easily linked, also to people. This means that object data may turn into personal data a posteriori. But besides that, we argue that central notions from the discussion of privacy may be helpful in the analysis of reconfigurability, in particular the notion of *context*.

Conceptual solution: reconfigurability in context

Reconfigurability puts the context of use and control of information—captured in notions such as 'spheres of justice' or 'spheres of access' (Hoven 1999; Nagenborg 2009) and 'contextual integrity', as used by Ackerman et al. (2001), Nissenbaum (2010)—even more crucially at the heart of the challenge put forward by privacy. For example, Nissenbaum understands privacy in terms of context-relative information norms, and distinguishes norms of appropriateness, and norms of distribution. She defines contexts as "structured social settings, characterized by canonical activities, roles, relationships, power structures, norms (or rules), and internal values (goals, ends, purposes)" (Nissenbaum 2010, pp. 132–134). Most relevant to the framework of Contextual Integrity are the roles, activities, norms and values. For reconfigurable systems there may be different roles, activities, norms and values that need to be combined in the design of one system, and its usage policies. How to deal with the composition of these different contexts for one system is a particular challenge.

Reconfigurability involves applicability of one system with flexible functionality in possibly distinct contexts. In the case of reconfigurable sensor networks, the challenge will be to formulate requirements that are both general and specific enough to cover each possible use. For example, how to balance privacy issues if the sensor system monitors individuals only in very few of its configurations? And how to go about changes in this configuration?

Practical solution: context dependent policies (values in context)

Nissenbaum's framework for Contextual Integrity provides explanation, evaluation and prescription, and thereby contributes to the design process. As Nissenbaum recognizes, it does not yet "support substantive descriptions for general families of technologies", and "the most fruitful assessments take place within particular contexts" (Nissenbaum 2010, p. 190). In the case of reconfigurable systems, the particular context may be underspecified, or only one of a vast number of possible contexts. Therefore, a specific challenge for VSD of reconfigurable technology, such as sensor networks, requires an analysis of the composition and interaction of different contexts, and its translation into policies.

We can conclude that in the case of the development of reconfigurable technology, or more accurately: technology that provides "function creep by design", the applicability of VSD is limited, and it may be more fruitful to focus on "values in context of use". This corresponds to a shift from the "technical" to the "social" component of socio-technical systems: technology developed to provide open affordances will not constrain the value choices, so this is left to the actors in the social context.

DISCUSSION

Reconfigurability of sensors in networks seems to be an attractive answer to the increasing and invariably changing demands in the security and crisis management domain, both in terms of economy and of effectiveness. The central aim behind the reconfigurability of the technology developed in STARS is to keep the use of the technology open to future functionalities, uses that are not explicitly envisaged yet in the design phase of the technology. In other words, the technology is designed to provide the affordance to address future, yet unknown, applications by making the technology reconfigurable. *Function creep* is replaced by the explicit goal of *function shift* towards yet undefined functionalities. Configurations could change overnight towards new functionality—but how do people subject to it or using it get to know this? Reconfigurability thereby implies that there is uncertainty about what the current normative framework is (which is an epistemic problem). This means the VSD approach has to start from a definition of a context of use.

Although at first sight, one could say that the sensor networks of STARS are intended to be closed systems, in the sense that the amount of user parties is limited and coordinated, reconfigurability gives the sensor networks open traits of a slightly different kind. The openness towards its functionality makes that systems' role based access models should also be reconfigured with the system. This contributes to the non-technological complexity of reconfigurable technology, an aspect which is not to be overlooked.

We expect the non-technological complexity of reconfigurable sensor technology to surface in particular around the ethical issue of privacy. Even if the STARS-sensors are not primarily intended for monitoring persons, privacy will inevitably become relevant in at least part of the usages of the STARS-technology over the coming years. One has to be aware that what counts as "personal data" is being stretched by connecting data from various sources, data gathered about objects are easily linked to (data about) people, and thereby transitively become personal data after all. Furthermore, the fact that privacy is recognized a human right in the UN Declaration of Human Rights, makes it always a juridical constraint. The European Union expects from companies and research consortia to take their own responsibility (*responsible innovation*): they should be able to justify how they dealt with constraint/secured values. For the case of reconfigurable sensor technology, with its *function-creep-by-design*, it follows that privacy issues should be accounted for, regardless whether the current, or currently intended, use deals with personal data.

CONCLUSION AND FURTHER WORK

In this paper, we have presented an initial, mostly conceptual reflection on challenges that reconfigurable technology raises for VSD. Functionality and even goals of the technology are deliberately left open in the design phase of the technology, so the assessment of the ethical and societal impact of the technology heavily depends on the definition of contexts of use, and the importance of policy design for the use of the technology in each such contexts increases.

In the coming years, with the progress of the STARS-project a more thorough analysis of the concept of reconfigurability will be possible. It will be interesting to see how reconfigurability can be analyzed from the perspective of the literature on function ascriptions and requirements engineering. Is (physical) reconfiguration essentially different from re-conception of the possible use of a piece of technology? We believe that a proper analysis and definition of context and spheres will be crucial in the VSD of such technology: it is essential both for understanding its potential effects and, in practice, for the formulation of usage policies.

ACKNOWLEDGMENTS

We are grateful to the audiences at the 17th International Conference of the Society for Philosophy and Technology (SPT), May 2011 in Denton (Texas, USA), and the Workshop on Values in Design, in Lisbon (Portugal), September 2011, for the discussions leading to this paper. Furthermore, we thank the anonymous referees for their constructive comments on earlier versions of this paper, and for the pointers to related work.

REFERENCES

1. Ackerman, M., Darrell, T., Weitzner, D. (2001). Privacy in context. *Human-Computer Interaction 16*, 167–176
2. Bellotti, V., & Sellen, A. (1993). Design for privacy in ubiquitous computing environments. In: *Proceedings of the third European conference on computer-supported cooperative work* (pp. 77–92). Kluwer: Kluwer Academic Publishers
3. Chan, H., & Perrig, A. (2003). Security and privacy in sensor networks. *IEEE Computer 36*(10), 103–105
4. Friedman, B., Kahn, P. Jr, & Borning, A. (2006). Value sensitive design and information systems. In P. Zhang, & D. Galletta (Eds.), *Human-computer interaction in management information systems: Foundations* (pp. 348–372). NY: M.E. Sharpe, Inc
5. Gaver, B., Moran, T., MacLean, A., Lövstrand, L., Dourish, P., Carter, K., & Buxton, B. (1992). Realising a video environment: EuroPARC's RAVE System. In *Proceedings of the ACM conference on human factors in computing systems, CHI 92* (pp. 27–35). New york: ACM Press
6. Gibson, J. J. (1986). The theory of affordances, In *The Ecological Approach to Visual Perception* (Chap 8, pp. 127–143). Hillsdale, New Jersey: Lawrence Erlbaum Associates, Inc., Publishers. (Originally published 1979
7. Guo, Y., (2006). Mapping applications to a coarse-grained reconfigurable architecture. PhD thesis, University of Twente, Enschede.
8. Houkes, W., & Vermaas, P. E. (2010). *Technical functions: On the use and design of artefacts (philosophy of engineering and technology)*. Berlin: Springer

9. Hoven, M. J. (1999). Privacy or informational injustice?. In L. Pourcia (Ed.), *Ethics and information in the twenty-first century* (pp. 140–150). West Lafayette, Indiana: Purdue University Press
10. Johnson, R. C. (11 Sept. 2006). Post-9/11, technology keeps us a step ahead, EE Times News & Analysis. http://www.eetimes.com/electronics-news/4064803/Post- 9-11-technology-keeps-us-a-step-ahead. Accessed October 12, 2012
11. King, R. S. (Sept. 2011). How 5 security technologies fared after 9/11—developed, deployed, and sometimes deep sixed. IEEE Spectrum: Biomedical /Devices. http://spectrum.ieee.org/biomedical/devices/how-5-security-technologies-fared-after-911. Accessed October 12, 2012
12. Moor, J. H., (1992). What is computer ethics?, Southern Connecticut State University, New Haven, CT, USA, pp. 1–11. Reprint of Moore, J. (1985). What is computer ethics?. Metaphilosophy, 16(4):266–275
13. Nagenborg, M. (2009). Designing spheres of informational justice. *Ethics and Information Technology 11*, 175–179
14. Nissenbaum, H. (2010). Privacy in Context. Stanford, CA: Stanford University Press
15. Ortmann, S., Langendörfer, P., & Maaser, M. (2007). A self-configuring privacy management architecture for pervasive systems. In: *Proceedings of the 5th ACM international workshop on mobility management and wireless access*. ACM, New York, NY, USA
16. Shi, E., & Perrig, A. (2004). Designing secure sensor networks. *Wireless Communication Magazine 11*(6), 38–43
17. Solove, D. J. (2008). Understanding privacy. Cambridge, MA: Harvard University Press
18. STARS-project. (2010). STARS project information, available at the STARS-Project website: http://starsproject.nl/
19. Zhao, W., Chellappa, R., Phillips, P., & Rosenfeld, A. (2003). Face recognition: A literature survey. *ACM Computing Surveys (CSUR) 35*(4), 399–458

CHAPTER 9

Rethinking Experiments in a Socio-Technical Perspective: The Case of Software Engineering

Viola Schiaffonati [1] and Mario Verdicchio [2]

[1] *DEIB, Politecnico di Milano, Piazza Leonardo Da Vinci 32, Milano 20133, Italy*

[2] *DIGIP, Università degli Studi di Bergamo, Via Salvecchio 19, Bergamo 24129, Italy*

ABSTRACT

Experiments in computing share many characteristics with the traditional experimental method, but also present significant differences from a practical perspective, due to their aim at producing software artifacts and the central role played by human actors and organizations (e.g., programmers, project teams, software houses) involved in the software development process. By analyzing some of the most significant experiments in the subfield of software engineering, we aim at showing how the conceptual framework that supports experimental methodology in this context needs an extension in a socio-technical perspective.

Keywords: experiments; empirical software engineering; socio-technical systems

1. INTRODUCTION

The investigation on the disciplinary nature of computing has been conducted from two different perspectives: the subject matter, on one side, and the methodology, on the other. They provide interesting suggestions on how to position computing with respect to traditional sciences (e.g.; physics) and engineering disciplines (e.g., electronic engineering). We shift the focus from the status of computing to what we consider a related critical issue, which is indeed connected with the methodological perspective: the lack of a proper conceptual framework dealing with the nature of experiments in computing. We have already argued that, although there is a general agreement on the fundamental features of the scientific method, the straightforward application of such a framework to computing can lead to oversimplifications that prevent us from achieving a complete picture of the practices in this discipline; see, for instance [1]. Experiments in computing indeed have much in common with more traditional sciences (e.g., the call for replication [2]), but also present significant differences from a practical perspective, due to their aim at producing artifacts (*i.e.*, computing systems, programs, *etc.*) and the key role played by human actors and organizations (e.g., programmers, project teams, software houses) in such production.

If the focus on artifacts in computing has shed light on the need to borrow conceptual instruments from the domain of engineering (besides those of science), the central role played by the human actors involved in the artifact creation process seems to call for a further extension of the relevant framework. Indeed, it has been recently pointed out by [3] that when it comes to socio-technical systems, that is systems whose function depends not only on technological devices, but also on human agents and social institutions, traditional taxonomies of experimental forms may be inadequate.

In this paper, we continue a discussion on computing and experiments narrowing our analysis down to how experiments are conducted in a subfield of computing, namely software engineering (SE), with the aim to show how the traditional experimental framework needs to be revised in an engineering context. We believe there are good reasons that justify our choice. Firstly, SE aims at the study and application of techniques for the design, development, operation and maintenance of software. As any computer system must rely on software, it is clear that SE is in its own right a vast discipline that intersects many, if not all other subfields in computing. Hence, we hope that any findings that we obtain in our endeavor in this context can be seen as having a significant connection to what is carried out in terms of experiments in the whole field of computing. Moreover, contrary to what was stated by some critics (e.g., "Evidence to the decline of scientific methods is found in textbooks on software engineering" [4] (p. 148), more and more attention has been devoted in SE to methodological and experimental issues, as shown by some general works by [5] and the birth of conferences and journals about these topics, like [6] and [7].

We analyzed the 50 most cited papers (according to [8]) from "Empirical Software Engineering: An International Journal" (ESE) in the 2003–2012 decade and selected the most representative papers of the characteristics that make experimentation in SE so interesting from a philosophical perspective, in that they call for a revision of the conceptual framework of the traditional experimental method (We are aware that this is a limited sample and that several experimental results in SE are currently presented in conference proceedings; still, the papers we analyzed suffice for the purposes of this work.). Our focus is on those endeavors that adopted the most rigorous methodology from the scientific tradition, namely the controlled experiment, so that we are able to observe the widest range of instruments for scientific investigation put to the test in the context of SE. Some of these papers [9,10,11,12] explicitly refer to the presented work as a "controlled experiment"; others [13,14,15] only speak of an "experiment", but due to the rigorous procedures adopted, can be considered equivalent from a methodological point of view. In particular, we selected these works because they present the results of controlled experiments performed with human subjects, which makes them examples of one of the most challenging methodological endeavors, where the most rigorous (and traditional) experimental procedures are applied in a context where the human factor plays a primary role and brings in its typical unforeseeable and disruptive effects.

As the aim of this work is to start a discussion on such effects, four papers that present controlled experiments without human subjects [16,17,18,19] are out of our scope and were not considered. The rest of the papers consists of 12 case studies [20,21,22,23, 24,25,26, 27,28,29,30,31], 11 reviews [32,33,34,35,36,37,38,39,40,41,42], 11 empirical analyzes [43,44,45,46,47,48,49,50,51,52,53], three comparative analyzes ([54,55,56]) and two field studies [57,58].

Obviously, this analysis is not to be considered exhaustive at all, but it is meant as the first step of a more comprehensive study on methodology in computing. We focus on the experimental method, and we argue that such a method is not sufficient to tackle all of the issues rising in this discipline from a socio-technical perspective. Our hope is to show that there exists a very interesting approach worth exploring, taking the various subfields of computing as starting points. The paper is structured as follows: In Section 2, we analyze a group of selected experiments illustrated in the ESE journal with a particular focus on the issues that may prevent a straightforward generalization of their results. In Section 3, we show how such issues are connected with the nature of SE as a discipline, which includes not only technological artifacts, but also features a significant social dimension. Finally, Section 4 concludes.

2. SOCIAL ISSUES IN THE LAB

Let us analyze the experiments presented in the papers we are considering, with a specific focus on social issues. In this context, we use the adjective "social" to indicate two different aspects that characterize all of the experiments: the context they refer to and the human actors they involve.

We must begin by taking the context of the software industry into account (by "industry" we mean the complex of organizations devoted to the production of software for commercial use), because it provides the main motivation for the researchers to set up their experiments in the first place: to discover the best practices to create quality software. The criteria that are used to judge such quality seem to refer mainly to what is currently practiced in the context of the software industry.

Here follows a brief description of the controlled experiments, which we present in decreasing order of the number of citations and refer to by the surname of the first author of the relevant paper, accompanied by the motivations for such endeavors, as declared by the authors themselves.

The Genero experiment aimed at investigating whether a collection of measures can be a good predictor of the maintainability of Unified Modeling Language (UML) class diagrams, which are meant to represent graphically the functional structure of a complex program. These indicators would allow software designers to make better decisions in the early stages of the software development life cycle, contributing to the creation of better quality products. The choice to focus the experimental endeavor specifically on the class diagram out of nine diagrams defined in the UML standard is motivated by the fact that surveys in the industry show that "[the class diagram] is perceived by practitioners as the most important diagram type." [10] (p. 518). Moreover, by looking for good predictors of class diagram maintainability, the proponents of the Genero experiment aim to "allow [object-oriented] software designers to make better decisions early in the software development life cycle, thus contributing to the development of better quality [object-oriented] software." [10] (p. 519). The Vokáč experiment replicated a previous experiment aimed at verifying the beneficial effects of design patterns (*i.e.*, proven solutions to design problems organized into reusable software modules). The main motivation is "to support the industrial use of design patterns." [12] (p. 149). The Shull experiment consisted of two replications on software requirements reviewing techniques to verify the advantages of introducing laboratory packages (*i.e.*, structured descriptions of experiments), with the long-term goal of building a "body of knowledge that identifies the benefits and costs of various techniques and tools to support the engineering of software." [15] (p. 112). The Karlsson experiment compared a number of techniques to introduce a priority hierarchy among different requirements when implementing a software system. The aim was to establish which technique showed the least time consumption, the greatest ease of use and the best accuracy. Furthermore, this effort on requirement prioritization is driven by considerations connected with the industry: "When requirements are elicited from several stakeholders, it often yields more requirements than can be implemented at once. (...) [The requirements] need to be prioritised so that the most significant ones are met by the earliest product releases." [11] (p. 4). The De Lucia experiment aimed at assessing the usefulness of an automatic tool in support of traceability recovery, that is the process of identifying functional links between different parts of a complex software project, such as requirements and implemented software modules. The assessment was based on the measurement of the time needed by software engineers to accomplish the assigned tasks with respect to manual identification. This experiment was carried out with two perspectives, one related to the researchers' interest to evaluate the benefit of a tool-based support in the task and the other belonging to "a project manager, who wants to evaluate the possibility of adopting the tool within his/her own organisation, depending on the skills of the involved human resources." [9] (p. 67). The Fenton experiment aimed at developing a causal model (in the form of a Bayesian net) for predicting the number of defects likely to be found in a piece of software during testing or usage. This endeavor is part of a larger research endeavor in software metrics, whose "ultimate goal (...) is to help project managers make decisions under uncertainty." [13] (p. 500). Finally, the Reinhartz–Berger experiment compared two modeling techniques, namely object-process methodology (OPM) and UML, to establish the level of comprehension and the quality of the constructed models in the context of web applications. The motivation for this effort came from "the exponential growth of the Web and the progress of Internet-based architectures, [which] have set the stage for the proliferation of a variety of Web applications." [14] (p. 57).

The experimenters care for what is carried out by the professionals in the context of the SE industry, because their aim is to improve their practice, and experiments seem to provide an argument for the adoption of some techniques rather than others. For such an argument to be more convincing, the experiments need to be as close as possible to the reality of SE practice. This means that whatever the experimental conditions, they must capture all of the significant factors that affect the processes of real industrial software development, if researchers want their experiments to be meaningful.

In the analyzed papers, we can find several factors that the researchers took into account while trying to recreate a realistic experimental environment.

The proponents of the Genero experiment have “used material that is more representative of real cases” [10] (p. 523) compared to previous endeavors of theirs. Moreover, they “improved the experimental tasks by trying to specify [tasks] that are similar to those required in real projects.” [10] (p. 523).

The Vokáč experiment was meant to improve on a previous experience by other researchers. The authors “replicated their experiment (...) using the same programs in a real programming environment, instead of pen and paper. This increases the experimental realism and, thereby, the applicability of the results.” [12] (p. 150).

The search for realism has found some obstacles in the intrinsic differences between a controlled experimental environment and the resource-constrained reality of the SE industry. From the Vokáč experiment: “[a] difference is that it is possible to walk away from a nonworking solution, as actually happened with several subjects. In industry that option is simply unavailable. When it happens, projects tend to become highly visible failures.” [12] (p. 190). Moreover, “the experimental design required all participants to be present at the same time, for two consecutive days. This made it much more difficult to get a sufficient number than for previous experiments that could be done on each company’s premises, one company at a time.” [12] (p. 185).

Sometimes the resource constraints are on the experiment’s side. Again, from the Vokáč experiment: “Real-world programs are much larger than those used in the experiment. With a restricted time and money budget, this is a limitation that is difficult to overcome.” [12] (p. 189). The De Lucia experiment presents similar issues: “Concerning the artefact repository used in the experiment, it is worth noting that it is not comparable to industrial projects (...) However, the type of experimentation has to be conducted in a controlled way and in a limited amount of time. For this reason, it is not easy to use repositories of larger size and even if larger repositories are used, there might anyway be the need to select a subset of (...) artefacts (...).” [9] (p. 82).

In some cases, like in the Karlsson experiment, such constraints are part of the motivation for the whole effort: "Since the major disadvantage of [a technique based on an exhaustive list of requirement comparisons] is the time-consumption for large problems, different investigations have been performed in order to decrease the number of comparisons, and thus the time needed." [11] (p. 7). Similarity to the industrial context plays such a primary role in this experiment that a technique that was "ranked rather low" in previous studies is still considered interesting to investigate, since it is used in a very active subgroup of the SE community.

In the effort to maintain similarity to actual SE practices, the proponents of these experiments must take several factors into account, but one plays a particularly important role: The human factor. SE research aims at helping make the best choices in terms of selecting, with respect to a given problem, the most suited requirement specification language, the most skilled programmer, the best fitting programming paradigm and language, *etc*. These choices have a social component in that they are strongly dependent on human actors: requirements consist of a call for a software solution to a problem expressed by groups of people or even entire organizations [59,60]; programmers are human beings endowed with honed technical skills, but also characterized by idiosyncrasies and subjective preferences with respect to programming paradigms and languages [61,62]. Thus, experiments intended to investigate these aspects of SE must make use of human subjects, and so did our chosen examples.

- The Genero experiment [10] was divided into two replicated rounds: The first involved 38 computer science students in the third year at a Spanish university, the second 23 computer science students in their final year at an Italian university.
- The Vokáč experiment [12] replicated a previous experiment, and according to the authors, experimental realism was increased by using a real programming environment instead of pen and paper; 44 paid professionals from multiple companies were the subjects instead of 29 volunteers from a single organization, like in the previous attempt.
- Each of the two replications in the Shull experiment [15] involved 18 undergraduate students from a different Brazilian university, all with an experience of slightly more than one year of software development in class.
- The Karlsson experiment [11] consisted of two sessions: one with a group of 16 subjects, including Ph.D. students, and one professor as part of a research methodology course, the other with 30 MSc

students in their final year, taking an optional requirements engineering course.

- The De Lucia experiment [9] was executed twice in an Italian university, with 20 first-year MSc students attending an advanced SE course and with 12 second-year MSc students from a course on software project management, respectively.
- The causal model used in the Fenton experiment [13] was comprised of variables and relationships identified by means of questionnaires distributed to a number of managers involved in 31 software projects in the consumer electronics industry.
- Finally, the Reinhartz–Berger experiment [14] involved 81 third-year students in a four-year information systems engineering program in the final four hours of a 13-week “Specification and Analysis of Information Systems” course.

Issues arise from the fact that experiments are conducted with groups of human subjects. People come with different prior knowledge, tastes, habits, skills, and so on, which means that they are likely to tackle the tasks in the experiments with results that may be very varied and make the researchers’ attempt to extract scientifically-proven facts more complicated. We have found that some of these issues derived from a less-than-ideal experiment design. In the two replications of the Genero experiment, for instance, the subjects in Spain were given material in their native language, whereas what was provided to their counterparts in Italy had been translated into English. The experimenters found out that not all of the subjects were sufficiently skilled in English and needed extra time to ask about the meaning of some statements. Another example is given by the Reinhartz–Berger experiment: some fatigue and order factors may have interfered with the results, since in all of the texts asking the subjects to create two models, one for a project management system and one for a book ordering system, the former always appeared first, so that it is likely that the subjects started working on it first.

Naturally, more uniform experimental conditions (possibly including the use of the subjects’ native language or a proper randomization in the sequences of experimental tasks) would enable experimenters to overcome such issues, but others are more intrinsic to the use of subjects and do not depend on experimental design.

One of the objectives of the Genero experiment was to verify that some structural properties of diagrams have an impact on their cognitive complexity, defined as the mental burden posed by the diagram on the persons who are studying it. Such complexity was measured in the experiment by means of the subjective judgment of the participants through a scale of five linguistic labels, ranging from "very simple" to "very complex". In the Karlsson experiment on requirement prioritization, the investigation is about the beneficial effect of the introduction of automated tools, but the decision making process on whether one requirement is more important than another ultimately rests on the subjects' shoulders. The causal model to predict software defects that was analyzed in the Fenton experiment was created on the basis of the expertise of professionals gathered by means of a questionnaire.

These examples show that a significant part of the results in the experiments relies on the personal point of view of the subjects involved. The very nature of this perspective strongly contrasts with the universality and objectivity expected of the results of scientific experimentation. An obvious, although not always possible, way to deal with this issue is to enlarge the pool of subjects involved in the experiments. Indeed, all of the experimenters are well aware of the issue of sample significance, especially since they relied on students attending an SE-related course, because in most cases, the experimenters are professors or lecturers at a university. Apart from concerns about the sample size, such a specific context would not be considered adequate for a general sociological or psychological study, but as clearly shown in the quotes below, all SE experimenters have indeed a more restricted aim: they perform experiments having the SE industry in mind: "Because of the difficulty of obtaining professional subjects, we used students from a software engineering course." [10] (p. 539); "Some [professional] maintainers may have more experience with design patterns than did our subjects." [12] (p. 189). "The generalisability of the study is limited due to the rather small sample and specific context. (...) the subjects may have opinions similar to decision makers in the industry." [11] (pp. 27–28). "All students are master students who either had some professional experiences or worked on industrial projects (...) This makes these students comparable to industry junior developers." [9] (p. 25). Let us explore the connection between experiments and the industry in the following section.

3. FROM THE LAB TO THE REAL WORLD

The generalization of experimental results is a highly desirable property for any experimenter, and SE practitioners are no exception in this respect. In most of the controlled experiments involving human subjects we have considered in our analysis, the authors express concern regarding the possibility of generalizing their results by discussing threats to external validity and possible solutions. External validity is the degree to which the results can be generalized and transferred to other situations.

In the Shull experiment, for instance, we can find a warning about "the sensitivity of experimental designs involving human subjects, where small variations in the execution of the experiment can have large effects on results." [15] (p. 132). More specifically, what is usually recognized in these articles as a common threat is the use of students (mostly MSc and/or Ph.D.) instead of real professionals as experimental subjects. As said in the De Lucia experiment, for instance, "this kind of threat is always present when experimenting with students." [9] (p. 81). Moreover, as remarked in the Karlsson experiment, although improving the validity of the conclusions, the homogeneity of a subject group (like students from a specific SE class) makes it difficult to generalize the results to a broader population. However, the experimenters rely on a previous study by Tichy [63] showing that if a student experiment shows that one technique is better than another, it is rather unlikely that professionals would come to the opposite conclusion. Although by means of specifically-designed courses, the students' modeling and programming skills can be brought up to the level of industry junior developers and strategies can be implemented to compensate for their limited familiarity with the application domains and the software systems used in the experiments (such as organizing meetings before the experiment to provide students with an acceptable knowledge level), the necessity to replicate the same experiment in an industrial context is clearly recognized. In the Genero experiment, for instance, further research is said to have to include "a family of experiments [which] should also use professionals as subjects and different experimental models and tasks." [10] (pp. 543–544).

A comprehensive analysis of the threats to external validity can be found in the Vokáč experiment, where several differences between the experimental environment and an actual software maintenance setting are considered, such as the fact that maintainers in the experiment were different from the programmers who created the software, so that the results are not applicable to maintenance by the original designers, who are normally "expected to remember not only the actual design, but also much of the motivation for it." [12] (p. 189). Another difference is the fact that some maintainers may have more experience than the experimental subjects. With respect to these and a number of other issues, the authors successfully showed that they do not constitute an actual threat to their experimental results: in these cases, for instance, if the use of design patterns has a beneficial effect on the maintainability of software programs (as proven by the Vokáč experiment), then a deeper knowledge of the artifact endowing the original designers or more experienced maintainers is very unlikely to make the situation worse. The authors "would expect the beneficial effects of patterns to be greater." [12] (p. 189). However, there are other issues that turn out to be more problematic, because the experimenters are not able to provide a solution to counter them. We have already mentioned that time and money budget restrictions force the experimenters to work with programs that are smaller than their real-world counterparts. Moreover, "real-world programs are sometimes less well documented, and changes may be larger and involve more than one [design] pattern. The effects of such differences are difficult to predict on a general basis. There are undoubtedly interaction effects that can occur between patterns, but would not be visible in such an experiment." [12] (p. 189).

The shift of from the lab to the real world (which means, in most of the cases we have considered, performing experiments in industrial contexts) presents several challenging issues from a methodological perspective. We have already argued in the past that taking into consideration how experimental methodologies are developed and used is a way of reflecting on the disciplinary status of computing that moves us away from the somehow too rigid dualism between science and engineering [1]. Here, we focus on how the conceptual categories of the traditional scientific method are challenged in the context of SE because of the social components that are involved. Keeping in mind the examples we have discussed so far, it is quite evident, for instance, that the notion of control as intended in the traditional experimental protocol (the experimentalist is not part of the experimental system and is able to intervene in and control the experimental system from the outside) does not fully apply. Not only is it problematic to draw a line between the experimental system and its environment, in particular if this environment is meant to include several subjects from different contexts, but since the experimental environment does not consist only of physical objects, also social components are to be taken into account.

We are aware that the social components at play in current SE experiments are only given by the human beings whose interactions with technical artifacts are under observation. However, we have also seen that other social factors exert a significant influence on the experimental practice in the context of SE: many, if not all, researchers are concerned about the external validity of their results, and the context they are referring to is the SE industry. The industrial dimension is not directly involved in SE experiments at the moment, but we believe that to move toward a more comprehensive account of future practices where experiments go out of the restricted environments of labs in universities and research centers, such a component should be taken into account, in terms of conditions, rules, organizations and companies.

Thus, with our work, we aim at showing that SE includes a very significant socio-technical dimension beside its long-studied technological dimension. The human factor brought in by the experimental subjects, as well as the whole organizational dimension given by the SE industry and its ramifications need to be taken into account if one wants to obtain a comprehensive analysis of the experiments in this discipline.

The work in [3] defines as socio-technical a system whose behavior is "significantly affected by [its] technical components but the functioning (...) as a whole depends as much on the functioning of these technical components as on the functioning of its social components (...) and it depends on the behavior of human actors." According to [64], a socio-technical system is a special sort of system, namely an entity that can be separated into parts, which are linked to each other in a specific way, including also components that are not tangible objects. In other words, a socio-technical system is a hybrid system formed by different components, some of which are described and analyzed using the natural sciences paradigm, while others must be described by drawing on the social sciences tools.

The example discussed at length in [64] is the world civil aviation system, composed of a great number of different technical artifacts (computers, airplanes, airports, baggage trolleys, *etc.*), people (cabin crew, personnel working at the check-in desks, air controllers, *etc.*), but also other components of a more abstract nature (air corridors, air companies, regulations having to be observed by pilots and airline companies, organizations enforcing such regulations, *etc.*). The articulation of these different components is what makes socio-technical systems different from even the most complex technical artifact, which, although incorporating characteristics of both physical and social objects, *i.e.*, its hardware and the function assigned to it by human designers and users, respectively, is a building block of systems of greater complexity.

The involvement of people, regulations and institutions beside technical artifacts is the reason why the natural scientific way of description and explanation represents only one way of formulating matters within the context of socio-technical systems and why a social-scientific approach is required. Although the hybrid character of socio-technical systems is what eminently characterizes them, their design, implementation and maintenance are still in the hands of engineers that have been mainly educated in the natural scientific way. However, the complexity of socio-technical systems consists precisely in the fact that they have many different users (unlike typical technical artifacts), and their functioning depends not only on the articulation of the technical components, but also on the coordination of the behavior of different users; the success of such coordination comes through habits, agreements, rules, laws, *etc.*, that is entities studied in the social sciences and not in the natural sciences.

The main aim of SE is creating technical artifacts in the form of software, but, as clearly shown by the examples in this work, technology alone does not cover everything that happens in SE: software is created by people to meet other people's needs within the context of communities, companies and markets, which do not simply work as a passive background, but pose several constraints, for example in terms of resource limitations, business strategies and societal values that shape and guide the software development process.

The co-existence of technological and social aspects in the conception, development and deployment of software indicates that the context of SE includes socio-technical systems. As a consequence, experiments in SE are significantly characterized by socio-technical features. Thus, we envision a development of SE experimentation in which the conceptual framework sustaining the methodological practice has been expanded in a socio-technical direction. After all, every aspect of SE can be and, in our opinion, should be interpreted from such a perspective.

Let us take a closer look at the variety of activities that are included in SE as a discipline. The construction of an SE artifact consists of a number of operations, which are carried out at different stages in the life cycle of the software that is being created. The context in which such a life cycle unfolds is notably vast, as there are several degrees of freedom depending on the choices made with respect to the entities involved at each stage. The process starts with the requirements, which describe what is required of the software; such a description can be provided in a natural or a formal language and at different levels of detail. A programmer or, more commonly, a team of programmers is responsible for the creation of the software, by first conceiving of an algorithm and then turning it into a program. The programming paradigm is the view that underlies the way the software is going to be written: with a functional paradigm, for instance, all operations that a program is to perform are modeled as functions, whereas the object-oriented paradigm views such operations as actions taken by different objects, that is independent subparts that constitute the program. Given a paradigm, a programmer can have more than one programming language to choose from: for example, C++, Java and Python are all object-oriented languages. The program itself can be written in different ways in the same language, depending on the structure by which the included operations are organized. A testing process must be carried out once a program is written to ensure that it fulfills the requirements it was created for without faults. Finally, if more programs are available to provide the same service, a benchmarking technique can be used to assess which one achieves the goal with the least resource consumption in terms of computing time and computer memory.

As said before, SE research aims at helping make the best choice with respect to every stage mentioned above. The decision making process comes with a significant social component, in that every choice is strongly dependent not only on human actors with their personal backgrounds and subjective preferences, but also on organizations and institutions with the relevant conditions and rules. When it comes to picking the programming language to develop software with, the decision is *de facto* taken by the market the engineers are aiming for: for instance, if they intend to develop software running on an iPhone, then the most natural choice is to work with the most used language for this specific platform, namely Objective-C, for which full support from the developer community and the relevant literature is ensured [65]. Programming languages are themselves the product of institutional decisions: the rules of the Java language, for instance, are determined by the Java Community Process (JCP), which is the sole official mechanism for developing standard technical specifications for this technology. Anyone from the SE community can sign up to become a JCP member and then join expert groups and participate in reviewing processes and providing feedback [66]. Testing is also an activity that is significantly influenced by organizations and, more specifically, by their economic conditions: the resources demanded by exhaustive program testing procedures are incompatible with any software company's economic strategy, so that a key task in testing is the selection of critical runs to detect as many software faults as possible to ensure that the final product reaches an acceptable quality level to be marketed [67]. Finally, even a highly technical practice, like benchmarking, is seen by SE researchers not only as a rigorous examination of contributions, but also a way to ensure greater communication and collaboration among researchers, leading to a stronger consensus in the community [68].

4. CONCLUSIONS

In this paper, we hope to have substantiated the importance of enlarging the context to socio-technical systems when analyzing SE experiments. The limits in applying the traditional experimental protocol to different fields of computing appears particularly significant in the case of SE experiments that we have systematically analyzed: not only the technical dimension, beside the scientific one, needs to be taken into account, but also the social issues, meaning both the human agents and the social institutions play a significant role in this framework. If SE experiments aim at leaving the limited contexts of the lab to move toward real industrial applications, it is important to recognize these issues, which we have explored here with a particular focus on the problem of external validity. Obviously, this social dimension has to be dealt with by the tools of the social sciences, as the functioning of complex socio-technical artifacts depends on the coordination amongst different users through rules, laws, habits and agreements, besides the proper working of the technical components (traditionally managed with the tools of natural sciences).

The acknowledgment of the social dimension of experiments within a socio-technical perspective calls for a paradigmatic shift that we have just initiated reflecting upon, considering SE experiments as a case study for the reasons we have discussed in the Introduction of this paper. We are confident, however, that an extension to other fields of computing, given the obvious differences, will show some very close similarities and strengthen the importance of extending the traditional notion of experiment in socio-technical systems. We believe, moreover, that this shift could have an impact on the philosophy of computing in general and in particular on those reflections aimed at clarifying the disciplinary status of computing from a methodological point of view.

Finally, to rethink experiments in a socio-technical perspective could represent also an important step for the methodological advancement of SE itself, where the attempts to reflect on experiments (see [5] for a comprehensive and influencing work on this topics) have generally been focused exclusively on the implementation of the traditional experimental categories, in particular the replication of experiments. Now, it is time to move beyond and construct the debate on experimental methods in SE upon the most recent developments in the philosophy of technology and in the philosophy of science to take into account a complex and articulated socio-technical practice that cannot be reduced to the categories of the traditional scientific disciplines.

REFERENCES

1. Schiaffonati, V.; Verdicchio, M. Computing and experiments. *Philos. Technol.* **2013**, *27*, 1–18.

2. Shull, F.J.; Carver, J.C.; Vegas, S.; Juristo, N. The role of replications in Empirical Software Engineering. *Empir. Softw. Eng.* **2008**, *13*, 211–218.
3. Kroes, P. Experiments on socio-technical systems: The problem of control. *Sci. Eng. Ethics* **2015**, 1–13.
4. Eden, A.H. Three paradigms of computer science. *Minds Mach.* **2007**, *17*, 135–167.
5. Juristo, N.; Moreno, A. *Basics of Software Engineering Experimentation*; Kluwer: Dordrecht, Germany, 2001.
6. Basili, V.R., Briand, L.C., Eds.; *Empirical Software Engineering*; Springer: Berlin, Germany, 1996.
7. Andersson, C., Runeson, P., Eds.; First International symposium on empirical software engineering and measurement. In Proceedings of the First International Symposium on Empirical Software Engineering and Measurement, Madrid, Spain, 20–21 September 2007.
8. Scopus. The Largest Abstract and Citation Database of Peer-reviewed Literature. 2014. Available online: http:// www.scopus.com (last accessed 10 February 2014).
9. De Lucia, A.; Oliveto, R.; Tortora, G. Assessing IR-based traceability recovery tools through controlled experiments. *Empir. Softw. Eng.* **2009**, *14*, 57–92.
10. Genero, M.; Manso, E.; Visaggio, A.; Canfora, G.; Piattini, M. Building measure-based prediction models for UML class diagram maintainability. *Empir. Softw. Eng.* **2007**, *12*, 517–549.
11. Karlsson, L.; Thelin, T.; Regnell, B.; Berander, P.; Wohlin, C. Pair-wise comparisons versus planning game partitioning–experiments on requirements prioritization techniques. *Empir. Softw. Eng.* **2006**, *12*, 3–33.
12. Vokáč, M.; Tichy, W.; Sjøberg, D.I.K.; Arisholm, E.; Aldrin, M. A controlled experiment comparing the maintainability of programs designed with and without design patterns—A replication in a real programming environment. *Empir. Softw. Eng.* **2004**, *9*, 149–195.
13. Fenton, N.; Neil, M.; Marsh, W.; Hearty, P.; Radliński, L.; Krause, P. On the effectiveness of early life cycle defect prediction with Bayesian Nets. *Empir. Softw. Eng.* **2008**, *13*, 499–537.
14. Reinhartz-Berger, I.; Dori, D. OPM *vs.* UML—Experimenting with comprehension and construction of web application models. *Empir. Softw. Eng.* **2005**, *10*, 57–79.
15. Shull, F.; Mendonc, M.G.; Basili, V.; Carver, J.; Maldonado, J.C.; Fabbri, S.; Travassos, G.H.; Ferreira, M.C. Knowledge-sharing issues in experimental software engineering. *Empir. Softw. Eng.* **2004**, *9*, 111–137.
16. D'Ambros, M.; Lanza, M.; Robbes, R. Evaluating defect prediction approaches: A benchmark and an extensive comparison. *Empir. Softw. Eng.* **2012**, *17*, 531–577.
17. Jiang, Y.; Cukic, B.; Ma, Y. Techniques for evaluating fault prediction models. *Empir. Softw. Eng.* **2008**, *13*, 561–595.
18. Pan, K.; Kim, S.; Whitehead, E.J., Jr. Toward an understanding of bug fix patterns. *Empir. Softw. Eng.* **2009**, *14*, 286–315.
19. Turhan, B.; Menzies, T.; Bener, A.B.; di Stefano, J. On the relative value of cross-company and within-company data for defect prediction. *Empir. Softw. Eng.* **2009**, *14*, 540–578.
20. Damian, D.; Chisan, J.; Vaidyanathasamy, L.; Pal, Y. Requirements engineering and downstream software development: Findings from a case study. *Empir. Softw. Eng.* **2005**, *10*, 255–283.
21. Gonzalez-Barahona, J.M.; Robles, G.; Michlmayr, M.; Amor, J.J.; German, D.M. Macro-level software evolution: A case study of a large software compilation. *Empir. Softw. Eng.* **2009**, *14*, 262–285.
22. Kapser, C.J.; Godfrey, M.W. "Cloning considered harmful" considered harmful: Patterns of cloning in software. *Empir. Softw. Eng.* **2008**, *13*, 645–692.
23. Karlström, D.; Runeson, P. Integrating agile software development into stage-gate managed product development. *Empir. Softw. Eng.* **2006**, *11*, 203–225.

24. Khoshgoftaar, T.M.; Seliya, N. Fault prediction modeling for software quality estimation: Comparing commonly used techniques. *Empir. Softw. Eng.* **2003**, *8*, 255–283.
25. Khoshgoftaar, T.M.; Seliya, N. Analogy-based practical classification rules for software quality estimation. *Empir. Softw. Eng.* **2003**, *8*, 325–350.
26. Khoshgoftaar, T.M.; Seliya, N. Comparative assessment of software quality classification techniques: An empirical case study. *Empir. Softw. Eng.* **2004**, *9*, 229–257.
27. Kommeren, R.; Parviainen, P. Philips experiences in global distributed software development. *Empir. Softw. Eng.***2007**, *12*, 647–660.
28. Mendes, E.; Watson, I.; Triggs, C.; Mosley, N.; Counsell, S. A comparative study of cost estimation models for web hypermedia applications. *Empir. Softw. Eng.* **2003**, *8*, 163–196.
29. Pikkarainen, M.; Haikara, J.; Salo, O.; Abrahamsson, P.; Still, J. The impact of agile practices on communication in software development. *Empir. Softw. Eng.* **2008**, *13*, 303–337.
30. Poshyvanyk, D.; Marcus, A.; Ferenc, R.; Gyimóthy, T. Using information retrieval based coupling measures for impact analysis. *Empir. Softw. Eng.* **2009**, *14*, 5–32.
31. Segal, J. When software engineers met research scientists: A case study. *Empir. Softw. Eng.* **2005**, *10*, 517–536.
32. Adolph, S.; Hall, W.; Kruchten, P. Using grounded theory to study the experience of software development. *Empir. Softw. Eng.* **2011**, *16*, 487–513.
33. Carver, J.C.; Jaccheri, L.; Morasca, S.; Shull, F. A checklist for integrating student empirical studies with research and teaching goals. *Empir. Softw. Eng.* **2010**, *15*, 35–59.
34. Dieste, O.; Grimán, A.; Juristo, N. Developing search strategies for detecting relevant experiments. *Empir. Softw. Eng.***2009**, *14*, 513–539.
35. Do, H.; Elbaum, S.; Rothermel, G. Supporting controlled experimentation with testing techniques: An infrastructure and its potential impact. *Empir. Softw. Eng.* **2005**, *10*, 405–435.
36. Falessi, D.; Babar, M.A.; Cantone, G.; Kruchten, P. Applying Empirical Software Engineering to software architecture: Challenges and lessons learned. *Empir. Softw. Eng.* **2010**, *15*, 250–276.
37. Ivarsson, M.; Gorschek, T. A method for evaluating rigor and industrial relevance of technology evaluations. *Empir. Softw. Eng.* **2011**, *16*, 365–395.
38. Juristo, N.; Moreno, A.M.; Vegas, S. Reviewing 25 years of testing technique experiments. *Empir. Softw. Eng.* **2004**, *9*, 7–44.
39. Lethbridge, T.C.; Sim, S.E.; Singer, J. Studying software engineers: Data collection techniques for software field studies. *Empir. Softw. Eng.* **2005**, *10*, 311–341.
40. Mohagheghi, P.; Conradi, R. Quality, productivity and economic benefits of software reuse: A review of industrial studies. *Empir. Softw. Eng.* **2007**, *12*, 471–516.
41. Runeson, P.; Hŏst, M. Guidelines for conducting and reporting case study research in software engineering. *Empir. Softw. Eng.* **2009**, *14*, 131–164.
42. Shull, F.J.; Carver, J.C.; Vegas, S.; Juristo, N. The role of replications in Empirical Software Engineering. *Empir. Softw. Eng.* **2008**, *13*, 211–218.
43. Beecham, S.; Hall, T.; Rainer, A. Software process improvement problems in twelve software companies: An empirical analysis. *Empir. Softw. Eng.* **2003**, *8*, 7–42.
44. Do, H.; Rothermel, G.; Kinneer, A. Prioritizing JUnit test cases: An empirical assessment and cost-benefits analysis. *Empir. Softw. Eng.* **2006**, *11*, 33–70.
45. German, D.M. An empirical study of fine-grained software modifications. *Empir. Softw. Eng.* **2006**, *11*, 369–393.
46. Mӑntylӑ, M.V.; Lassenius, C. Subjective evaluation of software evolvability using code smells: An empirical study. *Empir. Softw. Eng.* **2006**, *11*, 395–431.

47. Melton, H.; Tempero, E. An empirical study of cycles among classes in Java. *Empir. Softw. Eng.* **2007**, *12*, 389–415.
48. Nagappan, N.; Maximilien, E.M.; Bhat, T.; Williams, L. Realizing quality improvement through test driven development: Results and experiences of four industrial teams. *Empir. Softw. Eng.* **2008**, *13*, 289–302.
49. Schach, S.R.; Jin, B.; Yu, L.; Heller, G.Z.; Offutt, J. Determining the distribution of maintenance categories: Survey versus measurement. *Empir. Softw. Eng.* **2003**, *8*, 351–365.
50. Thummalapenta, S.; Cerulo, L.; Aversano, L.; Di Penta, M. An empirical study on the maintenance of source code clones. *Empir. Softw. Eng.* **2010**, *15*, 1–34.
51. Vegas, S.; Basili, V. A characterization schema for software testing techniques. *Empir. Softw. Eng.* **2005**, *10*, 437–466.
52. Weyuker, E.J.; Ostrand, T.J.; Bell, R.M. Do too many cooks spoil the broth? Using the number of developers to enhance defect prediction models. *Empir. Softw. Eng.* **2008**, *13*, 539–559.
53. Zou, X.; Settimi, R.; Cleland-Huang, J. Improving automated requirements trace retrieval: A study of term-based enhancement methods. *Empir. Softw. Eng.* **2010**, *15*, 119–146.
54. Azzeh, M.; Neagu, D.; Cowling, P.I. Fuzzy grey relational analysis for software effort estimation. *Empir. Softw. Eng.***2010**, *15*, 60–90.
55. Li, J.; Ruhe, G.; Al-Emran, A.; Richter, M.M. A flexible method for software effort estimation by analogy. *Empir. Softw. Eng.* **2007**, *12*, 65–106.
56. Li, Y.F.; Xie, M.; Goh, T.N. A study of the non-linear adjustment for analogy based software cost estimation. *Empir. Softw. Eng.* **2009**, *14*, 603–643.
57. Robillard, M.P.; DeLine, R. A field study of API learning obstacles. *Empir. Softw. Eng.* **2011**, *16*, 703–732.
58. Sharp, H.; Robinson, H. An ethnographic study of XP practice. *Empir. Softw. Eng.* **2004**, *9*, 353–375.
59. Sommerville, I.; Rodden, T.; Sawyer, P.; Bentley, R.; Twidale, M. Integrating ethnography into the requirements engineering process. In Proceedings of IEEE International Symposium on Requirements Engineering, San Diego, CA, USA, 4–6 January 1993; pp. 165–173.
60. Sutcliffe, A. *User-centred Requirements Engineering*; Springer: Berlin, Germany, 2002.
61. Naur, P. Understanding Turing's universal machine—Personal style in program description. *Comput. J.* **1993**, *36*, 351–372.
62. Harrison, R.; Smaraweera, L.G.; Dobie, M.R.; Lewis, P.H. Comparing programming paradigms: An evaluation of functional and object-oriented programs. *Softw. Eng. J.* **1996**, *11*, 247–254.
63. Tichy, W.F. Hints for reviewing empirical work in software engineering. *Empir. Softw. Eng.* **2000**, *5*, 309–312.
64. Vermaas, P.; Kroes, P.; van de Poel, I.; Franssen, M.; Houkes, W. *A Philosophy of Technology. From Technical Artefacts to Sociotechnical Systems*; Morgan and Claypool: San Rafael, CA, USA, 2011.
65. Skelton, G.W.; Jackson, J.; Dancer, C.F. Teaching software engineering through the use of mobile application development. *J. Comput. Sci. Coll.* **2013**, *28*, 39–44.
66. Java Community Process. Community Development of Java Technology SPecifications. 2014. Available online: http://www.jcp.org/ (last accessed 1 July 2014).
67. Myers, G.J.; Badgett, T.; Sandler, C. *The Art of Software Testing*, 3rd ed.; Wiley and Sons: Hoboken, NJ, USA, 2011.
68. Sim, S.E.; Easterbrook, S.; Holt, R.C. Using benchmarking to advance research: A challenge to software engineering. In Proceedings of the 25th International Conference on Software Engineering, Piscataway, NJ, USA, 3–10 May 2003; pp. 74–83.

CHAPTER 10

Why Virtual Friendship is No Genuine Friendship

Barbro Fröding • Martin Peterson

Division of Philosophy, Royal Institute of Technology KTH, Teknikringen 78B, 100 44 Stockholm, Sweden

Section for Philosophy and Ethics, Eindhoven University of Technology, P.O. Box 513, 5600 MB Eindhoven, The Netherlands

ABSTRACT

Based on a modern reading of Aristotle's theory of friendship, we argue that virtual friendship does not qualify as genuine friendship. By 'virtual friendship' we mean the type of friendship that exists on the internet, and seldom or never is combined with real life interaction. A 'traditional friendship' is, in contrast, the type of friendship that involves substantial real life interaction, and we claim that only this type can merit the label 'genuine friendship' and thus qualify as morally valuable. The upshot of our discussion is that virtual friendship is what Aristotle might have described as a lower and less valuable form of social exchange.

Keywords: Virtual friendship Aristotle Virtue ethics Facebook

INTRODUCTION

In the last few years, social community sites such as Facebook, MSN, and Hyves have gone from being driven by special interest groups to becoming basic social necessities of everyday life. If you do not exist online you simply do not 'exist'. The average user of Facebook in Europe has 139 friends, and it is not uncommon to have two or even three hundred online friends.[1]Although concerns have been raised about the lack of privacy, hacking, and the potentially tempting opportunity for service providers to use information stored on their sites for business purposes, many people seem happy to continue living their social lives online.

In this article we argue that social community sites are, contrary to what many users seem to think, not a key to meaningful social relationships. To be more precise, we argue that if we understand the notion of friendship in a broadly Aristotelian manner, virtual friendship does not qualify as genuine friendship. In our view, virtual friendship is what Aristotle might have described as a lower and less valuable form of social exchange. Further to this point, we argue that virtual friendship is analogous to certain, questionable, forms of alternative medicine: social community sites are potentially harmful since what is described as a route to social success may in fact turn out to be a toxic substance leading to isolation, just as some alternative medical substances harm rather than cure the patient. Furthermore, by opting for the alternative 'medicine' the individual may forego proven and functioning methods for achieving meaningful social interaction. In other words, what is flagged as a fast-track to meaningful social relationships and social inclusion is in fact an illusion as these relationships, whatever else they may be, do not contain the necessary components that go into genuine friendship.

In an early and attention-grabbing article on virtual friendship, Cocking and Matthews argue that, 'within a purely virtual context the establishment of close friendship is simply psychologically impossible'.[2] Cocking later developed this claim further and argued that, 'certain features of text-based online contexts largely rule out the development of close friendships exclusively in those contexts'.[3] However, in a comment on the original article by Cocking and Matthews, Briggle claimed that 'Cocking and Matthews are ... wrong about the possibility of friendship—as defined according to their own criteria—flourishing wholly online'.[4] Briggle's conclusion is, in opposition to Cocking and Matthews that, '[t]he increased distance and slowed pace of Internet relationships can foster friendships of equal or greater closeness than those in the offline world'.[5]

Although we agree with many of the views put forward by Cocking and Matthews, as well as with some of Briggle's criticism, there are also important differences between the claims we make in this article and theirs. First of all, in contrast to Cocking and Matthews, we do not argue that virtual friendship is *impossible*. Our claim is a more narrow claim about the moral value of virtual friendship; we do not question that virtual friendship counts as a form of friendship. All we seek to show is that from an Aristotelian point of view, virtual friendship is less valuable than other friendship relations. This also highlights an important difference between our view and that of Briggle: While he maintains that virtual friendships are not merely possible, but are often more valuable from a moral point of view, we of course deny this positive appraisal.

However, the main difference between our position and those mentioned above is that we explicitly relate our normative conclusion to Aristotle's theory as it is presented in the Nicomachean Ethics, and we do so on a relatively detailed level.[6] Cocking and Matthews briefly mention Aristotle's theory of friendship, but their main concern is their own analyses of various types of friendship relations. Our point of departure is thus different from theirs, as is the scope of our conclusion: All we believe to show is that given Aristotle's theoretical framework, virtual friendship does not qualify as genuine friendship, as defined above.

On the Aristotelian analysis, for a friendship to count as morally valuable, and hence virtuous, it must contain the following elements: it must be mutually recognised, the friends must engage in *theoria* (i.e., the contemplation that takes place between virtuous agents), and the love and admiration they feel for each other must be based on virtue. Our main thesis is that because virtual friendship cannot fully meet these criteria it does not qualify as genuine friendship. By 'virtual friendship' we mean the type of friendship that exists on the internet, and seldom or never is combined with real life interaction. To contrast this, we reserve the term 'traditional friendship' for the type of friendship that involves substantial real life interaction. The latter are the only type that we claim merit the label 'genuine friendship' and thus qualify as morally valuable.

All this said, we by no means object to initiating or maintaining friendships through social community sites. Nor do we have any general concerns about social community sites as such. Interaction on these sites can indeed be valuable in an instrumental sense. In the case of friendship, for example, the people that you first meet online can later turn into genuine friends or even life partners in the traditional sense. Our point is that for a relationship to qualify as genuine friendship it is not enough to merely interact online. Whatever goes on in the virtual world must always be supplemented by a substantive element of real life interaction.

The Aristotelian theory of friendship

Aristotle argues that friendship (*philia*) is key to human happiness. He claims that for any human to be happy she needs friends and other people close to her.[7] Generally speaking, the shared life is always superior and, as human wellbeing and social activity cannot be separated, it is better to engage in practical activities with a friend than to do it on one's own.

The paradigm case of friendship for Aristotle is a relationship that is mutually recognised and taking place between two adults of equal standing. While all other relationships are inferior to this one, Aristotle agrees that relationships between e.g. the non-virtuous may also be called friendship but of a lesser kind, as pointed out above. The most important aspect of friendship is spending time together, preferably engaging in *theoria* as this is the hallmark of the good friendship.[8]

Broadly speaking friendship helps us grow and become more virtuous as our friends inspire and help us. Both parties gain self-knowledge, 'we are able to observe our neighbours more than ourselves, and to observe actions more than our own'.[9] Further to that point Cooper writes that, 'the presumption is that even an intimate friend remains distinct enough to be studied objectively; yet because one intuitively knows to be fundamentally the same in character as he is, one obtains through him an objective view of oneself'.[10] So by watching our friend, our 'other self', we discover ourselves. Notably, this is an ongoing process: we change when going through life and therefore we must maintain our friendships not to lose track of ourselves. As Sherman succinctly puts it, 'friendship creates a context or arena for the expression of virtue and ultimately for happiness'.[11]

Good and true friends do things for one another and even though it might not be about counting and taking turns it is nonetheless vital that there is an overall balance which both parties are aware of. But what you do for your friend is not done to secure advantages for yourself, it is done simply because you see your friend as another self. Your friend is an extension of you in the sense that your happiness is to an extent dependent on him and, thus, that part of your fate lies in the hands of your friend(s).

A comprehensive definition of Aristotle's notion of *philia* is, 'the mutually acknowledged and reciprocal exchange of goodwill and affection that exists among individuals who share an interest in each other on the basis of virtue, pleasure or utility'.[12] In addition to voluntary associations of this sort, Aristotle also includes among friendships the non-chosen relations of affection and care that exists among family members and fellow citizens.[13]

Based on the above-mentioned definition of *philia*, Aristotle argues that there are three main qualities that determine whether someone qualifies as a friend: excellence, pleasantness and usefulness. He then moves on to saying that these translate into three types of friendships, which often overlap.[14]

1. friendship based on mutual admiration

2. friendship based on mutual pleasure
3. friendship based on mutual advantage

Aristotle claims that the first type of friendship is superior to the other two because it is based on excellence. What the two friends admire is the virtue of the other. It thus deals with the inner qualities of a person. In these situations we love our friend for intrinsic reasons and not solely as a road to pleasure and utility. You must not choose your friend because he makes you laugh or buys you expensive chocolates or has the right connections to secure you the best seats at the opera opening-nights. When you only love that which is useful and pleasant your friend becomes instrumental to securing those goods for you.[15]

Evidently, such behavior is not fitting for the virtuous agent. These intrinsic qualities are stable (contrary to e.g. fame, beauty and wealth) so even if your virtuous friend falls on hard times he will still have those personal qualities you admire and love. The foundations of such a friendship are good without qualification. Your friend and you like each other, share basic values and you admire each other for the right reasons. You see the virtue in one another and you are drawn to it and you wish each other good only for the sake of good.[16]

Although the three types of friendships overlap, it must always be the case that you see your friend as useful and pleasant *because* you love him and not the other way around.[17] That said, Aristotle also recognises that friends are important as instruments of happiness. He writes that, 'happiness also evidently needs external goods to be added, as we said, we cannot, or cannot easily, do fine actions if we lack the resources. For, first of all, in many actions we use friends, wealth, and political power just as we use instruments'.[18] In addition to this, friends are also intrinsic, necessary components of happiness: 'For we do not altogether have the character of happiness if we look utterly repulsive or are ill-born, solitary, or childless; and we have it even less, presumably, if our children or friends are totally bad, or were good but have died'.[19] In many cases people are friends in both senses because even the finest of friendships include pleasure and utility aspects and this does not taint them in any way.

It is worth stressing that the Greek term *philia* tends to be used in a broader sense by Aristotle and others than the English term 'friendship'. Aristotle's theory of friendship covers all the relationships we have with people around us, ranging from our family to our fellow statesmen. In addition to our modern notion of friendship, it includes a substantial chunk of all the other the members of society, for example the local cobbler[20] and one's political or business contacts.[21] Indeed, Aristotle writes that we even have a certain *philia* with all of mankind and that there is an ever so small element of care among all humans.[22]

Pakaluk claims that, 'since Aristotle uses the term [*philia*] for any affection that expects reciprocation, or that expects and finds reciprocation, no matter how extended or attenuated that affection, he applies it very widely: to families, clubs, clans, and even to reciprocal affections of loyalty and patriotism among citizens'.[23] A slightly more conservative approach can be seen in e.g. Price, Walker and Cooper who all (to various degrees) argue that the lesser kinds of friendship and relationship do not qualify as friendship proper.[24] We shall return to this idea about different kinds of friendships and its implication for virtual friendship towards the end of this article.

At this point it could perhaps be objected that since *philia* is used in such a broad sense by Aristotle, it seems that his theory of friendship does not imply any particular intimacy. If true, this could in turn be taken to speak against our claim that virtual friendship is no genuine friendship. If, for instance, business contacts count as friends, in the broad Aristotelian sense, it seems odd to maintain that a virtual friendship cannot count as genuine friendship. Our reply to this objection is that the Aristotelian theory of friendship emphasises the importance of *mutual* admiration and love among friends. A major problem with online friendship is that this is often not the case. Both parties have to be aware of the relationship, they must both harbour similar feelings for each other and there must be an overall balance.[25] This is one of the many reasons why virtual friendship is problematic. For the internet user it is often more difficult to ensure that the love and admiration is mutual, as we explain in the next section.

Friendship on the internet

In this section we analyse the difference between virtual and traditional friendship from an Aristotelian perspective. As explained above, there are three different types of friendships. While all qualify as worthwhile to some extent, the most valuable one is friendship based on mutual admiration. Aristotle maintains that for this kind of friendship to exist the following three conditions need to be satisfied.[26]

(i) the friendship is mutually recognised and takes place between two adult humans of equal standing;

(ii) the friends spend time together, principally engaging in *theoria*;

(iii) the admiration and love the friends feel for each other is based on the virtues they recognise in the other.

We concede that (i) can be satisfied in a virtual friendship and will therefore not discuss this condition any further. As for (ii), it is of course possible for the agents to engage in *theoria* in virtual reality. However, it is a mistake to assume that *theoria* only involves advanced or lofty ideas relating to the nature of science and the like. In fact, *theoria* requires the contemplation of a mixed bag of topics involving both the high and the low.[27] Arguably, agents sometimes withhold what they perceive as less than perfect character traits in themselves when given the practical opportunity to do so. This is problematic as this opens the door to pre-meditatated censorship with regard to the information one discloses about oneself. This threatens to compromise the variation required for *theoria.* Even base matters could, and sometimes should, feature in *theoria*. An excellent example of this is the passage about Heraclitos in the kitchen in Animals I.5. Here we meet a Greek philosopher and hero who does not only suggest that such lowly entities as animals are suitable material for philosophical contemplation of the higher orders but, also that this discussion should take place in a kitchen:

So one must not be childishly repelled by the examination of the humbler animals. For in all things of nature there is something wonderful. And just as Heraclitus is said to have spoken to the visitors who wanted to meet him and who stopped as they were approaching when they saw him warming himself by the oven he urged them to come in without fear, for there were gods there too so one must approach the inquiry about each animal without aversion, since in all of them there is something natural and beautiful.[28]

Two persons that spend time together in real life are more likely to face a wider spectrum of different situations, and consequently, encounter a larger range of topics meriting contemplation. This indicates that traditional friendship is more conducive to *theoria* than its virtual counterpart. In real life we stumble on situations that are both novel and unexpected and we have to deal with them *in promptu.* This seldom happens on the internet. In the online sphere agents can choose when to engage each other and are thus likely to select situations where they are in control, e.g. when they are in a good mood, not stressed, able to be private and so on. The result of this is that virtual interaction, by being subject to control, is too restricted and unlikely to bring about theoria as defined in the NE.

We concede that it might be possible to solve the problems detailed above through technological advances. Our concerns with respect to (ii) and the pursuit of *theoria* in a virtual friendship are based on empirical assumptions relating to the limits of our current technology. These limitations might be overcome in the future, and if so, these concerns would be silenced.

However, our main concern with virtual friendships pertains to (iii). Genuine admiration and love requires honesty. Both parties must tell the truth about themselves and, equally, be able to see the other as she is without embellishing or idealising the friend. Assuming that we are dealing with virtuous agents (or at the very least, agents who aspire to virtue, so called continent agents), blatant lying is less of a concern. Although it is easy to deceive people on the internet, those who aspire to virtue would simply not be tempted to provide false information about themselves or others. That said, this does not take care of the control element discussed in relation to (ii), which risks introducing another, and more subtle, form of vicious behaviour. Because social community sites allow friends to be selective as to the 'when', 'how' and 'for how long' aspects of the interaction in a different way to real life, they can (even unintentionally) choose to communicate only in certain situations. The price they pay is that they miss out on important, potentially problematic and complex, aspects of the friends' personality. Therefore the agent ends up admiring and loving parts of the friend rather than the whole of her. Of course the unknown parts can be just as virtuous as the known traits (although this seems somewhat implausible) but the mere fact that the friendship is based on limited information disqualifies virtual friendship from meeting (iii). Further to this point, this displays a lack of proper judgement and practical wisdom in the agent proving that she does not have a fully virtuous character.

To spell the problem out in even more detail, the novel possibilities for the agent to choose how they depict themselves online can give rise to two kinds of mistaken beliefs. Firstly, one or both of the agents may sometimes end up having less than full knowledge about the other, and thus poor foundations for her perception about the character and persona of her online friend. Secondly, each party would be unaware of this. Note here that the problem is not necessarily conditioned on the *actual* lack of virtue in either party but that the belief (about the goodness of the other) is based on incorrect or incomplete information. Withholding this type of relevant information is in itself vicious. The character traits hidden might well be exemplary and virtuous, but the fact that they are unknown to the friend is enough to give rise to problems for the Aristotelian. The complete and excellent friendship can only obtain when both agents are fine, noble and excellent in every aspect, and this is incompatible with the withholding or manipulation of relevant information.

To illustrate the problem, consider the following scenario. Alice and Betty met online and hit it off immediately. They now chat more or less every day and feel that they have become quite close. They have, for example, shared intimate secrets, laughed together and even exchanged holiday pictures. After a couple of months of intense communication, however, Alice begins to notice that Betty never seems to be available on Wednesday and Friday afternoons. Unfortunately this coincides with Alice's only free afternoons during the week, time which she would like to spend chatting to Betty. When asked about this Betty becomes evasive and snappy and Alice ends up confused and hurt. The following Wednesday afternoon when Alice is in town she swings by the Community Pool for a swim. As she is about to go in, she spots Betty, immediately recognising her from the pictures. Just as she is about to call out her name she sees that Betty is not alone. She is with her physiotherapist fully engaged in a session of rehab water gymnastics. It turns out that she suffers from an impairment caused by a traffic accident. Her condition is very painful and greatly restricts her ability to e.g. visit clubs, bars, restaurants and other social venues. She is very sad and embarrassed about this and did not want Alice to think of her as 'different' from any other woman their age and thus kept it secret. This is of course understandable from a human perspective, yet it is telling as it shows how Alice's love and admiration for Betty in fact was based on incomplete information. Regardless of what Alice would have made of the truth, the point is that she, unbeknownst to her, did not have access to all the relevant information. The judgment Alice reached about Betty was ill founded and, consequently, their friendship failed to meet condition (iii). This example shows that increased opportunities to withhold or distort information is in fact an element intrinsic to online life, and is as such morally problematic.

The internet and the possibility of lesser friendship

The highest form of friendship cannot be enjoyed solely by interacting online, as argued above. This type of friendship requires a real life component. However, as explained earlier, Aristotle recognises that friendship can come in more than one form and, further to that, be worthwhile even though it might to a limited extent only. What could be called the 'lesser versions' are neither useless nor without value. Consequently, they could indeed be worthwhile to pursue given that the agents involved do not confuse their relationship with the highest form, i.e. the genuine form, of friendship. This raises the following question: how ought we to think about the pros and cons offered by virtual 'lesser forms' of friendships.

To make the discussion concrete, consider the distinction between professional network sites (such as LinkedIn) and social community sites (such as Facebook and MSN). We believe that there is an important moral distinction to be made here. While the former offers a clear benefit to the users, e.g. in the way of mutually beneficial business relationships which the agents would not have been able to establish otherwise, the promise of the social network sites rings more hollow. Here the user is made to believe that she is likely to gain genuine friends and form meaningful and deep social relationships with other people. We showed in one of the earlier sections that this is false. However, that is not to say that social community sites are *entirely*without value, in all possible scenarios. If managed properly, they can of course offer very concrete benefits, in an instrumental sense. For example, a social community site might indeed be a very good place to meet people with whom you could become the friend of at a later stage as you advance from interacting online to meeting in real life. Further to this, it is a useful way to maintain already existing relationships, both when the friends are short on time or are geographically separated. None the less, there is an important moral distinction to be drawn between professional network sites and social network sites. Professional network sites, being of mutual advantage to the users, may well qualify as a good vehicle for establishing the 'lesser forms' of friendship, at least as long as no one using the professional network sites is led to believe that there is more to it than that (we take this to be a fairly uncontroversial claim and will thus not account for it in detail). The social network sites, on the other hand, do not even meet the criteria for 'lesser friendship'. Although we concede that these sites can sometimes be of mutual advantage to their users, our moral objection is that some users of social network sites are led to form false expectations and judgements about the true nature of their virtual (social) friendships.

Let us illustrate our argument in an example. Alice has two friends both of whom she has met online: Claudia and Daniella. Claudia and Alice first started talking in a chat forum on the professional networks site LinkedIn. They keep in touch regularly and as both are lawyers it has happened that they have recommended each other's services to potential clients. For all intents and purposes this is a mutually beneficial relationship and is thus a prime example of a valuable lesser friendship.

Alice's second friend, Daniella, shares Alice's keen interest in plants and Alice derives a lot of pleasure from discussing gardening on Facebook with her several times a week. On occasion Alice posts pictures of her garden on her Facebook wall. She is especially pleased about the compliments and positive feedback she receives from Daniella. Unbeknownst to Alice, however, Daniella's only motive for posting those comments is to encourage Alice to share gardening secrets with her. The information Daniella gleans has had a significant positive impact on Daniella's own garden, and allowed her to grow plants she would otherwise not have been able to.

From an Aristotelian point of view, there is a clear moral difference between the two cases. In the gardening scenario, Alice is deeply mistaken about the nature of her and Daniella's relationship. Alice thinks that Daniella's compliments are sincere and without agenda, something that is not the case. Our conclusion is that Alice's friendship with Daniella has no moral value what so ever, and may even be harmful to Alice, whereas her friendship with Claudia has at least some value, in virtue of being honest and mutually useful.

A metaphor might help to clarify the difference further. Alice's friendship with Daniella is like certain questionable forms of alternative medicine: The friendship is potentially harmful to Alice, since what she believes is a route to deep and meaningful interaction with Daniella in fact is a toxic substance leading to a feeling of betrayal, i.e. the friendship does harm to her rather than cure her social isolation. Furthermore, by opting for the alternative 'medicine' Alice also foregoes proven and functioning methods for achieving meaningful social interaction. In other words, what is flagged as a fast-track to meaningful social relationships and social inclusion is in fact an illusion as these relationships, whatever else they may be, do not contain the necessary components that go into genuine friendship.

To conclude, based on a modern reading of Aristotle's theory of friendship, we have shown that virtual friendship does not qualify as genuine friendship. Virtual friendship exists on the internet and is seldom or never combined with real life interaction, whereas traditional friendship involves substantial real life interaction. We have shown that only the latter type can merit the label 'genuine friendship' and thus qualify as morally valuable. This supports the idea that virtual friendship is what Aristotle might have described as a lower and less valuable form of social exchange.

FOOTNOTES

[1] Metro, The Netherlands, 1 December 2010.

[2] Cocking and Matthews (2000:224).

[3] Cocking (2008:124).

[4] Briggle (2008:72).

[5] *Ibid.*, p 73.

[6] We do not claim that our view is the 'true' or 'correct' exegetical analysis of Aristotle's position. All we claim is that it is a *possible* reading that is in line with the overall structure of his theory.

[7] NE 1169b10-15.

[8] NE 1157b19-24.

[9] NE 1169b33-35.

[10] Cooper (1977, p. 322).

[11] Sherman (1989:128).

[12] NE VIII.2.

[13] Cf. NE VIII.9, VIII. 12, IX. 6. See also Sherman (1989: 124).

[14] NE 1156a6-8.

[15] NE 1156a14-19.

[16] NE VIII.3-6.

[17] NE 1138a3-8, NE 1156b18-24, NE 1170a5-6, NE 1236b27-32 and NE 1237a26-33.

[18] NE 1099a31-1099b2.

[19] NE1099b2-6.

[20] NE 1163b35.

[21] NE 1158a28.

[22] See Book 8.1 of the NE.

[23] See Pakaluk (1998:264).

[24] See e.g. Cooper (1977, p. 316).

[25] NE 1155b26-56a5.

[26] NE 1157b25-30.

[27] See for example the passage on Heraclitus in the kitchen in Parts of Animals I.5.

[28] PA i 5.645a15-23.

REFERENCES

1. Aristotle. (1980). *The Nicomachean ethics* (trans: Ross, W. D.). Oxford: Oxford University Press.
2. Aristotle. (2002). *The Nicomachean ethics* (trans: Broadie, S., & Rowe, C.). Oxford: Oxford University Press.
3. Briggle, A. (2008). Real friends. *Ethics and Information Technology, 10*, 71–79.
4. Cocking, S. (2008). Plural selves and relational identity: Intimacy and privacy online. In J. van den Hoven & J. Weckert (Eds.), *Information technology and moral philosophy* (pp. 123–141). Cambridge: Cambridge University Press.
5. Cocking, S., & Matthews, S. (2000). Unreal friends. *Ethics and Information Technology, 2*, 223–231.
6. Cooper, J. M. (1977). Aristotle on friendship. In A. O. Rorty (Ed.), *Essays on Aristotle's ethics*. Berkeley: University of California Press.
7. Pakaluk, M. (Ed.). (1998). *Nicomachean ethics, books VIII and IX*. Oxford, England: Oxford University Press.

8. Sherman, N. (1989). *The fabric of character—Aristotle's theory of virtue*. Oxford: Clarendon.

CHAPTER 11

The Moral Implications of Software Piracy

Kyle Hamrick Olivet Nazarene University,

ABSTRACT

Computer software is integrated into almost every aspect of our professional and personal lives. Much of this software requires payment for use and is legally protected by the copyright system. This paper examines and analyzes the arguments pertaining to the moral use of protected software (digital piracy). The three arguments presented are the "victimless crime" argument, the "noble justification" argument, and the "willing but unable argument." These three arguments claim that piracy is morally justified in certain cases, and claim that software providers are not harmed in such situations. The three arguments are tested against counter-arguments, and it is discovered that each debate comes to a stalemate at a single question: is an action immoral because it is illegal? The paper also examines this question in relation to act-Utilitarian and ruleutilitarian ideology. The key research of Gary Santillanes and Ryan Marshall Felder is used to come to the conclusion that piracy is morally acceptable only in situations where there is sufficient need and no alternative. Rule-Utilitarian thought is found as insufficient to handle situations where such piracy is vindicated. The victimless crime argument is dismissed as failing to meet the two conditions, but the noble justification and willing but unable arguments are found as morally justified in certain situations.

THE MORAL IMPLICATIONS OF SOFTWARE PIRACY

In recent decades, software has facilitated research and business operations in virtually every modern institution. Large corporations such as Microsoft and Adobe cater to schools and businesses of every size. Most of this software requires purchased licenses from companies, and legal measures ensure that the software providers are compensated for their work. Gary Santillanes and Ryan Marshall Felder (2015) examine the moral implications of using the software without compensating the owner, and argue that such piracy can be morally justified in certain circumstances.

The digital world is so unlike the one we are born into that we are forced to analogize parts of it to understand it. While this allows us to easily conceptualize operations such as transferring and storing information, other concepts such as ownership, property, and rights become gray areas in both legality and morality. Is copying a file really stealing if the original owner never loses the original? If someone buys music online, can they make as many copies as they want? If someone buys a license to a software, can they let their friend use it? These issues would be either morally clear or impossible in the physical world, but the analogies we create through years of experience with information technology give people a variety of different moral standings and opinions. Where one person sees blatant thievery, another sees a sharing community. Where one person sees a justified use of authority, another sees a miscarriage of rights. In many cases, the use of analogies is actually detrimental to the ethical concept of the act. Because digital rights laws were created using these analogies, they are often unclear or insufficient to handle situations. Even where the laws are sufficient, the moral implications of an illegal act may not be enough to deter those who can so easily perform the illegal acts. Over the years, the people labeled as "pirates" have developed a number of justifications for using software without the express permission of its creator. These range from petty spite and unwillingness to support a particular company to a genuinely altruistic desire to promote knowledge and learning in an economically limited situation. The latter is the focus of Santillanes and Felder's research (2015).

The field of academics is no stranger to treating digital rights as more of a guideline than a law. While research on the subject is uncommon, studies show that piracy is as common in research institutions as it is among the general public. In one study, the majority (69%) of academics admitted to have knowingly used pirated software (Santillanes & Felder, 2015, p. 969). In another, about a quarter of individuals expressed the belief that it was acceptable to copy software for academic purposes. From this evidence it is easy to see an obvious disconnect with the analogy of piracy as synonymous with thievery. The failure of anti-piracy ads that use analogies to stealing a car are to be expected, when you consider that most pirates do not see piracy as comparable to stealing. The illegality of Piracy makes for very few pro-piracy arguments outside of internet forums. Reddit user faceyourfaces shares a mentality held by many pirates: "Most people I have seen argue against this go off on tangents about how piracy is immoral and hurts the content creator. Is piracy immoral? It certainly can be. Does piracy hurt the content creator? It definitely can. Does that make it theft? No" (Reddit, 2014).

Arguments for Piracy

Stealing a physical object can be a moral dilemma for many people because it deprives an owner of their rightful property, and damages the person. It is fairly easy for pirates to justify the decision with the following argument: Because the owner never lost their property, and because I was never going to purchase the software to begin with, I am causing no harm to the owner. Essentially, the owner is not affected at all whether the software is pirated or not, because the decision not to pirate would not lead to a purchase of the software. The decision to not purchase the software can stem from many reasons: an unwillingness to support a creator, no funds to do so, or an unwillingness to adopt the platform the content exists on. This victimless crime argument is another result of the range of worldviews that exist among users of information technology. The moral restraints created by the analogy to thievery are removed, and the ease at which the act is performed account in part for the high percentages of reported pirates.

A second argument that is the main focus of Santillanes and Felder regards software in relation to economics. Copyright law is credited with driving the progression of science and invention by rewarding individuals with accepted contributions to society. If individuals' work could be directly used by others with no compensation, there would be much less motivation to work in creative fields. The rewards that individuals receive are control over the usage of their work. In software, individuals or more likely corporations can charge customers for the use of their digital property. Santillanes and Felder (2015, p. 968) present a situation in which copyright law can act as a hindrance to the progression of science. A graduate who has recently received his doctorate in the field of biology wishes to complete his research that can benefit society. Difficulties in living in a capitalist environment force him to take low-paying jobs. His research can be performed alone, but requires the assistance of proprietary software that he is unable to afford. This is a clear example where copyright law prevents the advancement of knowledge by legal means. This noble justification argument is rarer than the victimless crime argument, but is much more attractive in such situations. Even software providers themselves may find it acceptable that such important work was performed on a pirated version of their product. Logic implies that the good the research will bring outweighs the relatively minor issue of using illegal software. While individuals would pay if they could, they are forced to use pirated software to complete their goals. Santillanes and Felder argue that these individuals are morally justified in pirating the software. While they are careful not to accuse the copyright system of being ineffective at promoting invention, they argue that in the copyright system limits the free exchange of ideas in certain situations (Santillanes & Felder, 2015, p. 974).

A third argument exists that has different roots than the first two. Many people simply cannot access software through legal means. For example: Company A offers a video-editing application that uses a proprietary file type that can only be legally created in that particular software. A recent college graduate named Jonah works from home for a private international news organization. The organization recently made the decision to switch to the proprietary file type, but the software from Company A is not licensed in Jonah's country. The price is very reasonable, but Jonah has no legal way to pay for the software. Jonah instead finds a cracked version online for people who suffer the same predicament. Because Company A does not seem to want his business, it doesn't seem wrong for Jonah to pirate the software. This "Willing but Unable" argument has many forms and is much more popular in media than software. People overseas often assert that they would gladly pay to watch American dramas at the same time as American audiences, but they would have to wait eight months for the shows to reach their home country. Other variations of this argument address unsatisfactory delivery methods. For example: Some software providers attempt to combat piracy by requiring users be constantly connected to the internet. This type of DRM does not satisfy those with spotty or limited internet access, and those users (who would gladly pay for the software otherwise) turn to piracy to fulfill their needs. A similar situation can be found in companies that go bankrupt or go are purchased and shut down. If the software is not available to be purchased anymore, is it wrong to pirate it? **Argument Analysis**

These three arguments, the victimless crime, the noble justification, and the willing but unable are three very attractive arguments for pirates of software because they morally justify (or at least remove moral barriers to) piracy using logic. The victimless crime argument maintains that whether the individual pirates or not, the creator/owner is unaffected. Therefore, the pirate is not harming the owner through piracy, nor benefitting the owner through inaction. The noble justification argument treats the issue like a simple mathematical equation: The benefits the research will bring are greater than the detriment of piracy. The willing but unable argument has many forms of varying strength, but even some that denounce piracy admit that those who are not presented with a legal way to use the software may be justified in doing so. There are many more arguments relating to the freedom of information or political ideologies, but these three arguments more often than not correspond to the brand of morality and the appreciation for logic that exist in those that denounce piracy. This allows both sides to debate with a similar mindset, and suggests that each argument can be understood by the opposite side. This is essential to explore the arguments in the dialectic method.

The victimless crime argument raises the most opposition of the three. People against piracy accuse the logic as an attempt to justify an immoral action after the fact. The decision not to purchase the software legally should come with pre-assumed consequences, one of which is the inability to use the software. The pirate's avoidance of the consequences means that the pirate actually benefits from the decision. This benefit raises serious doubts about the reason for the stance, and people against piracy argue that if the benefit had any influence on the pirate's decision to pirate, then the pirate's decision is a result from selfishness. Using this line of thought, the driving reason the pirate made the decision not to buy the software was because they could get it for free. This suggests that it would be possible that the pirate would purchase the software legally if there was no alternative. If this were true, then the decision to pirate would in fact be hurting the software provider because it is robbing them of business (even if it is the pirate's own). Of course, this cannot be proven and pirates are free to give any number of reasons why they made the choice not to buy the software. A very popular argument for piracy is that people pirate because they want to try the software out before they purchase it. If they like the software, then they will buy it to support the developer. Using this logic, a pirate can use the software as much as they want and payment is treated as optional. This argument has little affect on people against piracy. This difficulty in ascertaining the truth leads to a stalemate in arguments, but some find issues with the underlying assumption of the victimless crime argument. After the logic, the main assumption is that because nobody is being harmed by the piracy, it is morally acceptable to pirate. The filtered message is that it is morally acceptable to break laws if nobody is harmed by the action. The underlying assumption is that all laws are created to protect people from harm of any type. Arguments that disobeying the law is in itself an immoral act are shot down by pirates as legalism and irrelevant to morality. When applied to other laws, the moral content of the pirate's distilled message does not seem to change. Is it morally wrong to run a red light on a deserted street? Is it morally wrong for a large group of people to sing happy birthday (a licensed song) loudly, thus violating copyright law? The difficulty in arguing the morality of such actions usually results in a stalemate at legalism, with people against piracy holding that it is immoral because it is illegal, and pirates holding that such an argument is invalid. When discussing the idea that piracy is immoral because it is illegal, Reddit user Alterego9 believed "[t]his statement itself is of course a matter of ideological conviction and temperament, (and counterable with a belief in ideas such as civil disobedience), and not followed by most people who would rather believe that laws themselves are derived from moral axioms, and this morality is superior to legality" (Reddit, 2013). This disconnect seems to be the dividing factor for this argument.

The noble justification argument further divides parties along the lines of legalism. In this sense, piracy may not be morally neutral, but may be morally justified. When presented with the choice to pirate or not to pirate, should people choose the action that will bring the most benefit to the largest number of people, or should they assume that the laws in place exist to do that for them? This is a prime example of the philosophies of the two major types of Utilitarians: ruleUtilitarians and act-Utilitarians (Johnson, 2009, p. 38). Rule-Utilitarians argue that rules exist to handle such ethical issues and that individuals do not have the moral authority to decide which laws may or may not be broken for the sake of the greater good. Act-Utilitarians argue that rules are insufficient to decide the proper action in every conceivable situation, and that the appropriate action should be determined by analyzing each situation. Santillanes and Felder (2015) assume the latter ideology in their analysis of piracy in the academic field.

...notice that a specific set of circumstances must obtain for this moral justification to exist. There must be no other alternative open to the researcher than to pirate the software, and the researcher must have actively pursued every possible alternative. We do not believe that the vast majority of software piracy in research meets these standards for moral justification, but we think it is reasonable to expect that some such cases exist. (p. 974)

Santillanes and Felder argue that the copyright system (while adequate in the majority of cases) may be morally ignored in select cases that cannot be determined by preexisting laws. It is easier to argue against the act-Utilitarian standpoint than it is to argue against the moral content of this particular case of piracy. Opponents argue that even if there is moral justification for piracy, the individual(s) considering the act should not be vindicated in making the decision. Because every individual has a unique and personal set of moral values, there is no reliability in the outcome of the decision. Additionally, if such practice is accepted, anyone would be able to argue that any action corresponds with their moral code. Like victimless crime, arguments against the noble justification case usually result in an argument about whether it is or is not morally wrong because it is illegal. In this case however, proponents of academic piracy are able to argue for the moral value of the benefit such piracy will being.

The willing but unable argument presents an entirely different set of ethical deliberations. Because the variety of arguments this category contains has different strengths, the argument example presented (absolutely no legal method of payment) will be analyzed as the strongest. The field of business is constantly evolving and expanding. What is unavailable today may be within reach tomorrow. A weakness of Jonah's reasoning is assuming that Company A does not want his business. Company A may wish to expand to Jonah's country at the first opportunity, but may be experiencing difficulties doing so. That being said, is Jonah justified in pirating the software if he pays for it once it becomes available? In this case more than in the victimless crime case, an act of piracy by Jonah will not harm Company A until it is available in his country. This may be seen as the strongest version of the victimless crime argument, but the motives are different. In Jonah's case, the decision not to purchase the software is replaced with an inability to do so. The logic still applies: whether he pirates or not does not affect Company A. For a third time, the only grounds for immorality can be found in the legality of the action. The question becomes: Is Jonah's need for the software greater than the moral issue of piracy when there is no legal method of purchase? The argument can be further simplified: Does the exclusively moral (zero consequence) content of piracy have an effect on a situation, or does the act itself solely determine the morality?

CONCLUSIONS

In each case, the primary argument against piracy resulted from the legality of the action. While there are thousands of cases where arguments against piracy are simple and uncontested, these three cases best demonstrate examples where the act of piracy is moved away from harmful consequences. This allows the pure moral conduct of the act to be analyzed. When the recipient is not affected by the act (through pre-decision, lack of funds, or no access) the arguments against piracy skew towards the legality of the action: that it is immoral because it is illegal. While many would argue that such is the case with piracy, few would argue that all laws should be obeyed without question. In the United States of America, there are hundreds of documented cases of people violating laws because of a moral stance. The very foundation of America stems from an act that was considered illegal (rebellion against Great Britain), but morally justified. From this line of reasoning, it is tempting to follow the line of thought to a departure between morality and legality, but the intent behind the acts must be analyzed. In the foundation of America, and in the violations against laws, the acts were performed because the laws themselves were seen as immoral. The same cannot be said about copyright laws. While proponents of free information have opposite feelings and cannot be persuaded using such arguments, the individuals who pirate using either of the three justifications given cannot honestly give an argument that the laws prohibiting piracy are immoral. While it may be a fallacy to assume that disobeying laws is necessarily immoral, to do so without adequate justification is irresponsible to the society which one lives in. Taking into account Santillanes and Felder's conclusion, I agree that cases exist where it is morally acceptable to pirate software, but such cases are few and far between. There must be no alternative to piracy, meaning that the victimless crime argument does not apply. The precondition that an individual will not pay for the software is not enough to justify piracy. The noble justification and willing but unable arguments apply only when there is sufficient need, and no alternative. These two conditions must be satisfied to morally justify piracy. While "sufficient need" is too vague for ruleUtilitarians, the reach of human imagination is not enough to create a set of laws to address every situation where such piracy may be justified. The legality of the act is not enough to condemn piracy, but individuals who pirate without satisfying the two conditions are not morally justified in piracy according to my research and reasoning.

REFERENCES

1. Johnson, D. G. (2009). Computer ethics. Columbus, OH: Prentice Hall.
2. Reddit. (2013). The common moral arguments against piracy are ultimately all derived entirely from it's illegality, CMV. Retrieved December 7, 2016 from https://www.reddit.com/r/changemyview/comments/1kxoaj/the_common_moral_argume nts_against_piracy_are/
3. Reddit. (2014). CMV: Piracy is not theft. Retrieved December 7, 2016 from https://www.reddit.com/r/changemyview/comments/2ekiyk/cmv_piracy_is_not_theft/
4. Santillanes, G., & Felder, R. M. (2015). Software piracy in research: A moral analysis. Science and Engineering Ethics, 21(4), 967-977. Retrieved from http://olivet. summon.serialssolutions.com/#!/search?bookMark=ePnHCXMwdV1LSwM xEA4iWF38CxLo1YDJ7Gy7vRVRvAiF1nOY5gG9lFIF8d87s0nsyfuXSZiEWbIPO7UPUl u9vFrquGKqv4kTuVqHDFL6lRvRlyAwSup_FlYI5lS1xIYIRjo0c1axCTs eKNmT NP8 Vnp3q8yWjdM3nZPeHM4UfvThqFuK2kqv9buUtevWz6NTH68vu c3UcKmMCE1Rsi 9qpDxmH_hESIEXIcE3vu- 2XOS8IxsvUAZyG5BOwhuRgpEbgxDDFwQAedmhe5z Xr7e vGf3g 7ssgDKFJVO2YJqf OdPpYGEv7QKFoV4VogXhXiZ0PxQJU9sLeiaYABjwX Qvm0vECZ2XwQ664tEhsM_8IqUvdfljZYznGKGXx9gft4

CHAPTER 12

Service Robots, Care Ethics, and Design

A. van Wynsberghe[1]

[1] *Department of Philosophy, University of Twente, Drienerlolaan 5, 7522 NB Enschede, The Netherlands*

ABSTRACT

It should not be a surprise in the near future to encounter either a personal or a professional service robot in our homes and/or our work places: according to the International Federation for Robots, there will be approx 35 million service robots at work by 2018. Given that individuals will interact and even cooperate with these service robots, their design and development demand ethical attention. With this in mind I suggest the use of an approach for incorporating ethics into the design process of robots known as Care Centered Value Sensitive Design (CCVSD). Although this approach was originally and intentionally designed for the healthcare domain, the aim of this paper is to present a preliminary study of how personal and professional service robots might also be evaluated using the CCVSD approach. The normative foundations for CCVSD come from its reliance on the care ethics tradition and in particular the use of care practices for: (1) structuring the analysis and, (2) determining the values of ethical import. To apply CCVSD outside of healthcare one must show that the robot has been integrated into a care practice. Accordingly, the practice into which the robot is to be used must be assessed and shown to meet the conditions of a care practice. By investigating the foundations of the approach I hope to show why it may be applicable for service robots and further to give examples of current robot prototypes that can and cannot be evaluated using CCVSD.

Keywords: Robot ethics Care ethics Value-sensitive design Service robots Applied ethics

INTRODUCTION

In previous works I have argued in favor of a robust proactive framework for incorporating ethics into the design and implementation of robots named the Care Centered Value Sensitive Design (CCVSD) approach (van Wynsberghe 2012, 2013, 2015). This approach to design, to date, has been targeted at the design and implementation of care robots—robots used in a healthcare context to assist the nurse in his/her role—with the goal of explicitly and systematically including ethics into the design process of these robots. The approach is intended to help robot designers and ethicists in both the retrospective ethical evaluation of care robot design as well as the prospective design of future care robots. Through a series of steps for analyzing care practices (involving data collection, analysis and comparisons), the researcher is able to make an ethically grounded judgment concerning the design of a care robot that has the potential to contribute to good care.

Although robots in healthcare are a main area of development in robotics at the time of this paper (2016), according to the International Federation for Robotics (IFR), more than 4.7 million service robots were sold for personal and domestic use in 2014, including a 542 percent increase in assistive robots for the elderly and disabled.[1] There will be approx 35 million service robots at work by 2018. A service robot is a robot which "performs useful tasks for humans or equipment excluding industrial automation application."[2] Although healthcare robots fall within this definition, robots in other sectors like education, agriculture, policing, etc. have been and are currently being designed and implemented. Given these numbers one would estimate that a framework for the evaluation of robots in these sectors exists and is in testing at the moment. One would be wrong.

The CCVSD approach was originally, and intentionally, designed for the healthcare domain and seemed a natural fit given the already present role that care ethics plays in healthcare (Tronto 2010) and the inherently ethical context of nurses and doctors caring for patients. However, given the lack of a robust approach for robots in other domains one may wonder whether or not the same or a similar, comprehensive and systematic approach for incorporating ethics in care robot design can be applied to robots in other sectors. Such an approach is surely needed in virtue of the increasing role robots will play in our daily lives—at home, at work, and for play. This would be a welcome addition to the fields of robotics and robot ethics. I take the first steps in addressing this challenge in this paper and set the stage for the use of the CCVSD approach towards service robots both personal and professional. This approach fits into the study of robot ethics in that it appeals to the ethics not of the robots themselves (as machine ethics would have us consider) but to the ethics of the designers, users, regulators, and policy makers.

I begin by discussing what service robots are and why they deserve, and even demand, ethical attention. Following this I will present the CCVSD approach briefly for the reader. Afterwards I explore why the CCVSD approach can be used for service robots. In short, I suggest that the CCVSD approach can be used when a service robot is integrated into a care practice. In this paper, I present two necessary conditions for labeling a care practice as such. Given that the CCVSD approach relies on the concept of a care practice for its normative grounding, if these two conditions are met then the CCVSD approach ought to be used for evaluating the robot. The two conditions are as follows: (1) that the care practice is a response to the needs of another and (2) that the needs of the other are met through reciprocal interaction between the care giver and care receiver. If these two conditions are not met then the robot should not be normatively evaluated using the CCVSD approach. Using examples of current robot prototypes I will show why certain robots should or should not be evaluated using the CCVSD approach. To be clear, this paper is not meant as an exhaustive evaluation of different service robots as that will come in the future. Rather, this paper is meant to present a case for the use of the CCVSD approach outside of the healthcare domain.

Service robots: what and why?

According to the IFR between the years 2014 and 2018 there will be approx 35,000,000 personal service robots in use worldwide.[3] This is no small number. This number adds weight to the idea that robots in the homes and workplaces of people are no longer fuel for our imaginations or science fiction writers; rather, robots will soon greet us at the store, direct us to what we want to buy, check us out at the register, clean our floors while we are away, and read our kids a bedtime story.

As there is no universal definition of a robot it is also difficult to give a universal definition of a service robot. Service robots can have a range of capabilities (e.g. locomotion, infrared sensing), degrees of autonomy (e.g. amount of input from a human operator for the robot to fulfill its task), and appearances (e.g. creature like, machine like, humanoid) (Engelberger 1989). What the IFR, scholars, and roboticists can all agree on is that service robots function outside of the factory setting and in this way they are distinguished from industrial robots (Engelberger 1989; Lin et al. 2011; Veruggio and Operto 2006). For the IFR a service robot is: "a robot that performs useful tasks for humans or equipment excluding industrial automation application."[4]Service robots are then further classified as personal service robots, functioning in the homes of people, or professional service robots, functioning in a professional context for commercial use. For the former, examples include home cleaning robots (e.g. Roomba), or the latest Jibo[5](advertised to be the newest member of the family), and for the latter fall surveillance robots (e.g. Knightscope robot[6]), delivery robots (also referred to as logistic robotics[7]), robots to greet customers (e.g. Oshbot), and cleaning and disinfection robots (e.g. the UV disinfection robot[8]). An often neglected area of professional service robots are those in farming and agriculture. Sales in these areas are rising with little attention to the ways in which these robots change farming practices. Examples of such robots include but are not limited to: robots for milking cows (Driessen and Heutinck 2015), for crop maintenance (Evert et al. 2006), and for weeding (Pedersen et al. 2006).

This new context of functioning means that the robots must be able to co-exist, and even cooperate, with humans in the unpredictable, unstructured environments that humans live, work, and play in. To do this most often the robots must be embedded with a certain degree of artificial intelligence, e.g. machine learning, autonomy, advanced sensing etc. It is precisely this elevated level of intelligence added to the inclusion of robots into ethical contexts that presents society with a need to address the ethical issues related to these robots. In other words, society must consider that ethics with regards to these robots is not only about what happens when the robot is present and we must co-exist with such a technology—these are the questions that we constantly face as any new device enters our personal domain. Rather, it is about what happens when a technology can make decisions on our behalf as these robots most certainly will. What does it mean to delegate such a role to the robot? What does it mean to be in a relationship with such a technology (note that when I say be in a relationship I do not mean a romantic one; rather a relationship with the technology in which it takes on a vital role in our lives)?

There are many different layers of ethical issues to be discussed (Asaro 2006; Capurro 2009; Lin et al. 2011). There are fundamental issues such as: robot responsibility (Lokhorst and van den Hoven 2011; Sullins 2011; W. Wallach 2010), human responsibility (Allen and Wallach 2011; Asaro 2006; Wallach and Allen 2008), liability (Asaro 2011), agency (Sullins 2011; Wallach and Allen 2008), and well being (Sharkey 2014; Sharkey and Sharkey 2012; Sparrow and Sparrow 2006). There are also more applied ethical issues such as the potential impact on: privacy (Calo 2011; Denning et al. 2009), security (Denning et al. 2009), and so on. All of these issues must be dealt with; the only dilemma is when, how, and by whom?

Service robots touch on all of the ethical issues specified above. They are being designed to take over roles and responsibilities in a wide range of practices in our lives. There are ongoing studies attempting to understand what makes people trust, become emotionally attached to, and socialize with robots. Trusting, emotional attachment, and socializing are all ethically charged words—words we usually use in connection with other moral agents. Although I have doubts about whether a robot can be a moral agent, the *appearance* of moral agency is enough to warrant ethical evaluation. Furthermore, these robots are being placed into contexts in which they are responsible for children (educational robots and social robots) (Sharkey 2016), our grandparents (Sharkey and Sharkey 2012), and other vulnerable demographics. These contexts are, like healthcare, inherently ethical. For these two reasons (that robots are being built with at least the appearance of moral agency and that they are being placed into inherently ethical contexts), service robots demand ethical evaluation and reflection.

We do not want a situation in which society must deal with these issues in a piecemeal manner after the technology has become pervasive much like smart phones or online social networking sites today. Rather, we should put ourselves in a position in which ethical issues are identified and dealt with at an earlier stage in development during which time there may even be an opportunity to mitigate these concerns. The important question becomes how this can be achieved? I propose it is the robot ethicist who is tasked with the responsibility of finding a solution and the CCVSD approach, as will be argued in this paper, is a great place for a robot ethicist to start.

The care centered value sensitive design approach

The CCVSD approach brings together two traditions; Value Sensitive Design (VSD) and the care ethics perspective. These two traditions are brought together to answer the question: how can values be included in the design of (care) robots and which values ought to be included?

VSD is an approach to the design of computer systems that takes values of ethical and social impact into consideration throughout the design process (Friedman 1996; Friedman et al. 2002, 2003, 2015; Friedman and Kahn 2003; Spiekermann, 2015). The CCVSD approach is not a contradiction to VSD. It is an answer to the criticism that the values in VSD lack a normative foundation (Manders-Huits 2011). Care ethics, with its care values, is used to answer this challenge by providing a normative grounding to VSD theory and methods. More importantly, the CCVSD approach provides a systematic manner in which the methods of VSD can be carried out (i.e. through the steps of data collection and analysis that I later describe). CCVSD is just one way to provide normative grounding to VSD. One could, in principle, provide such grounding using another ethical theory like deontology or consequentialism. Therefore, VSD remains an important theory on its own. However, as argued elsewhere (Manders-Huits 2011), in ethical contexts there must be a normative grounding for the values which VSD is being used to promote.

For the purpose of this paper it is important to describe a focal point of the approach, the Care Centered framework (see Fig. 1), and its method of use. For an extensive theoretical account and justification of this framework along with the CCVSD approach please see previous publications (van Wynsberghe 2012, 2013, 2015).

Context – hospital (and ward) vs. nursing home vs. home setting…

Practice – lifting vs. bathing vs. feeding vs. delivery of food and/or sheets, social interaction, playing games…

Actors involved – nurse and patient and robot vs. patient and robot vs. nurse and robot…

Type of robot – assistive vs. enabling vs. replacement…

Manifestation of moral elements - Attentiveness, responsibility, competence, responsiveness

Figure. 1 The care centered framework for the ethical evaluation of robots

I speak of the framework and its method of use as being two separate things because the framework can be used in either a retrospective or a prospective manner. The way in which it is used in either case is similar but with differences in the: target audience, stage of development, and impact on the future design and/or implementation. For instance, if the ethical evaluation is done at a later stage in development only little tweaks here and there, in the hardware or software, may be possible whereas if the evaluation is done earlier on the entire interface or level of robot autonomy may be altered.

To use the framework, i.e. the methodology of the CCVSD approach, the ethicist and roboticist engage in a series of steps: data collection, value analyses, scenario comparisons, and recommendations based on the scenario comparisons. An outline of these steps is presented in Table 1. Note, however, that although these steps are presented in a linear fashion it is possible that they may be iterative depending on the stage of development and the needs of the design team, for example a team may complete steps 1–6 then return and complete steps 3–6 again with a new prototype. I must acknowledge that the design process is organic and thus there must be flexibility to go back to a previous step if necessary. In due time one would imagine that engaging in these steps and envisioning the future robot in its context of use before the prototype is made will ultimately reduce the number of prototypes made to meet the needs of the care practice in an ethical manner.

Table 1 Steps of the CCVSD methodology

Step	What Happens and How
1. Data collection of care practice prior to the robot	The researcher uses the components of the framework to paint a picture of the care practice by visiting the context in which the practice occurs and researching relevant literature [i.e. a thick description according to the nature of activities approach (for more on this see Sio and Wynsberghe 2015)]. This is done prior to the robot's introduction
2. Value Analysis of care practice prior to the robot to create a scenario for comparison	The researcher describes in great detail how the care values are manifest in the care practice as well as who has what role and responsibility. The researcher must also have an understanding of how the practice is linked with other practices and with an overall process, e.g. the practice of picking vegetables is linked with another practice of harvesting and/or crop maintenance and is also linked with the overall process of managing a farm
	The practice is not described in idealistic or utopic terms; rather, the practice is described as it occurs in reality at the time of description. Therefore, the practice is then open to criticism and scrutiny. In this way it is possible to observe when and how a robot may be a welcome, or necessary, addition to the practice for ensuring good care
3. Data collection of robot	The robot is described in terms of its capabilities and appearance, e.g. the robot has a machine-like appearance, the robot is capable of autonomous travel throughout a building but requires human input to complete its task, or the robot can complete its task without any input from a human operator
	If this step is done without a robot prototype in production it is possible that at this point the design team speculates on the kinds of capabilities and appearance the robot ought to have based on the results of the above two steps
4. Value analysis of the practice with the robot introduced to	The same practice is then described, in as much detail as it was originally described in step 1 (i.e. according to the CC framework), only this time once

create a scenario for comparison	the robot has been integrated. To do this it is optimal for the researcher to visit the context of use to observe the robot in its intended context of use Oftentimes this is not possible as the robot may still be in an early prototype phase (and as such considered an emerging technology); however, it is still possible to provide a detailed description of the care practice given the way in which the robot is intended to be used It is also possible that at this step there may be more than one robot prototype to enter the practice. Each of these prototypes ought to be integrated into a separate scenario to show the differing results to the resulting care practice
5. Scenario comparison	At this point one has an elaborate picture of the care practice, in terms of the manifestation of values, before the robot has been introduced and following the robot's inclusion. With this information one can now compare analyses according to the impact on: the distribution of roles and responsibilities, the impact on human–human relationships, the generation of new human-robot interactions, and the impact on the care values (either positive or negative) If there are multiple robot prototypes each of these scenarios will be compared with the scenario of the care practice prior to the robot's inclusion. If there is only one robot prototype then there ought to be two scenarios for inclusion
6. Recommendations for design, re-design, implementation, and/or policy	Depending on the stage of development, the analysis can yield recommendations for different stakeholders. If the analysis is done earlier on in the design process recommendations can be used to steer future prototypes If the evaluation is done later on recommendations may be used to steer implementation and policy to regulate the robot

The ethics of the CCVSD approach

I focus now on the ethical underpinning of the CCVSD approach as it plays a crucial role in the conditions for using the CCVSD approach which I will later present here in this paper. As I have stated earlier, the ethical tradition that serves as the normative foundation for this approach is care ethics. This tradition is neither consequentialist (i.e. the consequences of an action determine if it is right/wrong) nor deontological (i.e. adhering to a duty determines if an action is right/wrong). In fact care ethicists claim that the goal of care ethics is not to fit the traditional mold of either of these theories (Little 1998; Noddings 2002; Tronto 1993, 2010; Verkerk 2001). Instead, care ethics presents different elements that act as a starting point for uncovering and exploring a moral dilemma. Central to these elements are: relationships, roles, and responsibilities (Tronto 2010). In particular, the reciprocal nature of a relationship is highlighted which facilitates an active, rather than a passive, role of the care receiver (Ibid). From this standpoint then, the ethical dimension of the CC framework, and the CCVSD approach overall, is not to focus entirely on the consequences of the robot's actions nor to focus on certain duties that the engineer must abide by, or the robot must adhere to; rather, the approach echoes the care ethics perspective in that it focuses on promoting values inherent in the relationships, roles, and responsibilities of the practice at hand. Most importantly, it focuses on the relational nature of care activities.

What's more, care ethics argues that roles, relationships and responsibilities mark the starting point for the ethical analysis (rather than providing an equation or the like for solving an ethical dilemma). Accordingly, I suggest that the CC framework act as a starting point for identifying the ethical issues relevant to the robot in question. From the starting point of roles, relationships and responsibilities, both the consequences and duties must be weighed to come to an answer about the right thing to do, i.e. what is right/good.

Also of interest is that care ethics shares a likeness with the virtue ethics tradition in that there is a focus on character development. Of course this point may be contested if you consider that care ethics is relational at its roots rather than individual agent-based like virtue ethics (Noddings 2002); however, when we consider that the caring agent must fulfill both the caring action with a caring disposition we may suggest that care ethics is concerned not only with actions but also with character development (Tronto 1993; Vallor 2011). As such care ethics pays tribute to how a good care giver comes to be known as such, i.e. what is the good care giver, what are the characteristics he/she must embody and how does one arrive at becoming a good care giver? Consequently, the ethical nature of a situation in which a robot is involved is not entirely action based but must also pay tribute to how the robot contributes to, or hinders, the development of an individual as a good care giver. Thus, the ethical dimension of the robot, its goodness or badness, is not entirely based on how it may increase efficiency (i.e. consequence driven) or protect the privacy of users (i.e. duty driven) but also on how it will impact the ethical character development of users (e.g. will it have a long term effect of causing users to objectify other humans?).

Care ethics, and the CCVSD approach, in relying on care values and the relationship between care giver and care receiver was obviously suited to the healthcare context; however, I will show below that caring extends beyond the hospital—and, therefore, so does the CCVSD approach. Saying that the CCVSD approach can be used beyond the context of healthcare is unhelpful if there is no clear signs that a given context or practice involves *care*. In the next section I bring to light conditions that must be met in order for a robot to be considered to be a part of a care practice, and in so doing delineate under what conditions the CCVSD approach is applicable.

The applicability of the CCVSD approach outside of healthcare

As mentioned earlier, CCVSD brings together two approaches: the design approach of VSD for incorporating values in the design process and care ethics for normatively grounding the values used to evaluate the robot. The hurdle for this paper is not to show how VSD can be used outside of healthcare as its aim is in fact to be used in multiple and varying contexts. The hurdle is also not to show that care ethics can be used outside of healthcare as care ethicist Joan Tronto has already dedicated an entire book to this question (Tronto, 1993). The aim is to show that the CCVSD approach can be used for the design and evaluation of robots outside of healthcare contexts.

I will show that the CCVSD approach can be used in such cases *if* the robot in question meets certain conditions. The conditions I will arrive at are inherent to the normative foundation of the approach, namely that of care ethics. These conditions are also necessary within healthcare contexts and were implicit in my previous evaluations of care robots; however, because healthcare contexts so often meet these conditions, there was no need to spell them out explicitly. Given that the CCVSD approach revolves around the concept of a care practice for its normative grounding (i.e. for structuring the selection of values and the way in which values can be analyzed), the conditions for using CCVSD result from the conditions for labeling a care practice as such. To arrive at further conditions I take a closer look at the details and requirements of a care practice.

Condition #1: Care practices as a response to needs

Care practices are a concept used in the care ethics tradition to distinguish between caring activities and other types of activities. To label a practice as a *care* practice demands that one recognize the interrelatedness of: (1) the compassion and empathy required to care for another, (2) the thought that goes into understanding what can be done to care for one in need, and (3) the action to realize the practice of caring. By conceptualizing care practices in this way the aim is to view care as a complex and multi-layered activity rather than exclusively as an emotion or principle (Tronto 1993).

The most comprehensive articulation of a care practice, in this author's opinion, is that which is provided by Tronto in which she identifies stages and corresponding moral elements. The stages are as follows: *"caring about"* (recognizing a need for care), *"caring for"* (taking responsibility to meet that need), *"care giving"* (the actual physical work of providing are), and *"care receiving"* (the evaluation of how well the care provided has met the caring need) (Tronto 1993). The moral elements serve as a means for evaluating each of the stages and correspond to the four stages: attentiveness, responsibility, competence, and reciprocity respectively. For more information on a care practice please see Tronto (1993) and van Wynsberghe (2012, 2015).

The CCVSD approach relies on the concept of a care practice for its normative grounding in two ways: (1) It uses the stages of a care practice as the means of sketching the distribution of roles and responsibilities within the caring activity, and (2) It uses the moral elements of a care practice as the values of ethical importance (according to the requirements of VSD) to be included in the design of future care robots or to be used to evaluate the ethical nature of a current robot.

There are certain conditions necessary in order to label a care practice as such versus labeling it just a practice/activity (e.g. playing a sport or cooking a meal). First, care practices are considered a response to the needs of another. It is not the goal of this paper to restrict myself to one definition of a need but instead to highlight central features which serve as necessary conditions of the concept of 'need' from the care ethics perspective. It is helpful to distinguish needs from wants. Although we desire both, a need is something which is necessary for survival or well-being (Tronto 1993; van Wynsberghe 2015). Without a need one is lacking something preventing them from surviving or being well. A want that goes unfulfilled will not prevent one from surviving or being well. For example, one needs water in order to survive but one wants beer and will survive and be well without it. Although a thorough conceptual analysis of need is outside the scope of this paper, we can say, at the very least, that a need is something that without which one cannot be well or cannot survive. This leaves a lot of room for argument regarding what people need in order to be well—an argument to be had by competing theories of well being–which is precisely something for roboticists and robot ethicists to argue.

Many scholars have attempted to classify needs in different ways. The Psychologist Abraham Maslow developed a hierarchy of needs that typified, categorized and prioritized needs (Maslow 1970). According to this categorization, people moved from one type to another as their needs were being met. This presented a linear fashion to understand how needs can be met and how the more fundamental needs for survival must be met before any others. Many care ethicists would argue for a more particularistic vision of needs (Mol et al. 2010; Tronto 2010; Vanlaere and Gastmans 2011). Accordingly, 'needs' ought to be considered as individualistic in so far as they are related to the person as a unique multi-dimensional person, e.g. the high school student Sarah will have different needs from the University student Laura (Vanlaere and Gastmans 2011). Moreover, needs are not only specific to the individual but can change from one moment (year, month, day, hour, minute) to the next thereby giving needs a dynamic character, e.g. the elementary student Sarah will have different needs from the high school student Sarah (Mol et al. 2010). In fact, for Tronto "any agency or institution that presumes that needs are fixed is likely to be mistaken and to inflict harm in trying to meet such needs (Tronto 2010 pg. 164). Recognizing this individualistic and dynamic character of needs is pivotal for care ethics as it brings to light the active role of the care receiver in the care process—he/she must convey this to the care giver. This active role adds weight to the relational nature of care. In other words, because needs are particular and changing, care is not an activity bestowed on an individual without consideration for how it is received (uni-directional); rather, it is an activity (or series of activities) that two parties participate in with intention (it is bi-directional).

For Tronto (1993), every species (human and animal) is in a state of need at some point in their life and more often than not at multiple points in their life. Think of children or elderly persons in need of assistance with bathing, dressing, or mobility to function in their daily lives. Consider animals such as cows in need of being milked. This universality of needs lends itself to the idea that care is not restricted to the healthcare domain but is an inevitable fact of life; being in need and being cared for are conditions, that for humans at least, are necessary for being alive.

The first condition for labeling a care practice as such is that the practice must be a response to the needs of another. This is not to say that a condition for using CCVSD is that the (service) robot itself is responding to the needs of a care receiver; rather, it could be that the robot contributes to how a human or other technology responds to needs. Examples of robots contributing to the meeting of needs include: a feeding robot that responds to the need of an individual to eat, or a disinfecting robot responding to the needs of a pharmaceutical factory to be sterile and so on. An example of a robot that does not meet this description is a bar tending robot.

Condition #2: The reciprocal nature of needs

Because we have not restricted ourselves in the definition of need (in that I have not taken a stand regarding a theory of well-being) we may take needs to be broadly construed to include things such as entertainment or sexual intercourse—provided that these are included by one's preferred theory of well-being. The sex robot or the entertainment robot can then be considered as devices responding to needs. But as we have seen above it is the way in which these needs are met that bears significance. Because needs are individual and dynamic the care practice too is individual and dynamic which in turn reinforces the need for reciprocity within the care practice. Thus, needs must be met through a reciprocal interaction between the one providing relief (the care giver) and one in need (the care receiver). The dancing robot that merely dances for an audience is not engaged in a care practice in which it is paying attention to, or responding to, the needs of its audience in a reciprocal manner. Instead, the dancing robot performs its activity without concern for the particular needs of audience members (i.e. in a uni-directional manner).

In contrast to the dancing robot, a personal service robot for entertaining in the home which plays games with the human user is a different case. Such a robot interacts and responds to the actions of its user. Of greater interest is that such a robot could even be programmed to learn the behaviors of its users and ultimately become a partner in a reciprocal interaction. We will return to this idea later on ("Enter the robot" section).

What can we make of the disinfection robot mentioned above (in "Service roborts: what and why" section) with this vision of reciprocity in mind? It is true that the robot responds to a general need for sterility in the factory but it does so by taking cues from its environment (the air quality) in a passive manner; meaning, the environment does not act with intention to indicate its state of function. I would suggest that this is not in fact a reciprocal interaction and further that intention on the part of the care receiver is a necessary condition to consider that a reciprocal interaction has or is taking place.[9]

To expand further on this idea of intentionality on the part of the care receiver, think about a robot for picking vegetables. While it is true that the robot interacts with its environment we cannot say that the cucumber or tomato to be picked acts with intentionality to indicate to the robot that it is ripe and ready to be picked or that it must be left in the ground or on the vine. A second condition of labeling a care practice as such, in addition to it being a response to needs, is that: the entity in need of care acts with intentionality to engage in the reciprocal relationship.

A consequence of this is that although a technology may be considered in need of repair the activity of repairing the technology cannot be considered a care practice. It may indeed be considered a practice in which skill is exercised and a change of function occurs (i.e. the technology goes from broken to fixed/functioning); however, the practice is not one of care because the technology does not act with intention with its care giver.

What does this condition mean for the person in a coma or the infant? Both of these individuals are in need of care and are of course deserving of care but cannot interact with their care giver with intention to indicate whether their needs are met. This is not to say that such individuals were not at one time capable or will not be capable of such in the future; rather, that at that moment in time they are not capable of this kind of interaction. True, the person in a coma may exhibit a change in function and the infant may cease crying when fed but these reactions are quite different from an active care receiver who will knowingly tell its care giver they have had enough to eat or that they need to go to the bathroom and would like assistance with this. Again I repeat this does not mean that such individuals are not in need or deserving of care but rather that they are not capable at that moment in their life of a reciprocal interaction as I am defining here.

If we consider what such an interaction might look like with a robot, consider the way in which cows interact with the milkbot: the cow walks to where the milkbot resides to have itself milked by the robot. The cow, the care receiver, acts with intentionality to have its need (e.g. relief from swelling in its milk glands) met. The robot, the care giver, responds to the actions of the cow and in so doing meets its need.

Thus the second condition of a care practice is that the care receiver and care giver engage with intention in a reciprocal interaction to meet the needs of the care receiver. This reciprocity is essential for understanding the multi-layered needs of the care giver as well as their changing dynamic. It requires that the care receiver play an active role in the meeting of their own needs.

From this we may then ascertain certain practices for which the CCVSD may *not* be applicable. Consider the practices of weeding on a farm, vacuuming the floor, or assembling parts on an assembly line in a factory. The practice of weeding before any discussion of a robot occurs between a human weeder and the plant to be picked or pruned. Given that the plant to be picked or pruned is not capable of engaging in a reciprocal interaction with its care giver, the weeder, it is not possible to consider this a care practice. The cleaner that vacuums or mops the floor does not do so in a reciprocal manner with the floor. The floor can show a change in state, i.e. it appears clean after being mopped, but conveying this change of state or discussing its nature is not an intentional choice on the part of the floor. For the person working on the assembly line we may call this a practice for which it can be evaluated as good or bad and for which an interaction takes place between the person working the line and the parts he/she is working with; however, once again we cannot claim that the parts on the assembly line are capable of interacting in a reciprocal manner as has been outlined in this paper. True we may call all of these practices in which skills are developed and for which they may be evaluated as good or bad; however, provided that they do not meet the conditions we cannot call them care practices as such.

In the same vein, one may be hard pressed to show how a surveillance robot is integrated into a care practice when the original surveyor (the human security guard) was interacting with the environment around and not necessarily people. One can also think about a window washing robot; the practice for which the robot is being used originally occurred between the window washer and the window and although we may rightly refer to it as a practice it should not be considered a care practice.

Enter the robot

We have now arrived at two necessary conditions for labeling a care practice as such and it is a care practice that a robot must be involved in, in order to use the CCVSD for its evaluation. As we go further in our investigation of using CCVSD we will observe that the robot is involved in the meeting of needs as an actor but may not be solely responsible for doing so, i.e. the robot may enhance the ability of a human care giver to meet the needs of a care receiver by increasing the capability for the human care giver to be attentive or competent. Of particular interest will be the impact the robot bears on the element of reciprocity.

With this in mind there are various scenarios that we will then encounter, related explicitly to reciprocity, once a robot enters the care practice:

(i) The robot enters into a care practice in which the reciprocal relationship happens between two humans (or a human and an animal) and the robot does not diminish this interaction,
(ii) The robot enters into a care practice in which the reciprocal relationship happens between two humans (or a human and an animal) and the robot enhances the ability for reciprocity between actors,
(iii) The robot enters into a care practice in which the reciprocal relationship happens between two humans (or a human and an animal)

and the robot impedes, threatens, or abolishes the reciprocity between actors, and,

(iv) The robot enters into a care practice in which the reciprocal relationship happens between two humans (or a human and an animal) and the robot then becomes engaged in a reciprocal interaction with the human.

Each of these scenarios presents a picture in which a robot may be involved with the reciprocal aspect of the care interaction. To begin, it is possible that the robot will not have an impact on the reciprocal interaction between care giver and care receiver (i), for example a robot used to deliver materials throughout an office building. If we consider the robot to be integrated into the practice of delivery in which a care receiver can request certain materials to be delivered and the care giver (the person delivering items). When the robot is integrated the person in need (the care receiver) will still make requests to the care giver only now the materials will be delivered by the robot. If the use of the robot enhances efficiency of the office it may be considered an enhancement (ii); however, in many cases it will not have a direct impact on the reciprocal interaction between the original care giver and care receiver.

In cases of scenario (ii), consider any number of telepresence robots used to help maintain contact between employees and their employers when geographical distances separate the two. The care practice is considered the interaction between employee and employer for their daily activities, a need in order for the company to function, and the robot is used to enhance this interaction by enhancing the type (e.g. quality and or quantity) of communication between persons.

For scenario (iii), one might imagine the care practice of serving tables. The care giver, the server, and the care receiver, the customer, are engaged in a reciprocal interaction for the course of the meal. When a robot enters this care practice it may replace the server entirely thereby abolishing the reciprocal interaction between the server and customer. Alternatively, there may be instances in which the robot enters the care practice in a way that enhances the speed and/or efficiency with which the customer and server (or chef) can communicate (i.e. scenario ii).

Scenario (iv) presents a picture of what may happen in the future when the robot is embedded with enough artificial intelligence (AI) to be deemed capable of a level of attentiveness and competence for ascertaining when needs are changing as well as how to respond to these changing needs. In this way the robot may be considered a reciprocal partner for the care receiver. This presents the most interesting case to apply the CCVSD approach to and will be the task of future work when such robots begin to enter the market. Consider the same server robot in the paragraph above; if it were endowed with enough AI to engage in a reciprocal interaction with customers the scenario may be categorized as iv. Determining whether or not any of these scenarios are good or bad will be decided on a robot-by-robot basis when one follows the steps of the CCVSD approach (see Table 1).

Evaluating and designing service robots using CCVSD

In the above section my aim was to show that the CCVSD approach requires that certain conditions are met in order to make a retrospective evaluation. These conditions are derived from the conditions for labeling a care practice as such and are: (1) That the practice be a response to the needs of another and, (2) That the care giver and care receiver be engaged in a reciprocal interaction. Once this is accomplished, the steps for evaluating the (service) robot may be completed according to the CCVSD approach (see Table 1). In so doing, the roboticist and/or robot ethicist involved in the evaluation will be able to label the reciprocal interaction according to the scenarios presented above.

By specifying 'retrospective' evaluations I aim to indicate that one has a robot prototype in mind that they want to study. The goal then is to understand what the original practice was that the robot is integrated into. From this, the evaluation can take place. Current service robots in the research and design stages that fit the bill so to speak include but are not limited to: Jibo the personal service robot in the home for reading children bedtime stories, Autom the diet assist robot, assistive locator robots in a store (e.g. OshBot), server robots in a restaurant ("Mechanic masterchef" 2013; News 2014), pizza delivery robots (DRU from Dominoes), and milking robots to name a few. This list is only sure to increase in number in the coming years as the technology thrives and further develops. The next step will be to engage in systematic and comprehensive evaluations of these robots to show how CCVSD is applicable and what results its application yields. This will be the goal of the future work of this author.

When using the approach in a prospective manner the aim of the steps is to identify a care practice in which there exists a need for the assistance of a robot and to intentionally design it in a way that results in either scenario 'i' or 'ii' presented above. For the former, the example I presented in an earlier work, the wee-bot robot, shows how the robot ought to be designed to interact with the human user, the nurse, so as to maintain the presence, attentiveness, competence and responsibility of the nurse (van Wynsberghe 2013). The manner in which this is accomplished is through ensuring a reciprocal interaction between robot and nurse, i.e. that the nurse intentionally give the robot certain cues to allow the robot to move on to another portion of its task and/or to complete its task.

Also of particular interest will be the instance in which the creation of a robot results in a *new*care practice. In this instance a practice may not be considered as such until the robot is created and used. A robot used to guide an elderly person throughout a shopping mall or grocery store (Robovie II—the personal robotic shopper 2009) could appear to be the establishment of a new care practice provided that the robot and the person it is escorting are able to interact in a reciprocal manner. In other words, one would imagine such a case only when the robot has enough AI to engage as a reciprocal partner through verbal, auditory, visual and/or other cues. The CCVSD approach can also be used for evaluating these robots but a comparison between practices will not be possible. In this instance the practice itself can be evaluated using the moral elements in so far as they are a tool for evaluating a care practice on its own.

CONCLUSION

Within the coming years it should not be a surprise to encounter either a personal or a professional service robot in our homes and/or our work places. Since these robots will function in the unpredictable, unstructured environment that humans live and work in, they demand ethical reflection. I argue that the ethical evaluation (at least for the present time) should be specific to the robot (its capabilities and appearance) and the practice within which is has been placed. More specifically I mean to suggest that features of both the robot and the practice will help to decide if that robot is good or bad for the practice at hand. Thus, we may engage in the same *kind* of ethical evaluation for professional and personal service robots as we do for healthcare robots.

The approach created specifically for robots in healthcare remains to date the only one of its kind to systematically evaluate care robots and provide insights for ethicists, designers and policy makers. Given that robots are already in use and continue to be developed for contexts' outside healthcare it would seem to be of great benefit to show that CCVSD may also be used to evaluate robots outside of the healthcare domain. The normative foundations for CCVSD come from its reliance on the care ethics tradition and in particular the use of care practices for: (i) structuring the analysis and (ii) determining the values of ethical import. To apply CCVSD outside of healthcare one must show that the robot has been integrated into a care practice. Accordingly, the practice into which the robot is to be used must be assessed and shown to meet the conditions of a care practice. The two conditions are as follows, that: (1) The practice be a response to the needs of another, and (2) The care giver and care receiver be engaged in a reciprocal interaction. Provided these conditions are met, the (service) robot can then be evaluated according to the steps of the CCVSD approach (see Table 1).

This work was meant as a preliminary study investigating the applicability of the approach for robots outside of care, namely service robots. By investigating the foundations of the approach I hoped to show why it may be applicable for service robots. Added to this my aim was to provide initial reflections regarding the impact on the reciprocal interaction between care giver and care receiver and to use current robot prototypes as examples of robots that can and cannot be evaluated according to the CCVSD approach. This work was not intended to be an exhaustive evaluation of one or more service robots; rather, it argued that service robots can, and should, be evaluated using the CCVSD approach.

FOOTNOTES

1 http://www.ifr.org/service-robots/statistics/.

2 See http://www.ifr.org/service-robots/.

3 http://www.ifr.org/service-robots/statistics/.

4 http://www.ifr.org/service-robots/.

5 https://www.jibo.com/.

6 See http://knightscope.com/.

7 See http://www.blue-ocean-robotics.com/en/products.

8 See http://www.blue-ocean-robotics.com/en/products/production/cleaning-and-disinfection-robots/uv-disinfection-robot.

9 Tronto (1993) also claims that technologies may at times be in need and require care but I would disagree with her for this very point—intention on the part of the care-giver and care-receiver is necessary for a reciprocal interaction and the technology cannot provide this.

ACKNOWLEDGMENTS

Thank you to the reviewers of this text for their insightful comments and queries. Many thanks to my fellow robot ethicist Scott Robbins for his insight, constructive criticism, critical feedback, and support. A most gracious thank you to the National Science Foundation in the Netherlands for the funding to allow me to conduct this research.

REFERENCES

1. Allen, C., & Wallach, W. (2011). Moral machines: contradition in terms of abdication of human responsibility? In P. Lin, K. Abney, & G. A. Bekey (Eds.), *Robot ethics: The ethical and social implications of robotics* (pp. 55–68). Cambridge: MIT Press.
2. Asaro, P. (2006). What should we want from a robot ethic? *International Review of Information Ethics, 6*, 8–16.
3. Asaro, P. (2011). A body to kick, but still no soul to damn: Legal perspectives on robotics. In P. Lin, K. Abney, & G. A. Bekey (Eds.), *Robot ethics: The ethical and social implications of robotics* (pp. 169–186). Cambridge: MIT Press.
4. Calo, R. (2011). Robots and Privacy. In P. Lin, K. Abney, & G. A. Bekey (Eds.), *Robot ethics: The ethical and social implications of robotics* (pp. 187–202). Cambridge: MIT Press.
5. Capurro, R. (2009). Ethics and robotics. In R. Capurro & M. Nagenborg (Eds.), *Ethics and robotics* (pp. 117–123). Amsterdam: IOS Press.
6. Denning, T., Matuszek, C., Koscher, K., Smith, J. R., & Kohno, T. (2009). A spotlight on security and privacy risks with future household robots: attacks and lessons. *In Proceedings of the 11th international conference on ubiquitous computing* (pp. 105–114). New York: ACM. doi:10.1145/1620545.1620564.
7. Driessen, C., & Heutinck, L. (2015). Cows desiring to be milked? Milking robots and the co-evolution of ethics and technology on Dutch dairy farms. *Agriculture and Human Values,32*(1), 3–20.
8. Engelberger, J. F. (1989). *Robotics in service*. Cambridge: MIT Press.
9. Evert, F. K., et al. (2006). A mobile field robot with vision-based detection of volunteer potato plants in a corn crop. *Weed Technology, 20*(4), 853–861.
10. Friedman, B. (1996). Value-sensitive design. *Interactions, 3*(6), 16–23. doi:10.1145/242485.242493.
11. Friedman, B., Hendry, D., Huldtgren, A., Jonker, C., van den Hoven, J., & van Wynsberghe, A. (2015). Charting the next decade for value sensitive design. *Aarhus Series on Human Centered Computing, 1*(1), 4. doi:10.7146/aahcc.v1i1.21619.
12. Friedman, B., & Kahn, P. (2003). Human values, ethics, and design. In J. Jacko & A. Sears (Eds.), *The human-computer interaction handbook* (pp. 1177–1201). Hillsdale, NJ: L. Erlbaum. Retrieved from http://dl.acm.org/citation.cfm?id=772072.772147.
13. Friedman, B., Kahn, P., & Borning, A. (2002). *Value sensitive design: Theory and methods* (pp. 2–12). University of Washington technical report.
14. Friedman, B., Kahn, P., & Hagman, J. (2003). Hardware Companions?: What Online AIBO discussion forums reveal about the human-robotic relationship. In *Proceedings of the SIGCHI conference on human factors in computing systems* (pp. 273–280). New York, NY: ACM. doi:10.1145/642611.642660.
15. Lin, P., Abney, K., & Bekey, G. A. (2011). *Robot ethics: The ethical and social implications of robotics*. Cambridge: MIT Press.
16. Little, M. O. (1998). Care: From theory to orientation and back. *Journal of Medicine and Philosophy, 23*(2), 190–209. doi:10.1076/jmep.23.2.190.8922.
17. Lokhorst, G.-J., & van den Hoven, J. (2011). Responsibility for Military Robots. In P. Lin, K. Abney, & G. A. Bekey (Eds.), *Robot ethics: The ethical and social implications of robotics* (pp. 145–155). Cambridge: MIT Press.
18. Manders-Huits, N. (2011). What values in design? The challenge of incorporating moral values into design. *Science and Engineering Ethics, 17*(2), 271–287.

19. Maslow, A. H. (1970). *Motivation and personality*. New York: Harper & Row.
20. Mechanic masterchef: Robots cook dumplings, noodles and wait tables at restaurant in China. (2013, Jan 13). *Mail Online*. Retrieved from http://www. Dailymail .co.uk/news/article-2261767/Robot-Restaurant-Robots-cook-food-wait-tables-Harbin.html.
21. Mol, A., Moser, I., & Pols, J. (2010). *Care in practice: On tinkering in clinics, homes and farms*. Bielefeld; Piscataway, NJ: Transcript; Distributed in North America by Transaction Publishers.
22. News, F. (2014, Dec 4). This restaurant replaced humans with robot waiters. *New York Post*. Retrieved from http://nypost.com/2014/12/04/this-restaurant-replaced -waiters-with-robots/.
23. Noddings, N. (2002). *Educating moral people: A caring alternative to character education*.Teachers College Press, P.O. Box 20, Williston, VT 05495-0020 (paperback: ISBN-0-8077-4168-X, $21.95; cloth: ISBN-0-8077-4169-8, $48). Tel: 800-575-6566 (Toll Free); Fax: 802-864-7626; Web site: http://www. tcpress.com/index.html. Retrieved from http://eric.ed.gov/?id=ED468125.
24. Pedersen, S. M., Fountas, S., Have, H., & Blackmore, B. S. (2006). Agricultural robots—System analysis and economic feasibility. *Precision Agriculture, 7*(4), 295–308. doi: 10.1007/s11119-006-9014-9.
25. Robovie II—the personal robotic shopper. (2009, December 29). Retrieved from http:// www.gizmag.com/robovie-ii-robotic-shopping-assistant/13664/.
26. Sharkey, A. (2014). Robots and human dignity: A consideration of the effects of robot care on the dignity of older people. *Ethics and Information Technology, 16*(1), 63–75. doi:10.1007/s10676-014-9338-5.
27. Sharkey, A. (2016). Should we welcome robot teachers? *Ethics and Information Technology*, 1–15. doi:10.1007/s10676-016-9387-z.
28. Sharkey, A., & Sharkey, N. (2012). Granny and the robots: Ethical issues in robot care for the elderly. *Ethics and Information Technology, 14*(1), 27–40.
29. Sparrow, R., & Sparrow, L. (2006). In the hands of machines? The future of aged care. *Minds and Machines, 16*(2), 141–161. doi:10.1007/s11023-006-9030-6.
30. Spiekermann, S. (2015). *Ethical IT Innovation: A value-based system design approach*. Boca Raton: CRC Press.
31. Sullins, J. (2011). In M. Anderson & S. L. Anderson (Eds.), *Machine ethics*. Cambridge: Cambridge University Press.
32. Tronto, J. C. (1993). *Moral boundaries: A political argument for an ethic of care*. Abingdon: Psychology Press.
33. Tronto, J. C. (2010). Creating caring institutions: Politics, plurality, and purpose. *Ethics and Social Welfare, 4*(2), 158–171. doi:10.1080/17496535. 2010.484259.
34. Vallor, S. (2011). Carebots and caregivers: Sustaining the ethical ideal of care in the twenty-first century. *Philosophy and Technology, 24*(3), 251–268.
35. van Wynsberghe, A. (2012). Designing robots for care: Care centered value-sensitive design. *Science and Engineering Ethics, 19*(2), 407–433. doi:10.1007/s11948-011-9343-6.
36. van Wynsberghe, A. (2013). A method for integrating ethics into the design of robots. *Industrial Robot: An International Journal, 40*(5), 433–440. doi:10.1108/IR-12-2012-451.
37. van Wynsberghe, A. (2015). *Healthcare robots: Ethics, design and implementation*. Farnham: Ashgate Publishing Ltd.
38. Vanlaere, L., & Gastmans, C. (2011). A personalist approach to care ethics. *Nursing Ethics,18*(2), 161–173.
39. Verkerk, M. A. (2001). The care perspective and autonomy. *Medicine, Health Care and Philosophy, 4*(3), 289–294. doi:10.1023/A:1012048907443.
40. Veruggio, G., & Operto, F. (2006). Roboethics: A bottom-up interdisciplinary discourse in the field of applied ethics in robotics. *International Review of Information Ethics, 6*, 2–8.

41. Wallach, W. (2010). Robot minds and human ethics: The need for a comprehensive model of moral decision making. *Ethics and Information Technology, 12*(3), 243–250.

42. Wallach, W., & Allen, C. (2008). *Moral Machines: Teaching robots right from wrong*. Oxford: Oxford University Press.

CHAPTER 13

Robot Carers, Ethics, and Older People

Tom Sorell, Heather Draper

Politics and International Studies, University of Warwick, Social Sciences, Gibbet Hill Road, Coventry CV4 7AL, UK

Medicine, Ethics, Society and History (MESH), School of Health and Population Sciences, University of Birmingham, 90, Vincent Drive, Birmingham B152TT, UK

ABSTRACT

This paper offers an ethical framework for the development of robots as home companions that are intended to address the isolation and reduced physical functioning of frail older people with capacity, especially those living alone in a noninstitutional setting. Our ethical framework gives autonomy priority in a list of purposes served by assistive technology in general, and carebots in particular. It first introduces the notion of "presence" and draws a distinction between humanoid multi-function robots and non-humanoid robots to suggest that the former provide a more sophisticated presence than the latter. It then looks at the difference between lower-tech assistive technological support for older people and its benefits, and contrasts these with what robots can offer. This provides some context for the ethical assessment of robotic assistive technology. We then consider what might need to be added to presence to produce care from a companion robot that deals with older people's reduced functioning and isolation. Finally, we outline and explain our ethical framework. We discuss how it combines sometimes conflicting values that the design of a carebot might incorporate, if informed by an analysis of the different roles that can be served by a companion robot.

Keywords Robots, Ethics, Telecare, Assistive technology, Autonomy, Nature of care, ACCOMPANY.

What role might robots have in the future care at home of older people? That depends on the kind of robot and the aspects of ageing addressed. We argue that social robots—represented in our discussion by the Care-O-bot®—commendably integrate the functions of useful non-robotic telecare devices, and that they also have a claim to the potentially therapeutic property we call "presence". The relevant normative requirements that we discuss are mostly ethical, though they might also include value for money, and considerations about the range of physical environments in which the relevant technology can operate. The ethical framework we present in this paper was developed in the European Union-funded FP7 ACCOMPANY project, which has adapted the Care-O-bot® as a home companion for older people. ACCOMPANY scenarios primarily address the isolation and reduced physical functioning of some older people, especially those living alone in a non-institutional setting. Our ethical framework gives autonomy priority in a list of purposes served by assistive technology in general, and carebots in particular, for isolated and frail older people.

The paper consists of four sections. In the first, the capacity of robots to be "present" to their users is introduced. We argue that presence is a necessary condition for being a companion robot but a far from sufficient condition of being a carer or even of being a care-robot in a substantial sense of 'care'. In the second section we describe relatively low tech assistive technology and its benefits, and contrast these with what a robot can offer. The possible ethical problems of carebot design are then considered, on the basis of the review of recent carebot literature by Vallor (2011). This provides some context for the ethical assessment of robotic assistive technology. In the third section we consider what might need to be added to presence to produce care from a companion robot that deals with older people's reduced functioning and isolation. We distinguish between single-function and multi-function robots and what they can contribute to the care of older people in combination. Finally, we introduce an ethical framework that combines sometimes conflicting values that the design of a carebot might incorporate. The framework is informed by an analysis of the different roles that can be served by a companion robot, roles that are catered for in ACCOMPANY scenarios. Autonomy on the part of users is the pre-eminent value that we argue a carebot should promote. It is more important even than safety and social connectedness.

ROBOTS, 'PRESENCE' AND THE REQUIREMENTS OF CARE

Care that depends significantly on technology is already a reality in much of Western Europe, but the technology typically used is not robotic. On the contrary, some of it is decidedly low-tech. Assistive technology for older people in the UK includes wearable alarms for summoning help; smoke, CO_2 and flood sensors; pillboxes that are designed to help older people to take all of their medication on time; and fall sensors. Two-way visual contact through webcams and television monitors is also available, though less widely, and relatively inexpensively. This enables family or paid support workers to 'look in' on an older person and their home without travelling (Bayer et al. 2007; Bowes and McColgan 2006). If older people are comfortable with computers, virtual visiting is cheap and easily accessed—no more difficult to obtain than a Skype account—and there are virtual visiting systems simpler than Skype which operate using local broadband networks.

What, if anything, can robots add to this array of non-robotic assistive technology? There are a number of possible answers to this question, corresponding to differences between robots. A number of robots simulate the appearance, feel and behaviour of small domestic pets. These can have some of the beneficial effects of real pets—providing a facsimile of companionship and an outlet for a sort of affection for otherwise lonely and isolated people.[1] Such robots are relatively cheap, and their therapeutic benefits are in principle available not only to older people, but to younger people who are cognitively impaired, including children with autism. Considerably more expensive are multi-functioning humanoid robots that are able to move about in an apartment or bungalow, carrying out tasks for an older person, their carers, or both. An example of such technology which we will concentrate on in what follows is the Care-O-bot®.[2] Care-O-bot® platforms can be programmed to speak to the older person, to fetch and lift things, and to act as an audio-visual portal to the world outside the older person's home. In principle, the Care-O-bot® could help to steady an older person when moving from a sitting to a standing position or as they walk. It can also be programmed to learn and to remember an older person's daily routine, remind them of things, and summon help in the event of a fall or some other recognized mishap.

In what follows, we use the term 'humanoid robot' not for a robot that simulates human behaviour, including facial expressions and posture, but for an upright multi-functioning social robot that typically has facial features, usually quite schematic or cartoon-like features, some sort of communicative interface, and perhaps arms. Care-O-bot® is such a robot.

Without simulating the adult human body, it has the rough counterpart of such a body, much as a simple cartoon of a human being can be taken to resemble a human being. By 'non-humanoid' robot we will usually mean some single-function, often very compact, machine with no counterpart of a human body, though sometimes with the counterpart of an animal body.

Both humanoid and non-humanoid robots add to the potential benefits for older people of non-robotic assistive technology. A common benefit of some examples of both kinds of robots is that they have what we shall call 'presence'. This includes, but comes to more than, *being there* with the older person. Being there—in the minimal sense of being co-located with a person—is open to a mop, a broom, or a newspaper. What is meant by 'presence' is the kind of co-location of a thing with a person that brings it about that the person no longer feels alone. A child co-located with a bed will probably feel alone, even if the bed is comfortable and familiar. But a child co-located with a bed and a familiar cuddly toy will probably feel that they are in the presence of something or someone, even though the cuddly toy is inanimate and inert and has degenerated after years of handling to a lump of cloth. This illustrates what is meant by 'presence'.[3] In ACCOMPANY, the Care-O-bot® has sophisticated presence, since it not only acts against the feeling of being alone, but does so by moving around with the older person, appearing to take interest in activities in which the older person is engaged, prompting him or her undertake beneficial behaviours, communicating through a touch screen and reacting to the older person's commands. This is what we mean by 'sophisticated presence'.

Evidence gathered in ACCOMPANY[4] suggests that for older people exposed to it for short lengths of time in an experimental smart home, Care-O-bot® has presence. Some can even imagine forming what they term 'attachment' to a Care-O-bot®. Users may also give the robot gendered names. This shows that social robots are likely to be *more* than present to users, since attachment goes beyond not feeling alone. ACCOMPANY is adapting the Care-O-bot® for some scenarios illustrating what robot companionship for older people might mean. The scenarios call for the robot to be able to monitor the position of the user, sometimes remotely, and for fetching and carrying for the user in which the robot does not simply respond to user commands but prompts the user to do things with its aid, including moving between rooms. The ACCOMPANY robot is also in some sense a stand-in for another human presence, as it expresses a small range of emotions which aspire to make the relationship with the older person mutual and, in a straitened sense, 'empathic'. Another aspect of the robot design is that it aims to enable Care-O-bot® to distinguish its user from other human beings in a home setting (e.g. visitors or members of the extended family), and to respond differentially to the user.

The ACCOMPANY design captures some aspects of companionship. Since other kinds of presence can be provided more cheaply by non humanoid social robots and by real human beings making use of non-robotic assistive technology—especially Skype technology—we are not convinced that the Care-O-bot®—especially at its current cost and with its relatively meagre capabilities—is the right form for machine-assisted care to take.

On the contrary, we think it more likely that machine-assisted care will be some combination of non-robotic assistive technology and non-humanoid, social or single-function robots. It is true that humanoid, multi-function robots are more likely than non-humanoid, single-function robots and non-robotic technology to *combine* a relatively wide range of capabilities within a single, unified presence, but how important it is to combine capabilities, and how important the lifting, fetching and walk-assisting capabilities of the Care-O-bot® are, is unclear to us. It is possible that a humanoid robot has a role in assisting older people who are isolated to keep up their skills of social interaction. This is a possibility that the 'empathic' aspect of the ACCOMPANY design has tried to realize. But it is also possible that less is more in robotic technology in particular and assistive technology in general, and that a combination of different machines, some having presence and some not, may deliver more benefits.

Some comments on the ethical value of assistive technology for older people who are not cognitively impaired

Non-robotic technology already enables people with physical disabilities, including those who are older, to live relatively independently in homes of their own. Ramps, handholds, special kinds of lighting, highly legible telephone handsets and walkers are among the kinds of equipment that might be added to a conventionally designed house to adapt it to a resident with physical disabilities (Doughty et al. 2007; Alaszewski and Cappello 2006). Then there is a variety of sensors and alarms, some designed to alert the householder to a risk, and some designed to alert a source of outside help. This equipment is characteristic of telecare. Telecare can also make use of combinations of television and webcam equipment, enabling 'virtual' visiting by health care or social workers (Dixon and Stahl 2009; Percival and Hanson 2006). Telehealth is another use of assistive technology. Here the purpose of the technology is typically to monitor a medical condition that has required or could require hospitalization, and to facilitate timely interventions if the condition worsens (Rogers et al. 2011). Unlike telehealth, telecare does not necessarily monitor vital signs or other symptoms. It monitors, often unobtrusively, how an older person is using their house, and takes unusual patterns of movement or rest, or unusual events (like water overflows in bathrooms), as cues for intervention by people outside the home.

In considering the ethical value of assistive technology, it is useful to ask *whom* it assists, or assists primarily. Some technology primarily assists a householder or a patient; other technology may be designed primarily to reduce the burden on *carers* of patients or householders. To begin with patients or householders, healthy older people are one target group for non-robotic assistive technology. Frail older people are another; older people with dementia are a third. Younger people with some cognitive impairment are a further class.[5]Other technology assists family members or a care organization located at one remove from the householder or patient.

Cognitively able but physically not very capable older people are probably the central client group catered for by policy documents on telecare (Department of Health 2005, 2008, 2010).[6]These are people who meet the legal standards for making decisions about their welfare, entering into contracts, making wills and so on, but who may be much less mobile and agile than people in their 40 and 50 s. This class is representative of the growing proportion of ageing adults in the general population of the UK and other Western countries whose health and life expectancy are much better than those of previous generations, and whose eventual demands on public care provision are as yet unknown.[7] Telecare is, among other things, a way of extending the time that cognitively unimpaired and mobile older people are able to live in their own homes rather than hospital or residential care homes, publicly funded ones in particular. Those with physical and intellectual disabilities are in principle distinct from older users of telecare: although it is true that many older people suffer from dementia and physical disability, it is not clear that the ethical issues raised for telecare by the whole class of those with physically disability can readily be inferred from the ethical issues raised for telecare by older people with disabilities. What we have are overlapping groups with overlapping issues.

The decision of cognitively unimpaired but less able people to take assistive technology into their homes[8] is a decision to live with a certain loss of privacy[9] in return for effective emergency or care response or more effective management of long-term medical conditions. Sensors to detect falls, bed-wetting and chair-wetting, and the length of time people spend in bed or in the bathroom, are widely used. Sensors can also detect whether external doors have been opened at odd hours, or left open for an abnormally long time (Perry et al. 2009). Medical equipment that detects signs of deteriorating asthma, chronic heart and lung ailments or diabetes is now available (Garcia-Lizana and Sarria-Santamera 2007; Paré et al. 2007). Some of this equipment sends information automatically to central response centres or medical practices (Dang et al. 2009). Telehealth equipment can also send information about amounts of medication taken at different times.[10]

User-centred assistive technology can reduce the burden of carers, but it need not do so, especially if the user has habits that carry some risks, or has a lifestyle that is not risky but is disapproved of by family or carers. For example, if user-centred telecare facilitates private communication or visits to an older person from someone whom he or she, but not their family approves of, say a younger member of the opposite sex, whom the family suspects is only interested in obtaining the older person's money, then it might add to the worries and burdens of the carers (Draper and Sorell 2012).

This possibility notwithstanding, there are good moral reasons why user-centredness rather than carer-centredness is the appropriate default position for older people's care in general, including the design of assistive technology for them, and robot-assisted care in particular. One reason is that the older person is an adult, with a life of his or her own to lead. This point is crucial to the approach taken in this paper, which is to extend the norms of fully able-bodied, socialized adult life to older people who are frail and isolated, when there is a technology available to them that compensates for isolation and physical decline.

Why apply norms for able-bodied adults in their prime to older people? The older person is no less an adult than someone much younger whose choice of, for example, sexual or other companions would normally not be anyone else's business, not even the business of the younger person's friends or family. Likewise, a middle-aged person can form relationships that his or her family disapproves of, but their disapproval is not normally taken to be decisive for what the middle-aged person should do, even if the middle-aged person is dependent on e.g. his daughter for meals, laundry and general organization.

If the older adult is to be treated differently from the younger adult, and if that difference in treatment is to be justified morally, then there has to be something about being older that makes one less able to lead one's own life. Although there may be some facts of older-age that are relevant here—declining mobility and worsening memory, for instance—these do not mean that one is not able to make decisions about how one's life should be run. After all, many much younger people who are forgetful, or who are not very agile or mobile through some accident, are not taken to be incapable of decision-making, and the quality of their decision-making may be no better than that of an older person who has age-related forgetfulness. Unless an older person is cognitively impaired—and cognitive impairment need not attend older-age even when older people are no longer mobile—

there is every reason to treat older people as all other unimpaired adults are treated, namely as able to make their own decisions, including exercising the right to make their own mistakes and take the consequences.

From a moral point of view, the decision-making of the user may even have to be allowed to extend to the question of what counts as an emergency. Under some telecare and telehealth regimes, it can be the judgment of telecare *monitors* that triggers an emergency intervention. Why should not people who are medically needy but perfectly competent trigger a rescue or medical intervention themselves?[11] This would be more consistent with the autonomy-promoting aims of assistive technology than the intervention triggered by the outsider. Some assistive technology does work in this autonomy-promoting way. Among the longest-established assistive technology devices are alarms that older people can set off if they fall or are in distress (Fisk 2003). These alarm devices can be worn, and so need never be out of reach. Alarms belong to a wider class of assistive technology devices in the control of the person receiving the care. These can include sensors for over-running baths and smoke that alert the user rather than a remote telecare hub, and that can prompt the user to set off their alarm if they feel they cannot cope. In the event that users feel that they can cope, a mishap such as a fall need not be regarded as an emergency, still less an emergency so overwhelming that outsiders need to come to the rescue.

In the same way, sensors monitoring medication use might alert healthy older *users,* rather than a remote carer, that tablets had not been taken. The user would then be free to take the tablets if he or she chose. Instead of outsider-controlled assistive technology, we have here the technology of the 'smart home', that is, technology that keeps track of hazards *for the householder* rather than an outsider, that does not keep track of the householder him or herself, and that does not make *decisions* for the householder or involve others in decision-making independently of the householder. It is smart home technology along these lines that should probably be preferred to outsider-controlled technology in the care of the competent but capable elderly, as it better respects their autonomy.[12]

User control can sometimes be the other side of the coin of user isolation. The older householder who autonomously manages her life, whether or not she has a chronic illness, may often do so because she is alone. Assistive technology sometimes reduces regular human contact, in particular visits of, or to, care assistants and other carers (Clark et al. 2007; Lim et al. 2007). Whether telecare *must* increase isolation in order to serve its standard twin purposes of promoting independence and reducing the public expenditure involved in hospitalization or maintaining state-run care homes is not always easy to determine (Sorell and Draper 2012;

Murray et al. 2011; Pols 2010).[13] Relatively inexpensive virtual visiting can make available important facets of human contact. Contact can even be enhanced, since technology reduces the size of the obstacle posed by distance and the familiar problem of having to be in two places at once. When one visits virtually, with good, real time audio/visual access to the older person's home, the virtual visitor *is* in two places at once, for most purposes. Similarly for the older people visited. The equipment enables older people who are virtually visited to see not only the people they are conversing with but their home environments in real-time. Earphones can even make virtual visitors much more audible for the hard of hearing than being co-present with the same people in a room with a great deal of background noise.[14] It is true that equipment for virtual visiting also creates opportunities for intrusion—but this may be no greater than home visiting. When it is user-controlled and combined with the usual array of sensors and monitors, however, it may at least in principle provide health and social gains for older people, cost-savings for health and social services, and a more manageable care burden for friends and family living separately from users.

Carebots versus low tech assistive technology

What do robots actually add, and what can they potentially add, to non-robotic assistive technology? What in particular can the Care-O-bot® in ACCOMPANY add? ACCOMPANY is user- centred rather than carer-centred. Its robot has not been conceptualized as a piece of technology that helps the older person's family or care providers by being their representative in the older person's house, doing for the older person what the family or carers would otherwise do, and allowing carers more time of their own. Instead, it has been designed to serve as a multi-functioning, humanoid presence in an older person's home, capable of acting as companion, helper and enabler, and sensitive to the wishes of the user before the wishes of others.

Multi-functioning, humanoid robot companions are at one end of a spectrum of robotic products with care or enablement functions.[15] At the companion end of the robotic spectrum can also be found non-humanoid companion devices such as dolls and simulations of small animals. There are also hybrids of humanoid robots and companion animals. Riken have developed the Riba robot (Robot for Interactive Body Assistance): an adult-sized robot designed to look like a streamlined white teddy bear that can pick up and carry humans from a bed to a wheelchair.[16] At the other end of the spectrum are devices that are more like clothing, and, when worn, markedly increase the strength of users.[17] In between are robotic devices[18]whose functions, including monitoring and functioning as an interactive portal, are carried out separately by different kinds of telecare devices. Other robots are single function machines that enable the user to be more independent, such as My Spoon[19] and vacuuming and floor-washing robots.[20]

The literature on the ethics of robotics sometimes identifies tensions between the design and use of carebots—including some that are similar to the Care-O-bot® used in ACCOMPANY—and the morally desirable treatment of older persons as autonomous adults. Vallor (2011) has identified the following concerns from her review of the literature:

1. The objectification of the elderly as "problems" to be solved by technological means (Sparrow and Sparrow 2006; Sharkey and Sharkey 2010)
2. The potential for carebots to either enhance or restrict the capabilities, freedom, autonomy, and/or dignity of cared-fors (Borenstein and Pearson 2010; Sharkey and Sharkey 2010; Decker 2008)
3. The potential of carebots to enhance or reduce engagement of cared-fors with their surroundings (Borenstein and Pearson 2010; Sharkey and Sharkey 2010)
4. The potential of carebots to enhance or intrude upon the privacy of cared-fors (Sharkey and Sharkey 2010)
5. The quality of physical and psychological care robots can realistically be expected to supply (Coeckelbergh 2010; Sparrow and Sparrow 2006)
6. The potential of carebots to either reduce or enhance cared-fors' levels of human contact with families and other human caregivers (Sparrow and Sparrow 2006)
7. The potential of carebot relations to be inherently deceptive or infantilizing (Sparrow and Sparrow 2006; Sharkey and Sharkey 2010; Turkle 2006)

The second of these seven concerns is the one that ACCOMPANY principally tries to address. Far from restricting the capabilities, freedom, autonomy and dignity of users, ACCOMPANY aims at maximizing those things. We shall first explain *how* ACCOMPANY sets out to enhance or at least preserve the autonomy and capabilities of users.[21] We shall consider whether ACCOMPANY addresses other issues on Vallor's list. Finally, we shall return to the question of whether the ACCOMPANY Care-O-bot® and carebots generally do better than lower tech solutions in assistive technology.

The ACCOMPANY Care-O-bot® is designed to be stationed in the home of the older person. It is not brought into the older person's home only from time to time. Nor does it divide its time between a number of different rooms in an institutional care-home corridor. Instead, it is a semi-permanent fixture in a home occupied by one person. Its role in that person's home is partly that of helper, partly that of enabler, partly that of co-learner, and partly that of companion.

The Care-O-bot® is able to lift and carry household objects, and also to fetch objects at the command of the user. It is also able to remind the user of different scheduled or routine events, such as a visit, or the need to place a shopping order, or to go to a doctor's appointment. The robot is able to keep track of the user's position in the house and to register falls or other signs of harmful incidents. It is able to distinguish the user from other people who might enter the home, and it is able to treat the orders or requests of the user as having more authority than those of visitors. The co-learner role is a matter of the process by which the robot and the user accommodate themselves to one another. The user's routine will shape many interactions between the user and the robot. For example, if the user often wakes at 9 am rather than 7 a.m., the robot can learn that and be ready to fetch a morning drink at the right time. If a routine-disrupting appointment is made, the robot can register that and prompt the user to get dressed for a visit or to go out at a certain time. The robot can also take part in recreational pastimes—providing music to go with a song, or perhaps staying in the same room as the user as he or she watches television, and making gestures which suggest it is giving attention to the user or what the user is doing. Through its tray/tablet it can offer a video and internet portal, making possible virtual visiting.

Although the Care-O-bot® is in most respects at the service of its older user, the Siena roboticists in the ACCOMPANY consortium are seeking to endow it with capabilities that contribute to maintaining the social skills of the user—by making the obedience of the robot less than absolute. Sometimes the 'displeasure' of the robot is communicated to the user through a tablet interface showing a diagrammatic "empathic mask" (Stienstra and Marti 2012). This empathic mask can signal the robot's 'worry' about the older person, its 'pleasure' at his or her co-operation or 'displeasure' when he or she is over-insistent or impatient. The effect of this is to remind a user who may be interacting socially mainly with the robot that it, as the local representative of the social world, needs to be treated with a kind of consideration, and that the social skills of responding to gentleness with gentleness need to be kept up. The Siena work also makes possible a high degree of sensitivity to tactile interactions between the user and the robot. The tablet interface between the user and the robot can register the pressure exerted on it when the user squeezes it, and can process a hard squeeze as an indication of urgency in a command. Again, the robot can simulate 'emotional synchronization' with the user, apparently taking up a position near the user and directing its attention where the user does when, for example, the user watches television.

In what ways does the Care-O-bot® promote the autonomy and independence of the older person? To begin with, the Care-O-bot® takes the user's routine and the preferences embodied in this routine as its frame of reference. This means that the user's choices are foremost and, other things being equal, are implemented unquestioningly. So the agent's autonomy is not at all impaired. It *would* be impaired if the choices of others started to supplant the user's choices. But in ACCOMPANY scenarios this does not happen. Far from introducing new choices, still less choices at variance with those of the user, the Care-O-bot® takes its cue from the user's choices. The robot is also able to co-ordinate those choices through scheduling and prompting abilities. In both of these ways the ACCOMPANY Care-O-bot® promotes the autonomy of the user.

User safety constrains the autonomy-respecting features of carebots. Sharkey and Sharkey (2010) ask what would happen if a user instructed a lifting and carrying robot to release him or her over the side of a high balcony in an apartment building. Even if that suicidal request were competent and highly autonomous—not the result of treatable depression, for example—there would be a good moral reason for programming a carebot not to comply with it, namely that the robot would not normally be able to *tell* whether such a request was competent and autonomous. Likewise, one would expect a carebot with monitoring abilities to abort an activity requested by the user if, while undertaking that activity, it registered that the user had fallen or that his or her vital signs had suddenly changed. But notice that we would expect the same behaviour of an autonomy-respecting *human* helper as well, since loss of life and sudden medical emergencies disrupt or undermine the exercise of autonomy.

Autonomy—choosing for oneself and acting in accordance with one's choices—is different from independence. Independence is being able to act on one's choices without depending on the consent or co-operation or resources of others. If someone chooses to live the life of a sailor, for example, but can only succeed in doing so if a shipping company offers employment, then there is a clear sense in which the agent is not an independent sailor, however much his choice of sailing as an occupation is autonomous. Independence might only be achieved if he owned a suitable boat or ship for as long as he wanted to live the life of a sailor. In the case of older people, autonomy can sometimes co-exist more or less constantly with dependence rather than independence. In other words, the choices of older people can often need to be realized through the efforts of *others*.

Although an older person can still make his or her own choices, he or she may not be physically strong enough, or mobile enough, or rich enough to see them through. Carebots can act against some sources of dependence, like physical weakness, but not others, such as lack of wealth.[22]Sometimes in discussions about older people, independence is not understood with respect to one's command of the means for realizing choices in general, but rather in terms of one's ability to keep oneself clean, fed, sheltered, unharmed and legally occupied without the constant assistance of other people. Carebots readily promote independence in this limited sense, but not independence in all of the respects in which it affects autonomous choices.

Carebots in general and the ACCOMPANY Care-O-bot® in particular also address issues other than autonomy and independence on Vallor's list. The sixth concern ('The potential of carebots to either reduce or enhance cared-fors' levels of human contact with families and other human caregivers') highlights the potential isolation of older people whose main interactions are with machines. We have already seen that, in telecare, technology can be a gateway rather than a barrier to human contact. This is because it can act as a portal for two-way communication and more. What about carebots? Although they also afford the possibility of virtual visits, there is a clear difference between having a two-way television in one's house and having a Care-O-bot®. The Care-O-bot® is a *presence* in its own right in a way a television is not. To put it in another way, the television is not a subject of interaction but at most a medium of interaction. On the other hand, a Care-O-bot®*is* a subject of interaction—a possible 'conversation' partner and a possible participant in synchronised activity.

Admittedly, the Care-O-bot®'s current communicative abilities are so limited as to be conversational only in an inverted commas sense; admittedly, the extent to which its presence fills a gap left by human contact is very restricted. Still, there is a sense in which the Care-O-bot® in ACCOMPANY and other carebots fulfil the conditions for what, at the beginning of this paper, we called 'presence': when carebots are co-present with people, those people can feel that they are not alone. This does not seem to be a power of television sets. It is of course possible for an image on a two-way television to produce the feeling of not being alone. But a Care-O-bot® resident in one's home can create this feeling even when it is not functioning as a virtual visiting portal. This is an important difference between non-robotic assistive technology and robots.

Can *only* a carebot—typically an adult-sized humanoid fetcher-carrier-lifter with restricted communicative capacities—have presence, or is presence open to much simpler robots? Much simpler non-humanoid social robots undoubtedly *can* have presence. The Paro [23]—a small fur-covered robot that looks like a seal—is specifically designed for therapeutic uses with older people. It is programmed to exhibit a range of responses to being petted, including moving its tail and opening and closing its eyes. It also 'learns' actions that the user likes or dislikes (it responds to being hit as well as petted), can respond to voice direction and tone, and detects the difference between day and night and is more or less active accordingly. Although the evidence regarding Paro's actual effectiveness as a companion is open to more than one interpretation, some studies have concluded that animal robots have some of the therapeutic powers of real domestic animals,[24] but without the care-burdens of owning a real domestic animal. Again, it is plausible that Paro serves the purposes of the oldest old better than the somewhat younger intended users of the ACCOMPANY carebot.

If the isolation and loneliness of older people were the only or the main problem that robots were being introduced to solve, then, a multi-functioning humanoid robot might not be needed. And there would be lower-tech solutions to the loneliness problem if there were regular two-way virtual visiting between actual human beings. Or a combination of simpler robots and two-way television may be the best of both the high-tech and low-tech worlds for combating loneliness, providing both permanent and low-maintenance presence but also regular human contact. (No tech visits would of course be possible and might be best of all therapeutically, but could be expensive in time and money).

An ethical framework for judging carebots for older people

When what is in question is the promotion of autonomy, independence and some form of human contact, what, if anything, recommends a carebot solution to providing care for older people over a telecare, or single-function and simple companion robot solution, or a combination of telecare and single-function and simple companion robots? If the money cost of a multi-function, humanoid carebot is taken into account, the answer may be "Nothing". On the other hand, if financial costs are disregarded, then the answer on the basis of the previous discussion may be that the carebot solution delivers physical help, and the ability in principle to *integrate* telecare and sophisticated presence. By 'sophisticated presence', as explained in the section "Robots, 'presence' and the requirements of care", we mean that the carebot interacts and can even initiate interaction with the user. Moreover, the quality of interaction is far more sensitive and far more challenging than the passive twitches and facial expressions of Paro. By taking over some of the functions of telecare, the ACCOMPANY Care-O-bot® can keep track of the location and condition of the user. It does so, however, from close at hand, potentially enabling quicker intervention or emergency response than conventional telecare devices relaying data to a remote information hub (assuming that the carebot is not itself programmed to summon help from a similar hub). In other words, cost considerations apart, a carebot may give us in a single package a highly desirable embodiment of assistive technology alongside practical help with lifting, carrying and fetching.

The previous discussion, however, may be inadequate for a full answer to the question of the comparative value of low-tech and robotic assistive technology. So far we have been guided by a list of ethical issues raised by philosophers and technologists who have reflected on the capabilities of robots designed or used to provide care for older people and meet the needs of older people as they present themselves in ordinary experience. But perhaps the common sense of philosophers and technologists is a bad guide to the needs or preferences of the elderly. The preferences and needs of the current population of older people are not representative of future older people, and are subject to cultural variation even in the present. Moreover the list of ethical issues depends on the assumption that the experience of older people, especially in the West, is more or less uniform (Parks 2010).[25]

Any development of an ethical framework for evaluation of carebots must be informed by the attitudes of older people themselves, with allowances being made for big variations in technophobia between people who currently are 60 years old and people who are currently over 80 years of age. The importance to an ethical framework of taking into account user-attitudes is connected with the value of autonomy. If carebot use is to take its cue from the wishes of individual older users of carebots, and if surveys of older people reveal a range of design-relevant preferences which do not correlate with the design features of the carebot that engineers intend to realize (Van der Plas et al. 2010), this may suggest that engineers think they know better than their potential older users what carebots should be like, or that they do not know and have not bothered to find out what older users of carebots might be looking for. Either way, the potential of the engineer-designed carebot to promote the autonomy of older users might be compromised.

ACCOMPANY has conducted research among panels of older people in the UK, the Netherlands and France. The project is investigating what users might want from a carebot, and has found that mobility, self-care and isolation are major preoccupations, while co-learning seems not to be.[26] Does this finding mean that ACCOMPANY should drop co-learning from its designs for robots? Not necessarily. Co-learning may have other effects that older people could benefit from and that they want, even if they want other effects more. There could be a therapeutic rationale for some design features that older people don't want or don't want much, so long as on balance groups of older people have been consulted and listened to in relation to design, and so long as the ACCOMPANY Care-O-bot® accommodates itself to individual users rather than coming up with an agenda of its own. To go back to Siena's methods of keeping up older people's social skills by adjusting its behaviour to the user's tone of voice, this might have what is broadly speaking a therapeutic benefit even if the older person doesn't like it much.

Vallor's list of ethical concerns indeed anticipates the way that the Siena design might be justified. It in effect asks philosophers and technologists to think about:

3. The potential of carebots to enhance or reduce engagement of cared-fors with their surroundings

and

5. The quality of physical and psychological care that robots can realistically be expected to supply.

The Siena innovations try to improve social skills and, indirectly, the psychological well-being of older users. They also introduce companionship into such routine ways of engaging with one's surroundings such as watching television and helping with such tasks as moving objects from one room to another, which promotes living in orderly and clean surroundings.

Even when the attitudes of users are taken into account, there may be conflicts within the range of ethical values that are individually relevant to providing care for older people. We have already seen that autonomy can conflict with safety: a carebot that is otherwise dedicated to fulfilling the wishes of its older user ought not to comply with a request that is suicidal. Similarly, although older autonomous people have a right to privacy at least as extensive as that of younger people, there may be occasions when a carebot should report a fall to a non-resident carer or a medical assistance hub, even if that is against the wishes of the older person himself or herself.

Against this background, what sort of ethical framework should be proposed for the design of carebots. The framework must identify and define values that should be promoted or at least respected by carebot design and use in relation to older people, and it must say which value is, or which values are, overriding when there is a conflict. The ACCOMPANY project addresses isolation and declining physical capacity in older people who continue to live and want to live in their own homes. If a robotic companion is to be a solution, its design must promote the following:

- autonomy—being able to set goals in life and choose means;
- independence—being able to implement one's goals without the permission, assistance or material resources of others;
- enablement—having or having access to means of realizing goals and choices;
- safety—being able readily to avoid pain or harm;
- privacy—being able to pursue and realize one's goals and implement one's choices unobserved;
- social connectedness—having regular contact with friends and loved ones, and safe access to strangers one can choose to meet.

All of these values lie in the background of most able-bodied, independent adult life, and our approach is to extend these values to later life unless there are reasons not to do so. Isolation and physical decline might be thought to be such reasons—unless a technology can compensate for them. The ACCOMPANY scenarios animate these reasons. And a particular design of robot companion compensates for them.

It is, however, inevitable that circumstances will arise where these values are in tension. When this happens one value is likely to be given priority over another. The preceding discussion has suggested that autonomy is a crucial value but that it can be outweighed when respecting it would threaten a user's life or physical well-being. It might be thought to follow, then, that of the six values, safety is supreme, trumping even autonomy.

This seems to be a mistake. Not every threat to safety, even when realized, produces major injury. When the worst that the exercise of autonomy produces is minor harm, or not-so-minor but tolerable and survivable harm, autonomy might win out over safety. Admittedly, the meaning of 'major harm' and 'minor harm' varies over a life-course. Falls that are tolerable at 45 years of age and classifiable as minor then would not be classifiable as minor at 90, but the threshold has to be quite high if the older person's autonomy is not to be in danger of being entirely undermined by too conservative a safety regime. In other words, autonomy, not safety, should normally be the ruling value in carebot design. For example, if an older person prefers being bruised for a week to staying seated or using a walker, not interfering with a decision to get up and be active seems to be consistent with the discretion usually allowed to middle-aged and younger adults with respect to their health and safety, even when minor harm results. Allowing the older person the same discretion might mean designing a carebot so that its prompts to use a walking frame etc. can be disabled (and perhaps later re-enabled) by the user.

Because privacy promotes autonomy by allowing users to discover when unobserved what their limits or vulnerabilities are, and to factor those into their plans, carebots should not normally be able to report information about users to outsiders or let anyone into an older person's home without permission. On the other hand, acting on some kinds of information without reporting to outsiders might be valuable. Thus, if the carebot has or is connected to flood sensors in a smart home, there is no reason why it or the smart home technology cannot trigger a cut in the water supply and then ask the user what they next want done. This is in keeping with autonomy. Cutting the water supply and asking an *outsider* for subsequent instructions would undermine user autonomy unless the user was incapacitated.

Social connectedness is desirable, other things being equal, because of its potential benefits to physical and mental health. But the 'other things being equal' is important: it is possible for social connectedness to empower busybodies, without any benefit to the user. Instead of social connectedness full stop, *chosen* social connectedness with *chosen* people seems desirable, with the user deciding, as most adults routinely do, whom to include and whom *not* to include in their social circle. A user who disliked all eligible social connections might intelligibly choose isolation, but, given the reach of social networks afforded by the World Wide Web, the number of eligible social connections is likely to be much larger than the number of people the user has good reasons or any reasons for shunning.

Enablement might also be in tension with autonomy, since enablement may require individuals to do things for themselves that they might prefer were done for them, or that they might prefer not to do at all. Robotic devices are being developed to help with physical rehabilitation following stroke, accident or amputation. Physiotherapy of this kind often requires patients to be coaxed, persuaded and even paternalistically coerced into repeating movements by physiotherapists, who may themselves move or position the patients in ways that although initially uncomfortable are necessary for rehabilitation. Returning someone to a state of greater independence is certainly compatible with autonomy; the question is whether it is compatible with autonomy for a carebot to coerce someone to adhere to regimes that will return them to greater independence.

The answer to this question may lie in what is agreed with the older person at the time a rehabilitation device or robot with enabling capabilities is placed with that person's consent in their home. In the case of single-purpose device, there would be no objections to removing a state-funded device that was lying unused or not being used properly. Carebots pose a different challenge because they are designed to be multi-functioning and these other functions would also be lost if they were removed. Enablement functions are not quite the same as those providing potential social interaction. Disliking social interaction and preferring isolation is a matter of taste. Working against a carebot programmed to maintain independence is not simply an expression of taste, but a kind of resistance to independence. Again, the robot and its developers would not necessarily be working against the autonomy of older users if the robot refused to do things that the older person could reasonably do for herself, or which it might be good for her to do for herself. Indeed, we can envisage something of a spectrum of mutual accommodation. At one extreme might be a user's refusal to co-operate with the robot in maintaining his or her mobility. At the other extreme might be automatic robot compliance with all user requests, even the request to be thrown off the balcony. Between the extremes might be cases where the robot enables the user to eat, or drink or smoke excessively. In this respect choices about the programming of carebots reflect the ethical issues raised more generally in health promotion and public health, where what people want is not necessary what is good for them, and satisfying their desires can be in tension with health interests.

One of the challenges for the ethical framework in ACCOMPANY is that the Care-O-bot® can play a variety of roles (companion, helper and enabler), each of which is subject to different norms in human-to-human service provision.

To take companionship first, we can assume that the Care-O-bot® is not designed to simulate a family member but rather to counteract the experience of being always or mostly alone. The Care-O-bot® might therefore play a role similar to that of a paid companion in late eighteenth and early nineteenth century England. The companion was paid to provide constant company, usually for single people, and shared their employer's home. This was a role that struck a balance between friend and servant. The companion could be a confidante, but, unlike the friend, was an employee who had very little autonomy and could be called upon to help with 'light' duties—such as sewing or playing sport. As in the case of the Care-O-bot®, the relationship was one-sided, with the feelings, wishes and whims of the employer (or older user in the Care-O-bot® case) having most of the weight and those of the companion having little or none. However, it was considered unseemly to be unduly rude to or rough with the companion—which corresponds with the concerns for 'respectful' interaction being worked on in by the Siena partner in ACCOMPANY.

A helper may be a servant, professional or volunteer, and these three roles will now be considered in turn. Servants are paid to do their employer's bidding, usually without question. As it operates in ACCOMPANY, Care-O-bot® does not quite take on the traditional role of the servant, because it is intended to perform tasks that users are physically unable, rather than unwilling, to do for themselves. On the other hand, to place Care-O-bot® in the servant role suggests, appropriately enough, that the older user is controlling the robot rather than the robot controlling the older user. It also suggests that the robot should be discreet, keeping household matters private.

To the extent that it is designed for the frail and those with physical impairments, the Care-O-bot® could be associated with caring roles filled by nurses, healthcare assistants and doctors, especially when they are equipped with interfaces for telehealth interventions. Human carers are not necessarily obedient servants. On the contrary, they are likely to have their own ideas about how much help to give and when, what constitutes help and what form it should take from occasion to occasion. So there may be a tension between placing Care-O-bot® in the caring role and placing it in a servant role. In one the older person is the boss, and in the other the older person sometimes needs to accommodate the carer. Informal, voluntary care such as that which might be provided by a friend, incorporates both the care element and that of companionship. It reinforces the idea that that whilst the robot is present at the invitation of the older user, it should not be exploited or ordered about. It is also more of a relationship between equals, even though the older user retains the upper hand and the robot has only limited capacity to withdraw from unsympathetic behaviour or tone.

'Enabler' may suggest superiority over the enabled: the human enabler is the one with the knowledge, skills, abilities and powers to enable. This may also raise questions about who is deferred to when older person and their enabler are in conflict. There is a corresponding tension between enablement and autonomy.

When autonomy conflicts with other values that govern the possible roles of Care-O-bot®, which should prevail? A way of summarizing much of the foregoing is by saying that autonomy should. Autonomy can make sense as the organizing value of the ethical framework for the design of carebots. Being the organizing value, autonomy also constrains additions to the value framework: other potential values would be consistent with autonomy or else have some independent moral grounding. Should further values be added to those already introduced?

One source of further values is the interests of carers connected to the older person. Carers enter the ethical framework developed so far through its values of safety and social connectedness, in turn constrained by the value of keeping the older person autonomous for as long as possible. This may not be the right way for carers to enter the framework. It might be thought that by putting older people and their choices at the centre of things, the framework denies the dependence of older people on carers and is in any case too individualistic. For example, the framework recognises threats to the autonomy of older people from carers but not the sheer hard work and sometimes sacrifice of their carers. Perhaps the framework needs to reduce the value of autonomy in interactions with the older person the more other people have *their* choices reduced by their caring role. Concretely, this might mean that the ability of the older person to judge and take risks that could lead to injury and greater dependence might be restricted the more dependent they are on others. It might also justify more monitoring and more reporting to carers.

We are not persuaded that autonomous older persons necessarily overburden carers, even when they are dependent. But it helps to remind ourselves that we are not concerned with the general question of the best way of being fair to carers. We are only concerned with the way that carers' interests should be represented in a framework for the design of carebots. Since carebots of the kind being developed in the ACCOMPANY project assume only moderate physical disability and near complete cognitive functioning in the older people who would be living with the Care-O-bot®, the question of trade-offs between autonomy and *high* dependence does not arise. That does not mean that there are no difficult questions about what carers have a right to know about in the lives of older people and what decisions of older people they have a right to veto, but in general the burden of proof will be on carers rather than the other way round.[27]

FOOTNOTES

[1] See for instance Paro the seal http://www.parorobots.com/index.asp.

[2] See http://www.care-o-bot.de/english/.

[3] 'Presence' in this sense is an effect of the robot on its companion. This is not to be confused with telepresence, where someone is *not* co-located with a robot, but is able to get audio-visual experience of the place where a remote robot is located, or where it is moving.

[4] The authors are members of the ACCOMPANY project team. The research leading to these results has received funding from the [European Union's] Seventh Framework Programme (FP7/2007-2013) under grant agreement no. [287624]. Some ACCOMPANY deliverables (including those referred to below, are public and can be found at the project website: http://www.accompanyproject.eu/.

[5] But as ACCOMPANY is aimed at older people, this group of potential users will not be discussed here.

[6]These documents tend not to distinguish between those over 65 and the much older. The 'oldest old' may pose special problems for assistive technology [see Misselhorn et al. (2013)]. The ACCOMPANY robot is not designed specifically for those in the Fourth Age, and may not be well-suited to the oldest old.

[7]Though predictions for care needs based on current demand, extrapolated to the expected increase in the proportion of older people in the population, suggest something of an explosion in the need for care; e.g. OECD (2011).

[8]It is important to remember that all assistive technology is introduced into a person's *home*. The significance of its being a home and the significance of 'home' to individuals means, as we explain below, that it should only be introduced with appropriate consent from the autonomous person whose home it is.

[9]Users of telecare and telehealth equipment are sometimes conscious of being under surveillance and dislike the Orwellian 'Big Brother' aspects of telecare and telehealth (Percival and Hanson 2006; Robinson et al. 2007). Do these aspects count against telecare morally? Whether they do depends on whether the purpose of the monitoring is sufficiently similar to that of state security surveillance, and whether sensors are necessary for that purpose. State security surveillance seeks to collect evidence of behaviour that is illegal or damaging to the security of the state. It is not done for the benefit of the person surveilled, but for the wider public, or for the maintenance of the state. Telecare and telehealth, on the other hand, *are* operated for the benefit of the person monitored, and, in Western countries, are only introduced into people's homes with their consent. They are an early warning system for a health problem or a health emergency. Far from leading to arrest, prosecution and imprisonment, they trigger rescue or medical intervention for a particular person identified in advance as being medically needy. Beyond that, telecare is at least often claimed by its promoters to maintain the independence of its users (Woolham 2006). This purpose is often given greater weight, at least officially, than relieving the burden on a carer (Brownsell et al. 2003; Poole 2006; Greenhalgh et al. 2012). For a full discussion of the ethics of surveillance in telecare, see Sorell and Draper (2012).

[10]The US Food and Drugs Administration recently approved the use of an ingestible sensor in pills to monitor the medication intake of patients: http://proteusdigitalhealth.com/proteus-digital-health-announces-fda-clearance-of-ingestible-sensor/.

[11]We assume here responsible use of emergency facilities (see Draper and Sorell (2002)) and also note that there are anecdotal reports on older

users using alarm devices because they feel lonely and regard this as one means, perhaps their only means, of getting human contact.

[12]Locating control with the user fits in with one finding from surveys of users of telecare, namely that they dislike the way it takes away their control over *revealing* increasing frailty or disability (Sorell and Draper 2012). This desire to retain control was also reflected in the data collected by MADoPA and reported in ACCOMPANY deliverable D1.2 Report on user and system requirements and first outline of system functionality.

[13]It is also possible the developments in technology and human interactions with technology will change the meaning of isolation. People who, even when young, maintain social relations online may feel far less lonely when they lose physical mobility than older people for whom computer contact by itself does not relieve feelings of isolation. In the same way, playing with a robot dog might well come to feel just as satisfying for future older people as playing with a real dog might today. For more on the way that the evolving *use* of technology can reconstitute such concepts as isolation, see Coeckelbergh (2012).

[14]There is considerable literature on tele-consultation and within this, evidence that focussing on a single item can improve concentration, see for instance Sävenstedt et al (2005)

[15]ACCOMPANY deliverable D1.1 Status of elderly care in Europe and the potential for service robots. Available http://accompanyproject.eu/ [Accessed 19/1/2014].

[16]http://rtc.nagoya.riken.jp/RIBA/index-e.html.

[17]For a video of the Cyberdene HAL "power suit", see http://www.youtube.com/watch?v=fy7ipDAyXtI&feature=related.

[18]These are the so-called "Remote Presence" robots manufactured by intouch. See http://www.intouchhealth.com/products-and-services/products/rp-7i-robot/

[19]My Spoon helps users to feed themselves. http://www.secom.co.jp/english/myspoon/usage.html

[20]Such as those produced by iRobot http://www.irobot.com/uk/home.aspx.

[21]'Capabilities' is used here in its ordinary sense, as a synonym for 'abilities', and not in the senses assigned to the term by Sen, and Sen and Nussbaum, in their work on the capability approach to development. Although we realize that this approach (primarily in the Nussbaum version) has been extended to robots (Coeckelbergh 2012), we think that it

is ill-suited to different stages of the life course, in particular older age. In Nussbaum's hands, the capability approach is essentially a repackaging of Aristotle, which is in its turn ill-suited to questions about the flourishing of human beings with disabilities and the possibility of flourishing at stages of life in which there is a natural decline in capacities. A proper discussion of these issues is beyond the scope of the paper, but an important point would be that flourishing for a species (central to Aristotle) question-beggingly privileges certain points in the life course of members of that species.

[22]ACCOMPANY deliverable D1.1 'Status of elderly care in Europe and the potential for service robots' noted that there are inconsistencies across the EU in how assistive technologies for older people may be funded, and not all EU citizens in need of help can be sure of receiving state funded assistance. Available http://accompanyproject.eu/[Accessed 19/1/2014].

[23]http://www.parorobots.com/index.asp

[24]See Misselhorn et al. op. cit.

[25]Jennifer Parks, for instance, notes that cultural difference may be significant in terms of both how people respond to robots versus humans and different robotic persona. Discussion about the paucity of data about, and understanding of, cultural differences in perceptions and approaches to ageing in general can be found in University of Birmingham Policy Commission 2014 'Health Ageing in the 21st Century: the best is yet to come.' Available http://www.birmingham.ac.uk/research/impact/policy-commissions/healthy-ageing/index.aspx [Accessed 6/3/2014].

[26]See ACCOMPANY deliverables D1.1 Status of elderly care in Europe and the potential for service robots and D1.2 Report on user and system requirements and first outline of system functionality. Both available http://accompanyproject.eu/ [Accessed 19/1/2014].

[27]A further and extremely important consideration is monetary cost. Until the price of Care-O-bot® is much lower than it is now, the presumption in favour of lower tech and cheap assistive technology must remain strong.

ACKNOWLEDGMENTS

The work in this paper was partially funded by the European project ACCOMPANY (Acceptable robotics COMPanions for AgeiNgYears). Grant agreement no.: 287624.

REFERENCES

1. Alaszewski, A., & Cappello, R. (2006). *Piloting telecare in Kent County council: The key lessons. Final report.* Canterbury: University of Kent, Centre for Health Services Studies.
2. Bayer, S., Barlow, J., & Curry, R. (2007). Assessing the impact of a care innovation: Telecare. *System Dynamics Review, 23*(1), 61–80.
3. Borenstein, J., & Pearson, Y. (2010). Robotic caregivers: Harbingers of expanded freedom for all?'. *Ethics and Information Technology, 12*(3), 277–288.
4. Bowes, A., & McColgan, G. (2006). *Smart technology and community care for older people: innovation in West Lothian, Scotland.* Edinburgh: Age Concern Scotland.
5. Brownsell, S., Bradley, D., & Porteous, J. (2003). *Assistive technology and telecare: Forging solutions for independent living.* Bristol: The Policy Press.
6. Clark, R. A., Inglis, S. C., McAlister, F. A., et al. (2007). Telemonitoring or structured telephone support programmes for patients with chronic heart failure: Systematic review and meta-analysis. *British Medical Journal, 333*, 942.
7. Coeckelbergh, M. (2010). Personal robots, appearance and human good: A methodological reflection on robots. *International Journal of Social Robots, 1*(3), 217–221.
8. Coeckelbergh, M. (2012). How I learned to love the robot: Capabilities, information technology and elderly care. In I. Oosterlaken & J. van den Hoven (Eds.), *The capabilities approach, technology and design* (pp. 77–85). Dordrecht: Springer.
9. Dang, S., Dimmick, S., & Kelkar, G. (2009). Evaluating the evidence base for the use of home telehealth remote monitoring in elderly with heart failure. *Telemedicine and e-Health, 15*(8), 783–796.
10. Decker, M. (2008). Caring robots and ethical reflection: The perspective of interdisciplinary technology assessment. *AI & SOCIETY, 22*(3), 315–330.
11. Department of Health. (2005). *Building telecare in England.* London: Crown.
12. Department of Health. (2008). *High quality care for all: NHS Next stage review.* London: Crown.
13. Department of Health. (2010). *Building the National Care Service.* London: Crown.
14. Dixon, R. F., & Stahl, J. E. (2009). A randomized trial of virtual visits in a general medicine practice. *Journal of Telemedicine and Telecare, 15*(3), 115–117.
15. Doughty, K., Monk, A., Bayliss, C., et al. (2007). Telecare, telehealth and assistive technologies—do we know what we're talking about? *Journal of Assistive Technologies, 1*(2), 6–10.
16. Draper, H., & Sorell, T. (2002). Patients' responsibilities in medical ethics. *Bioethics, 16*(4), 335–353.
17. Draper, H. & Sorell, T. (2012). 'Telecare and Care' Bioethics 16th April. doi:10.1111/j.1467-8519.2012.01961.x.
18. Fisk, M. (2003). *Social alarms to telecare: Older people's services in transitions.* Bristol: The Policy Press.
19. Garćia-Lizana, F., & Sarrĩa-Santamera, A. (2007). New technologies for chronic disease management and control: A systematic review. *Journal of Telemedicine and Telecare, 13*, 62–68.
20. Greenhalgh, T., Procter, R., Wherton, J., et al. (2012). The organising vision for telehealth and telecare: Discourse analysis. *BMJ Open, 2*, e001574.
21. Lim, F. S., Foo, M., Kanagalingam, D., et al. (2007). Enhancing chronic disease management through telecare—the Singapore Health Services Experience. *Journal of Telemedicine and Telecare, 13*(Suppl. 3), 73–75.
22. Misselhorn, C., Pompe, U., & Stapelton, M. (2013). Ethical considerations regarding the use of social robots in the fourth age. *The Journal of Gerontopsychology and Geriatric Psychiatry,26*(2), 121–133.

23. Murray, E. et al. (2011). Why is it difficult to implement e-health initiatives? A qualitative study. *Implementation Science, 6*(6), 1–11.
24. OECD (2011). *Health at a Glance 2011: OECD Indicators.* OECD Publishing. http://dx.doi.org/10.1787/health_glance-2011-en.
25. Paré, G., Jaana, M., & Sicotte, C. (2007). Systematic review of home telemonitoring for chronic diseases: The evidence base. *Journal of the American Medical Informatics Association, 14*(3), 269–277.
26. Parks, J. (2010). Lifting the burden of women's care work: Should robots replace the "human touch"? *Hypatia, 25*(1), 100–120.
27. Percival, J., & Hanson, J. (2006). Big brother or brave new world? Telecare and its Implications for older people's independence and social inclusion. *Critical Social Policy, 26*, 888–909. see esp. p. 898.
28. Perry, J., Beyer, S., & Holm, S. (2009). Assistive technology, telecare and people with intellectual disabilities: Ethical considerations. *Journal of Medical Ethics, 35*, 81–86.
29. Pols, J. (2010). The heart of the matter: About good nursing and telecare. *Health Care Analysis, 18*(4), 374–388.
30. Poole, T. (2006). *Wanless social care review: Telecare and older people.* London: The King's Fund.
31. Robinson, L., et al. (2007). Balancing rights and risks: Conflicting perspectives in the management of wandering in dementia. *Health, Risk and Society, 9*, 389–406.
32. Rogers, A., et al. (2011). Established users and the making of telecare work in long-term condition management: Implications for health policy. *Social Science and Medicine, 72*, 1077–1084.
33. Sävenstedt, S., Zingmark, K., Hydén, L.-C., & Brulin, C. (2005). Establishing joint attention in remote talks with the elderly about health: A study of nurses' conversation with elderly persons in teleconsultations. *Scandinavian Journal of Caring Sciences, 19*, 317–324.
34. Sharkey, A., & Sharkey, N. (2010). Granny and the robots: Ethical issues in robotic care for the elderly. *Ethics and Information Technology,*. doi:10.1007/s10676-010-9234-6.
35. Sorell, T., & Draper, H. (2012). Telecare, surveillance and the welfare state. *The American Journal of Bioethics, 12*(9), 36–44.
36. Sparrow, R., & Sparrow, L. (2006). In the hands of machines? The future of aged care. *Minds and Machines, 16*(2), 141–161.
37. Stienstra, J., & Marti, P. (2012). Squeeze me: gently please. Paper presented to NordiCHI 2012 conference in October 2012.
38. Turkle, S. (2006) A nascent robotics culture: new complexities for companionship. AAAI Technology Report Series Available online http://web .mit.edu/sturkle/www/pdfsforstwebpage/ST_Nascent%20Robotics%20Culture.pdf accessed 5.9.2012.
39. Vallor, S. (2011). Carebots and caregivers: Sustaining the ethical ideal of care in the 21st century. *Philosophy and Technology, 24*, 254.
40. Van der Plas, A., Smits, M., & Werhmann, C. (2010). Beyond speculative robot ethics: A vision assessment study on the future of the robotic caretaker. *Accountability in Research, 17*(6), 299–315.
41. Woolham, J. (2006). *Safe at home—the effectiveness of assistive technology in supporting the independence of people with dementia: The safe at home project.* London: Hawker Publications.

CHAPTER 14

Can we Program or Train Robots to be Good?

Amanda Sharkey[1]

[1] *Department of Computer Science, University of Sheffield, Sheffield S1 4DP, UK*

ABSTRACT

As robots are deployed in a widening range of situations, it is necessary to develop a clearer position about whether or not they can be trusted to make good moral decisions. In this paper, we take a realistic look at recent attempts to program and to train robots to develop some form of moral competence. Examples of implemented robot behaviours that have been described as 'ethical', or 'minimally ethical' are considered, although they are found to operate only in quite constrained and limited application domains. There is a general recognition that current robots cannot be described as full moral agents, but it is less clear whether will always be the case. Concerns are raised about the insufficiently justified use of terms such as 'moral' and 'ethical' to describe the behaviours of robots that are often more related to safety considerations than to moral ones. Given the current state of the art, two possible responses are identified. The first involves continued efforts to develop robots that are capable of ethical behaviour. The second is to argue against, and to attempt to avoid, placing robots in situations that demand moral competence and an understanding of the surrounding social situation. There is something to be gained from both responses, but it is argued here that the second is the more responsible choice.

Keywords: Ethics, Moral competence, Robot, Decisionmaking, Minimally ethical

INTRODUCTION

Our increasing deployment of and reliance on robots means that there is a pressing need for a clear position on the possibility of developing robots that can be described as 'good' or 'ethical'. High profile concerns have been raised about the potential impact of artificially intelligent systems on humans, and arguments have been made about the need to constrain the behaviour of such systems (e.g. Bostrom 2014; Russell 2016). Two areas in which there is a growing awareness of the extent to which robotics can directly impinge on the health and safety of humans are those involving (i) autonomous vehicles and (ii) robotic weapons, especially 'autonomous' robot weapons. It is apparent that autonomous cars are likely to encounter situations in which it is necessary to make life or death decisions about whether to protect themselves, or other humans (Lin 2013, 2015). And autonomous robotic weapons could be deployed in situations in which they make decisions about when to use lethal force, and who to kill (Sharkey 2012; Asaro 2012; Altmann et al. 2013). The stakes in such domains are high, and the issues important.

Both self-driving cars, and lethal autonomous weapons, would directly affect the physical safety of human beings. But life or death decisions are not the only ways in which robots could affect human lives: their potential effects are not limited to physical damage. As discussed by Sharkey (2016), a robot deployed in a classroom as a teacher or as a teacher's assistant, could be required to make decisions about what children's behaviour was acceptable or punishable. A robot 'carer' of vulnerable older people might have to make decisions about which of its charge's activities should be facilitated, or prevented (Sharkey and Sharkey 2012; Sorrell and Draper 2014). Similarly, to be effective, a robot 'nanny' or minder of children would need to make decisions about when to stop children from doing something, and when to encourage them (Sharkey and Sharkey 2010).

If robots are to be placed in situations in which they will make decisions that have a direct impact on human well being, or on human physical safety, it is only sensible to try to ensure that they make the right decisions. The aim of this paper is to examine the various approaches that have been taken to answering the question about whether robots can be programmed to be good, and to assess their current level of success. In doing so, any examples of actual implementations, as opposed to abstract discussions of what might be possible in principle, will be highlighted. This examination will form the basis for a consideration of the best response to the current situation, and a discussion of the circumstances in which robot use should be encouraged, discouraged, or even banned. This in turn will contribute to the ongoing debate about what is meant by taking a 'responsible' approach to robotics.

Programming robots to be good

There have been various attempts to program robots to be 'good' and to make decisions that might be described as ethical or moral. Famously, the science fiction writer Isaac Asimov proposed the 3 laws of robotics (Asimov 1942)

1. A robot may not injure a human being or, through inaction, allow a human being to come to harm.
2. A robot must obey the orders given it by human beings except where such orders would conflict with the first law.
3. A robot must protect its own existence as long as such protection does not conflict with the first or second laws.

However, many of Asimov's stories illustrated the unintended problems that could occur as a result of following these rules. The rules are of course fictional, and there is no simple way of translating them into implementable code. How could you write a program to ensure that a robot's action or inaction did not lead to a human being coming to harm? How could the robot foresee all the possible consequences of its actions, and their interaction with human behaviour? How could a robot even recognise harm? The rules seem more focused on short-term physical safety, when clearly there are other ways in which humans could be harmed. A robot's actions might indirectly cause future long-term physical harm. Its actions could also lead to other kinds of damage such as psychological trauma or emotional upset. The rules also imply robots that can understand the orders given to them by humans (and the extent to which they conflict with the first law). As Murphy and Woods (2009) point out, such robust natural language understanding has not yet been achieved.

There have been some practical attempts to program robots to be 'good', or to make decisions that have been described as ethical. Winfield et al. (2014) report experiments in which robots are programmed to stop other robots (designated as proxy humans) from coming to harm. The robot is placed in an environment in which it has a goal to reach, but in which there is also a 'hole' or dangerous area that is a risk both to it, and to the other proxy human robots. They propose an internal-modelling based architecture for what they describe as 'a minimally ethical robot'. The robot has access to an internal model, or simulator, through which it can assess all possible actions by looking at their consequences: in particular their consequences in terms of the dangerous hole area. The robots used the internal model to anticipate the consequences of different trajectories of movement for themselves or other robots. These anticipated consequences, combined with pre-set preferences, are used as the basis for determining which action to undertake. Possible actions include moving towards (and falling in) the hole, or blocking the path of another robot (the proxy human) in order to prevent it from falling into the hole. The robot's predetermined 'preferences' are set by the human programmer. Winfield et al. (2014) describe a situation in which there are two proxy human robots, both of which are following a trajectory that would lead them to enter the dangerous area. The main actor robot is programmed to try to intercept the path of both robots, but given no way of prioritising which one to rescue first. As a consequence, the robot was sometimes found to dither between two possible trajectories, as if it were unsure of which proxy human to save.

Winfield et al. (2014, p. 5) write 'What we have set out here appears to match remarkably well with Asimov's first law of robotics: A robot may not injure a human being or, through inaction, allow a human being to come to harm'. The work has stimulated discussion in the media (e.g. Rutkin 2014) of whether or not the robots should be described as ethical. However the robots in question have been programmed to behave as they do. Although they appear to hesitate about what to do when faced by the dilemma of two proxy human robots that both need rescuing, this hesitation is a consequence of their programming. One of the main reasons that this work has stimulated discussion is that it describes the main robot as being 'minimally ethical'. This use of the term 'ethical' is controversial, as will be discussed later. Nonetheless a strength of the study is that it provides an implemented and practical example of research into issues related to robots and ethical decision-making.

Ron Arkin has argued that robots and computational agents could be more ethical and moral than flawed and emotional humans. In a paper about implementing an 'ethical governor' for autonomous military robots, he writes, 'It is not my belief that an unmanned system will be able to be perfectly ethical in the battlefield, but I am convinced that they can perform more ethically than human soldiers are capable of' (Arkin 2007, p. 4). His reasons are: (i) the robots will not need to protect themselves and could be self-sacrificing; (ii) they might have better sensors for battlefield observation than humans; (iii) they could be designed without emotions that could affect their judgment; (iv) unlike humans, they would not be vulnerable to 'scenario fulfillment', and to interpreting a situation in the light of prior expectations; (v) they could integrate information from several sources faster than humans can; (vi) they could independently monitor (and report) the behavior of those in the battlefield.

The ethical governor proposed by Arkin et al. (2009) is part of a system architecture that is described as 'potentially capable of adhering to the

International laws of war (LOW) and rules of engagement (ROE) to ensure that these systems conform to the legal requirements of a civilized nation' (Arkin 2009, p. 1). The ethical governor would be introduced as a bottleneck to evaluate the actions proposed by the reasoning subsystems of the overall system, permitting only those actions that were deemed ethically acceptable. The acceptability of actions would be determined based on a set of constraints, which themselves would be based on stored representations of the International laws of war, and the specific rules of engagement. Actions could be deemed unethical and prohibited if they did not conform to the laws of war, or if they were not recommended as appropriate, ('obligated' in their terminology). A further check would be carried out to ensure that potential collateral damage would be minimized, based on a table indicating acceptable levels of collateral damage given the military necessity associated with the target. Arkin (2009) describes an evaluation of the architecture undertaken within the MissionLab simulation environment, in which the decisions made as a result of the interaction between the ethical governor and the behavioural control system are examined in a number of test scenarios. The simulated tests indicate that the system, together with the ethical governor, would make decisions about the use of lethal force that would limit collateral damage with reference to the levels of military necessity (as determined by the military).

We could see the ethical governor as constituting an approach to programming robots 'to be good', or ethical. At the same time, autonomous military robots programmed in this way would have no choice about what actions they would perform. Their action choices, in this case about deploying lethal force, would be determined by the system and the set of constraints, which are set up and decided upon by the programmers and those using the system. As in Winfield's et al. (2014) experiments described above, the programmers of the system effectively determine the action choices.

A number of objections have been raised to Arkin's proposals. Matthias (2011) discusses the paper in detail, and points to a number of difficulties. One of these is that many of the rules on which the system is based are unclear and contradictory. For example, the rules of engagement for use in Kosovo stated "You may use minimum force, including opening fire, against an individual who unlawfully commits or is about to commit an act which endangers life, in circumstances where there is no other way to prevent the act" (Arkin 2007, p. 37, cited by Matthias). Adhering to this rule would require considerable interpretation, and knowledge and understanding of individuals' intentions. Matthias (2011) also points out that the military can adjust or override the ethical governor if military necessity is considered to be high, and that it should therefore be described as an ethical *advisor*, rather than as an ethical governor.

Matthias characterizes Arkin's view of a moral agent as one that follows rules. The ethical governor performs its actions 'according to a pre-installed program, with no possibility of dissent or of questioning the commands issued to it' (Matthias 2011), unlike the case of a soldier who could refuse to carry out an immoral command. Crucially, it lacks the autonomy that Matthias considers to be 'a key ingredient of moral agency'.

A similar objection could be made to the idea that Winfield's robots are 'minimally ethical'. Interestingly, in a recent paper, (Vanderelst and Winfield 2016), the point is made that if a robot can be programmed to make 'ethical' choices, it can also be programmed to make ones that are 'unethical'. In a 'shell game', in which the desired action was either to approach the shell on the left or the right, Vanderelst and Winfield (2016) used a robot that was able to detect whether or not another robot (again designated a proxy human) was moving towards the correct shell, or heading in the wrong direction. They programmed the robot to indicate to the human when they were heading in the wrong direction. They also programmed two other versions of the robot: a competitive version which headed to the goal first and prevented the proxy human from reaching it, and an 'aggressive' version that deceived the proxy human and sent it in the wrong direction. They conclude from their experiments that it is just as possible to program a robot to be unethical as it is to program it to be ethical.

Moor (2006) developed a typology of ethical agents, and it is interesting to consider how it would apply to Arkin's 'ethical governor', or to Winfield's 'minimally ethical' robots. Moor identified and defined four types of moral agent: ethical impact agents, implicit ethical agents, explicit ethical agents and full ethical agents. *Ethical impact agents* are computers or robots that 'do our bidding as surrogate agents and impact ethical decisions such as privacy, property and power' (ibid p. 19). Moor gives the example of the robot camel jockeys in Qattar that have reduced the use of young boys as slaves to ride the camels. *Implicit ethical agents*, by contrast, act ethically because they are programmed, or have internal functions, which promote ethical behavior, or avoid unethical behavior. Moor gives the example of automatic teller machines (ATMs) that are programmed to deliver the right amount of money. An *explicit ethical agent*can 'represent ethics explicitly and then operate effectively on the basis of this knowledge' (ibid p. 20). A *full ethical agent* can both make ethical judgments and justify them. A human with consciousness, intentionality and free will is a full ethical agent. Moor points out that some would argue that computational artifacts (computers and robots) will never be full ethical agents whilst lacking consciousness, intentionality and free will. He disputes this claim, on the basis that 'we can't say with certainty that future machines will lack these abilities' (ibid p. 20). Rather than engaging in this debate, Moor argues that it is important to examine the other categories, and in particular to research the possibility of developing explicit ethical agents.

Moor wants to encourage efforts to develop explicit ethical agents because (i) we want machines to treat us well (ii) machines are becoming more powerful and need a more powerful ethics, and (iii) programming or teaching ethics to a machine will increase our understanding of ethics. He suggests that a major barrier to creating explicit ethical agents will be their lack of common sense and world knowledge. For example, a robot could only refrain from harming humans if it had a good knowledge of what possible harms there were.

Arkin's ethical governor could be considered to fit in the Moor's implicit ethical agent category since it contains internal ethical functions that promote ethical behavior. However its operation is more sophisticated than the ATMs that Moor offers as an example of implicit ethical agents, since its assessments of possible actions are based on a combination of constraints (from Laws of War, and specific Rules of Engagement) and considerations of collateral damage and military necessity. As such it might be considered to be an explicit ethical agent in Moor's typology, even though its explicit representation of ethics takes the form of constraints specific to military situations rather than more general ethical principles. However Moor seems to expect more of an explicit ethical agent, and wrote, in 2007, that 'an explicit ethical agent is futuristic at the moment' (Moor 2007, p. 12). His argument seems to be that explicit ethical agents are an appropriate goal to aim for, even though they may not be fully achieved. An explicit ethical agent should be one that 'can identify and process ethical information about a variety of situations and make sensitive determinations about what should be done in those situations'(ibid p. 12), working out resolutions when principles conflict. It should also be able to give persuasive justifications for its decisions. It is not clear that the ethical governor is able to do all of this. In particular, the range of situations to which it can be applied is limited to the battlefield, and its determinations are largely predetermined by the way it is set up.

The 'minimally ethical' robots of Winfield et al. (2014), that can prevent other robots from entering an area designated as dangerous, are able to make judgments that Winfield et al. describe as ethical. As such they might be considered to be explicit ethical agents in Moor's terms. However, their behavior is quite specific to one situation, and they are not able to process information about a variety of situations. Nor are they able to offer justifications for their decisions. Perhaps they, and Arkin's ethical governor, would be better described as implicit ethical agents, more akin to the ATMs that Moor uses as an example.

Susan and Michael Anderson (Anderson and Anderson 2007) also write about ethical agents, with the aim of developing an explicit ethical agent as defined by Moor (2006): able to represent particular ethical principals and to operate effectively on the basis of that representation. They contrast this with the idea of *ad hoc* programming of a machine to behave correctly in certain circumstances (implicit ethical agents). Interestingly, the Andersons make a distinction between moral responsibility, which implies intentionality and free will, and performing the morally correct action in a given situation.

The Andersons make use of an ethical theory based on *prima facie* duties (duties or obligations which individuals should try to satisfy but which could be overridden by stronger obligations), developed by Ross (1930). They (Anderson et al. 2006) use inductive logic programming to learn the relationships between these duties, which often give conflicting advice. Their work resides in the domain of medical ethics and is based on Beauchamp and Childress's four principles of medical ethics (Beauchamp and Childress 1979): respect for autonomy, and the principles of beneficence, non-maleficence, and justice. The particular dilemma they focus on is one where a health-care worker has recommended a treatment for a competent adult patient, and the patient has rejected the treatment. Should the health care worker accept the patient's decision, or attempt to change their mind? Anderson et al. (2006) implemented a machine learning approach using inductive logic programming to learn the relationships between the principles and the two possible actions. The system (MedEthEx) has access to a representative set of cases in which humans have made 'ethically correct' decisions, and uses inductive logic to abstract ethical principles from them. They claim that the system discovered a new principle: 'a health-care worker should challenge a patient's decision if it isn't fully autonomous and there's either any violation of non-maleficence or a severe violation of beneficence', (ibid p. 1764). They have also implemented another version of the system, EthEl on a Nao robot that can decide whether or not to remind the patient to take the medicine, and whether or not to report the patient to an overseer for not taking the medicine. They admit that the system at present is limited, but suggest that it could be scaled up to make a wider range of ethical decisions.

A strength of the Andersons' work, and that of Winfield and his colleagues is that the decision mechanisms have been implemented and shown to work with actual robots. Nonetheless, the examples seem disappointingly limited in their scope: the implemented systems can only make 'ethical' decisions in quite specific scenarios; either about preventing others from entering a dangerous area, or deciding whether or not to insist a patient take their medicine. The ethical governor seems to have a wider scope of application that covers varying battle scenarios, but in practice has only been tested in quite constrained simulations of military situations.

Wallach and Allen (2009) in their book, 'Moral Machines', distinguish between top down and bottom up approaches to the development of Artificial Moral Agents (AMAs). The approaches considered in this section are similar to their top down approach, which they define as 'any approach that takes a specific ethical theory and analyses its computational requirement to guide the design of algorithms and subsystems capable of implementing that theory' (pp. 80). The top-down approaches they discuss include that of the Andersons (Anderson et al. 2006), but they focus more on the difficulty of getting a machine to apply the sets of moral principles that constitute deontological or consequential ethics. Major problems with developing an ethical system for a robot-based utilitarian ethics lie in the need to anticipate the effects of undertaking action, and even more so in the need to evaluate the goodness or desirability of such effects. Any implementation of Kant's categorical imperative raises another set of seemingly intractable problems (ibid pp. 95–97).

Wallach and Allen contrast this top down approach to a bottom up one, in which an emphasis is placed on 'creating an environment where an agent explores courses of action and learns and is rewarded for behavior that is morally praiseworthy' (ibid p. 80), with the idea that any ethical principles will be discovered or constructed, rather than imposed in a top down manner. As will become apparent in the next section, there have also been various attempts to train or evolve robots to be ethical.

Training robots to be ethical

Malle (2015) takes the approach of outlining what is required for moral competence, and considering how this could be achieved in robots. For him, the requirements for moral competence are: a moral vocabulary; a system of norms; moral cognition and affect; moral decision-making and action; and moral communication. A moral vocabulary would include terms referring to norms (e.g. 'fairness', 'honesty'), their violations (e.g. 'wrong', 'thief'), and responses to violations (e.g. 'blame', 'forgiveness'). Knowledge of such terms could help a robot detect when humans refer to morally significant situations. A system of norms forms the basis of morality in humans, and is built up over time, initially on the basis of the moral judgments that adults make about concrete behaviors: 'that was naughty!'; 'He did something wrong'. Malle argues that it would not be possible to preprogram such norms into a robot since they are too subtle and context dependent. Instead he suggests that, 'a more promising direction is to mix unsupervised and supervised learning, "practice" through constant browsing of existing data (e.g. novels, conversations, movies) along with feedback about inferences (e.g. through crowdsourcing of "inquiries" the robot can make) and teaching through interaction' (Malle 2015, p. 9).

As well as discussing how norms could be acquired, Malle also considers how they could be represented in robots: suggesting a flexible network activated by features of the environment. In humans, knowledge of norms forms the basis for moral judgments since it enables them to recognize when norms have been violated and to allocate blame to responsible individuals, depending on the intentionality behind an act. A robot would need to be able to identify the aspects of an event that violated social and moral norms, via some mechanism that did not require comparison to every stored norm. Malle suggests that it would be able to do so even if it had no affect or emotion.

According to Malle, moral decision-making and action, a prominent component of human moral competence, would not necessarily require free will on the part of the robot, but rather the ability to receive blame and take it into account in its future actions. In human moral decision-making, there is a tension between the human's own goals and social-moral norms that is balanced by empathy for others. But Malle suggests that robots will have less need for empathy since they will not have a tendency for selfish behavior. At the same time, in order to be trusted by humans, robots might need to at least behave in a caring empathetic way towards others. Moral communication, the last component, is also required for moral competence so that moral judgments can be made, and moral decisions explained.

Malle's (2015) paper provides a useful account of what might be required for moral competence in robots. He does not say that such competence has yet been achieved, nor does he suggest a timeline for it. However, he makes the argument that creating a morally competent robot would be a good thing, since such robots 'could be trustworthy and productive partners, caretakers, educators, and members of the human community' (Malle 2015, p. 19). He suggests that if a robot was to become morally competent it should have some rights, and should not necessarily have to obey human commands. Controversially he suggests it should even be allowed to kill humans in certain circumstances. However his account of moral competence is an analysis of what would be required, not an implementation, and it does not provide any working simulations or examples of moral robots. How might the components of moral competence be acquired by robots? Malle and Scheutz (2014) suggest that it might be necessary to raise robots in human environments, since this 'may be the only way to expose them to the wealth of human moral situations and communicative interactions' (ibid p. 34).

Russell (2016) also advocates a related training method when he considers the possibility of super intelligent machines, and the need to ensure that their goals do not conflict with those of humans. He suggests that by following three principles this should be possible: (i) 'the machine's purpose must be 'to maximize the realization of human values' (ibid p. 59) (ii) the machine must be 'initially uncertain about what those human values are' (ibid p. 59) and (iii) the machine must 'be able to learn about human values by observing the choices that we humans make' (ibid p. 59). He suggests that Inverse Reinforcement learning could be used to allow the machine to infer human values from observations of human actions. He does admit that some humans would form poor role models, and that humans exhibit diverse sets of values. As well as directly observing human behavior, he also suggests that machines could be given access to 'vast amounts of written and filmed information about people doing things (and others reacting)' (ibid p. 59).

Suggestions such as these, that robots could be trained or 'raised' to develop human values, tend to be made in very general terms. There are very few examples where some form of training or evolution has been used to train a robot, or a computer, to develop some aspect of moral competence. Riedl and Harrison's paper (Riedl and Harrison 2015) is an exception that presents preliminary results from a study exploring the possibility of a machine learning the norms of moral behavior from stories. They describe their goal as being one of ensuring 'value alignment', which they define as 'a property of an intelligent agent indicating that it can only pursue goals that are beneficial to humans' (ibid p. 1). They argue that rather than programming such values into a computational system, value alignment could be better achieved by reading stories, and reverse engineering the values that underlie them. They admit that 'how to extract sociocultural values from narratives and construct a value-aligned reward system remains an open research problem' (ibid p. 4), but report a study in which stories are generated via crowd sourcing that pertain to the situation and behavior that they want their virtual agent to perform.

In their preliminary study, a plot graph is learnt from the generated stories, and then a trajectory tree is developed that indicates all legal transitions from one plot point to another. The story-reading agent receives a reward every time it performs an action in the environment that is a successor of the current node in the trajectory tree, and a punishment for any action that is not a successor of the current node. The situation they consider is one in which an agent must acquire a drug to cure an illness and return home, in a scenario they term 'Pharmacy World'. In this story world, the computer essentially learns to avoid the bad action of stealing the drugs instead of obtaining them by legitimate means as a result of the rewards associated with following the steps in the trajectory tree. They acknowledge that their system is some way from being one that could be scaled up for a system of general artificial intelligence, and that it is dependent on the content of the generated stories. They suggest that a more general solution to value alignment could be achieved by using all the stories associated with a given culture, assuming that subversive texts will be washed out by those that conform to social and cultural norms.

Riedl and Harrison's work provides an indication of how value alignment might be achieved by reading stories. At the same time, as with many of the examples considered so far, the actual progress towards this goal that is evident in their paper is extremely limited: One scenario, some automated learning, but also a dependence on human intervention to select the scenario and to determine the reward schedule. The question of whether it would be at all practical to scale this up to a general system for learning moral value is not given a clear answer here.

Can robots be ethical?

As well as efforts to program, or to develop moral competence in robots and machines, another way of approaching the issue is to consider whether, or to what extent, robots could ever be full ethical agents. Peter Asaro's (Asaro 2006) contribution here is to reject any strict division between full moral agents and other agents. He proposes that 'it will be helpful to think of moral agency as a continuum from amorality to fully autonomous morality.' (ibid p. 11). He suggests that the simplest way of getting robots to make moral decisions would be for them to randomly choose between a number of alternatives. Or they could be programmed to make decisions on the basis of a set of moral principles instantiated in the form of rules. Or, at another level of sophistication, they could be programmed to learn such a set of principles, and even to evolve their own ethical systems. None of these would mean that they should be considered to be fully autonomous moral agents. That, he argues, would require them to have further abilities such as 'consciousness, self-awareness, the ability to feel pain or fear death, reflexive deliberation and evaluation of its own ethical system and moral judgments' (ibid p. 11).

Wallach and Allen (2009) also divide up the space of artificial moral agents (AMAs), distinguishing between 'operational morality' and 'functional morality'. Operationally moral systems depend on their designers and users for any moral significance, and have little autonomy or sensitivity to morally relevant facts. As machines become more sophisticated they may achieve 'functional morality', and have the capacity to assess and respond to moral challenges. The distinction between the two is not clear-cut. They refer to two dimensions of AMA development: autonomy and ethical sensitivity. Systems corresponding to operational morality are lower in autonomy and ethical sensitivity than those corresponding to functional morality. The highest levels of autonomy and ethical sensitivity belong to systems with full moral agency, and Wallach and Allen (2009) are clear 'that humanity does not have such a technology'. They seem uncertain whether or not it will in the future, although they state that there are no proven limits to the abilities of AMAs. They write that 'whether computer understanding will ever be adequate to support full moral agency remains an open question' (ibid p. 69).

John Sullins seems happier to accept the notion that robots could be full moral agents. He (Sullins 2006) claims that a robot can be a full moral agent if (i) the robot is 'significantly autonomous' (ii) the robot's behavior is intentional and (iii) the robot is in a position of responsibility. His requirements for autonomy and intentionality are uncomplicated. By autonomous, Sullins means that the robot should not be under the direct control of a human, and that it should have a practical independent agency. For intentionality, he refers to behaviour that is 'complex enough that one is forced to reply on standard folk psychological notions of predisposition or 'intention' to do good or harm', (ibid p. 28) and where the interaction between the robot's programming and the environment results in actions that are seemingly 'deliberate and calculated'. Sullins considers a robot to be a moral agent when it 'behaves in such a way that we can only make sense of that behaviour by assuming it has responsibility to some other moral agent(s)' (ibid p. 28). His argument is based on Floridi and Sanders (2004) and their assertion that when viewed at the appropriate level of abstraction an artificial agent can be considered a moral agent. Sullins does not consider current robots to be the moral equals of humans, but advocates paying attention to on-going developments in this area.

Johnson and Miller (2008), by contrast, do not consider Floridi and Sanders' arguments about levels of abstraction to be decisive. For them, there is 'no preexisting right answer to the question whether computer systems are (or could ever be considered to be) moral agents; there is no truth to be uncovered, no test that involves identifying whether a system meets or does not meet a set of criteria' (ibid p. 123); they write here of computer systems, but their discussions apply equally as well to robots. Instead they frame the debate as an argument between two distinct groups of scholars or researchers with different underlying motivations. The first group they call 'Computational Modelers'. They characterize Computational Modelers as being committed to establishing the validity of computational modeling. Those in the computational modeling camp believe that giving computer systems (or robots) the status of moral agents will further endorse the approach.

Johnson and Miller (ibid*)* distinguish the Computational Modelers from the 'Computers-in-Society' group. According to them, this group is against ascribing the status of moral agent to any computer system on the grounds that doing so is dangerous. It is dangerous because it distances human developers, owners, and users, from their responsibility for the robots or computer systems that they have developed or deploy. For those in this group it is important to emphasize the connection between humans and the technology they develop.

Johnson (2006) also argued that computer systems (and robots) should be viewed as moral entities, but not as moral agents. Her argument is extensive and based on the idea that, although the actions of computer systems can have moral consequences, these necessarily involve the intentions of humans. A computer system does not have the same freedom to act based on intentions that humans have. She gives the example of a landmine, which once placed in the field, is distant from the humans who designed and built it, and from those who placed it there. The landmine will be triggered when stepped on. Nonetheless, the landmine is only there as the result of human activity, and the humans involved in its deployment are morally responsible. Even if the landmine were replaced by something more sophisticated that made a decision about whether or not to detonate based on an assessment of the surrounding situation, humans would still be implicated in developing the rules that determine that decision. For Johnson, the artifact itself, the artifact designer, and the artifact user, together form a moral entity that can be morally evaluated. A related argument is made by Hew (2014), who argues that 'For an artificial agent to be morally praiseworthy, its rules for behavior and the mechanisms for supplying those rules must not be supplied entirely by external humans' (ibid p. 197). He claims that this is not technologically feasible for foreseeable artificial agents, and that systems based on technologies such as machine learning, evolutionary computing and self- organisation are all dependent on rules supplied by humans.

Johnson and Miller's argument rings true to the present author, and provides some explanation of the reasons why some writers and researchers are keen to believe that robots are, or could become, full moral agents, and why some are against the idea. In a similar way, there are those who insist that there is no in principle reason why it should not be possible one day to build robots that can feel pain and have emotions. And there are those who do not believe in this possibility. It does seem to come down to a belief, since there is certainly little in the way of tangible evidence that this will ever happen.

One reason for being skeptical about the likelihood that non-living, non-biological machines could develop a sense of morality at some point in the future is their lack of a biological substrate. A case can be made for the grounding of morality in biology. For instance, Churchland (2011) argues that morality in humans and mammals depends on their biology and is grounded in their ability to care for kith and kin; their recognition of other's psychological states; problem solving in a social context; and learning social practices. The basis for caring, she writes, lies in the neurochemistry of attachment and bonding in mammals. Humans and other mammals extend their self-maintenance and avoidance of pain in mammals to their immediate kin; feeling 'anxious and awful' when either their own well-being is threatened, or the well-being of their loved ones. They also feel pleasure when their infants are safe, and when they are in the company of others. These emotions form the basis for more complex social relationships, grounded in the rewarding pleasures of approval and belonging, and the 'generalised pain of shunning and disapproval' (Churchland 2011, p. 131), and the internalization of social standards.

An argument for a biological basis for morality implies that existing robots lack the biological basis for the development of morality. Current robots, lacking living bodies, cannot feel pain, or even care about themselves, let alone extend that concern to others. How can they empathise with a human's pain or distress if they are unable to experience either emotion? Similarly, without the ability to experience guilt or regret, how could they reflect on the effects of their actions, modify their behavior, and build their own moral framework?

How crucial are emotions and empathy for the development of morality? Docherty (2016) has argued that robots should not be allowed to make an autonomous kill decision in battle because they lack both empathy and emotion. Robots, she claims 'lack real emotions, including compassion', and 'could not truly understand the value of any human life they chose to take'. By contrast, because they possess empathy, 'people can feel the emotional weight of harming another individual', and refrain from unjustified killing. She argues that humans are able to make the judgments about proportionality that are required by the laws of war. Humans can apply judgment based on their past experience and moral consideration to assess the necessity of an attack, but Docherty thinks it unlikely that robots could be preprogrammed to do so, or that they would be able to reason about unanticipated scenarios.

Docherty (2016) describes robots and robot weapons as lacking both emotions and empathy. However Prinz (2011) questions the extent to which empathy is required for morality, and claims that empathy itself is not very motivating, and that it is subject to bias. According to Prinz, empathy is not necessary for making moral judgments, or for moral development, or for motivating moral conduct. Sentiments such as disapprobation, or emotions such as anger are more likely to form the basis for moral judgments about offensive behavior. The point is sometimes made that psychopaths lack empathy, and that they are also deficient in moral reasoning. But, as Prinz points out, psychopaths are also characterized by other emotional deficiencies, such as a low level of guilt, and an indifference to punishment, and their lack of empathy does not demonstrate its necessity for morality. Prinz's arguments rest on the careful distinctions he makes between empathy and emotions. Nonetheless, Prinz (2011) is clear that moral judgment, moral development and moral motivation do require emotions. And his discussion about the role of empathy does not refute the arguments made by Docherty (2016), since her objections to robots making life or death decision are primarily based on their general lack of emotions in general, and their inability to understand the value of human life.

Responding to the current situation

In the research we have looked at here, there is general agreement that current robots are not yet full moral agents. There is some disagreement about whether they could ever become so in the future. The situation is complicated by developments in robotics that make it increasingly possible to develop robots that look and behave in ways that create and encourage the illusion that they are able to understand and relate to humans. There is also a strong tendency to use terms to describe robots and computational agents that strongly imply that they are already ethical beings and moral agents.

Van Wynsberghe (2016) expresses her concern 'that robots are being built with at least the appearance of moral agency and that they are being placed into inherently ethical contexts', (ibid p. 313), and argues that such robots (in her case, service robots) demand ethical evaluation and reflection. She is also concerned about the use of ethically charged words (e.g. trusting, emotional attachment, socialising) to describe robots and their behaviour.

Noel Sharkey expresses similar concerns (Sharkey 2012) when he writes about the application of ethical terms to robots and machines. He talks about the way in which applying terms such as 'ethical' or 'humane' to machines leads to false attributions of abilities to them: 'They act as linguistic Trojan horses that smuggle in a rich interconnected web of human concepts that are not part of a computer system or how it operates. Once the reader has accepted a seemingly innocent Trojan term .... it opens the gate to other meanings associated with the natural language use of the term' (ibid p. 793). For instance, when Arkin writes that robots 'can be more humane in the battlefield than humans' (Arkin 2009, p. 30), his language implies that robots are capable of kindness, mercy and compassion.

Miller et al. (2016) were disturbed by the use of the word 'ethical' to describe the hole-rescuing robots of Winfield et al. (2014), and by the subsequent journalistic reporting of the study in the New Scientist by Rutkin (2014). As well as pointing out that inaccurately describing robot behaviour as ethical decision making is 'more likely to confuse than educate', (Miller et al. 2016, p. 392) they also set out some requirements for ethical decision making. They propose that ethical decision-making requires an 'openness to self-doubt' that they term an 'elenchus experience' with reference to Reed, (2013). They argue that for a machine to be considered to be capable of ethical decision-making, it needs to have '(i) a capacity to sense some aspects of the outside world (ii) an implementation of a function of merit that quantifies the acceptability of the current situation and (iii) a capacity to reprogram itself in order to improve performance in future situations.' (Miller et al. 2016, p. 393).

The elenchus experience of a machine may not be the same as a human elenchus experience. But according to Miller et al. (2016), if the machine or robot cannot reconsider a decision after it has been made in order to lead to better decision making in the future, it should not be considered to have made an ethical decision. To be described as ethical, a machine should be developing, or trying to develop, ethical expertise.

Miller et al. (2016) apply this definition of ethical decision making to the hole avoiding robots described by Winfield et al. (2014), pointing out that the robots cannot reconsider their actions given their effects, and cannot reprogram or adjust them in order to achieve a better outcome in the future. Winfield's robots have no ability to maintain a record of past decisions and their outcomes, or to reflect on these. The 'decisions' made by the robots are clearly the result of their programming, as we have already discussed. Miller et al. argue that terms such as 'ethical' should not be used without better justification to describe the apparent behaviour of a robot. Instead any description should be 'as simple as possible, making as few assumptions as possible about the capabilities of the AA [artificial agent]'(Miller et al. 2016, p. 400).

Of course, if robots are to be used in situations that impinge on humans, it is important to take steps to ensure that they will not harm them. But given these points about the misuse of ethically charged words, perhaps the robot programming undertaken by Winfield would be better described as addressing safety concerns, rather than as creating 'minimally ethical robots'. Similarly, the Andersons describe their systems as involving explicit ethical agents, but their work might be better, and more prosaically, described as being about patient reminders. In the case of Arkin's ethical governor, which involves harm to humans, perhaps his system would be better described as a military situation advisor.

An important reason for being careful about the use of language to describe the operations and underlying mechanisms of robots and computers is the need to remain aware of the unavoidable human involvement and responsibility highlighted by some (Johnson 2006; Johnson and

Miller 2008; Hew 2014). As Johnson (2006) convincingly argues, even if a computational artefact is placed in a situation in which it is required to make decisions with moral consequences, the responsibility for such decisions still rests with the humans and the society that developed them and decided to deploy them there. Describing such machines as being moral, ethical, or humane, risks increasing the tendency for humans to fail to acknowledge their ultimate responsibility for the actions of these artefacts. It could encourage the use of inappropriate use of machines to make morally sensitive decisions that affect humans when they lack the moral competence that such decisions require.

An important component of undertaking a responsible approach to the deployment of robots in sensitive areas then is to avoid the careless application of words and terms used to describe human behaviour and decision making to robots. If those writing about robots were to eschew, or at least limit, the use of terms such as 'moral', 'ethical', 'humane', and 'caring' in their accounts, it would be easier to clearly assess their current abilities.

It is relevant at this point to question here the difference between the idea of robots making decisions in circumstances that require moral competence, and our increasing reliance on automatic and algorithmic decision-making. There are many crucial issues about this reliance that need to be addressed but that are beyond the scope of the present article (see Carr 2015; Susskind and Susskind 2015). However there are some important differences between the uses of robots in social roles, and the use of non-embodied computational systems. The robot in the classroom, or the robot on the battlefield, may be required to make decisions that require an understanding of the surrounding human social context. The inputs to its decision-making would be based on the information gathered from its sensors, and on its interpretation of the social meaning of that information. This is quite different from a medical decision maker or advisor that is fed information off-line, and that does not have to rely on real-time interpretation of a social situation. Circumstances that require morally competent decision makers are those for which there is some ambiguity, and a need for a contextual understanding: situations in which judgment is required and there is not a single correct answer.

The idea that the circumstances in which morally competent decision makers are required are those in which there is some ambiguity about what the right decision is raises some questions about some of the examples we have considered here that are described as requiring *ethical*decisions. Moor (2006) described an ATM that dispenses the right amount of money as an implicit ethical agent: but is there any ambiguity here about whether or not the right amount of money should be dispensed? Likewise, when a robot prevents a proxy human robot from falling into a hole, is this an ethical, or a safety decision? Is the auto-pilot of an aeroplane making ethical decisions when it oversees a smooth take-off and landing? None of these examples involve the interpretation of a human social situation: instead they involve an understanding of the physical surroundings (in the case of hole-avoidance and flying), or of accurate data checking (in the case of the ATM). The situations that require a morally competent decision maker seem different to these.

Given these deliberations, and given what seems to be a general agreement that robots are not yet full moral agents, we turn now to a consideration of what would be the responsible way to respond. There seem to be two main alternative responses. Response 1 advocates the need to work towards the development of robots that have *some* level of ethical ability. Response 2 is to make efforts to prevent or dissuade people from deploying robots in situations and roles in which moral decisions are required. We will examine and evaluate both of these responses in turn.

Response 1: building 'ethical' robots

There are some authors who consider it both important, and possible, to develop robots and machines with some degree of ethical behaviour. For instance, Wallach (2010) advocates the building of 'moral machines' as a practical goal, motivated by 'the need to ensure that increasingly autonomous machines will not cause harm to humans and other entities worthy of moral consideration' (ibid p. 243). He suggests that artificial moral agents (AMAs) will continue to be developed for practical applications over a long period of time, and that testing these systems will enable an understanding of the limits of the implemented mechanisms. For instance, he proposes that the limitations of a system that lacks specific mechanisms such as emotions, a theory of mind, or consciousness will become apparent to engineers when they are tested and found not to be 'sufficiently sensitive to moral considerations essential for making judgments in certain situations'.

The idea that the limitations of such systems should be found by testing them seems unconvincing to the present author. Apart from anything else, the developer of an AMA is likely to be more concerned with showing its strengths than in finding its limitations. The limitations of a given system or robot, or of robots in general, are also usually easily identifiable without the need for actual testing. For example, Winfield's robots are able to prevent other robots from falling into a hole. But it would make little sense to test the robots to see if they were able to prevent other robots from, for instance, running out of energy, because that is not what they were programmed to do. It also seems unnecessary to actually build a childcare robot without phenomenal consciousness or emotions in order to demonstrate that the children left in its care for long periods start to exhibit dysfunctional behaviour and attachment problems. The question of whether or not robot Nannies are a good idea is one that can be considered and answered on the basis of knowledge about the current abilities of robots; it doesn't need a practical (and potentially risky) demonstration.

Moor (2006) was also, as we have seen, keen on the idea of developing ethical agents even if they always fall short of what is needed for a full ethical agent. He argued that it is important to examine the other types of moral agent he identified (ethical impact agents, implicit ethical agents and explicit ethical agents), and especially encouraged the development of explicit ethical agents because of the need to ensure that 'machines treat us well'. He also conjectured that programming or teaching ethics to a machine would improve our understanding of ethics.

Our understanding of ethics is indeed likely to be improved as a consequence of attempts to teach or program ethics into machines. This is also the case in many other domains of computational modeling where improvements in understanding have been gained as a result of having to be more explicit about the assumed underlying mechanisms. However, if we look at the current state of AMAs and 'minimally ethical' robots, and compare it to what is known about human moral abilities there seems to be an insurmountable gap between the two. As reported earlier, Malle (2015) provided a useful outline of what is required for moral competence (a moral vocabulary; a system of norms; moral cognition and affect; moral decision-making and action; and moral communication). There is no evidence of an artificial system that has come any way close to achieving such competence. In addition, the systems that have been developed to date are all severely limited in scope to a particular domain.

Despite the current level of progress and achievements in this area, there are still those who continue to advocate the deployment of robots in situations in which moral decision-making will be required. For example, Arkin (2009) has argued that robots will be able to make more ethical decisions in the fog of war than humans. When it is pointed out that current systems do not have sufficient understanding of the human situation and context, the argument is sometimes made that there is no reason in principle to expect that they will not be able to develop this, and that given time, they will.

Response 2: limiting robot use

While some like to believe that at some time in the future robots and machines will become sentient, conscious, and able to understand the human world, there are others (including the present author) who prefer to focus instead on their actual capabilities. Currently existing robots are neither sentient or conscious, nor capable of understanding the complexities of social situations involving humans. They are also unlikely to become so in the near future: a statement that cannot be proved, but for which there is little convincing counter evidence. Given this, it is argued here that the responsible approach should be to identify those situations in which robots should not be deployed, and the social roles that they should not be given. This again is the contention of the present author.

There are other writers who are beginning to suggest this. Most prominent are those who are writing about the use of robots in warfare, and arguing against the deployment of lethal autonomous weapons, where the robot, machine, or weapon, makes 'decisions' about who to kill without human supervision. For example, Christof Heyns (2013), the UN Special Rapporteur on extrajudicial, summary or arbitrary executions has argued that robots should not be allowed to make lethal decisions on the battlefield, on the basis that they lack 'human judgment, common sense, appreciation of the larger picture, understanding of the intentions behind people's actions, and understanding of the values and anticipation of the direction in which events are unfolding' (2013, A/HRC/23/47).

Similar concerns have also been raised about the use of robots by the police (Sharkey 2016). Then there are authors who have looked at the use of robots for the care of older people (Sharkey and Sharkey 2012; Sparrow and Sparrow 2006; Coeckelberg 2010; Vallor 2011), and raised concerns about the extent to which robots can care for and respond to them in the way that human carers do. Likewise, concerns have been raised about the use of robots as nannies, carers, and teachers of children (Sharkey and Sharkey 2010; Sharkey 2016). In a similar vein, Sherry Turkle has written persuasively about her concerns about people developing and being encouraged to develop, relationships with computational artifacts that 'cannot love you back' (Turkle 2011).

Robots that are tasked with killing people are clearly treading on ethical territory. It is less obvious that this is the case for robots that are developed for the care and supervision of children, older people, or as companions. But how could a robot make appropriate decisions about when to praise a child, or when to restrict his or her activities, without a moral understanding? Similarly how could a robot provide good care for an older person without an understanding of their needs, and of the effects of its actions? Even a bar-tending robot might be placed in a situation in which decisions have to be made about who should or should not be served, and what is and is not acceptable behaviour. All of these seem to the present author to require both moral understanding and moral competence.

Saying that there are some situations in which robots should *not* be used is not the same as being overly negative about robot use. There are many situations in which robots can offer people something that would not otherwise be available (Sharkey 2014). The challenge is to find the right path to steer between capitalising on and benefitting from the unique opportunities that robots can offer, and avoiding a future in which robots are placed in positions and roles that require a moral understanding that they do not have.

CONCLUSIONS

In this paper, we have examined the progress made towards developing moral robots. We have seen how some have taken the route of attempting to program robots to be good. Others have proposed training, or raising, robots to develop moral understanding, or moral competence. The progress along both roads has been limited. Systems that have been either programmed or trained have so far been successfully applied only in quite narrow and specific domains.

We have considered some of the debates about the extent to which robots could ever be full moral agents. There are those who are skeptical about the possibility that robots could ever be said to be moral. Nonetheless, there are several writers (Asaro 2006; Moor 2006; and; Wallach and Allen 2009) who have looked at the possibility of developing robots, which, while not being full ethical agents, exhibit some level of ethicality. For instance, Moor distinguishes between ethical impact agents, implicit ethical agents, explicit ethical agents and full ethical agents.

The need to limit the unjustified use of terms such as moral and ethical to robots has been highlighted here. The circumstances in which morally competent decision-makers are needed have also been discussed. In addition, in the light of the current progress, (or *lack* of progress) towards the development of robots that are full moral agents, or explicit ethical agents able to reason about, and reflect on their decisions, two responses to the current situation are identified here: (1) building 'ethical' robots and (2) limiting robot use.

Those advocating the first response who are interested in working towards the development of 'minimally ethical' robots, or explicit ethical agents, do not necessarily fall in the camp identified by Johnson and Miller (2008) as 'Computational Modellers'. Indeed, they are often motivated by the need to ensure the safety of humans as robots increasingly work near them, or with them, or are even placed in charge of them. An advantage of this response is that it is likely to advance our understanding of moral decision making in general, even if the ultimate goal of an artificial moral agent is never achieved. However it is argued here that their work could often be better described as having the goal of developing safe robots, than as developing ethical robots.

Those commending the second response of Limiting robot use, are likely to feel an affinity with others in the Computers-in-Society group identified by Johnson and Miller (ibid). Given the gap between current robot abilities, and those required for full moral agency, it is important to recognise that humans remain responsible for any deployments of robots in morally sensitive domains. Humans should not offload their responsibility for the effects of robot actions onto the robots that carry them out. It is also crucial that, recognising this responsibility, steps are taken to anticipate the potential negative effects of placing robots in situations where moral decisions are required, and that efforts are made to restrict their use. Appropriately developed and deployed robots have the potential to bring many benefits to human society, but the responsible robotics approach should have the aim of limiting their incursions into morally sensitive situations before it is too late.

REFERENCES

1. Altmann, J., Asaro, P., Sharkey, N., & Sparrow, R. (2013). Armed Military Robots: Editorial. *Ethics and Information Technology, 15*(2), 73–76.
2. Anderson, M., Anderson, S., Armen, C. (2006). MedEthEx: A prototype medical ethics advisor. In *Proceedings of the eighteenth conference on innovative applications of artificial intelligence*. Menlo Park, CA: AAAI Press.
3. Anderson, S., & Anderson, M. (2007). Machine ethics: creating an ethical intelligent agent. *AI Magazine, 28*(4), 15–26.
4. Arkin, R. C. (2007). *Governing lethal behaviour: Embedding ethics in a hybrid deliberative/reactive robot architecture*. Atlanta: Georgia Institute of Technology.
5. Arkin, R. C. (2009). *Governing lethal behavior in autonomous robots*. Boca Raton, FL: CRC Press.
6. Asaro, P. (2012). On banning autonomous lethal systems: human rights, automation and the dehumanizing of lethal decision-making. *Special Issue on New Technologies and Warfare, International Review of the Red Cross 94*(886), 687–709.
7. Asaro, P. M. (2006). What should we want from a robot ethic? *International Review of Information Ethics, 6*, 9–16.
8. Asimov, I., (1942, March) Runaround. *Astounding Science Fiction 29*, 94–103.
9. Beauchamp, T. L., & Childress, J. F. (1979). *Principles of Biomedical Ethics*. Oxford: Oxford University Press.
10. Bostrom, N. (2014) *Superintelligence: Paths, dangers, strategies*. Oxford: Oxford University Press.
11. Carr, N. (2015). *The glass cage: Where automation is taking us*. London: Bodley Head.
12. Churchland, P. S. (2011). *Braintrust: What neuroscience tells us about morality*. Oxford: Princeton University Press.
13. Coeckelbergh, M. (2010). Health care, capabilities, and AI assistive technologies. *Ethical Theory and Moral Practice, 13*(2), 181–190.

14. Docherty, B. (2016) Losing control: The dangers of killer robots. *The Conversation*, June 16th, 2016.
15. Floridi, L., & Sanders, J. W. (2004). On the Morality of Artificial Agents. *Minds and Machines, 14*(3), 349–379.
16. Hew, P. C. (2014). Artificial moral agents are infeasible with foreseeable technologies. *Ethics and Information Technology, 16*, 197–206.
17. Heyns, C. (2013). Report of the Special Rapporteur on extrajudicial, summary or arbitrary executions, A/HRC/23/47. NewYork: United Nations
18. Johnson, D. G. (2006) Computer Systems: Moral entities but not moral agents. *Ethics and Information Technology*, 8(4), 195–204
19. Johnson, D. G., & Miller, K. W. (2008). Un-making artificial moral agents. *Ethics and Information Technology, 10*, 123–133.
20. Lin, P. (2013). The ethics of autonomous cars. *The Atlantic*, October 8th 2013. http://www.theatlantic.com/technology/archive/2013/10/the-ethics-of-autonomous-cars/280360.
21. Lin, P. (2015). Why ethics matters for autonomous cars. In M. Maurer, J. C. Gerdes, B. Lenz, H. Winner (Eds.), *Autonomes Fahren: Technische, rechtiliche und gesellschaftliche aspekte* (pp. 69–85). Berlin Heidelberg: Springer.
22. Malle, B. F. (2015). Integrating robot ethics and machine morality: the study and design of moral competence in robots. *Ethics and Information Technology*. DOI:10.1007/s10676-015-9367-8.
23. Malle, B. F., & Scheutz, M. (2014, June). Moral competence in social robots. IEEE international symposium on ethics in engineering, science, and technology. *Presented at the IEEE international symposium on ethics in engineering, science, and technology* (pp. 30–35). Chicago, IL: IEEE.
24. Matthias, A. (2011). Algorithmic moral control of war robots: Philosophical questions. *Law, Innovation and Technology, 3*(2), 279–301.
25. Miller, K. W., Wolf, M. J., & Godzinsky, F. (2016). This "ethical trap" is for roboticists, not robots: on the issue of artificial agent ethical decision-making. *Science and Engineering Ethics*. doi:10.1007/s11948-016-9785-y.
26. Moor, J. H. (2006). The nature, importance, and difficulty of machine ethics. *IEEE Intelligent Systems*, 21, 18–21. doi:10.1109/MIS.2006.80.
27. Moor, J. H. (2007). Four kinds of ethical robot. *Philosophy Now, 72*, 12–14.
28. Murphy, R. R., & Woods, D. D. (2009). Beyond Asimov: The Three Laws of Responsible Robotics. *IEE Intelligent Systems, 24*, 14–20.
29. Prinz, J. J. (2011) Is empathy necessary for morality? In A. Coplan & P. Goldie (Eds.) *Empathy: Philosophical and psychological perspectives*. Oxford: Oxford University Press, pp. 211–229.
30. Reed, R. C. (2013). Euthyphro's elenchus experience: Ethical expertise and self-knowledge. *Ethical Theory and Moral Practice, 16*, 245–259. doi:10.1007/s10677-012-9335-x.
31. Riedl, M.O., & Harrison, B. (2015) Using stories to teach human values to artificial agents. Paper presented at the *2nd international workshop on AI, ethics, and society*. http://www.aaai.org.
32. Ross, W. D. (1930). *The right and the good*. Oxford: Clarendon Press.
33. Russell, S. (2016, June) Should we fear supersmart robots? *Scientific American*, 314, 58–9.
34. Rutkin, A. (2014, September) Ethical trap: robot paralyzed by choice of who to save, *New Scientist*. Amsterdam: Elsevier.
35. Sharkey, A. (2014). Robots and human dignity: The effects of robot care on the dignity of older people. *Ethics and Information Technology, 16*(1), 53–75.
36. Sharkey, A. (2016). Should we welcome robot teachers? *Ethics and Information Technology, 18*(4), 283–297.
37. Sharkey, A. J. C., & Sharkey, N. E. (2012). Granny and the robots: Ethical issues in robot care for the elderly. *Ethics and Information Technology, 14*(1), 27–40.

38. Sharkey, N. (2012). The evitability of autonomous robot warfare. *International Review of the Red Cross, 94*(886), 787–799.
39. Sharkey, N. (2016) Policing with Robots, *Open Rights Group (Ed) 10×10: Digital rights, the next decade*.
40. Sharkey, N. E., & Sharkey, A. J. C. (2010). The crying shame of robot nannies: An ethical appraisal. *Interaction Studies, 11*(2), 161–190.
41. Sorrell, T., & Draper, H. (2014). Robot carers, ethics and older people. *Ethics and Information Technology, 16*(3), 183–195.
42. Sparrow, R., & Sparrow, L. (2006). In the hands of machines? The future of aged care. *Mind and Machine, 16*, 141–161.
43. Sullins, J. P. (2006). When is a robot a moral agent? *International Review of Information Ethics, 6*(12), 23–30.
44. Susskind, R., & Susskind, D. (2015) *The Future of Professions: How technology will transform the work of human experts*. Oxford: Oxford University Press.
45. Turkle, S. (2011). *Alone together*. New York: Basic Books.
46. Vallor, S. (2011). Carebots and caregivers: Sustaining the ethical ideal of care in the twenty-first century. *Philosophy and Technology, 24*(3), 251–268.
47. Van Wynsberghe, A. (2016). Service robots, care ethics, and design. *Ethics and Information Technology, 18*(4), 311–321.
48. Vanderelst, D. & Winfield, A. (2016) The dark side of ethical robots, ARXiv:1606.02583v1 [cs.RO} 8 June 2016.
49. Wallach, W., & Allen, C. (2009). *Moral machines: Teaching robots right from wrong*. New York: Oxford University Press.
50. Wallach, W. (2010). Robot minds and human ethics: The need for a comprehensive model of moral decision making. *Ethics and Information Technology, 12*, 243–250. doi:10.1007/s10676-010-9232-8.
51. Winfield, A. F. T., Blum, C., & Liu, W. (2014). Towards an ethical robot: Internal models, consequences and ethical action selection. In M. Mistry, A. Leonardis, M. Witkowski & C. Melhuish (Eds.), *Advances in autonomous robotics systems* (pp. 85–96). Berlin: Springer International Publishing.

CHAPTER 15

The Evolution of Intellectual Property Protection in China

Lu Chen

School of Management, Jinan University, Guangzhou, China

ABSTRACT

This paper developed an in-depth understanding of China's IPP strategies, regulations, and implementation, collected and analyzed China's policy documents, laws, statistics and reports relevant to IPP, and reviewed the IPP evolution from 1970s to 2010s. Then we introduced four main enforcement options in China. At last, we evaluated the IPP level of China based on Park's method from 1985-2012. The quantitative result shows that the strength of IPP in China is growing year by year, and has reached a level close to developed countries.

Keywords: Intellectual Property Protection, Evolution, Assessment, China

1. INTRODUCTION

The intellectual property protection (IPP) in China aroused heated discussion among people all around the world. Intellectual property owners around the world have long complained about the inefficacy of intellectual property protection in China and have gone so far as to accuse the Chinese government.

It is questionable that why Chinese people are resistant to the concept of IPP? Is that correct? Will it change ever time?[1] This paper gives a descriptive review and assessment of China's intellectual property policies, in particular relating to implementation.

China is often criticized by its poor record in the protection of IP rights [1] - [4]. However, an examination of China's IP laws and the international treaties it has acceded to would put China at more developed economies.

Effective protection of intellectual property rights depends on both the existence of IP laws and the enforcement of the laws. Although much has been written about the IP rules and laws [5] [6] , there is little empirical evidence of the importance of the enforcement. One reason is that studies of IP protection are generally performed at country level.

Country level analysis does not allow researchers to separate the confounding effect of the existence of IP laws and the effectiveness of enforcement. Unlike dealing with cross country variations in IP laws, Ang, Cheng and Wu analyzed the impact of the local enforcement of IP laws on financing of and investing in R & D by firms in provinces throughout China [7]. They found that enforcement in the protection of intellectual properties did matter in China. High tech firms in provinces with better IP rights enforcement have greater access to external debt (formal and informal).

The rest of this paper proceeds as follows: Section 2 traces the development of IPP in China from 1970s to 2010s; Section 3 represents four main enforcement options in China; Section 4 measures the IPP of China from 1985-2010 based on the method of Park (2008); Section 5 makes a short conclusion.

2. THE FOUR PERIODS OF IP "HOT WAVES" IN CHINA

The rush establishment and rapid development of the Chinese IP system is strongly related to the four periods of IP "hot waves" in China [8] [9]. Initially, during the negotiations of the Sino-US Trade Agreement and the Sino- US High Energy Physics Agreement, IP became an objectionable issue and negotiations reached an impasse. The US negotiators believed that IPP should be an integral part of bilateral agreements in science, technology and trade, as negotiators would be only allowed to sign an agreement on the condition of adequate IPP required by the US president.

The process of development of IPP in China is quite tortuous. What follow in this part are the four period of IP wave and the revolution of the IPP in China from 1973-2014.

2.1. The First "IP Fever" (1979)

The Chinese government had little knowledge and experience about the IP clauses in the early days; therefore, they were very laith to sign the agreements with US. This period has been described as the first "IP fever" because it was the start point of China to conduct very intensive research in this special area [9]. The current authors name this as the first "hot wave" on IP in China—the formation wave.

1973, Premier Minister Zhou's visit to WIPO

In 1973, the approval of Premier Zhou Enlai and the China Council For The Promotion Of International Trade (CCPIT)'s Director of Legal Affairs visited WIPO. It was the first time that the Chinese government allowed an officer to visit an international intellectual property institution since the foundation of the state.

1974, the first formal activity of IPR

The first formal activity of international intellectual property of China occurred in 1974 when the governments of China and Australia signed an exchange of notes regard the registration of trademark between both nations.

1979 Agreement

Intellectual property rights (IPRs) have been acknowledged and protected in the People's Republic of China since 1979. The Agreement on Trade Relations Between the United States of America and the People's Republic of China of 1979 ("1979 Agreement") was the starts of intellectual property protection by Western.

1980, access of WIPO

China joined WIPO as its 90rd member in 1980. Only one year after the acceding to WIPO, in 1981, the Director of WIPO Dr. Arpad Bogsch visited China, convened the first patent agents course personally at the CCPIT Headquarters in Beijing.

1984, the first patent law

In March 1984, China's National People's Congress promulgated the first Chinese patent law, which is an epoch-making event which opened a new chapter of the protection of intellectual property rights.

1985, SIPO's foundation

The State Intellectual Property Office of the People's Republic of China (SIPO), also known as the Chinese Patent Office, is the patent office of the People's Republic of China (PRC). It was founded on 1980, as the Patent Office of the People's Republic of China, the predecessor of SIPO[2]. It is responsible "for patent work and comprehensively coordination of the foreign related affairs in the field of intellectual property"[3].

1989, joined the Madrid Agreement Concerning the International Registration of Marks

China joined the Madrid Agreement Concerning the International Registration of Marks in 1989[4] the Madrid Agreement Concerning the International Registration of Marks, which dates from 1891, and the Protocol Relating to the Madrid Agreement, which was adopted in 1989, entered into force on December 1, 1995, and came into operation on April 1, 1996. Common Regulations under the Agreement and Protocol also came into force on that date.

2.2. The Second Hot Wave (1991-1994)

1992, Signed MOU with US

Urged by American business executives, the USTR placed China on the "Priority Watch List" in 1989. By doing so, the United States gained leverage in negotiations with China while it did not need to initiate a Section 301 investigation.

In response to the Priority Watch List designation, China passed a new copyright law[5] and issued new implementing regulations in 1990. A separate set of computer software regulations[6] followed in 1991.

Hours before the deadline for imposing sanctions, both countries averted a potential trade war by signing the Memorandum of Understanding Between China and the United States on the Protection of Intellectual Property ("1992 MOU").

2.3. The Third Hot Wave—The Improvement Wave

The third hot wave of IP occurred in 1995 when China became the signatory countries to the TRIPS agreement—"the improvement wave". The third wave reflected a very different picture of IP in China. In the first two waves, China was mainly influenced by the US under its one-sided threats and the treaties and conventions of WIPO. Nevertheless, in the third wave, one-sided threats have been confined under the WTO agreement. TRIPS became the most important measurement on international trade and IP.

1995, agreement between China and United States

The 1995 National Trade Estimate Report estimated that U.S. industries suffered almost $850 million in losses due to copyright theft alone. On June 30 1994, the USTR again designated China a Priority Foreign Country and immediately initiated a Special 301 investigation. According to the Xinhua News Agency, China's international news service, China needed to take such retaliatory measures "to protect its sovereignty and national dignity."[7] Ending up with the signing of "1995 Agreement".

2.4. The Forth Hot Wave—The Enforcement Wave

The fourth wave—the enforcement wave, appeared in 2001 when China access into the WTO after 15 years of consultations. The principal character of this period has been the enhancement of legal enforcement.

2001, access of WTO, TRIPs

In order to gain the access to WTO, China enhanced the protection of IPRs by designing a number of strategies in 1996 APEC Individual Action Plan, including improve the enforcement of administration, enhance the public awareness of IPRs protection and strengthen judicial tools [2].

Finally, China has gained the entry into the WTO in 2001, complying with the minimum standards of the Agreement on Trade-Related Aspects of Intellectual Property Rights (TRIPs). The TRIPs requires a number of countries to strengthen their patent and other intellectual property rights (IPR) systems. Even though policy makers have committed to significant reforms, the implementation of this agreement remains contentious.

2004, SIPO's progress

In 2004, China has become the fourth largest patent office in number of patent filings in the world, the top three countries are, by rank, Japan, the United States, and South Korea[2].

2014, the Beijing WIPO office

To improve cooperation on protection of intellectual property rights (IPR), the WIPO Director-General Francis Gurry came to Beijing in July for opening a WIPO office[8], one of only four globally. Gurry spoke highly of China's support for WIPO, the high importance China attached to innovation and IPR protection and the remarkable achievements it made.

2014, the new strategy

The Promotion Plan for the Implementation of the National Intellectual Property Strategy in 2014 was published on May 5th. This plan is promulgated to implement the requirements of the 18th National Congress of the Communist Party of China (CPC) and the 3rd Plenary Session of the 18th CPC Central Committee, carry out the requirements of the national television and telephone conference on implementing intellectual property strategy, promote the deep and efficient implementation of the national intellectual property strategy, identify the primary goals and measures for the implementation of the strategy in 2014[9].

3. THE ENFORCEMENT OF IPP IN CHINA

The registered intellectual property rights available are invention patents, utility patents, design patents and trade-marks. Copyright can also be recorded[10]. There are four main enforcement options in China [8] :

3.1. Administrative Actions

The key Chinese administrative bodies are the Intellectual Property Offices (IPOs), the Administrations for Industry and Commerce (AICs), the Copyright Office, and the Quality and Technical Supervision Bureaus (QTSBs) (local divisions of the Administration for Quality Supervision Inspection and Quarantine). These are empowered to take certain actions against companies infringing intellectual property rights. These bodies between them have the power to:

- Raid defendants' premises and to seize and destroy infringing items;
- Impose injunctions to force the infringing party to desist;
- Levy fines on the infringing party for TM infringement, copyright infringement and counterfeiting patent certificates.

3.2. Civil Litigation

Civil litigation is equivalent to a court case. The usual remedies sought are injunctions, damages, delivery up and destruction (of tools/products). A civil action will generally take six to 12 months from the issuance of proceedings until handing down of the judgment. Infringing acts under Chinese Patent law include sale, offer for sale, use, manufacture and export. It is usually easiest to obtain evidence of sale or offer for sale, but even this is not entirely straightforward because Chinese courts require this type of evidence of infringement to be given by a notary public.

3.3. Criminal Sanctions

Criminal sanctions are only used in relation to patents where the counterfeiting of the patent certificates themselves has taken place. Such actions are rare; criminal proceedings are more common in relation to trade mark and copyright infringements. There are three methods of bringing criminal sanctions:

- IP owner reports to the Public Security Bureau (PSB);
- An administrative agency transfers its case to a criminal agency when it comes to suspect the damage inflicted by the defendant exceeds certain thresholds;
- A trade mark owner can choose to file a criminal lawsuit with the court known as a private prosecution.

The first two approaches are common whilst the final approach is less so because claimants have no compulsory powers to obtain evidence, unlike the authorities.

3.4. Using Customs to Halt Exports and Gather Evidence

While practice varies across China, many Chinese customs authorities will proactively enforce trademarks registered with them. However, generally they will not do the same for patents nor copyright and therefore patent/copyright owners wishing to have infringing goods seized by customs must inform the customs officers of the precise details of each shipment to be seized (including the container number). In practice, obtaining this information entails extensive use of private investigators.

4. THE MEASUREMENT OF IPP OF CHINA BASED ON THE METHOD OF G. PARK

There are a lot of studies on patent systems, but only few of them have measured the overall strength of the system in each country [10] - [12]. The few available studies are based either on a "dummy variables" approach or on surveys of firms. Examples of the "dummy variables" approach are Bosworth [13] and Ferrantino [14]. Their approach is to develop various indicators of whether certain features of patent laws exist.

They do not provide an integrated index of those indicators. Rapp and Rozek [15] aggregate their indicators (which too are dummy variables: 1 if a feature exists, 0 otherwise). The survey of Mansfield [16] is a typical example by using Rapp's approach, which samples the views of 94 U.S. multinationals of patent rights in 16 countries during 1991.

Ginarte and Park [17] present an index of patent rights for 110 countries for the period 1960-1990. The index differs greatly from previous measures in a number of ways for three reasons. First, it provides information about national patent rights for more countries and periods than do the surveys, which otherwise offer many insights. Secondly, broader categories of the patent system are considered, particularly the treatment of foreigners. Thirdly, the measures obtained in the dummy variable approaches exhibit little variability across countries. Park [18] has updated the index of patent rights to the year 2005 and included more countries (such as China and the East European countries, which were originally excluded because laws protecting industrial property were either non-existent or based on a different system, such as inventor certificates).

Five categories of the patent laws were examined: 1) extent of coverage; 2) membership in international patent agreements; 3) provisions for loss of protection; 4) enforcement mechanisms; and 5) duration of protection. [5] Each of these categories (per country, per time period) was scored a value ranging from 0 to 1, as discussed below. The unweighted sum of these five values constitutes the overall value of the patent rights index.

What follows in this part are the measurement of IPP in China from 1985 to 2010 for easy comparison with Park's result[11].

4.1. Coverage

The value of this category indicates the fraction of 8 elements (pharmaceuticals; chemicals; food; surgical products; microorganisms;

utility models; software; plant and animal) that were specified as being patentable in the law or not.

As the government revised the patent law in 1992, plants and animals are excluded from the scope of patentable subject matters under the patent law of China since Jan. 1st, 1993 (Article 25)[12].

So the value of "Coverage" is 1 from 1985 to 1992, and it is 0.875 since 1993.

4.2. Membership in International Treaties

Countries which are signatories to all 5 international agreements receive a value of 1 in this category, those which are signatories to just one receive a value of 1/5. Table 1 shows the join time and the effective time of each treaty of China.

The Paris convention and revisions is one of the most early comprehensive international treaty which have the most extensive memberships, and it's the most basic and significant global multilateral international conventions in today's world as well.

The main objective of the Patent cooperation treaty is to coordinate and simplify the administrative procedures, helps patent Offices with their patent granting decisions, and facilitates public access to a wealth of technical information relating to those inventions.

Table 1. The effective time of each treaty of China.

Treaty	Join time	Effective time
Paris convention and revisions	12/19/1984	3/19/1985
Patent cooperation treaty	1/1/1994	1/1/1994
UPOV Convention	3/23/1999	4/23/1999
Budapest Treaty	4/1/1995	7/1/1995
TRIPS	12/11/2001	12/11/2001

Source: World Intellectual Property Organization (WIPO)[13].

The International Union for the Protection of New Varieties of Plants (UPOV) is an intergovernmental organization which was established by the International Convention for the Protection of New Varieties of Plants, with the mission of providing and promoting an effective system of plant variety protection.

The Budapest treaty allows "deposits of microorganisms at an international depositary authority to be recognized for the purposes of patent procedure".

The Agreement on Trade-Related Aspects of Intellectual Property Rights (TRIPS) is an international agreement administered by the WTO that sets down minimum standards for many forms of intellectual property regulation as applied to nationals of other WTO Members.

As is shown in Table 1, it can be concluded that the value of "Membership in international treaties" is "0.2" from 1985 to 1993, "0.4" in 1994, 0.6 from 1995 to 1998, "0.8" in 1999 and 2000, and it has reached the full mark "1" since 2001.

4.3. Restrictions on Patent Rights

Patent holders may also face risks of losing their patent rights. This category measures protection against losses arising from three sources: working requirements; compulsory licensing; and revocation of patents. A country that protects against all three sources receives a value of 1 in this category.

4.3.1. Working Requirements

Working requirements refer to the utilization of inventions, if a country doesn't require working at any point during the patent term, it receives a value of 1/3.

Under Article 22 of the Patents Law 1984, "Inventions and utility models for which patent rights are to be granted shall be ones which are novel, creative and of practical use." this entry get "0" points.

4.3.2. Compulsory Licensing

Compulsory licensing requires patentees share the invention with third parties, if a country does not impose compulsory licensing within 3 or 4 years from the date of patent grant or application, it receives a 1/3 value.

According to Section 6 of the Patents Law 1984, the patent administration department under the State Council may, upon application made by any unit or individual that possesses the conditions for exploitation, grant a compulsory license for exploitation of an invention patent or utility model patent. Thus this entry gets "0" points.

4.3.3. Revocation of Patents

Revocation of patents indicates that if a country does not revoke patents for non-working receives a value of 1/3. This entry can get 1/3 point as there's no relevant provision in the Patent law.

As a result, the value of "Restrictions" is "1/3".

4.4. Enforcement

4.4.1. Preliminary Injunctions

Preliminary injunctions are pre-trial actions that require individuals to set-back an so-called infringement in order to protect the patentee from infringement until a final decision is made in a trial.

Under Article 61 of the Patents Law 2000 (take effect on July 1st, 2001), "If the patentee or interested party has evidence to prove that another person is committing or is about to commit a patent infringement, which, unless being checked in time, may cause irreparable harm to his lawful rights and interests, he may, before taking legal action, file an application to request that the people's court order to have such act ceased." This entry gets 1/3 point since 2001.

4.4.2. Contributory Infringement

Contributory infringement means infringement actions that are not caused by themselves but result in infringement by others.

Under Article 130 of the Civil Law 1986 (take effect on Jan. 1st. 1987)[14], "If two or more persons jointly infringe upon another person's rights and cause him damage, they shall bear joint liability." This entry gets 1/3 since 1987.

4.4.3. Burden-of-Proof

Burden-of-proof is the imperative on a party in a trial to produce the evidence that will shift the conclusion away from the patentee to the alleged infringer.

Under Article 60 of the Patents Law 1992 (take effect on Jan. 1st. 1993), "If a dispute over patent infringement involves an invention patent for the method of manufacturing a new product, the unit or individual manufacturing the same product shall provide evidence to show that the manufacturing method of their own product is different from the patented method." This entry gets 1/3 since 1993.

Finally, the value of "Enforcement" is "1/3" from 1987 to 1992, "2/3" from 1993 to 2000, gets the full mark "1" since 2001.

4.5. Duration of Protection

According to the Patent Law, the duration of patent right for inventions shall be fifteen years counted from the date of filing since April 1st, 1985, and the duration shall be twenty years since Jan. 1st, 1993.

4.6. Results and Discussion

From the above we can calculate the level of IPP yearly. As shown in Figure 1, the result is quite different from Park's. The source of the error may be either case laws which said no (not available) or the fact that the documentation Park got were not accurate[15].

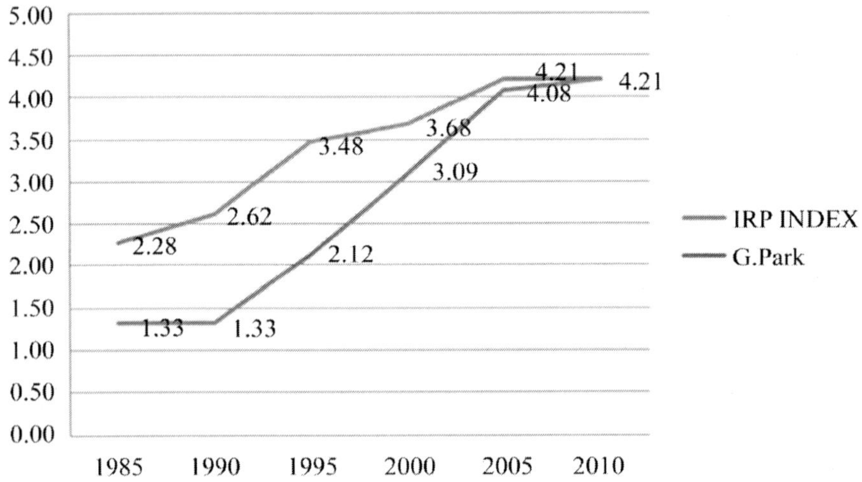

Figure 1. IPP of China based on the method of G. Park.

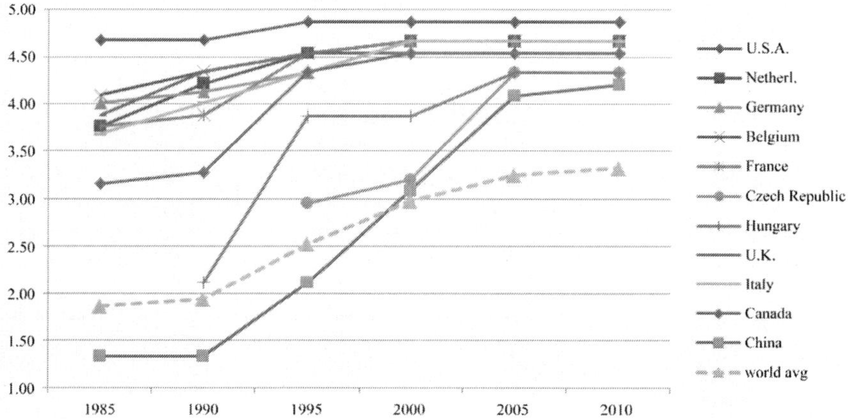

Figure 2. IPP indexes of China and other developed countries. Source: Park's personal page (2014).

For purposes of comparison we choose some developed countries (top-ten countries ranked by the average in past 25 years). From Figure 2 we can find that the level of IPP in China are growing year by year, and has reached a level close to some developed country, but it does not mean the higher the better, Chinese policy makers should make clear of the fundamental purpose of strengthening intellectual property protection in China is to promote domestic industrial upgrading, encourage innovation, and promote China's economic development, however, China cannot afford this under the existing economic level.

5. CONCLUSION

China's drive to modernization through improving intellectual property protection has been pushed forward by constant efforts to improve formal intellectual property laws and institutions for acquiring, maintaining, and enforcing intellectual property rights. Chinese policymakers and government officials have worked diligently for thirty years in a consistent, unwavering drive to create and improve the country's public intellectual property institutions. It's for sure that both the management and enforcement of IPP in China will make a progress over time.

FOOTNOTES

[1]Am. Chamber of commerce in the P.R.C., Am. Chamber of commerce in Shanghai & Am. Chamber of commerce in S. China, 2008 White Paper: American Business in China. pp. 55 fig. 24 (2008) [hereinafter White Paper], available at http://web.resource.amchamchina.org/wysiwyg/WP2008-PartI.pdf (follow "Part One: The Business Climate for American Firms in China" hyperlink).

[2]http://www.sipo.gov.cn, History, 2002-03-05. Consulted on April 26, 2007.

[3]http://www.sipo.gov.cn, Responsibilities of the State Intellectual Property Office of the People's Republic of China (SIPO), 2002-04-27. Consulted on April 26, 2007.

[4]See, e.g., WIPO, Status of Madrid Agreement Concerning the International Registration of Marks (July. 15, 2009).

[5]Copyright Law of the People's Republic of China 1990, translated in2 China L. Foreign Bus. (CCH) ¶11-700 (1993) [hereinafter Copyright Law]

[6]Computer Software Protection Rules 1991, translated in 2 China L. Foreign Bus. (CCH) ¶ 11-704 (1993).

[7]Hamilton,U.S. to Hit China with Stiff Tariffs, supranote 77, at A1.

[8]http://news.xinhuanet.com/english/china/2014-07/11/c_133477844.htm

[9]http://english.sipo.gov.cn/laws/developing/201405/t20140505_944778.html

[10]China IPR SME Helpdesk. (2013) Enforcement of Intellectual Property Rights in China: http://www.wipo.int/sme/en/best_practices/china_sme_helpdesk.html

[11]The IP index of China is "0" until 1985 for there exist no laws or regulations concerning about IPR.

[12]Patent Law of the People's Republic of China (1985, 1992, 2000, 2008)

[13]http://www.wipo.int/treaties/en

[14]The Civil Law of China (1986).

[15]The author sent an email to Park, and his reply was as follow:

"I only update the index every five years, but I hope to work on 2015 to make the index more timely. The source of the error may be either case laws which said no (not available) or the fact that our documentation here is not up-to-date and current, since we have to translate Chinese laws into English and there is a lag. You should for now use your own scores to represent China until I figure out the difference."(From Park, March 2014).

REFERENCES

1. Beebe, B. (2014) Shanzhai, Sumptuary Law, and Intellectual Property Law in Contemporary China. Vol. 47, New York University, New York, 849-872.
2. Wang, L. (2004) Intellectual Property Protection in China. International Information & Library Review, 36, 253-261. http://dx.doi.org/10.1016/j.iilr.2003.10.007
3. Godinhoa, M.M. and Ferreirac, V. (2012) Analyzing the Evidence of an IPR Take-Off in China and India. Research Policy, 41, 499-511. http://dx.doi .org/10 .1016/j.respol.2011.09.009
4. Maskus, K., Dougherty, S. and Mertha, A. (2005) Intellectual Property Rights and Economic Development in China. In: Maskus, K.E. and Fink, C., Eds., Intellectual Property and Development: Lessons from Recent Economic Research, Oxford University Press, Oxford, 295-331. http://dx.doi.org/10.1596/0-8213-5772-7
5. Gould, D. and Gruben, W. (1996) The Role of Intellectual Property Rights in Economic Growth. Journal of Development Economics, 48, 323-350. http://dx.doi.org/10.1016/0304-3878(95)00039-9
6. Moser, P. (2005) How Do Patent Laws Influence Innovation? Evidence from Nineteenth-Century World's Fairs. American Economic Review, 95, 1214-1236. http://dx.doi.org/10.1257/0002828054825501
7. Ang, J.S., Cheng, Y. and Wu, C. (2014) Does Enforcement of Intellectual Property Rights Matter in China? Evidence from Financing and Investment Choices in the High-Tech Industry. Review of Economics and Statistics, 96, 332-348. http://dx.doi.org/10.1162/REST_a_00372
8. Europe Chamber (2013) Enforcement of Intellectual Property Rights in China. China IPR SME Helpdesk 2013, Beijing, 1-8.
9. Yang, D. and Clarke, P. (2005) Globalisation and Intellectual Property in China. Technovation, 25, 545-555. http://dx.doi.org/10.1016/j.technovation.2003.10.008
10. Xu, C.M. and Shan, X.G. (2008) Constructing of the Index System and Verification for the Intensity of Intellectual Property Protection in China. Studies in Science of Science, 26, 715-723.
11. Li, W. (2011) The Improvement of China's Intellectual Property Rights Protection. China Collective Economy, 30, 125-126.
12. Rai, V., Schultz, K. and Funkhouser, E. (2014) International Low Carbon Technology Transfer: Do Intellectual Property Regimes Matter? Global Environmental Change-Human and Policy Dimensions, 24, 60-74. http://dx.doi.org/10.1016/ j.gloenvcha.2013.10.004
13. Bosworth, D.L. (1980) The Transfer of U.S. Technology Abroad. Research Policy, 9, 378-388. http://dx.doi.org/10.1016/0048-7333(80)90033-5
14. Ferrantino, M.J. (1993) The Effect of Intellectual Property Rights on International Trade and Investment. Weltwirtschaftliches Archiv, 129, 300-331. http://dx.doi.org/10.1007/BF02707699
15. Rapp, R.T. and Rozek, R.P. (1990) Benefits and Costs of Intellectual Property Protection in Developing Countries. Journal of World Trade, 24, 75-102.
16. Mansfield, E. (1994) Intellectual Property Protection, Foreign Direct Investment, and Technology Transfer, International Finance Corporation Discussion Paper, World Bank, Washington, DC, 19.

17. Ginarte, J.C. and Park, W.G. (1997) Determinants of Patent Rights: A Cross-National Study. Research Policy, 26, 283-301. http://dx.doi.org/10.1016/S0048-7333(97)00022-X

18. Park, W.G. (2008) International Patent Protection: 1960-2005. Research Policy, 37, 761-766.

CHAPTER 16

Quantitative Evaluation for the Level of Intellectual Property Protection in China

Liping Xu[1,2]

[1]University of Shanghai for Science and Technology, Shanghai, China

[2]Shanghai Publishing and Printing College, Shanghai, China

ABSTRACT

Making a review for the quantitative evaluation research of Intellectual Property Protection Level's in china, we get the research status, research perspective and the research method of the study from the review. After analyzing the existed problem and deficiency of those studies, we set forth the prerequisite of Intellectual Property Protection Level's study, and some new study thoughts, which are based on performance theory and value chain theory. We suggest a new study method should be used in the quantitative evaluation study of intellectual property protection level, which is a comprehensive evaluation method based on experts information. In the end, the principle of constructing parameter system will be elaborated, which will be used in quantitative evaluation study of intellectual property protection.

Keywords: Intellectual Property Protection, Ginarte-Park Method, Value Chain, Comprehensive Evaluation Method

1. INTRODUCTION

Intellectual property protection has been an international policy concern. It is a legal confirmation to the right, which comes from intellectual activities in the field of science, technology, culture, art, etc. It is a complex problem, which is related with economic, science and technology, culture, education and others. More and more scholars have engaged in the research of intellectual property protection since China's accession to the WTO. There are different opinions about the protection level of intellectual property in China. Some scholars believe that Chinese existing intellectual property protection level is higher than what can be overwhelming, which is not conducive to economic development. Some scholars think that Chinese intellectual property protection level is lower. It is not suitable to rapid economic growth, and adverse to the national development.

Foreign scholars' research shows choosing appropriate strategy of intellectual property protection is helpful to economic grow [1] [2]. So, we must make clear of intellectual property protection level and do some adjust to the protection strategy. It is a key point to quantitative evaluation for the level of intellectual property protection. In this review, we make careful analysis comparison of previous research on quantitative evaluation of intellectual property rights in China, point out the shortage of existing research on the basis of summarizing research ideas and research methods. We put forward a new research perspective and research method, combined with the existing scholar's research and full consideration of actual economic and social conditions. West rive to make quantitative evaluation research of the intellectual property protection level has more realistic guiding significance that can give some specific advice to corresponding indus- try development or give some efficiency suggestion to adjust policy, but not give theory description only.

2. THE QUANTITATIVE EVALUATION RESEARCH STATUS

2.1. Research Review

After intellectual property protection being the basic framework of the international trade in WTO, the research related to it quickly became a hot topic. The protection issue of intellectual property is complicated, which affected by such factors as the legislative, judicial and law enforcement. It is more difficult to directly evaluate the level of intellectual property protection. [3] At present, most scholars have conducted qualitative studies or analysis of the theoretical model [4] , because of lacking of quantitative measurement method. After joining the WTO, Chinese scholars have begun to use quantitative analysis methods to measure the level of intellectual property protection.

In 2005, Yuxiong Han and Huaizu Li pointed out that intellectual property protection level cannot be accurately measured by using the Rapp-Rozek method or Ginarte-Park method in China. They put forward a revised method by adding parameters related law enforcement variables to Ginarte-Park method, and apply it to measure the level of intellectual property protection level [3]. In 2008, Chunming Xu and Xiaoguang Shan revised the Ginarte-Park method by adding intensity of law enforcement as a parameter, and evaluate intellectual property protection level of China from 1985 to 2004. They pointed that intellectual property protection level will be affected by five key factors, such as judicial protection level, administrative protection level, economic development level, the social public consciousness and international supervision [5]. Hui Peng, Jiejing Yao has adopted similar approach to evaluate the IP protection level of 58 countries [4]. Xuebing Shen and Jia Liu put forwards that that the economic development level, degree of governed by law and enforcement of law are the three main factors to influent the intensity of actual IP protection, after measuring IP protection level from of 1995 to 2007, which results between that getting by G-P method and Han's method [4]. Xue-bing Dong believe less strict protection of intellectual property will more advantageous to our country economic growth in a short term. He get it from experiment result [4]. Hefa Song, Yuguang Li and Wan Qudo the experiment to respectively measure four kinds of ability, including intellectual property creation ability, intellectual property application ability, intellectual property protection ability and intellectual property management ability. They set up a parameters system, which includes 20 parameters. Research shows it is important to encourage intellectual property creation ability [6].

Since 2013, researches about IP protection have improved. Jing Yang, Xue-zhong Zhu's research covered a wide field, concluding patent protection, copyright protection, trademark protection and geographical indication protection. Research objects are more comprehensive than previous research, which make research results more scientific and valid. They pointed out that high strength of the intellectual property protection in developing countries can lead to limited domestic policy system, the problems such as system establishment and operation cost increase. It is very important to do intellectual property protection of quantitative research. We should explore system establishment on the basis of protection economy, law and social influence with objective evaluation of intellectual property rights [7]. In 2013 YingZhan, revised the method of

Ginarte- Park by adding law enforcement indexes such as national intellectual property rights protection implementation index, law enforcement index of global rule of law and PC software piracy rate. He evaluated the influence of economic development level on IP protection in the world 122 countries. He judged whether the level of intellectual property rights is adapted to the current level of economic development and pointed out that China's legislative level is higher and the intellectual property protection level corresponded to the current economic developing level. [4] In 2012, Mei-ling Yang finished the evaluation index system of enterprise's intellectual property management research. [8] She put the four procedures of intellectual property rights into the concrete implementation of the enterprise to build the evaluation index system in order to improve the enterprise value maximization. [9] In 2012, HualinTan, Junqiang Zhang finished the evaluation research of judicial protection of intellectual property rights. They determined the evaluated objects of judicial protection of intellectual property rights and selected the key indicators of the performance evaluation. [10] In 2013, Lv Min, Ya-bin Zhang in the study of the actual intensity measurement for the protection of intellectual property, with the entropy method to determine the weighting parameters. They measured the actual strength for the protection of intellectual property in China from 1985 to 2010. They pointed out that the government strengthened the construction of related intellectual property law system, participated in the international intellectual property rules and other mea- sures, can obviously accelerate and promote the protection of intellectual property rights [11]. In 2014, Chunlin Chen, Yanqiu Song and Yunting Wang evaluated the performance of intellectual property management. They took intellectual property rights management of government funding research project as the research object. They set up the evaluation index system with the laws and regulations. They proposed a fuzzy comprehensive evaluation model based on expert opinion. They determined the evaluation index system with the expert opinion [12]. Sun He quantified and analyzed the law enforcement level of intellectual property of our country. They used the normalization method and the grey system model on data processing and pointed out that the administrative law enforcement protection is the important means to realize the protection of intellectual property rights protection from the level of judicial protection of intellectual property and administrative protection law enforcement level, social and enterprise and individual protection level [13].

2.2. The Characteristics of the Existing Research

In recent years, many Chinese scholars have carried out quantitative evaluation research of intellectual property protection in the field of law, economics, science and technology. In the field of law, the study focuses on quantitative evaluate relevant legislation ability, judicial protection ability and IP protection law enforcement ability. Most scholars applied the revised G-P method to measure the intellectual property protection level, emphasizes IP protection system security and order, social values and public interests. Researcher in the field of science and technology Focus on the development and application of advanced science and technology. They research all kinds of intellectual property protection technology and method.

The commonly used research methods includes: 1) the questionnaire method; 2) Ginerate-Park method; 3) the information processing method. Ginerate-Park method is an evaluation method for the strength of intellectual property legislation, is not for the protection level of intellectual property. In some studies, scholars set the equal weights for different lever parameters, ignoring its different impact degree on the intellectual property protection, which will probably make results less scientific, less valid. In recent studies, some scholars have been aware of the shortage, gradually introduced the data analysis method, such as fuzzy mathematics theory and entropy theory to set parameter weight.

3. THE LIMITATION OF THE EXISTING RESEARCH

1) The regional differences were not been considered. People's protection consciousness varies from different economic conditions. If not considering regional economic differences, but blindly implement the intellectual property policies, it will be frustrated. In China, there is unbalanced development and the economic condition is difference in various regions. So, regional differences must be considered in the research.

2) Applicable Conditions of research is not been illustrated. Intellectual property protection is a complicated issue related with the legislative, judicial, economy, technology, culture, education, social environment and etc. [4]. In the study, Most scholars choose these factors, which relevant to the economic, technical, and social environment, as parameters that have an effect on enforcement ability. The results of that study will be limited by the condition of research environment. So, applicable Conditions of research should be illustrated. In reality, if condition of the legislative level, economic condition and social environment does not meet the applicable environment, the application of research is meaningless. I think that study of intellectual property is valuable only in specific conditions: 1) the complete legal system; 2) the higher degree of governed by law; 3) the stability of the economic level; 4) adaptable level of law enforcement to the current economic.

3) Main types of intellectual property were not involved comprehensively in study, the results of study has limitations. At present, most study about intellectual property protection mainly do macro-research to make clear that effect of Intellectual property protection on whole society. Patent data is chosen to represent the whole intellectual property category, while intellectual property includes patent, trademark, copyright. There are limitations to research only using patent data. Foreign scholars have set about to conducted studies according to the categories of intellectual property [13].

4) Proxy variable for same parameter is not same in different study. After choosing parameters, proxy variable will be used to replace the parameter values, such as the number of lawyers to measure the judicial protection level, legislative years measure whether the legal system is complete, the per capita GDP to measure economic development level (some studies using urban per capita disposable income to measure economic development level), the trademark administrative enforcement cases measure administrative law enforcement efficiency (some measured by the copyright administrative authorities investigate case level administrative law enforcement efficiency) [11] [14] [15]. That would lead to the results are inconsistent.

5) Lack of corresponding evaluation. The existing evaluation study of intellectual property protection focus on evaluating a country's overall strength of intellectual property protection, [7] there is less targeted evaluation. There is no quantitative evaluation research have been conducted to the administrative department of industry, industry and different policy.

4. THE NEW PERSPECTIVE OF QUANTITATIVE EVALUATION RESEARCH OF INTELLECTUAL PROPERTY RIGHTS PROTECTION

With the development of the quantitative evaluation research of intellectual property protection, many scholars put forward the new idea. In order to break through the bottleneck of the present research, it is imperative to introduce new perspective and new method to research.

4.1. Perspective Based on the Theory of the Performance

Performance reflects the efficiency of organization in complete or exact fulfillment of the terms of plan for achieving goals [16]. Performance targets of the intellectual property protection is to attach importance to on both social public interests and the balance between fairness and efficiency, while intellectual property right were protected effectively and properly. The object of performance evaluation is a series of activities, which can be evaluated, in four process of intellectual property work.

The performance evaluation parameters are diverse, and evaluation criterion is dynamic, but not static [17]. The parameters are difference in various stages of development [16]. We select the performance evaluation parameters from the objects and users of four processes in intellectual property work. It will make evaluation parameter set comprehensive.

We do the quantitative evaluation research of intellectual property protection will be helpful to make the intellectual property system healthy, orderly and sustainable.

4.2. Comprehensive Evaluation Idea Based on Value Chain

In 1985, Michael porter of Harvard University introduced analysis method based on value chain in book: competitive advantage. [18] The enterprise and relevant unit seemed to be a whole body, it create the same value. [8] In recent years, foreign scholars have done the research of intellectual property protection using that theory based on supply chain management [19]. In China, Heng Tang, Yang Zhang and Jinyan Chai was proposed to do intellectual property research based on value chain. They qualitatively expounded and analyzed the theory [8] , but no quantitative evaluation.

Intellectual property value chain is the combination of Michael porter's value chain theory and the intellectual property management [20]. The value of intellectual property including creator's value and social value, it covers four process of intellectual property work: intellectual property creation, intellectual property utilization and intellectual property protection, and intellectual property management. It organically combines all processes together, and it makes a more complete intellectual property research. comprehensive evaluation idea based on value chain make four process connection and integration effectively to raise their ability, maximizing the value of intellectual property.

4.3. Commercial Perspective Evaluation, Comprehensive Evaluation by Using Parameters Covered Field of Technology, Economy and Law

In China, the research of intellectual property protection mainly concentrated on legal in the field of law, concentrated on new technology in the field of science & technology, or concentrated on economic income distribution in the field of economics. There is no combination research of law, economy and technology. It is significant to build analysis frame work with "technology- economy-aw", study the effect of intellectual property of on market structure and competition from the view of commerce. It can be more comprehensive quantitative evaluation of intellectual property protection level.

4.4. Classification Evaluation According to the Kind of Intellectual Property

There are three important content of Intellectual property: the patent, trademark and copyright. They have the following characteristics in China: 1) administrative department of intellectual property is varied from different categories of intellectual property. There is difference in aspect of law enforcement strength [14]. 2) private rights and public rights is different, resulting in different social value; 3) protection consciousness is different between different categories of intellectual property; 4) corresponding value chain is different, and the objects in the value chain is also different.

In the existing research of intellectual property protection, the patent data has been used to instead of three kind of intellectual property, it lead to that result has a lot of restrictions. Especially in the environment of culture industry rapid development, copyright of digital publishing is very important. I think the research results should be more effective in realistic guiding significance, if the research have been done in accordance with the patent, trademark, copyright (copyright), respectively.

5. NEW RESEARCH METHODS OF QUANTITATIVE EVALUATION

5.1. Information Technology Related with E-Government Performance Evaluation

Chu Zhang introduces the concept of performance evaluation to quantitative evaluation research of intellectual property protection first time in the paper: "On the Constitution of the Performance Evaluation System of Intellectual Property Protection". He introduced the foreign government performance control target system, and put forward three kinds of e-government control goal system: COBIT system, FEA system and EAF system. In his research, Chu Zhang points out that those evaluation model system have consult value for evaluation of intellectual property protection in our country [19]. EAF system has strong Indian aboriginality, so we temporarily not to consider this approach.

COBIT is internationally recognized standards for management and control of security and information technology, it is the integrated information control system, in which the performance evaluation is an important component. COBIT system is helpful to quantitative evaluation of intellectual property protection. We can do the research follow steps: 1) defining the high-level performance evaluation goal of intellectual property protection; 2) describing the evaluation criteria; 3) in a series of activities, defining evaluation goals in detail according to the high-level goal; 4) give scores to parameters according to the evaluation criteria; 5) applying three kinds of assessment tools to research: the key success factors, key performance indicators and key target index.

5.2. System Comprehensive Evaluation Method Based on Expert Selection

In recent years, scholars have applied comprehensive evaluation methods to quantitative evaluation research of intellectual property protection level, such as fuzzy theory, gray model, entropy value method and so on [9]. There is broader research idea and better research methods than before. At present research, most scholars resolve problem in a certain way, which lead to inconsistent evaluate conclusion because of different methods, or random error caused by single quantitative method. Seeking for the new evaluation method is necessary to make the research effectively, On the premise of fully recognize evaluation objects, external environment and the evaluation requirements. Professor Qian Xue- sen proposed a method of processing complex giant system, which is comprehensive integration method from qualitative to the quantitative thinking. It is be used to combination evaluation research based on the method set. Under the guidance of the basic evaluation principles, several methods will be concentrated extracted from the set of basic evaluation method according to certain standards and rules, and then the object will be evaluated by those methods, which will eliminate the system error and random error caused by single method. At the same time, it can solve the inconsistency problem evaluation conclusion. In the study, if we can study with combined method based on expert suggestion, the research conclusion will be more scientific and more effective.

6. RULES OF BUILDING A QUANTITATIVE EVALUATION INDEX SET

It is an importance step for research of intellectual property protection quantitative evaluation to building a relevant evaluation index set. There are some advi- ces about choosing parameter, it follows.

Index source: index selected exist in the process of intellectual property, it contains four aspects: intellectual property creation, intellectual property application, intellectual property protection and intellectual property management. The system is not complete if one of the four aspects is absent [6].

The range of data: we need experiment data corresponding to index, which reflect the real situation. China joined in the WTO in 2001. After 2001, economic condition, social environment, degree of legislation and law enforcement has met the essential condition for the research of intellectual property protection in China. Making use of data from 2001 to 2015 to do research, the research results will have more scientific and realistic significance for the future work.

The basic principle of index selection: it is key point for the guarantee of results is scientific and validity to establish the index system. It is complex and difficult, we should follow principles: 1) the index set should containboth quantitative parameters and qualitative parameters; 2) the scientific principle; 3) comprehensive, systemic principle; 4) operational principle; 5) the principle of stability.

7. CONCLUSIONS

Quantitative research of intellectual property protection came up in the 1980's. Scholars adopted various quantitative or qualitative analysis methods to measure the level of intellectual property protection. The most famous is the Rapp-Rozek method and Ginarte-Park method. In China, people get the concept of intellectual property later than other countries. People lack protection consciousness of private property. Until 2001, after the accession to the WTO, Chinese scholars actively have begun to do research on the protection of intellectual property, but the method is limited, mostly is revised method but not technique innovation.

Intellectual property, which is core elements of international competitiveness and resource of national development strategic, has become an important means of development and competition in many countries. View from the angle of eco- nomics and management, it would have more practical value to do research of intellectual property protection based on the industrial chain. Intellectual property protection issue is complex, related with many field. Research on intellectual property protection should be system, comprehensive.

In China, economic development is different from different region. The quantitative evaluation study of intellectual property protection should fully consider the actual situation of unbalanced development in various regions. According to industry characteristics and intellectual property categories, targeted research will be better. The process of research needs to follow certain principles: firstly, choose the better research perspectives and target orientation; secondly, establishing a comprehensive & systematic index system, and get the effective data that is the best response to actual situation; Last, choosing the appropriate research methods. That will ensure research process scientific and research conclusion effectiveness.

REFERENCES

1. Kim, Y.K., Lee, K., Park, W.G. and Choo, K. (2012) Appropriate Intellectual Property Protection and Economic Growth in Countries at Different Levels of Development. Research Policy, 41, 358-375
2. Gu, Z. and Shen, Y. (2015) Intellectual Property Rights Protection, Technology Innovation and Technology Transfer: From the Perspective of Developing Countries. Industrial Economics Research, No. 3, 474-493
3. Han, Y. and Li, H. (2005) Quantitative Analysis for Intellectual Property Protection Level of China. Studies in Science of Science, 23, 377-382. (In Chinese)

4. Zhang, Y. (2013) Empirical Research on the Protection Level of Intellectual Property Right in China—International Compare and the Appropriateness Evaluation. Studies in Science of Science, 31, 1347-1354. (In Chinese)
5. Xu, C.-M. and Shan, X.-G. (2008) Constructing of the Index System and Verification for the Intensity of Intellectual Property Protection in China. Studies in Science of Science, 26, 715-723. (In Chinese)
6. Song, H., Li, Y. and Qu, W. (2013) The Intellectual Property Capacity Measurement Indicator System and Method Based on Equivalent Relationship of Kinds of Intellectual Property and Kinds of Research Institute. Studies in Science of Science, 31, 1826-1834. (In Chinese)
7. Yang, J. and Zhu, X.-Z. (2013) The Index System of FTA Intellectual Property Strength and Measurement. Studies in Science of Science, 31, 883-891. (In Chinese)
8. Tang, H. and Zhang, Y. (2013) A Study of the Enterprise Intellectual Property Standardization Management Based on the Value Chain. Science and Technology Management Research, No. 15, 175-180. (In Chinese)
9. Yang, M.-L. (2012) The Evaluation Index System Research of High-Tech Enterprise Intellectual Property Management. Thesis, Guangdong University of Technology, Guangzhou. (in Chinese)
10. Tan, H. and Zhang, J. (2012) Research on Performance Evaluation of the Judicial Protection of Intellectual Property. Journal of Social Sciences, No. 5, 109-116. (In Chinese)
11. Lv, M. and Zhang, Y. (2013) The Strength Measurement for Intellectual Property Protection in China. Science & Technology Progress and Policy, 30, 113-118. (In Chinese)
12. Chen, J., Song, Y., Wang, Y., et al. (2014) Evaluation for Performance of Intellectual Property Management: A Fuzzy Comprehensive Evaluation Model Based on the Specialists' Opinion Selection. China Science & Technology Resources Review, 46, 15-22. (In Chinese)
13. Park, W.G. (2008) International Patent Protection: 1960-2005. Research Policy, 37, 761-766
14. Sun, H. (2015) Measurement and Analysis of the Level of Law Enforcement of the Intellectual Property Protection in China. Studies in Science of Science, 33, 1372-1380. (In Chinese)
15. Zhao, S.-M. (2003) The Conflict and Coordination between Information Resources Sharing and Intellectual Property Protection. Nanning, Guangxi University. (In Chinese)
16. Zhang, C., Bai, J.-Y. and Dai, C. (2011) On the Constitution of the Performance Evaluation System of Intellectual Property Protection. Journal of Shandong Teachers' University (Humanities and Social Sciences), 56, 120-127. (In Chinese)
17. Thomas, D.J. and Griffin, P.M. (1996) Coordinated Supply Chain Management. European Journal of Operation Research, 94, 1-15
18. Tsay, A. and Nahmias, S. (2000) Modeling Supply Chain Contracts—A Review. In: Quantitative Models for Supply Chain Management, Kluwer Academic Publisher, Amsterdam, 299-366
19. Lawrence, D., Letmathe, P. and Uebe-Emden, N. (2016) Supply Chain Management Practices and Intellectual Property Protection in China: Perceptions of Mittelstand managers. International Journal of Operations & Production Management, 36, 135-163. https://doi.org/10.1108/IJOPM-12-2013-0526
20. Chai, J.-Y. (2009) The Cultivation of Intellectual Property Competitive Advantage of Enterprises Based on Value Chain. Science & Technology Progress and Policy, 26, 53-56. (In Chinese)

CHAPTER 17

Virtuous Reality: Moral Theory and Research Into Cyber-Bullying

Tom Harrison[1]

[1] *Jubilee Centre for Character and Virtues, School of Education, University of Birmingham, Birmingham B15 2TT, UK*

ABSTRACT

This article draws on a study investigating how 11–14 year olds growing up in England understand cyber-bullying as a moral concern. Three prominent moral theories: deontology, utilitarianism and virtue ethics, informed the development of a semi-structured interview schedule which enabled young people, in their own words, to describe their experiences of online and offline bullying. Sixty 11–14 year olds from six schools across England were involved with the research. Themes emerging from the interviews included anonymity; the absence of rules, monitoring and guidance and, the challenges associated with determining the consequences of online actions. The findings demonstrate the advantages of adopting a character-based moral theory to compliment rules and/or consequence based moral theories as the basis for future research into cyber-bullying. The findings evoke some wider implications for future research into cyber-bullying that might equally be applied to investigations into other Internet related moral concerns.

Keywords: Virtue ethics Cyber-bullying Character Moral concern

INTRODUCTION

Research into the impact of the Internet on children and young peoples' morality has struggled to keep up with technology's rapid development and innovation. The existing body of research on how the Internet influences young people does not paint a clear picture and is often polemical (Harrison 2014; Livingston 2009). Interpretations oscillate between extremes; young people online are depicted as either predominantly truthful or dishonest, compassionate or callous, selfless or selfish, altruistic or egotistical, courageous or cowardly, or as modest or vain. What is clear is that the Internet appears to influence young people in a number of complex ways, and that it is not obvious if such influence is predominantly positive or negative. The only common ground in the literature is a broad agreement that the Internet is, in some way, having an effect on the morality of children and young people (Harrison 2014).

What is missing from many empirical studies attempting to determine what this effect might be is a philosophical foundation on which to ground the research. Recent articles have explored the Internet from deontological, utilitarian and virtue ethical lines of normative enquiry (see, for example Plaisance 2013; Vallor 2010; Lievens 2011), although the arguments expressed in them have not been put to an empirical test. The research reported in this article describes an attempt to do exactly that. It draws on an empirical study that prioritised young peoples' perspectives of cyber-bullying, with the aim of suggesting what the best moral philosophical framework for researching the pervasive issue might be. The findings of this study raise some wider implications for future research into cyber-bullying, which might equally be applied to investigations into other moral concerns found on the Internet.

Cyber-bullying as a moral concern

Despite varying definitions of bullying and more specifically cyber-bullying (Tokunaga 2010), it is understood by many to be deliberate and hostile behaviour that is intended to harm others and as such, a moral concern (Belsey 2005; Menesini et al. 2011). The concern is great due to the high rates of cyber-bullying (Hasebrink et al. 2009) and also because of the emotional, psychological and at times, physical damage it causes its victims (Layard and Dunn 2009). Despite this, moral development and cyber-bullying research have largely been undertaken independently. Although studies have shown that the processes and reasons behind cyber-bullying do not differ much from face-to-face bullying (Froese-Germain 2008; Lenhart 2007), there is also a specific body of research that draws on moral disengagement theory (Bandura 2002) to explain why young people might bully others on the Internet. Empirical research supports a claim that the Internet acts as a 'moral disengagement mechanism' so that young people feel they are doing less moral wrong if they bully online than if they bully face-to-face (Gini 2006; Perren and Gutzwiller-Helfenfinger 2012; Menesini et al. 2013). Studies have also shown that cyber-bullies have less shame and guilt than offline bullies (Menesini and Camodeca 2008) and a corresponding lack of moral compassion (Pozzoli and Gini 2010).

Despite such promising research, none of these studies have sought to ground their findings in any established moral theory. This state of affairs persists despite accounts drawn from wider research into the Internet, demonstrating why such an approach might be beneficial. Several recent articles consider the rights and duties of Internet users, adopting largely deontological lines of enquiry (see, for example, Lievens 2011; Granitz and Loewy 2007; Zhang 2010; Lyu 2012). The preponderance of deontological approaches is not surprising, given that Kantianism (in some form of another) has been the moral theory of choice, until recently at least, in many areas of applied ethics (Arthur et al. 2015). Articles that explore pleasure, happiness or subjective wellbeing as a consequence of web-based interactions, adopting consequentialist lines of enquiry, are also common in the literature (see, for example, Valkenburg et al. 2006; Ali 2013; Cross et al. 2012). There has been growing awareness of virtue ethics as a serious moral rival to deontological and utilitarian theories ever since the publication of Elizabeth Anscombe's influential article Modern Moral Philosophy (1958) and Alasdair MacIntyre's book *After Virtue* (1981). Examples of recent articles that apply a virtue ethics theory to the Internet include: Elder (2014) who presents an Aristotelian defence of friendships online and argues that the concept of a shared life can equally be applied to online and offline friendships; and McFall (2012) who also adopts an Aristotelian conception of friendship to argue that online friendships are 'character friendships'.

In short, virtue ethics is a character based moral theory that offers an alternative framework for research into the Internet to the more directly action-guiding theories of deontology and utilitarianism. Vallor (2010) believes that several distinctive features of virtue ethics make it uniquely applicable to the domain of Internet ethics—and, furthermore, that a virtue-based perspective is needed to balance a strong utilitarian bias in the literature. Couldry (2010) agrees and argues that virtue ethics offers a more compelling and useful basis from which to study the Internet, than deontology. In particular, since rules are hard to establish and uphold online, and consequences are hard to predict, an approach to morality that is based on individual own character virtues has immediate appeal. As Plaisance (2013: 92) argues, 'rather than getting mired in the philosophical thicket regarding the motives and duties of actors in an online world we should focus on what behaviour and guidelines contribute to the flourishing of digital lives'. In this view, a virtue ethical approach offers a promising way to explore the specific moral issues found online. It is also useful to those interested in developing practical solutions for dealing with such moral issues, as it provides general guidance regarding appropriate virtue- or character-based behaviour, without reducing it to a set of codifiable rules that tend to be insensitive to particularities and contexts.

All the same, recognising Williams (1985) call for a non-reductionist approach to moral philosophy, there are merits in considering how each of the classical moral theories might be applied to a specific problem such as cyber-bullying. In doing so, the basis for pluralistic models that integrate notions of duty, consequence and virtue into future moral enquiries might well be sought. The research reported on here attempts to consider the potential interplay between different moral theories by attending to young peoples' own perspectives on cyber bullying. It holds that by taking into account the actual views of young people on the realities of their daily experience we might come to a more informed position that may be useful for further studies into the issue. Still, this study has a particular interest in exploring the potential of virtue ethics to better understand the influence of the Internet on young people, regarding it as providing an attractive and fruitful lens though which to question why people do good and/or bad things online.

Research methodology

The research reported on here is part of a larger project that investigated the influence of the Internet on the character and virtues of 11–14 year olds. This study is reported on in more general terms elsewhere (Harrison 2014) whereas this paper focuses on one section of the qualitative findings of that piece in greater detail and depth.

The study described here sought to learn from 11 to 14 year olds, growing up in England and how they conceived the motivations of cyber-bullies. It was deemed important to hear the actual voice of the young people at both the data collection and data analysis stages. Semi-structured group interviews are widely used with young people for this purpose. Unlike individual interviews, group interviews reduce adult influence in data collection methods (Heary and Hennessy 2002) by providing children of like age with the opportunity to share perceptions and experiences with peers. Horner (2000: 510) suggests that children 'are more relaxed and willing to share perceptions when discussions are held with a group of peers'. When conducting group interviews with young people, it is important to try to avoid 'answer dialogue' (Heath et al. 2009) in which young people try to answer questions in the 'expected' way, for fear of negative consequences if they get things wrong. The aim was therefore to encourage participants to speak freely and honestly about their cyber-bullying experiences.

Twelve group interviews were undertaken in six schools between September and November 2013. In total, sixty 11 to 14 year olds were interviewed, of whom thirty-five were female and twenty-five male. The group interviews lasted between 30 and 45 min. On average, there were six students in each group (smallest number was 4, largest 7), since a group of this size is considered small enough to be manageable, whilst large enough to enable a genuine group dynamic (Krueger 1994). Group interviews followed the ethical procedures outlined by the British Research Association (2011). All participants were informed about the nature of the research and were asked to give consent to participation. Parents also gave consent for their children to be involved. The interviews were recorded and transcribed.

The interview schedule mostly consisted of open questions divided into four sections. Section one addressed the question of how young people experience and/or understand cyber-bullying in their everyday lives; section two explored the difference between face-to-face and online bullying; section three sought a better understanding of the motivation of cyber-bullies; and, section four contained questions aimed at determining effective educational strategies for reducing cyber-bullying. The group interview schedule was pre-piloted and piloted to ensure that it was fit for purpose.

The data collected during interviews was analysed thematically using a deductive approach of qualitative interpretation, since the themes informing the data had already been shaped by the prior literature review. Qualitative interviews permit access to unanticipated types of data, insofar as their purpose is to allow participants to recount their own narratives and to identify the themes that best reflect their own experiences of cyber-bullying. Thematic analysis is also a way of ensuring that the key issues are highlighted and addressed (Boyatzis 1998).

Contextual details: the participants and schools

The participants were selected through cluster sampling, a method widely used by educational researchers working with schools (Robson 2011). The aim of cluster sampling is to identify a mixture of schools representing as far as possible the wider school population. The present research sites were selected on the basis that they contained a diverse range of participants within the 11–14 age bracket, in terms of gender, ethnicity, religion and rural and urban settings. Table 1 below provides details of the six schools involved.

Table 1 Group interview research sites

School	No. of pupils on roll	Age	Type M/F	Setting	% free school meals	Ethnicity of pupils	5 or more A*–C Grades at GCSE (in Maths and English) (2013)
School A	1074	11–18	Mixed	Urban	17.5	94 % White-British heritage Remainder: minority ethnic groups	64 %
School B	1195	4–19	Mixed	Urban	29.2	Proportion of ethnic minority groups is higher than national average.	46 %
School C	1607	11–18	Mixed	Urban	10	Majority White-British heritage; remainder being from ethnic backgrounds, mostly Polish	78 %
School D	482	9–13	Mixed	Urban	21.5	Most pupils White-British heritage.	n/a
School E	179	11–19	Mixed	Urban	12.3	Not Known	n/a
School F	1922	11–18	Mixed	Urban	26	Proportion from minority ethnic backgrounds: quarter of the national figure	71 %

The contact teachers in each school were asked to select a representative sample of young people to take part in the interviews. However, such selection may have been limited by availability of students at the time of the interviews, their willingness to participate, and also whether permission to participate had been granted by parents. Thus, it should be conceded that it is unlikely that representative samples were achieved in all interview groups. In addition, since participants were selected by the schools themselves, it is possible that more articulate students, and/or those of higher academic aptitude were selected to 'represent' their school.

Insofar, the participants in this research might not be considered a probability sample. Still, though the selected groups cannot be said to be representative of all 11–14 year olds living in England, they may still provide useful indications of the attitudes of young people of this age group using the Internet. The results therefore need to be treated with caution, especially when attempting to draw more general conclusions from them.

The findings from the research are detailed in the section below. The quotations used in this article have been extracted from the group interviews and are generally representative of emerging themes. Genders of participants are indicated by an f (female) or m (male).

FINDINGS

Deontological challenges

A deontological ethic regards moral conduct as the observance of rules or principles usually defined in terms of duties and/or obligations. In the group interviews questions were asked about how 'rules' and 'duties' on the Internet informs conduct, with particular regard to cyber-bullying. Two themes emerged from the data. The first was a perceived absence of clear rules online, leading to an increased sense of freedom. The second was that Internet users may easily conceal their identities, enabling them to bypass their ordinary duties.

Perceived absence of rules leading to an increased sense of freedom online

I don't know anyone who follows rules on the Internet, I don't even know where they are (m).

Many of the participants identified an increased sense of freedom from restrictions as one of the biggest differences between their online and offline lives. In schools it was said that students know what is expected of them, and furthermore that rules were enforced. A distinction was drawn between schools, which is 'full of rules' (m) and the Internet, which is a 'free for all' (m). One participant stated 'There are rules at school about no bullying, no fighting. The teachers sort this out. They don't online' (f). Participants noted that although websites used by young people might or should have more rules and regulations they do not enforce these 'as young people won't use their sites' (m). As one student put it: 'young people want freedom online' (f). Moreover, while many participants did acknowledge there are terms and conditions when they sign up to a website, they said that they did not read them and also that no one respects rules not issued by a person. This position was summed up by the observations that: 'rules in real life are more intimidating because you know that they have authority, whereas on Facebook there is no person enforcing them' (f); In addition, the scale of the Internet led many participants to doubt whether rules could ever be enforced. For example, one participant questioned how Facebook could monitor its 1 billion users.

The lack of any 'central power' or 'hierarchy' controlling the Internet has encouraged the belief that there are no rules in cyberspace and it seems to be such sense of freedom that makes the Internet so attractive to many young people. Research has shown, moreover, that such increased sense of freedom (and by association, lack of rules) also makes cyber-bullying and similar abuses more likely. For example, a study by Mishnan et al. (2009) found that cyber-bullying mostly takes place when young people are alone and unsupervised. This is pertinent given that studies show that the majority of the young people use the Internet alone (Harrison 2014) although evidence suggests that the majority of parents do monitor their child's online use (Ofcom 2014). Furthermore, young people who use their computers privately and unsupervised are more likely to be bullied (Anirban and Anoshua 2011). Shariff and Hoff (2007) have also found that schools have a difficult time supervising online activities and knowing when to involve law enforcement, or when to distinguish freedom of expression from harassment.

Internet users may easily conceal their identities

Some people use the Internet to reinvent themselves to be like a completely new person. But it can actually lead to cyber-bullying, what they say behind screens might be different to what they say to your face. (m)

A difference between online and offline forms of bullying, raised more consistently, and forcefully by participants, was that the Internet encouraged or at least allowed for anonymity. This meant that young people were able to bypass any rules and duties expected of them by operating online as someone else. It was observed that some of the websites commonly used, such as ASK FM, actually had 'anonymous' buttons on them, where it was possible to hide profile and name.

Participants held there to be a difference between open bullying in (say) a playground and being anonymously bullied through the Internet. One participant described the Internet as a 'mask' (m) insofar as 'you never know who you are talking to'. Another described it as a 'costume' (f) that bullies put on to avoid being detected. One participant stated: 'I think cyber-bullies are like cowards, they are hiding in their shells which is the Internet' (m). As such, cyber-bullying was often described as being 'darker' (m) and more 'underground' (m) then playground bullying.

The issue of anonymity was raised regularly in the interviews and is also discussed at length in the literature. It is considered to be one of the biggest challenges by those concerned with finding solutions to cyber-bullying (Ackers 2012; Mishnan et al. 2009). The possibility for anonymous action whilst in cyber-space also presents a challenge to deontological lines of moral thinking as it makes both rule setting as well as tracking the observance of rules difficult. Research participants thought that acting anonymously made people more likely to do things they would not normally do offline, as in face-to-face bullying there is often a more obvious power differential usually where the stronger bully torments the weaker victim, whereas in cyber-bullying the power also lies in anonymity (see also Brydolf 2007; Winter and Leneway 2008).

Perhaps even more significant is that anonymity online diminishes accountability. Accountability to others constrains conduct and without it the more anti-social aspects of human nature may come to the fore. Mishnan et al. (2009) describe the situation as bullies hiding behind the keyboard with little fear of exposure or repercussions. Weisband and Keisler (1996) found that when Internet use reduces fear of censure. Anonymity also reduces sensitivity towards their victims (Ang and Goh 2010).

Suler (2004: 322) described this mindset as 'dis-associative anonymity'. This is where the bully does not 'own their behaviour' and distances themselves from their actions. Many interviewees cited example of people they knew who acted differently online deciding to adopt a different Internet lifestyle and identity. If young people believe they cannot be found out, and that what they do online cannot be traced to other social dimensions of their lives, they are more likely to act in socially dislocated ways.

Utilitarian challenges

they think it's [cyber-bullying] a victimless crime, nothing is going to happen to anyone, they don't see people getting hurt, so why not (m).

Utilitarianism holds that the right course of action is that which maximises human wellbeing and/or reduces suffering. In this regard, evidence from interviews was clear: that participants thought that cyber-bullying was wrong and caused widespread unhappiness. However, there was also evidence that while utilitarianism provides some guide to right and wrong moral actions, features of the Internet make enforcing this principle challenging. It was commonly held that: 'seeing the consequences is a good learning tool for getting better' (f). However, it was also commonly held that due to the nature of the technology, it is hard to determine the consequences of actions online, with some consequent erosion of empathy (as empathy requires emotional identification with particular others; see Hoffman 2000).

The Internet was described by the participants variously, but words such as 'impersonal' (m) and 'faceless' (f) were used regularly. For many, the main concern was a lack of 'visual clues' (m and f), which meant communication could often 'go wrong' (m and f). This suggests that there is much scope for misinterpretation of intended meaning in communication and that it is often hard to determine the consequences of their online communications. They felt that that this meant some people might cyber-bully inadvertently. As one participant put it: 'in real life you can see it and will say sorry about things. Many don't realise their mistakes and that they are hurting you' (f). The concern appeared to be for a diminished empathy insofar as for some: being unkind 'does not mean as much online' (m); or because it is 'hard to feel as bad about online actions as those you see first-hand' (m). One participant stated that: 'because you can't see others you don't think about their pain' (f).

The absence of visual clues, so that 'people are communicating behind screens' (f) meant that many of the common ways that people make sense and bring meaning to words appeared to get lost. There was concern that: 'you can't see people's reactions, so you might not be able to tell when you are offending them'(f). Some students felt that the impersonal black and white text may also make some messages seem more threatening than intended.

The literature on cyber-bullying has also shown that victims are sometimes unintended (Shariff 2008) insofar as people are more likely to cyber-bully when they do not think they are causing harm to the victim (Willard 2002; Cross et al. 2009). Wider research has shown that cyber-bullies often do not see the painful implications of their actions insofar as they are often removed from them (Campbell 2005). If the direct consequences of any particular online action are not seen, it is less likely that guilt will be felt when someone is hurt. The time/space distance of online messages may leave young people morally disengaged as the Internet enables young people to avoid feelings of guilt (Arsenio et al. 2009; Malti et al. 2010). Menesini et al. (2013) believe that the Internet encourages individuals to participate in 'egocentric reasoning', insofar as they are more concerned with gaining credibility from their peers than the outcomes of their actions. Research has shown that participants bullied others online because it made them feel amusing, popular, or powerful, although many indicated feeling guilty afterwards (Mishnan et al. 2010).

Utilitarianism supposes that we can weigh up moral dilemmas by calculating the consequence of our actions. However, the Internet by its very nature makes it harder for young people to determine the consequences of their actions. Both the interview data and research described in the literature show that while consequences are sometimes unintentional, they are also hard to determine. Such considerations raise doubts about the usefulness of utilitarianism for addressing problems of cyber-bullying or other web-based abuse.

Virtue ethics

When you're little you learn right and wrong. The Internet should not change the way you have been thinking and been brought up. It is still down to you what you do (f)

Virtue ethics is a moral theory that emphasises the place of character and virtues as the determining factor in why people feel and do the morally 'right thing'. The role of character in moral life was also a topic central to the group discussions of interviewees. It was generally agreed by participants that bullying in either online or offline forms seems to be due to an absence of certain virtues on the part of the bully. Virtues mentioned by the participants included care, self-discipline, compassion, humility and trust. Many interviewees held that while it might be 'easier' to bully online, a large factor determining why someone might choose to send nasty messages or not is the quality of their character. For these, the decision of whether or not to bully depended on the 'kind of person they are' (m).

As previously discussed, many participants thought that their peers were more likely to bully online because of the absence of constraining principles. For many, this placed responsibility on individuals themselves to regulate their online activities. As one student put it: everyone must 'police themselves' (f). Some looked for external guidance to help them do this, such as the student who observed that they had 'their mum's voice in their head whenever they are online' (f).

Virtues considered by interviewees to be important when using the Internet included 'being kind and compassionate' (m and f); whilst another observed that: you have to take responsibility for yourself and be honest' (m). Self-discipline was also considered important insofar as the Internet is 'always on', permitting communication with others at any time of the day and from any location. Feelings and emotions were also given as reasons why young users do good and bad things online. For example, empathy was also raised as an important emotion as there is a need to think about the consequences of actions before sending online messages. A female participant stated:

You have to understand how what you're going to say is going to affect them and their perspectives so you have to imagine how they are going to be feeling, if you have said something to them. So you would have to be like cautious of like the consequences of your actions and be responsible for what you're going to say? (f)

The importance of making wise judgments when online was discussed by some of the participants. In order to make wise judgements they explained they had to draw on past experiences to work out the best course of action. For example, one female student commented: 'my friends all act the same online as they understand what it is like to be backstabbed' (f); and another, that 'sometimes you know afterwards you have done wrong – you read back and think why did I write that' (f). It was this process of learning through experience, and building up a bank of knowledge to guide action that the participants felt helped them make wise decisions when communicating online.

Although, as previously expressed, many participants felt that features of the Internet made it more likely that people would be bullied, there was also an understanding that some features might actually inhibit such actions. These features were: its connective properties; access to large audience; and the fact that written communication have a permanency that oral communications do not. Students felt that the larger audiences and connectivity made them more careful about what they wrote online, insofar as they were concerned about where their messages might go and who might see them. Likewise, some students observed that online communications are hard to erase and often have a life of their own. They explained that it is more difficult to prove if something was said verbally in the playground than online where the evidence can be seen in black and white.

A moral theory that prioritises human character over consequences and/or duties is helpful when seeking courses of action for dealing with cyber-bullying. This is because emphasis regarding educating good digital citizens can be firmly placed on the development of character as a guide to conduct online. This will be discussed in more detail below.

DISCUSSION

The research reported on in this article has sought to break some new ground in the field. The merits of virtue ethics, by contrast with utilitarianism and deontology, have been examined through an analysis of the interview data. Deontology has been found to be a somewhat imperfect basis for online morality, with many of the participants admitting that they have a reduced sense of duty when they are in cyberspace. This is not meant as a general condemnation of deontology; only as an indication of the fact that cyberspace might carve up an area of 'moral space' that is not easily amenable to a rule-and-code based approach. Likewise, although utilitarianism offered some sound principles for moral behaviour on the Internet, it is not so easy to apply it in practice. The research findings show that calculating the consequences of any particular online interaction is difficult at best, and often impossible. Moreover, utilitarianism assumes empathy with the plight of other people, but the emotional identification underlying empathy seems to require (psychologically if not necessarily logically) the existence of particular rather than non-specific others. That said, utilitarianism and deontology might continue to provide useful frameworks for future research into morality and the Internet, in various partial and specifically targeted ways. However, the findings from this research show that virtue ethics has some advantages over the other two theories.

One distinct advantage of virtue ethics is that it places the onus on young people to make wise decisions on the Internet when it might not be obvious what the right thing is to do. As the Internet is described as 'free' and often involves new experiences, young users are often faced with situations where there is no categorical imperative, or right course of action based on rational principles. In these situations users are required to make judgements about the right course of action, based on previous experience as well as their effective emotions and feelings. Virtue ethics allows for young users of the Internet to become wiser over time through the development of online practical wisdom, or what we could call 'cyber-*phronesis*'. While a deontological approach might continue to be seen as helpful to bolster general rules and guidance governing what is good and bad online conduct, and a utilitarian one to draw greater attention to harmful online consequences, a virtue ethical perspective would prioritise the formation of wise and virtuous online citizens. Such an approach might provide young people with resources to negotiate the inevitable moral dangers and challenges of using the Internet.

Adopting a virtue ethical theoretical base for research into the impact of the Internet on young people requires us to ask different questions. Perhaps most importantly is to ask *how the character virtues of young people influence their use the Internet,* rather than simply to assume that because the technology exists, immoral behaviour will inevitably ensue. Unnecessarily deterministic accounts should be avoided. The interviews show that young people perceive their relationship with the Internet to be reciprocal, rather than unidirectional. They recognised that it was how agents used the Internet that largely determined the consequences of their online actions: that there was nothing about the technology as such that *made* people cruel, or dishonest. From a virtue ethical perspective, technological determinism is implausible insofar as it is the character virtues of the users of the technology, rather than the technology itself, that determines online behaviour. From this viewpoint, the technology is neither good nor bad (although it has features that might encourage good or bad behaviour): it is the user that determines whether it is put to good or bad use.

CONCLUSION

It would not be desirable to draw any simple picture of user experience from the data presented in this article. Given the limited nature of the present research, any attempt to make universal claims is not possible. However, the findings discussed above may provide a useful basis upon which to construct a moral theoretical framework for future research into cyber-bullying in particular as well as perhaps other moral issues found online. The new opportunities that the Internet has opened up for young people require them more than ever to 'do the right thing'; not so much motivated by rules, duties or consequences (since these may not always be explicit), but by having the character to choose wisely between alternatives. An important question to ask in future research is how best to educate digitally virtuous citizens to help them make good and wise decisions. On the basis of the present study, virtue ethics arguably offers not only a promising theoretical basis for researching the influence of the Internet on character virtues, but also provides a promising practical strategy for dealing with online moral concerns. The educational implications of this research will be returned to in a later paper.

REFERENCES

1. Ackers, M. (2012). Cyberbullying: Through the eyes of children and young people. *Educational Psychology in Practice, 28*(2), 141–157.
2. Ali, A. (2013). Cyber-bullying taxonomies: Definition, forms, consequences and mitigation strategies. In *2013 5th International conference on computer science and information technology (CSIT)* (p. 126–133).
3. Ang, R. P., & Goh, D. H. (2010). Cyberbullying among adolescents: The role of affective and cognitive empathy, and gender. *Child Psychiatry and Human Development, 41*(4), 387–397.
4. Anirban, S., & Anoshua, C. (2011). Are social networking sites a source of online harassment for teens? Evidence from survey data. *Children and Youth Services Review, 33*(2), 284–290.

5. Anscombe, G. E. M. (1958). Modern moral philosophy. *Philosophy, 33*(124), 1–19.
6. Arsenio, W., Adams, E., & Gold, J. (2009). Social information processing, moral reasoning and emotion attributions. Relations with adolescents reactive and proactive aggression. *Child Development, 80*, 1739–1755.
7. Arthur, J., Kristjansson, K., Thomas, H., Kotzee, B., Ignatowitcz, A., & Qiu, I. (2015). *Virtuous medical practice*. Birmingham: University of Birmingham. Accessed July 20, 2015 from http://www.jubileecentre.ac.uk/ userfiles/jubileecentre /pdf/Research%20Reports/Virtuous_Medical_Practice.pdf.
8. Bandura, A. (2002). Selective moral disengagement in the exercise of moral agency. *Journal of Moral Education, 31*(2), 101–119.
9. Belsey, B. (2005). *Internet usage: Facts and news*. Accessed September 17, 2012 from www.cyberbullying.ca.
10. BERA. (2011). *Ethical guidelines for educational research*. London: BERA.
11. Boyatzis, R. (1998). *Transforming qualitative information: Thematic analysis and code development*. Thousand Oaks, CA: Sage Publications.
12. Brydolf, C. (2007). Minding myspace: Balancing the benefits and risks of students' online social networks. *Education Digest, 73*(2), 4–8.
13. Campbell, M. (2005). Cyber-bullying: An old problem in a new guise? *Australian Journal of Guidance and Counseling, 15*(1), 68–76.
14. Couldry, N. (2010). Media ethics: Towards a framework for media producers and media consumers. In S. J. Eards & H. Wasserman (Eds.), *Media ethics beyond boarders: A global perspective*. New York: Routledge.
15. Cross, D., Li, Q., Smith, P. K., & Monks, H. (2012). Understanding and preventing cyberbullying. Where have we been and where should we be going? In Q. Li, D. Cross, & P. K. Smith (Eds.), *Cyberbullying in the global playground. Research from international perspectives* (pp. 287–305). West Sussex: Wiley.
16. Cross, E., Richardson, B., Thaddaeus, D., & Jessica, V. (2009). *Protecting young people from cyberbullying*. London: Beatbullying.
17. Elder, A. (2014). Excellent online friendships: An Aristotelian defense of social media. *Ethics and Information Technology, 14*(16), 287–297.
18. Froese-Germain, B. (2008). Bullying gets digital shot-in-the-arm. *Education Canada, 48*(4), 44–47.
19. Gini, G. (2006). Social cognition and moral cognition in bullying: What's wrong? *Aggressive Behaviour, 32*, 528–539.
20. Granitz, N., & Loewy, D. (2007). Applying ethical theories: Interpreting and responding to student plagiarism. *Journal of Business Ethics, 72*(3), 293–306.
21. Harrison, T. (2014). *The influence of the Internet on the character virtues of 11–14 year olds*. Birmingham: University of Birmingham.
22. Hasebrink, U., Livingstone, S., Haddon, L., & Olafsson, K. (2009). *Comparing children's online opportunities and risks across Europe: Cross-national comparisons for EU Kids Online* (2nd ed.). Deliverable D3.2, LSE, London, EU Kids Online. Accessed June 17, 2014 from http://eprints.lse.ac.uk/24368/1/D3.2_Report-Cross_national_ comparisons- 2nd-edition.pdf.
23. Heary, C., & Hennessy, E. (2002). The use of focus group interviews in paediatric health care research. *Journal of Pediatric Psychology, 27*(1), 47–57.
24. Heath, S., Brooks, R., Cleaver, E., & Ireland, E. (2009). *Researching Young people's lives*. London: Sage.
25. Hoffman, M. L. (2000). *Empathy and moral development: Implications for caring and justice*. Cambridge: Cambridge University Press.
26. Horner, S. D. (2000). Focus on research methods: Using focus group methods with middle school children. *Research in Nursing & Health, 23*(6), 510–517.
27. Krueger, R. (1994). *Focus groups: A practical guide for applied research* (2nd ed.). London: Sage Publications.
28. Layard, R., & Dunn, J. (2009). *A good childhood, searching for values in a competitive age*. London: Penguin.

29. Lenhart, A. (2007). *Cyberbullying*. Washington, DC: Pew Research Center. Accessed November 20, 2013 from www.pewinternet.org/Reports/2007/Cyberbullying.aspx.
30. Lievens, E. (2011). Risk-reducing regulatory strategies for protecting minors in social networks. *Strategy for Telecommunications, Information and Media, 13*(6), 43–54.
31. Livingston, S. (2009). *Children and the Internet*. Cambridge: Polity Press.
32. Lyu, H. (2012). Internet policy in Korea: A preliminary framework for assigning moral and legal responsibility to agents in Internet activities. *Government Information Quarterly, 29*(3), 394–402.
33. MacIntyre, A. (1981). *After virtue*. Notre Dame: University of Notre Dame Press.
34. Malti, T., Gasser, L., & Gutzwiller-Helfenfinger, E. (2010). Children's interpretive understanding, moral judgments, and emotion attributions: Relations to social behaviour. *British Journal of Developmental Psychology, 28*, 275–292.
35. McFall, M. T. (2012). Real character-friends: Aristotelian friendship, living together, and technology. *Ethics and Information Technology, 14*(3), 221–230.
36. Menesini, E., & Camodeca, M. (2008). Shame and guilt as behaviour regulators: Relationships with bullying, victimization and prosocial behaviour. *British Journal of Developmental Psychology, 26*(2), 183–196.
37. Menesini, E., Nocentini, A., & Calussi, P. (2011). The measurement of cyberbullying: Dimensional structure and relative item severity and discrimination. *Cyberpsychology, Behaviour and Social Networking, 14*(5), 267–274.
38. Menesini, E., Nocentini, A., & Camodeca, M. (2013). Morality, values, traditional bullying, and cyberbullying in adolescence. *The British Journal of Developmental Psychology, 31*(1), 1–14.
39. Mishnan, F., Cook, C., Gadalla, T., Daciuk, J., & Solomon, S. (2010). Cyber-bullying behaviours among middle and high school students. *The American Journal of Orthopsychiatry, 80*(3), 362–374.
40. Mishnan, F., Michael, S., & Solomon, S. (2009). Ongoing and online: Children and youth's perceptions of cyberbullying. *Children and Youth Services Review, 3*(12), 1222–1228.
41. Ofcom. (2014). *Children and parents: Media use and attitudes report*. Accessed July 10, 2015 from http://stakeholders.ofcom.org.uk/binaries/research/media-literacy/media-use-attitudes-14/Childrens_2014_Report.pdf.
42. Perren, S., & Gutzwiller-Helfenfinger, E. (2012). Cyberbullying and traditional bullying in adolescence: Differential roles of moral disengagement, moral emotions, and moral values. *European Journal of Developmental Psychology, 9*, 195–209.
43. Plaisance, P. (2013). Virtue ethics and digital flourishing: An application of Philippa foot to life online. *Journal of Mass Media Ethics, 28*(3), 91–102.
44. Pozzoli, T., & Gini, G. (2010). Active defending and passive bystanding behavior in bullying: The role of personal characteristics and perceived peer pressure. *Journal of Abnormal Child Psychology, 38*, 815–827.
45. Robson, C. (2011). *Real world research*. Oxford: Blackwell Publishers.
46. Shariff, S. (2008). *Cyber-bullying: Issues and solutions for the school the classroom and the home*. Oxon: Routledge.
47. Shariff, S., & Hoff, D. L. (2007). Cyberbullying: Clarifying legal boundaries for school supervision in cyberspace. *International Journal of Cyber Criminology*. http://www.cybercrimjournal.co.nr. Accessed 12 Oct 2013.
48. Suler, J. (2004). The online disinhibition effect. *Cyberpsychology and Behaviour, 7*(3), 321–326.
49. Tokunaga, R. S. (2010). Following you home from school: A critical review and synthesis of research on cyberbullying victimisation. *Computers in Human Behaviour, 26*(3), 277–287.
50. Valkenburg, P. M., Peter, J. S., & Schouten, A. P. (2006). Friend networking sites and their relationship to adolescents' well-being and social self-esteem. *Cyberpsychology Behavior, 9*, 584–590.

51. Vallor, S. (2010). Social networking technology and the virtues. *Ethics and Information Technology, 12*(2), 157–170.
52. Weisband, S., & Kiesler, S. (1996). Self-disclosure on computer forms: Meta-analysis and implications. In *Proceedings of the SIGCHI conference on human factors in computing systems* (pp. 3–10).
53. Willard, N. (2002). *Computer ethics, etiquette, and safety for the 21st-century student.* Eugene, OR: ISTE.
54. Williams, B. (1985). *Ethics and the limitations of philosophy*. Cambridge: Harvard University Press.
55. Winter, R. E., & Leneway, R. J. (2008). *Cyberbullying—Terror in the classroom: What can be done?* Accessed January 15, 2013 from http://www.chaminade.org/cyberbullying.htm.
56. Zhang, X. (2010). Charging children with child pornography—Using the legal system to handle the problem of sexting. *Computer Law & Security Review, 26*(3), 251–259.

CHAPTER 18

Is There an Ethics of Algorithms?

Felicitas Kraemer, Kees van Overveld, Martin Peterson

ABSTRACT

We argue that some algorithms are value-laden, and that two or more persons who accept different value-judgments may have a rational reason to design such algorithms differently. We exemplify our claim by discussing a set of algorithms used in medical image analysis: In these algorithms it is often necessary to set certain thresholds for whether e.g. a cell should count as diseased or not, and the chosen threshold will partly depend on the software designer's preference between avoiding false positives and false negatives. This preference ultimately depends on a number of value-judgments. In the last section of the paper we discuss some general principles for dealing with ethical issues in algorithm-design.

Keywords: Algorithm, Image analysis, Medical technology, False positive, False negative

INTRODUCTION

The focus of this article is on ethical aspects of algorithms. An algorithm is, roughly speaking, a finite sequence of well-defined instructions that describe in sufficiently great detail how to solve a problem. Both computers and humans use algorithms for solving a wide range of problems. However, in this paper we shall be exclusively concerned with algorithms implemented in computers.

At first glance it might be tempting to conclude that algorithms are value-free entities that do not, at least not in their most abstract form, have any ethical dimensions. However, in this article we argue that this commonsense view about algorithms is false. Many, but not all, algorithms implicitly or explicitly comprise essential value-judgments. By an 'essential value-judgment' we mean the following: If two algorithms are designed to perform the same task, such as classifying a cell as diseased or non-diseased, these algorithms are essentially value-laden if one cannot rationally choose between them without explicitly or implicitly taking ethical concerns into account. Another way of saying this is that the algorithm cannot be designed without implicitly or explicitly taking a stand on ethical issues, some of which may be highly controversial.

If true, our claim about essentially value-laden algorithms has to be taken seriously by software engineers who design algorithms. If some algorithms are essentially value-laden, i.e. if people who design algorithms cannot avoid making ethical judgments about what is good and bad, then it is reasonable to maintain that software designers are morally responsible for the algorithms they design.[1] Although the term 'ethics of algorithms' might have far-reaching connotations, it nevertheless captures what is at stake here. If our claim about essentially value-laden algorithms is correct, there will indeed be an ethics of algorithms.

The issue that we address has to some extent already been touched upon in the literature. Turilli seeks to develop a solution to "the problem of specifying computational systems that behave in accordance with a given set of ethical principles".[2] Allen et al. argue that, "we need to integrate artificial moral agents into ... new technologies to manage their complexity".[3] Finally, Arkin argues that there is a need for, "immediate investment in ... machine/robot ethics".[4]Although we do not necessarily disagree with these authors, we wish to point out that their papers are concerned with questions that are slightly different from the issue discussed here. Turilli, Allen et al., and Arkin discuss how we can construct algorithms and systems that behave in *accordance with* ethical principles. As we argue in the next section, such algorithms need not be essentially value-laden themselves. Indeed, an algorithm or system that behaves in *accordance with* some ethical principles can itself be ethically neutral. What we wish to discuss in this article is the ethical features of the algorithm itself.

The structure of this article is as follows. In "Value-laden algorithms and the design process" we characterise some of the central concepts used in the paper and argue that algorithm design is in many respects similar to designing other technological artefacts. In "A real example" we discuss a real example of an essentially value-laden algorithm used in medical image analysis, and in "The precautionary principle and ethical theories" we discuss some ways of dealing with the ethical problems raised by this algorithm. Finally, in "Should ethical values be user-defined?", we argue that software designers should as far as possible leave ethical choices to users, and when this is not possible the ethical assumptions underlying the algorithm should at least be made transparent. This may lead to new or refined procedures in the process of software engineering.

Value-laden algorithms and the design process

It is notoriously difficult to give a precise characterisation of what an algorithm is. Numerous definitions have been proposed in the literature.[5] However, nothing in our reasoning hinges on how exactly algorithms are formally defined. The definition of an algorithm suggested in "Introduction" will suffice for our present purposes.

Arguably, the claim that many algorithms implicitly or explicitly comprise essential value-judgments cannot be accurately assessed without first defining some of its central concepts. We take a value-judgment to be any proposition expressing a view on how things ought to be or not to be, or what is good or bad, or desirable or undesirable. It takes little reflection to see that not all algorithms express value-judgments. Consider, for instance, algorithms used for calculating the lexicographical order of a finite set of strings, such as the ones used to put words in alphabetical order. Given the input to this algorithm, there is only one possible correct output. Therefore, this task is fully specified in terms of various logical relationships, which require no approximation or interpretation.

That said, some algorithms clearly produce genuine value-judgments. Consider, for example, algorithms used in decision support programs, i.e. systems that help decision makers to make better decisions by ranking a set of alternative actions with respect to some predefined criteria. A typical outcome of an algorithm used in such a program is a verdict like "Alternative X is the best option" or "Alternative X is better than alternative Y with respect to criterion Z". It would be pointless to deny that these sentences express genuine value-judgments. This is acknowledged by Turilli, who discusses how to specify "systems that behave in accordance with a given set of ethical principles".[6] However, the fact that an algorithm or system yields a value-judgment as its output does not prove that the algorithms used for producing the value-judgments are essentially value-laden, as pointed out in "Introduction". The string of sentences (or data) that is produced by running through an algorithm might very well be value-laden even if the algorithm itself is not value-laden.

Let us take a closer look at this important point. The general features of the kind of value-judgments generated by decision support programs can be clarified by invoking Kant's famous distinction between hypothetical and categorical imperatives. A hypothetical imperative is a statement about what you ought to do, given that you wish to respect some exogenously defined goal. "Don't study philosophy if you wish to become rich" is a hypothetical imperative. A categorical imperative is a value-judgment that tells us what we ought to do irrespective of our desires or goals. The distinction between hypothetical and categorical imperatives does not exhaust the logical space of all possible value-judgments, but it illuminates the nature of the kind of value-judgments that are most likely to be encountered in computer programs used for aiding decisions. Such value-judgments are best characterised as hypothetical value-judgments, because they tell you what you ought to do given that you wish to achieve a specific aim. However, as pointed out above, the presence of a hypothetical value-judgment does not entail that the algorithm used for generating the recommendations comprises value-judgments. The underlying algorithm can, when taken in isolation, be completely value-free.

So what, then, could it possibly mean to say that an algorithm comprises a value-judgment and therefore has an essential ethical dimension? Consider the following suggestion.

ALGORITHM COMPRISING AN ESSENTIAL VALUE-JUDGMENT

An algorithm comprises an essential value-judgment if and only if, everything else being equal, software designers who accept different value-judgments would have a rational reason to design the algorithm differently (or choose different algorithms for solving the same problem).

Note that this criterion does not make the definition circular. It is indeed true that the term 'value-judgment' occurs both in the left-hand part and in the right-hand part of the criterion, but the term that is to be explained is 'algorithm comprising an essential value-judgment', not 'value-judgment'. (Recall that we have already explained above how we understand the term 'value-judgment'.) It is also worth noting that our necessary and sufficient criterion explains why an algorithm producing hypothetical imperatives, such as the decision aid program discussed above, does not comprise essential value-judgments. If two or more software designers are asked to write a decision aid program it does not matter what ethical principles (or other value-judgments) the software designers themselves subscribe to.

Since the aim of the program is to produce hypothetical imperatives, the algorithm itself can be entirely value-free.[7]It is certainly true that value-judgments are produced by a successful implementation of the program, but it is not true that two or more software designers accepting different value-judgments would ever have a rational reason to design the underlying algorithm differently.

We now come to a fundamental assumption in this paper, viz. that a software designer designing an algorithm is facing a decision process that is in many respects similar to that faced by people designing other technological artifacts. If true, this entails that value-judgments that are built into an algorithm will in many respects be similar to value judgments that feature in other design processes. For this reason it is helpful to elaborate a bit on the design processes from which algorithms and other technological artifacts are derived.

Let us first explain what we mean by 'design' by proposing a necessary condition we believe all such activities have to fulfill. (For present purposes, we can do without a sufficient condition.)

DESIGN ACTIVITY: Something is a design activity only if it is an activity where a designer (or a team of designers) make decisions in order to reach a pre-defined goal.

We use the term 'variable' both for characterising what it means to 'make decisions' and for 'reaching a goal'. We assume that a variable stores exactly one piece of information. For instance, when designing a bicycle, the designer can decide that the front wheel shall have a diameter of 60 cm; this 'diameter of the front wheel' is a variable, and '60 cm' is the value stored in this variable. However, we must carefully distinguish the (four) different ways in which values can be stored in a variable.

First, the designer can *decide* that a variable shall have a particular value. For the rest of this paper, this settles our definitions of 'making a decision' as 'assigning a value to a variable', assuming that the decision-maker has the authority to assign the chosen value to the variable in question. For instance, no designer can decide that 'this bicycle will have a maximum velocity of 250 km/hour', since the maximum velocity of a bicycle is the result of a range of other decisions and empirical circumstances. Variables that may occur in a decision are called *decision variables* or *category-I* variables.

Second, a variable can take a value as part of the assessment of the degree to which a design-objective was met. For instance, suppose the bicycle we are designing should be as light as possible; then the value of the variable 'weight of the bicycle' helps to assess if we were successful. Such variables will be called *objective variables* or *category-II* variables. Notice that any category-II variable receives a value as the final consequence of assigning values to category-I variables, plus a number of (physical, economical, and social) mechanisms that causally propagate the category-I values through the designed artifact and its usage.

From the account sketched above, we see that category-II variables are (in a mathematical sense) a function of category-I variables: the design function. However, they are also a function of another class of variables. In order to see this, note that the weight of the bicycle depends on e.g. the density of the chosen material. If we choose carbon fiber (a possible value for the category-I variable 'material of the bicycle'), we have to accept that the density of carbon fiber occurs as an argument in our design function. Such variables cannot be controlled by the designer. We call them *context variables* or *category-III* variables. From the perspective of the designer, they are constants.

In all but trivial cases, the design function is immensely complicated. It is next to impossible to write down the values of category-II variables even if we know the values of category-I and category-III variables. We can regard it, however, as a composition of a large number of much simpler functions. The weight, for instance, is the sum of the weights of the two wheels plus the weight of the frame. The weight of one wheel is not a category-II variable (because the design is not necessarily better if this weight is less); it is also not a category-I variable (if we decide the diameter, the weight is no longer independent); finally it is not in category-III because it is not a constant. It is what we call an *auxiliary variable* or a *category-IV* variable.

This completes our general outline of the design process. What it amounts to is the establishment of a network of functional relationships between variables in categories I, II, III, and IV and the assignment of values to the category-I variables; the 'quality' of the design is assessed by inspecting the resulting values for the category-II variables. Consider Fig. 1.

The four categories characterised in Fig. 1 suggest an important link between ethics and the freedom of the designer: The choice of category-II variables may reflect, among other things, the designer's ethical view. For example, the designer might stipulate that the bicycle be constructed from recycled materials (reflecting the value of sustainability), and that it must be manufacturable in third-world countries (reflecting the value of geopolitical justice). In this respect we should observe that category-II variables come in two versions: *requirements*, that is: predicates that must be true (say, 'the bicycle shall be lighter than 15 kg'), and *desires* that relate to variables subject to optimisation (say, 'the bicycle shall be as light as possible'). Among other things, the framework of four categories allows us to clarify trade-offs in the design, it helps us to pinpoint compromises, and it provides insight into possible design alternatives. We will use it here to decouple, on the one hand, the acceptance of an ethical position (such as the categorical imperative) from the implementation of the consequences of this position in a design (a hypothetical imperative): The first consists of deciding which ethical values should be represented; the second of representing these ethical values in terms of category-II variables.

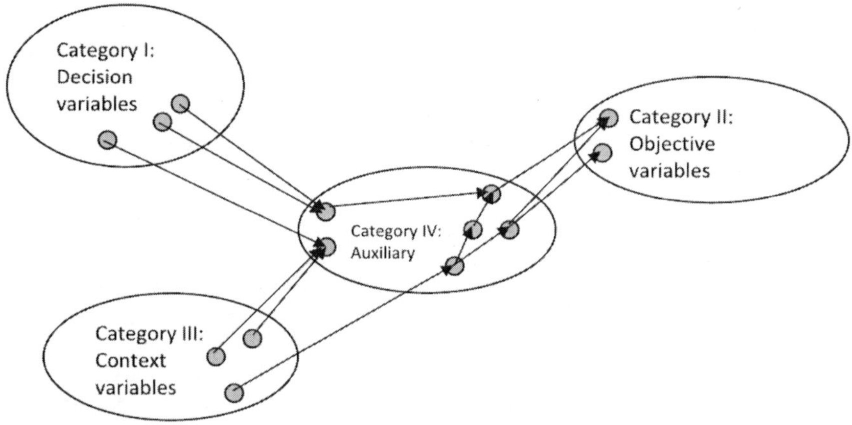

Figure 1 Designing as a process of connecting 4 types of variables. *Arrows* indicate functional dependencies

A real example

In this section we discuss some examples of algorithms that are essentially value-laden. All our examples, one of which is real, are concerned with a particular kind of algorithm used in medical image technologies. Very briefly put, medical image technologies aim at representing human biological structures in computers in an accurate way, such as human organs and cells in blood samples, and thereby improve the diagnostic or therapeutic prospects of diseases affecting the biological structures in question. One of the many ethical issues raised by such algorithms is the risk of using algorithms that produce false positive and false negative results. By definition, a false positive result occurs whenever the algorithm triggers the system to count something (a cell, a symptom of a disease) in a digital image that is not actually there. A false negative occurs if an algorithm in a similar vein fails to identify a structure in the picture that is actually there.

For all practical means, it is virtually impossible to totally eliminate the risk of getting false positives and false negatives. However, if one is willing to accept a large number of false positive results one will typically get a smaller number of false negatives. In order to understand why, imagine you are asked by an eccentric millionaire to build a device that automatically counts the number of tigers in the jungle who passes through a certain spot. Naturally, if you build a device that simply counts everything yellow that passes through the jungle you will get a rather large number of false positive results, since e.g. bees, bananas, and yellow flowers will trigger the device. However, if on the other hand, you impose more conservative criteria on what is to count as a positive result and design the device such that only very large yellow items are detected, then you can expect the number of false negatives to rise (since e.g. new-born tigers babies will not be detected by the device).

Clearly, software designers designing toy algorithms for detecting tigers, as well as real-life algorithms used in medical image technologies, have to make a trade-off between minimising the number of false positives or the number of false negative results. This trade-off will inevitably be based on a value-judgment. There is simply no objective fact of the matter about whether it is more desirable to avoid a false positive or a false negative. Different users may have different preferences, and several conflicting preferences appear to be equally rational. It therefore seems hard to deny that an algorithm used for, say, counting the percentage of cells infected by a virus in relation to the number of non-infected cells will invariably contain a value-judgment about how many false positive results are tolerable in relation to the number of false negative ones. That said, both false positive and false negative results may give rise to severe negative effects for individual patients. Doctors base diagnostic as well as therapeutic decisions on what they come to believe about e.g. the number of infected cells in relation to the number of non-infected cells.

We can also illustrate our point about essentially value-laden algorithms by way of a real example. To start with, recall that in "Value-laden algorithms and the design process" we discussed some of the choices faced by designers by discussing the design of a material object (a bicycle). However, as pointed out in that section, the design process is roughly the same when designing immaterial artifacts such as algorithms. We can therefore use the general design framework introduced in "Value-laden algorithms and the design process" for illustrating how values for category-I variables propagate in the direction of category-II variables.

In the image to the right we see an MR-scan depicting a cross section of a human heart. For the purpose of diagnosing a variety of possible pathologies, it is necessary to accurately estimate the blood volume of the heart during various stages of a heart-beat cycle. The difference between blood and heart muscle tissue occurs as a difference in grey values in MR images. Estimating the blood volume starts with establishing which part of the image is colored lighter grey, and counting the number of pixels in this light grey area. This is called *segmentation*between light and dark, as depicted schematically in the diagram below.

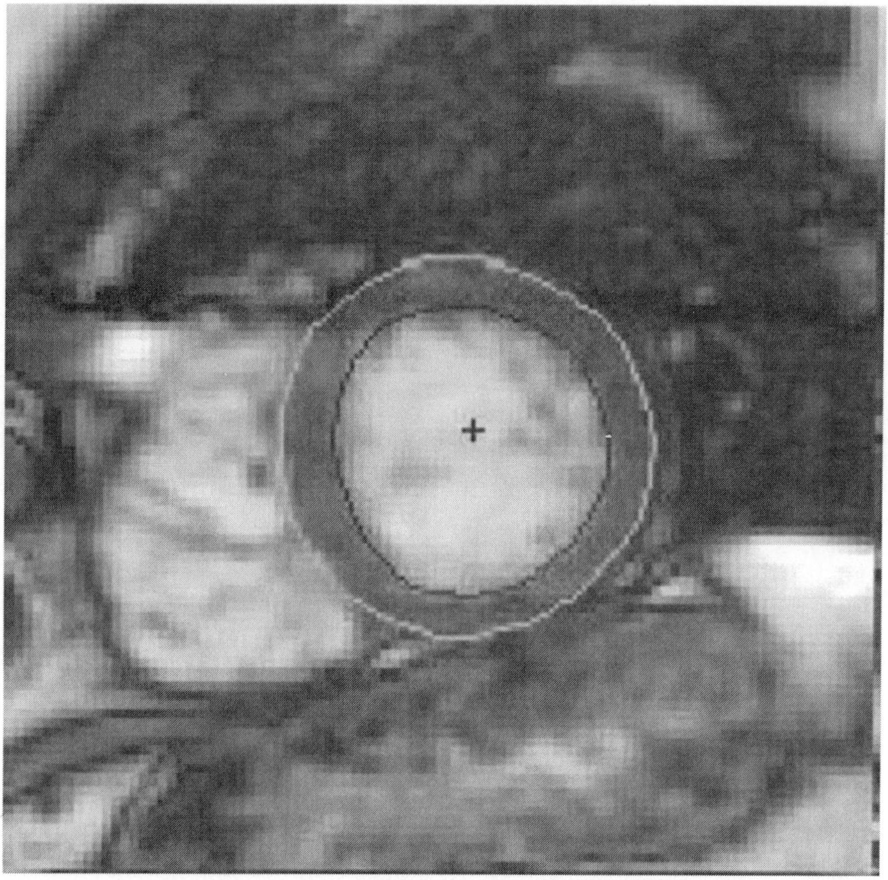

One way of carrying out a segmentation is to use a numerical threshold, say T. This means that pixels with a lightness value above T are labeled 'light' and pixels with a lightness value below T are labeled 'dark'. The border between light and dark areas is, however, not sharp (Fig. 2).[8]

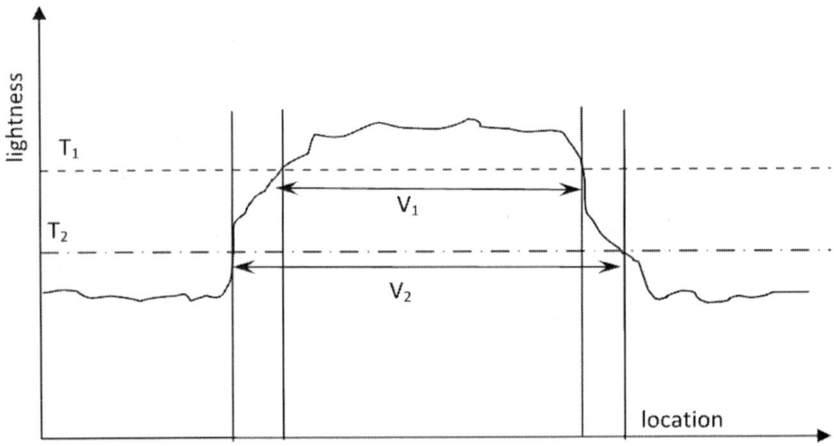

Figure 2 Segmentation in a noisy image. For a larger threshold value, T1, the estimated area V1 of the 'blood' segment will be lower than the estimated area V2 which is found with a lower threshold T2. It is not a priori clear, however, which of the two threshold values is the 'correct' one. The software engineer chooses a threshold without real argument—thereby biasing the outcome of the algorithm when it is used on patient data. This may statistically influence the change of false positive diagnostic errors in favour of false negatives, or vice versa

It is quite common to introduce thresholds in segmentation-like algorithms. Such thresholds serve as nice examples of category-I variables in the design process of the algorithm. There is no *a priori* correct value for such thresholds. There is no rigorous first principles from which they can be derived, since the noise in the image is an inevitable artifact of the MR measuring process, caused by numerous non-modeled sources. Software engineers typically choose a value that 'seems reasonable' for thresholds. In the example above we see how the choice for this threshold influences the estimated blood volume. This estimate will in turn affect further values, and eventually the diagnosis. So in borderline cases the diagnosis will (indirectly) depend on the value of the threshold T.

Of course, if the anatomical configuration is far from a borderline case (either 'very' healthy or 'very' pathological), there is obviously no problem. For the majority of pathological conditions, however, there is a continuum between 'healthy' and 'pathological'. Somewhere in this continuum there is a grey zone where the diagnostic outcome will critically depend on thresholds somewhere along the computational pipeline.[9]

Now, given that we take the computational output of a medical diagnosis support system to be a designed artifact, we can ask ourselves what the category-II variables are in this example. Apart from trivial ones, such as patient safety during clinical measurement, and operational comfort for the medical specialist, we may want the quality of the (automated) diagnosis to be 'good'. But the notion of goodness must of course be rendered more precise, and this is exactly where our observations concerning false positives and false negatives come in. In borderline cases the bias towards false positives or false negatives may to a large extent be determined by the values of thresholds or other category-I variables.

At this point it could be objected that the best way forward is to design the algorithm such that it totally eliminates all false positive and false negative results. However, a closer analysis of the risks for getting false positives and false negatives in medical images shows that this is not likely to be a feasible option. First, all algorithms that are applied to a large set of measurements will be sensitive to stochastic effects in the data set, and this will automatically yield false positive and negative effects. Second, it should be noted that the images are in most cases numerical (re-)constructions of large amounts of physical measurements. This point has important consequences: Nowadays the quality of such computer generated images is getting close to being photo realistic. That is, the data is presented (rendered) as if it is an actual photo of some 3-D internal organ or tissue structure, perhaps even isolated from its environment. Such rendering or isolation, however, requires complicated algorithms that include segmentation (as mentioned above), as well as much other subtle image processing. So rather than being objective photographs, the 3D medical images that specialists look at (and base their diagnosis on) are the result of an elaborate algorithmic process, depending on arbitrary thresholds in a difficult-to-predict manner. Now, medical specialists, like any other humans, are accustomed to interpret plausible 3-D images as accurate projections of corresponding 3-D objects. It is very difficult *not* to interpret a realistically looking 3-D image as a trustworthy projection of a 3-D object. This introduces the risk that one will forget that in order to generate these 3-D images, a number of decisions about thresholds had to be taken. Had some decisions been taken differently, the image could have looked (very) different—and, as we argue, there is no a priori way to give the correct value for such thresholds.

Understanding the relations between what is seen in the image and what really exists in a patient's body (or in the microscopic slice of tissue being examined) therefore requires a thorough understanding of the various algorithmic steps that are applied to the physical data in their transformation towards a collection of colored pixels. Such transformations include a choice of filters (involving variables with values that may significantly affect the eventual 'looks' of the image – and therefore the conspicuousness of certain types of anomalies) and perhaps geometrical algorithms (such as contouring, segmentation, tracking, etcetera). These filters and algorithms have to be chosen with some goals in mind – but these goals cannot typically, for practical reasons, be formulated in terms of minimising the risk of acquiring false negatives.

The third and last reason why it is not feasible to design algorithms that avoid all errors can be formulated as follows. Many (semi)automated diagnostic tools have been developed in recent years, and these diagnostic tools typically incorporate implied hypotheses of expected pathologies. For instance: a probe to identify a potential aneurism, stenosis or tumor can only do so because it *expects* certain characteristic features of, say the shape or volume of such anomalies. An experienced medical doctor realises when she sees something 'unexpected'; a software program usually cannot. Again, these diagnostic tools may not have been developed with the goal of minimising the percentage of false negatives in mind. Finally, there is one further problem. Software for complicated tasks such as medical diagnosis is often immensely complicated. Much if it consists of components that may have been developed earlier for 'general purposes'; these components are re-used in order to make the software production process economically feasible. Segmentation algorithms are an example of such components. Components are preferably treated as black boxes: based on their formal, functional specifications, their behavior can be assumed to be 'correct'. The ethical position, however, that was adopted during the construction of such a component when either choosing a more conservative or more liberal threshold, is typically not part of their formal specification.[10] That means that the same segmentation algorithm, applied in two different systems, will behave equally with respect to its formal, functional requirements, but at the same time it may behave oppositely with respect to its tendency to produce false positive or false negative judgments.

The total complexity of highly interrelated physical techniques, algorithms, heuristics and visualisation methods is immense, and there is little hope to optimise the entire, integrated chain in terms of minimising the chance of false negatives. What may be feasible, however, it to take this chain apart in a systematic way, and to analyse the various stages as if they were (to some extent) independent. This route may be facilitated to use a framework such as the 8-layers model developed by one of the authors to argue about the information contents in images. This model separates this information into coherent chunks (e.g., the shape-related information is separated from the texture-related information, which is separated from the 3-D surface related information, etcetera.)

The precautionary principle and ethical theories

In this section we discuss two different ways of addressing ethical issues raised by the risk of obtaining false positive and false negative results when using medical image algorithms. In the next section we shall identify and discuss some ethical issues that apply to algorithm design in general.

Briefly put, we propose that a possible way of managing false positive and false negative results obtained from medical image algorithms is to adopt an epistemic interpretation of the precautionary principle.[11] The second option, which we will also explore in depth, is to approach this choice as an ethical judgment to be determined by an ethical theory.

Let us first consider the precautionary-oriented, epistemic strategy. The precautionary principle was originally invoked by policy makers for addressing environmental issues, such as global warming, toxic waste disposal, and marine pollution. In recent years it has also been suggested that the precautionary principle may also be applied to medical issues. David B. Resnik argues that "properly understood, the [Precautionary Principle] can provide physicians and patients with a useful approach to medical decision making." The precautionary principle can, however, be interpreted in many different ways. According to the epistemic (belief-guiding) interpretation outlined in Peterson (2007), the precautionary principle should be characterised in terms of what it urges us to believe. To put it more precisely, the epistemic version of the precautionary principle holds that it is better to get a false positive rather than a false negative result, contrary to what is commonly accepted in science, since this will prevent doctors and patients from falsely believing that someone who is in fact ill is healthy.

Scientists generally agree that it is more important to avoid false positives than false negatives. This is because scientific knowledge tends to be cumulative: We add new beliefs about what the world is like to a set of already existing beliefs; and the justification for the new beliefs typically depends in more or less intricate ways on the old beliefs. Therefore, if we start to accept too many false positive beliefs we may end up in a situation in which future research is directed in the wrong direction by our false positives. A more conservative approach, in which false negatives are preferred over false positives, is more likely to be successful in the long run, since it makes it more likely that we will not base new beliefs on old but false ones.

However, when discussing medical image algorithms it is far from clear that we should adopt the same set of (epistemic) values that guide, or at least ought to guide, scientific research.[12]The reason for this is that the consequences of falsely believing something to be safe when it is not might be disastrous. If the algorithm is designed such that doctors come to believe that patients who are actually diseased are not, then the doctors may indirectly cause indirect harm to patients by failing to treat them. We therefore conclude that when addressing ethical aspects of medical image algorithms it is far from clear that medical decision makers should prefer algorithms that make them believe what is most likely to be true. On the contrary, a strong case can be made that medical image algorithms should be designed such that they are more likely to produce false positive rather than false negative results. That said, it is of course essential to make sure that an increased number of false positives does not lead to too many unnecessary and potentially dangerous operations. The computer image is just a tool. The final responsibility for deciding whether a surgical intervention is appropriate has to be taken by the doctor.

It is important to bear in mind that our conclusion about the importance of avoiding false negatives in medical image algorithms is a value-judgment, not a factual one. No observations or other purely empirical methods can be appealed to for backing up this conclusion. This is part of the explanation of why medical image algorithms are essentially value-laden. However, as with nearly all value-judgments, good reasons can also be given against this value-judgment. For example, it could be claimed that the software designer designing the algorithm ought to perform a detailed cost-benefit analysis of the pros and cons of accepting different trade-off rates between false positives and false negatives. This is an approach that, broadly speaking, tallies well with traditional consequentialist intuitions. It may turn out, for instance that the total utility of accepting some false negatives but avoiding a very large number of false positives may actually be optimal, because of various empirical circumstances. It is beyond the scope of this paper to take a definitive stand on whether the consequentialist view outlined above is at least as plausible (or even more plausible) than the precautionary approach. Here we just wish to highlight that the choice of a trade-off rate between false positives and false negatives is a genuine value-judgment, and that this makes medical image algorithms essentially value-laden.

Let us now consider an alternative, slightly more theoretical approach to the choice between false negatives and false positives. The ethical assumption underlying this approach is in short the following: A doctor's decision about whether avoiding false negatives is better than avoiding false positives should depend in part on the answer to the question of how dangerous the suspected disease is for the patient. The more dangerous it is, the more important it will be to detect it, and the more acceptable it is to take the risk of mistakenly identifying it in a patient (i.e. to produce a false positive). The same holds, secondly, for contagious diseases that could jeopardise others. Third, a rising number of false negatives is less acceptable if the faulty diagnosis is likely to bring about drastic side-effects such as e.g. a dramatically diminished quality of life after diagnosis, or even the triggering of suicidal tendencies in those who are mistakenly informed of a positive result in the testing of a disease. For instance, the mass screening for breast-cancer in women has recently raised severe criticism due to the fact that it bears a high risk of false-positives. Being wrongly diagnosed with breast cancer is usually extremely disturbing for women and their families.[13]

The choice for a certain threshold in an algorithm is a decision that is a judgment about which there is, or at least could be, a controversy between advocates of the major theories of normative ethics. This supports our claim that algorithms manifest or reflect certain ethical judgments. This point can be illustrated as a choice between deontological and consequentialist or utilitarian theories of normative ethics.

Deontologically minded users of the software are most likely to focus on the physical and mental integrity of the individual patient. Therefore, in a Kantian vein, medical doctors will most probably opt for the implementation of some more liberal algorithm when it comes to computerised image analysis of severe diseases, aiming at more false positives and less false negatives. In such a Kantian vein, it is pre-eminent to protect the individual patient, i.e. to avoid doing harm to him or her. Such harm would be most likely to be done if a severe disease remains undetected and therefore untreated. For this reason, the deontologist is willing to put up with problematic consequences of more false positives that are brought about by her choice.

In contrast, a scientist who tries to gather statistical data about a population of patients may wish to use the same software, but she will typically prefer more strictly or conservatively designed algorithms. This will bring about results that are more in line with consequentialist or utilitarian decision-making: What is important in this perspective is that the body of scientific data as a whole remains valid and intact, because this will typically bring about better consequences in the long run, although some individual patients may have to suffer along the road. This is in accordance with the consequentialist or utilitarian view that overall well-being in society should be our focus of attention, rather than the rights or needs of single individuals. Therefore, on this view, aiming at scientific insights that benefit the greatest number of people is the highest priority. This means that within the consequentialist approach, scientific results should be free from false positives as far as possible. It is therefore acceptable that a relatively large amount of false negatives will arise.

It is beyond the scope of this paper to take a stand on the controversy over deontological and consequentialist normative theories. However, it is important to note that the choice of a threshold value for a certain algorithm goes hand in hand with some normative background assumptions specified by the software designer. Problems arise if it is unforeseeable whether the background assumptions of the designer will be in accordance with those of future users (i.e. a doctor or a scientist). If the ethical assumptions made by the software designer remain implicit and differ from those preferred by the user, we face a rather severe problem: Without knowing it, the user who bases her decisions on the output of the software will base those normative choices on ethical assumptions that are in conflict with the implicit ethical assumptions made by the software designer.

Should ethical values be user-defined?

Our general view is that software designers should, as far as possible, leave ethical decisions to users (say, with an external switch to choose between 'preference for false positives' or 'preference for false negatives'). Moreover, when this is not possible the ethical assumptions in the algorithm should at least be transparent and easy to identify by users. In the case of medical image algorithms the users amount to medical doctors and scientists. For the reasons outlined in "The precautionary principle and ethical theories", physicians who use a software will typically work on the assumption that it is better to get a false positive than a false negative. But when scientists use exactly the same software they are likely to have the opposite ethical view. Such users will typically avoid false positives in the first instance. These divergent ethical judgments could, of course, easily be implemented into the software by including some code that allows the user to specify her preference between false positive and negatives.

We are aware that such a leave-it-to-the-user approach may sometimes be impossible to implement, as indicated above. There are cases in which it may just not be feasible to directly ask users to express the required kind of ethical judgments in numerical or quasi-numerical terms. Having to take a stand on fundamental ethical issues before using a software may be too difficult, and it may also be too impractical to ask users to make the relevant ethical choices. In such cases the underlying ethical assumptions should at least be made transparent. However, in what follows we shall set such cases aside and focus on cases in which it may actually be feasible to eliminate ethical judgments as far as possible from the algorithms underlying the software.

At the heart of the matter is the fact that it is not just a practical decision that must be made when designing an algorithm, but a genuinely ethical one. To opt for either a consequentialist or utilitarian or a deontological attitude towards e.g. the choice between false positives and false negatives amounts to a value judgment that brings about ethically relevant implications for the patient's health, for those who are in touch with her, and for the progress of science as a whole. As explained above, we therefore advocate that the software designer should design the algorithm such that it remains flexible and applicable to the requirements of different ethical settings. To be more precise, we propose that algorithms as far as possible remain open to the user's ethical preferences. The design of the algorithm must allow the user to choose the circumstances in which she situates herself. We therefore think that it is not enough that the designer makes his or her assumptions transparent by letting the user know what the ethical assumptions in the design stance were. Rather, it is necessary that the designer leaves it to the user to specify what ethical parameters to choose.

Having said this, we are of course aware that we now go beyond what is usually required in the literature on the ethics of image analysis. One of the few comments on the ethics of image analysis stems from Gert-Jan Lokhorst who was interviewed by van Strien (2008). He suggests that software has to be trained on image materials that resemble the ones it will have to analyse in the clinic, because the software will recognise and identify only those matters on which it was explicitly trained before. Moreover, Lokhorst suggests that doctors have to undergo proper training before they use the software. This will ensure that users become aware of which sorts of image analysis samples the system is familiar with, and this in turn ensures that users will be able to properly assess the competences of the software. Lokhorst's final suggestion is that users and other relevant parties should be provided with some basic knowledge about the situations in which the algorithms were developed by the designer, because (i) this will enable them to decide in which clinical situations the program should be applied and in which not, and (ii) it must also be made clear in advance who is responsible for which step in the process of image analysis.

Lokhorst thus implicitly agrees with us that it is essential to make the software designer's ethical choices transparent to the user. However, our way of formulating this ethical requirement is more far-reaching than Lokhorst's. As explained above, we maintain that the software designer must leave the responsibility for defining the default state of the software to the user herself. This is required for avoiding possible discrepancies between the ethical background assumptions of the software designer and users.

The example of the stenosis in "A real example" shows that there is no a priori "correct" threshold value. However, the very first choice made by the software designer will influence all further steps of diagnosis. In borderline-cases, those choices may even affect the treatment suggested to the patient by the doctor. This is disturbing, especially in face of the fact that the software designer may lack the proper medical background of a fully trained physician.

According to our argument for implementing user-defined ethical values, the doctor treating the individual patient should be enabled to make the relevant choice herself, as far as possible. If the patient is facing a severe condition, the threshold value should perhaps be set more liberally, to make sure that at any sign of deterioration, treatment is initiated. However, if the health condition of the patient is estimated as overall good, the threshold value should be set more conservatively to avoid "false alarms" that could lead to surgical interventions causing unnecessary harm to the patient.

CONCLUSION

We conclude that a strong case can be made for the claim that some algorithms are essentially value-laden. Some algorithms, such as those used for classifying cells as diseased or non-diseased, forces the designer of the algorithm to take a stand on controversial ethical issues, e.g. whether it is more desirable to prefer false positive errors over false negative ones. This is a controversial ethical issue, on which there is a lot of disagreement among ethicists. We propose that designers of algorithms should, as far as possible, leave ethical issues to users, and when this is not possible, the ethical assumptions in the algorithm should at least be transparent and easy to identify by users.

FOOTNOTES

[1] Software designers can, of course, be morally responsible also for algorithms that are not essentially value-laden. A fatal accident caused by a faulty algorithm can be (partially) blamed on the software designer, irrespective of whether the algorithm is essentially value-laden or not.

[2] Turilli (2007: 49).

[3] Allen et al. (2006: 13).

[4] Arkin (2009: 2).

[5] See e.g. Sipser (1997).

[6] Turilli (2007: 49).

[7] It is worth noticing that some decision support systems may apply a reasoning algorithm that, although deterministic, is too complicated to be

followed by human users (e.g., since it would take too long to be practical). In that case, the user cannot do anything else but follow up on the algorithm's suggestion *or* ignore the suggestion altogether. This renders the decision support algorithm value-laden.

[8] The algorithm outlined here is a deliberate simplification of actual, modern segmentation algorithms. In particular, the algorithms demonstrated in the cited website are much more sophisticated. For our argument, however, this is not relevant. Even advanced algorithms typically involve parameters, the values of which are to be chosen in order to decide, eventually, between 'normal' and 'pathological' cases.

[9] In principle, an algorithm can detect when it bases a decision on a 'near-borderline'-case. Sophisticated algorithms sometimes give an estimate of the reliability of their conclusions (for instance: long-term weather predictions, e.g. in the form of error-bandwidths). In this way, an algorithm can become less value-laden, since it moves part of the responsibility in the direction of the (human) user.

[10] By 'ethical position' we mean the decision at which level the threshold should lie. We will deal with this in detail in the next section.

[11] This section is partly based on some ideas first presented in Peterson (2007).

[12] To be precise, we should distinguish between on the one hand, medical imaging algorithms used exclusively in interaction with a medical specialist, and on the other hand, algorithms used to screen large volumes of data to seek for pathologies. In the latter type of cases the data volume is too large for (human) medical staff to handle in given clinical contexts (consider e.g. yearly screenings of ten thousands of women for early signs of breast cancer). In such cases, the risk for statistically significant biases in the results (that are ethically value-laden) is much larger than in the first case.

[13] See e.g. Gigerenzer et al. (2009) about these problems of screening and their statistical implications.

ACKNOWLEDGMENT

The authors wish to thank Robin van der Sligte for extremely helpful discussions, and Sven Diekmann and Rosemary Lowry for helpful comments on earlier drafts.

REFERENCES

1. Allen, C., Wallach, W., & Smit, I. (2006). Why machine ethics? *IEEE Intelligent Systems, 21*, 12–17.
2. Arkin, R. C. (2009). *Accountable autonomous agents: The next level.* Position paper for the DARPA Complete Intelligence Workshop, Feb. 2009.
3. Gigerenzer, G., Mata, J., & Frank, R. (2009). Public knowledge of benefits of breast and prostate cancer screening in Europe. *Journal of the National Cancer Institute*. doi:10.1093/jnci/djp1237.
4. Peterson, M. (2007). Should the precautionary principle guide our actions or our beliefs? *Journal of Medical Ethics, 33*(1), 5–10.
5. Resnik, D. B. (2004). The precautionary principle and medical decision making. *Journal of Medicine and Philosophy, 29*(3), 281–299.
6. Sipser, M. (1997). *Introduction to the theory of computation*, PWS Publishing Company.
7. Turilli, M. (2007). Ethical protocols design. *Ethics and Information Technology, 9*, 49–62.
8. van Strien, W. (2008). Opvangen zwakheden maakt beeldtechnieken waardevoller (interview met G.J.C. Lokhorst). *Ethiek, Onderzoek en Bestuur*, pp. 20–25. March 2008.

CHAPTER 19

Reliability of NAND Flash Memories: Planar Cells and Emerging Issues in 3D Devices

Alessandro S. Spinelli*, Christian Monzio Compagnoni and Andrea L. Lacaita

Dipartimento di Elettronica, Informazione e Bioingegneria, Politecnico di Milano, 20133 Milano, Italy

ABSTRACT

We review the state-of-the-art in the understanding of planar NAND Flash memory reliability and discuss how the recent move to three-dimensional (3D) devices has affected this field. Particular emphasis is placed on mechanisms developing along the lifetime of the memory array, as opposed to time-zero or technological issues, and the viewpoint is focused on the understanding of the root causes. The impressive amount of published work demonstrates that Flash reliability is a complex yet well-understood field, where nonetheless tighter and tighter constraints are set by device scaling. Three-dimensional NAND have offset the traditional scaling scenario, leading to an improvement in performance and reliability while raising new issues to be dealt with, determined by the newer and more complex cell and array architectures as well as operation modes. A thorough understanding of the complex phenomena involved in the operation and reliability of NAND cells remains vital for the development of future technology nodes.

Keywords: Flash memory; NAND Flash reliability; NAND Flash scaling; 3D NAND Flash

1. INTRODUCTION

The concept of electrically programming and erasing a semiconductor memory cell dates back to the 1970s, when a floating-gate device with a thinned oxide layer in a portion of the channel area was successfully demonstrated [1] as a viable alternative to ultra-violet erasable cells. However, it was not until the development of the one-transistor Flash memory device [2], in which the entire chip could be erased at once, that solid-state non-volatile code and data storage began to experience a tremendous growth. Data storage, in particular, was fostered by the invention of the NAND Flash [3], which is possibly one of the most successful semiconductor devices ever developed. Thanks to its extremely compact area occupation and to clever system design and optimization, it benefited from the market explosion of digital cameras first, followed by smartphones and tablets and culminating in the development of NAND solid-state drives (SSDs) [4], that have recently overtaken hard-disk drives (HDDs) in terms of areal storage density [5]. Although the discussion on whether SSDs will eventually replace HDDs aside from small consumer devices is far from settling, it is indubitable that more and more applications are falling within reach of SSDs, that are now targeting enterprise-class storage [6,7].

The NAND success story is the result of continuous scaling of the planar cell dimensions, that has led all major NAND Flash manufacturers to reach the 15 nm planar node [8,9,10,11]. However, several constraints have severely challenged the latest nodes development [12,13,14], eventually putting an end to the scaling of the planar Flash technology. Some of these constraints are process-based and reflect the increasing difficulty in patterning smaller and smaller features with the ArF immersion lithography. Then, traditional device design issues arise, such as the need to curb short-channel effects (exacerbated by the thick gate oxide, which has very limited scaling possibilities) and avoid degradation of the subthreshold slope (reducing the operation margin) while limiting junction leakage (which impairs channel boosting during program). Finally, there are issues that are specific to memories, such as cell-to-cell interference and program noise, as discussed in detail in [15]. On top of all this, reliability issues, to be discussed in the following, played a key role in challenging further scaling.

This increasingly dramatic scenario prompted all major NAND Flash manufacturers to seek an alternative way to continue to deliver increasing storage capacity, that has resulted in the demise of the planar technology and the pursuit of three-dimensional (3D) memory arrays. In this field, a pioneering proposal was put forward in 2001 [16], evolving into early proposals based on layer stacking [17,18,19] and reaching the first cost-effective solution of a 3D-integrated array in 2007 [20]. From there on, 3D NAND based on vertical channels have taken the lead in the quest for higher density, resulting in the first commercial product in 2013 [21,22] and reaching a capacity of 786 Gb with a density of 4.29 Gb/mm2 [23] while stacking up to 64 layers [23,24,25]. Chip capacity was boosted not only by 3D integration, but also by multi-bit storage: while two bit-per-cell (i.e., multi-level cells, MLCs) were used in the first prototype [21], today's technology relies on storing three bit-per-cell (triple-level cells or TLCs) and quadruple-level cells (QLCs) are under active development. Recent reviews of the different 3D NAND technologies and architectures can be found in [26,27,28].

The revolutionary move to 3D memories was not without problems, as discussed in [29]. In fact, while traditional device reliability issues must still be accounted for, new phenomena came to prominence, which are not merely an extension of what has already been observed on planar devices, but rather a consequence of the different device and array fabrication. The aim of this paper is then to review the main reliability issues that play a key role in planar NAND Flash devices and discuss the main experimental data and current understanding. Then, 3D NAND reliability is addressed, highlighting the role of such "traditional" mechanisms in the new cell size and geometry and introducing issues that are specific to the new memory cell. The analysis of the literature—though selected—reveals on the one hand the complexity of Flash reliability and, on the other hand, the huge efforts undertaken by the scientific community to understand the root causes of the experimental observations and develop effective predictive models that allowed a successful product design. Three-dimensional NAND bring further challenges to the table, with tighter requirements in an increasingly complex device, meaning that Flash reliability will offer an exciting field for scientific investigation for many years to come.

2. THE NAND MEMORY ARRAY

This Section contains a brief discussion on the planar NAND memory array architecture and program operations, with the aim of providing the basic knowledge to understand the rest of the paper. We will not discuss in detail the different array operations but rather restrict ourselves to the general principles, referring to specific literature for more details [27,30,31].

2.1. Array Architecture and Layout

The basic element at the heart of the NAND memory is the floating-gate (FG) transistor, in which a polysilicon island is sandwiched between a gate (or tunnel) oxide and an interpoly dielectric, in turn connected to a control gate. Whenever a charge Q is stored in the floating gate, simple electrostatics [32] shows that the threshold voltage VT of the cell becomes

$$V_T = V_{T0} - \frac{Q}{C_{pp}}, \tag{1}$$

where V_{T0} is the so-called *neutral* threshold voltage, corresponding to zero charge on the floating gate, and C_{pp} is the interpoly dielectric capacitance. As charge cannot easily leak from the isolated floating gate, a non-volatile memory element is achieved. In the NAND array, transistors are connected in series along *strings*, while their control gates, connected across different strings, constitute the wordlines (WLs), each making a *page*, as schematically depicted on the left of Figure 1 (more than one page can be associated to the same WL, but this is irrelevant for our purposes). Strings are connected to the bitlines (BLs) and the common sourceline via select transistors driven by a drain select line (DSL) or a source select line (SSL). This array constitutes a *block*, whose size is a key element in NAND array organization.

Cross-sections of a current planar NAND array are shown on the right of Figure 1: Figure 1a depicts a typical cut along the WL (green = silicon, red = floating gate, magenta = WL, white = silicon oxide), highlighting the shallow trench isolation used to separate the active area of adjacent devices [33] and the planar structure of the cell, without any wrapping of the control gate along the floating gate sidewalls [34]. Figure 1b depicts instead the cross-section along the string direction. Figure 1c shows the layout of the memory array, which becomes extremely compact thanks to these solutions. In particular, given the feature size F of a given technology, i.e., the size of the smallest feature that can be created on silicon, the elementary memory cell (shown as a dashed square in Figure 1c) occupies the ideal area of 4F2. However, the *effective* cell area, i.e., total chip area divided by the number of cells, is somewhat larger than that because of the overhead circuitry, string contacts with BLs, dummy cells and select transistors. To decrease such overhead, the number of cells in a NAND string has reached the impressive number of 150 in the latest planar nodes [10].

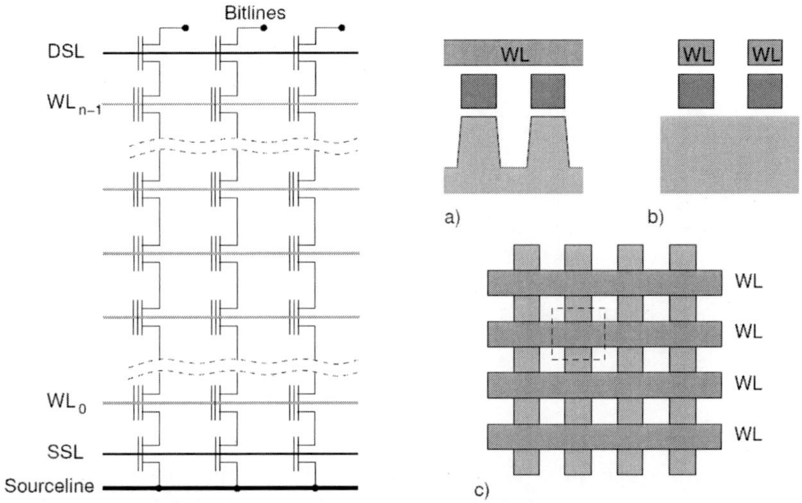

Figure 1. (**Left**) Schematic view of a NAND Flash array. Vertical strings of series-connected devices are attached to the bitlines and the sourceline via select transistors driven by a drain select line (DSL) and a source select line (SSL); (**Right**) schematic cross-sections of the array along the WL direction (**a**) and the string direction (**b**) (green = silicon, red = floating gate, magenta = WL, white = silicon oxide). Inset (**c**) shows the layout of the array with the elementary cell (dashed square).

2.2. Array Operation

As mentioned above, digital information is stored as charge onto the floating gate of a NAND cell. While the first devices relied on two discrete charge levels to store a single bit (the so-called single-level cells, SLCs), it was soon realized that the linear relation between floating-gate charge and cell V_T could be exploited to store more than one bit per cell [35,36]. Nowadays, TLCs are in production and used in SSDs [6], while quadruple-level cells (QLCs) are under active development [37]. Figure 2 shows the V_T distribution of a TLC NAND array: eight different levels are necessary to encode three bits. In the NAND architecture, all but one V_T level are positive (shown in different shades of blue) and are referred to as the *programmed* levels. The negative (red) VT level is instead the *erased* one.

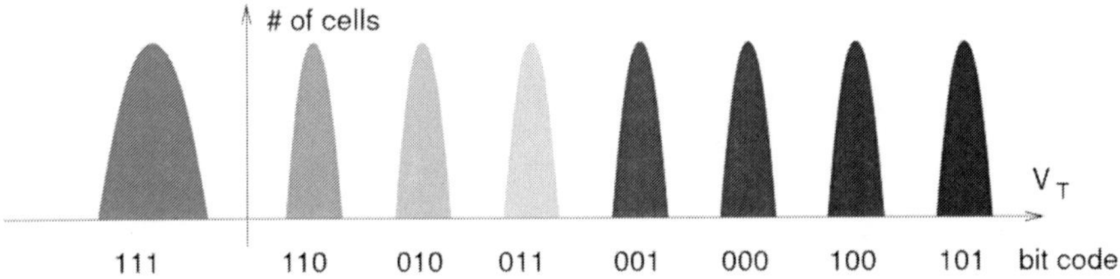

Figure 2. Pictorial view of the threshold voltage (V_T) distribution in a TLC cell. A three-bit code is associated to each of the V_T levels.

2.2.1. Read

Reading the state of a cell means discriminating one V_T level against another and irrespective of the V_T levels of all other cells in the string. To accomplish this, the string has first to be connected to the sensing circuit, i.e., the select lines are high and the BL is biased; therefore, a current can flow to the (grounded) sourceline. Unselected cells are then placed in a pass-transistor mode, biasing their WLs at a *pass voltage* larger than the maximum V_T. The actual read operation is then performed, sensing the cell current and comparing it against fixed reference values to assess the corresponding binary code [38]. The operation is performed more than one time for MLC and TLC cells. To increase parallelism, strings are read in parallel across a page (typically 8–16 kB), resulting in a throughput higher than 150 MB s^{-1}.

2.2.2. Program

NAND Flash cells adopt channel Fowler–Nordheim (FN) tunneling to store electrons onto their floating gates [39]. To this aim, the selected cell is first connected to the external circuitry by biasing the unselected WLs to a pass voltage and turning on the DSL, while the BL is grounded and the SSL is left off. A high voltage is then applied to the desired WL, resulting in tunnel electron injection from the inverted substrate to the floating gate. In MLCs or TLCs, fine placement of the V_T levels is needed because of the reduced distance between adjacent distributions (see Figure 2). This is achieved via an *incremental step-pulse programming* (ISPP) technique [40], in which fast pulses of increasing amplitude are used to inject a constant, controlled amount of charge onto the floating gate at each step, interleaved with *program verify* phases in which V_T is monitored. Several refinements are applied to this basic concept in 8- or 16-level cells [38].

Since the program operation is controlled by the WL bias, it acts on a page of the array. However, not all cells must be programmed at the same level (or programmed at all), which means that a program-inhibit strategy must be implemented for those cells. Inhibiting programming means reducing the gate-channel bias of such cells, thereby lowering their tunnel oxide field responsible for the electron injection. In NAND arrays, this is achieved via a *self-boosting* procedure [36], where the inhibited strings are first floated and their potential is temporarily raised via capacitive coupling to the WLs, that are raised at the passing voltage [38]. Program throughput in excess of 50 MB s−1 has been reported [10].

2.2.3. Erase

Erase operation is performed on blocks, i.e., a set of pages that can be 4–8 MB in size, exploiting the same FN mechanism used in program. As NAND chips do not allow the use of high negative voltages, WLs are grounded and a high positive voltage is applied to the *p* and *n* wells of the array. A verify phase follows, aimed at ensuring that all cells have negative V_T, which triggers another erase pulse if the condition is not met: the technique is similar to what is used during program, and is called *incremental step-pulse erasing*. Floating WLs are used to inhibit erase on unselected blocks.

3. NAND FLASH MEMORY RELIABILITY

From a general standpoint, the reliability of a product is related to its capability to continue to work as expected as a function of time. Reliability is enhanced by screening, testing and qualification and, eventually, by error management procedures. In all these respects, Flash memories are peculiarly different from other semiconductor devices, as they intrinsically operate under high-field conditions that other devices purposely avoid. Moreover, reliability is usually extrapolated from accelerated tests, that should replicate as close as possible the real operating conditions. Unfortunately, the development of such tests is complicated by the many different patterns of program/erase and read cycles under which Flash memories usually operate.

As a matter of fact, there are a number of failure modes and mechanisms that are specific to Flash, not to mention error-management procedures that are put forward by the manufacturers to increase reliability and yield. These depend on the type of failure, which can be recoverable (i.e., it can be fixed by erase and reprogram) or not. Recoverable errors are usually dealt with via error-correction codes (ECC). Initially, when SLCs were prevalent, simple Hamming algorithms were employed, which could correct single-bit errors and detect two-bit ones. Later, more efficient BCH (Bose, Ray-Chaudhuri and Hocquenghem) algorithms were implemented in the Flash controller [41]. However, as Flash capacity continued to rise and MLCs gave way to TLCs, more efficient ECCs were needed, resulting in a transition to soft decoding schemes. Low-density parity-check (LDPC) ECCs [42,43,44,45] are gaining traction in modern SSDs, while concatenated codes for NAND Flash have been also proposed [46,47,48,49].

Non-recoverable errors, or errors exceeding the ECC capabilities, result in *bad blocks* [50], i.e., blocks that are excluded from the memory operation and are replaced by available valid blocks. A small number of initial bad blocks is allowed, which can increase with use. Eventually, when the number of bad blocks exceeds the device specifications, the entire chip fails. It is worth mentioning that both ECC and bad blocks management require extra cells/blocks, thereby contributing to an increase of the cell effective area as discussed in Section 2.1, i.e., a reduction in the array efficiency.

Reliability specification by the manufacturers typically includes the number of bad blocks, the endurance and retention properties of the array and the number of error-correction code bits. In the following, however, we will go beyond the manufacturer specifications and look at the basic physical mechanisms at the heart of the reliability issues. Since this is an extremely large field, we restrict our discussion to issues that develop with memory usage, neglecting for example defect-related issues, time-zero problems such as VT placement, cell-to-cell interference and many others, for which we refer the reader to other overviews of Flash memory reliability that are available in the literature, leaning either toward a device/technology [27,51,52,53,54,55,56,57,58,59] or user [60,61,62] viewpoint.

3.1. Endurance

This parameter specifies the maximum number of program/erase (P/E) cycles NC that the memory chip can tolerate. The onset of this limitation can be observed by repeating simple, single-pulse P/E operations on a cell and monitoring the resulting VT. An example of such a dependence in a NAND device is shown in Figure 3 [63]: the V_T levels experience an increase for high values of NC, eventually leading to a read failure. Higher increase is usually observed on the erased level.

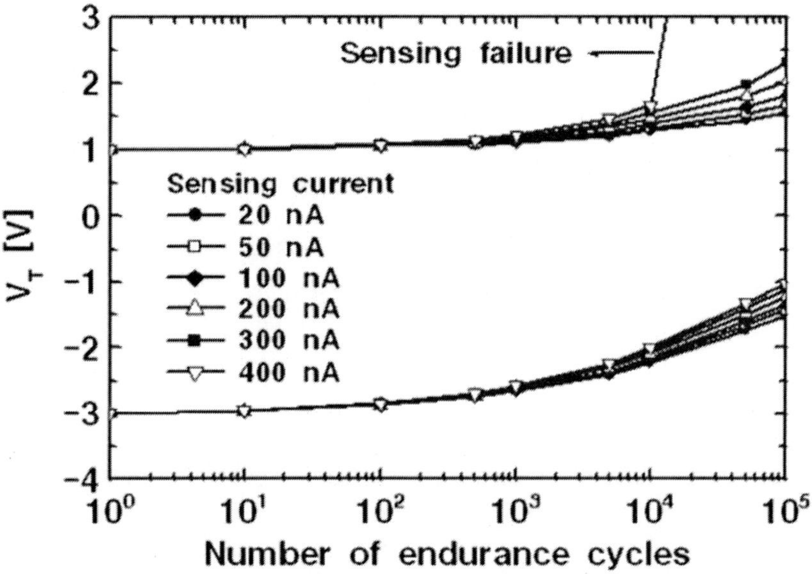

Figure 3. Endurance characteristics of a NAND cell [63], © 2004 IEEE.

Like most reliability concerns for Flash memories, this phenomenon is one consequence of the electron flow through the oxide during FN tunneling, which has been observed to generate oxide traps and fixed charge in MOS (Metal-Oxide-Semiconductor) devices since the 1980s [64,65,66,67,68,69]. At first glance, such charges should result in a constant shift of both V_T levels, while the actual behavior features a milder dependence on NC by the programmed V_T with respect to the erased one, pointing toward an interplay among different mechanisms. This picture is confirmed by early results on different non-volatile memory technologies, where different P/E schemes (i.e., different electron injection conditions into the oxide) resulted in different V_T dependences on NC [70].

A thorough analysis of the phenomenon was first conducted in [71], using charge-pumping and V_T measurements to assess the roles of interface and oxide charges, and was later corroborated by subsequent studies [72]. The authors' conclusion was that trapped oxide charge should not influence the programmed V_T value because tunnel injection from the channel during the program pulse is controlled by the oxide field at the cathode interface, meaning that the final condition reached by the cell will be of equal field, irrespective of the starting V_T and trapped charge (if not too close to the cathode interface). In turn, the same oxide field at the channel interface means the same final V_T, explaining the experimental behavior of a nearly constant programmed V_T level. Simply put, as trapped negative oxide charge builds up, it reduces injection, giving a negative V_T shift that is balanced by the positive VT shift given by the charge itself. This "converging" behavior of the FN tunneling operation has been in fact exploited to tighten the V_T distribution after erase [73].

The same reasoning holds for the erase operation, but now the final condition is the one with the same field at the oxide-floating gate interface (the cathode during erase), which does not translate into the same V_T, now dictated by the oxide trapped charge: a larger increase in V_T vs. NC is then expected. This model was then refined [63] to include the effect of charged stress-generated interface states, that provide an additional positive V_T shift for both levels (while not influencing injection) for high cycling conditions, explaining the slight increase in the programmed V_T. Finally, the effect of non-uniform tunneling current flow and oxide charge distribution over the area of scaled NAND devices have been investigated in [74,75], pointing out their effect on the endurance curve and on the assessment of the endurance degradation with scaling. A recent technique for separating the endurance degradation components, including subthreshold slope and transconductance, which also influence the V_T values, has been reported in [76].

From the product viewpoint, the V_T window degradation is managed by the program and verify ISPP algorithm, meaning that any variation in the P/E characteristics of the cell will result in a change in the number of ISPP pulses. Eventually, when—say—the number of erase pulses will become higher than the maximum allowed by the erase time specifications, an erase failure will be signaled. This is an unrecoverable error, resulting in the block being marked as *bad*(overprovisioning ensures that failed blocks do not reduce the device capacity). When the number of bad blocks exceeds the device specifications, the entire chip fails. Several techniques of *wear leveling* [77] are adopted by current NAND suppliers to distribute the number of P/E cycles among the blocks and make degradation uniform. Typical endurance values for SLC cells are in the 105 P/E cycles range, which drops by an order of magnitude or even more when moving to MLC [62]. An even lower value (around 1000) can be expected for TLCs. For SSDs, this specification translates into the total amount of writable data in—say—five years for a given workload.

3.2. Random Telegraph Noise (RTN)

3.2.1. RTN Amplitude

Random telegraph noise (RTN) in MOS transistors is an abrupt two-level fluctuation in drain current ID following trapping and release of a single electron by an oxide defect. This well-known phenomenon has been observed and studied since the 1980s [78,79,80,81,82,83,84,85,86], resulting in a prediction for the RTN amplitude [79] equal to

$$\frac{\Delta I_D}{I_D} = \frac{q}{C_{ox}WL(V_G - V_T)} + \frac{\Delta\mu_n}{\mu_n}\frac{\Delta A}{WL}, \tag{2}$$

where q is the electron charge, C_{ox} the oxide capacitance per unit area, W and L the channel width and length, respectively, and μ_n the electron mobility in the inversion channel, which experiences a fluctuation $\Delta\mu_n$ over an area ΔA as a consequence of the additional scattering with the trapped electron. Typical values for the first term are smaller than 10%, while the scattering part is typically negligible in the subthreshold regime, where NAND cells are usually read; therefore, RTN has long been regarded as a curious scientific phenomenon not constituting a noticeable reliability threat. This held true even if unusually large values for the RTN amplitude had been consistently reported on some devices, together with complex time behaviors [79,82,87,88,89], that seemed to defy Equation (2). When Flash memories made a large quantity of data readily available, making it possible to pinpoint and study anomalous behaviors occurring with extremely low probability, larger and larger fluctuation amplitudes emerged [90,91], and RTN became a serious reliability constraint in static random-access [92,93,94] and Flash memories [90,91,95,96,97,98,99], prompting an intensive research effort to understand its root cause and impact on memory array operation.

An example of the relevance of RTN is shown in Figure 4, where results obtained on a selected decananometer Flash cell are reported [91]: current fluctuations up to 60% were detected, which cannot be explained with the above theory. Though early explanations pointed to multiple carrier emissions [88,89], change in defect properties [87] or interacting quantum wells [82], a different approach based on the percolative nature of the current conduction around the charged trap site had been mentioned in [100] and thoroughly revisited in [101], exploiting a concept applied to amorphous semiconductors [102]. The authors pointed out that no uniform conduction in a MOSFET (Metal-Oxide-Semiconductor Field-Effect Transistor) channel could account for the anomalous RTN amplitudes. Rather, a non-uniform conduction with filamentary current flow was needed as a starting point; therefore, an RTN trap could effectively modulate a current-crowding path, resulting in a large current fluctuation. Fixed oxide and interface charges were proposed as sources of percolation, following earlier studies [103,104]. This is now regarded as the correct interpretation of the RTN amplitudes, with the caution to add the actual main source of percolation: random dopant fluctuation (RDF). In fact, statistical variation in number and position of dopants was known to affect MOSFET parameters such as VT and subthreshold slope [105,106,107,108], and was shown to have a significant impact on RTN in [109] and in [110,111], where large amplitude values compatible with experimental data were reported.

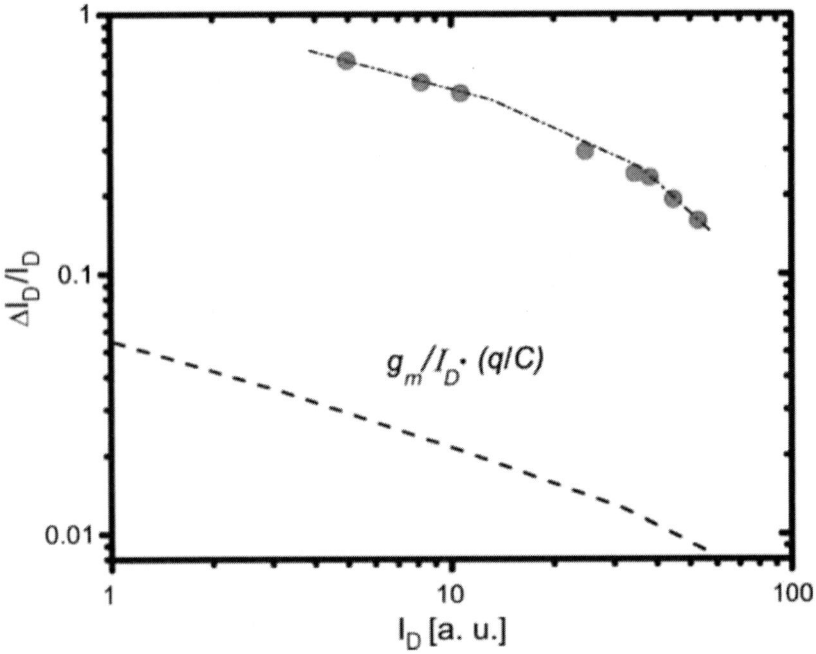

Figure 4. RTN fluctuation amplitude measured on a decananometer Flash cell (points = experimental data, dashed line = predictions from (2) when mobility modulations are neglected) [91], © 2007 IEEE.

For the evaluation of the RTN effects in Flash memories, the statistical distribution of its amplitude is needed, accounting for realistic device geometry and doping. This task was undertaken in [112,113], where 3D Monte Carlo simulations of electron transport were carried out, accounting for randomized doping and trap positions. Simulations were conducted within the drift-diffusion framework, neglecting the field dependence of mobility to avoid unphysical effects close to the ionized dopants and RTN trap. The impact of Coulomb scattering on the RTN and variability characteristic of such MOSFETs has been studied in [114,115,116,117], showing that those effects are barely relevant for NAND Flash, that are read in the subthreshold regime. Charge quantization was instead accounted for via the density-gradient correction to the drift-diffusion formalism [118,119]. Its relevance from the variability viewpoint was discussed in [120].

Figure 5 shows results for the current density profile in the channel of a decananometer Flash memory biased around a threshold. The two figures on the left-hand side correspond to a case of high V_T shift: transport in the empty trap case (upper figure) has a percolative nature because of RDF and field intensification around the active area corners. The RTN trap (black square in the lower figure) is located right above one main percolation path, close to one edge: its charging will shut off such path, leading to a large variation in current and V_T. On the other hand, if the trap is located over a region in which the current density is already negligible, its charging and discharging will barely affect the overall current flow, as shown by the figures on the right-hand side.

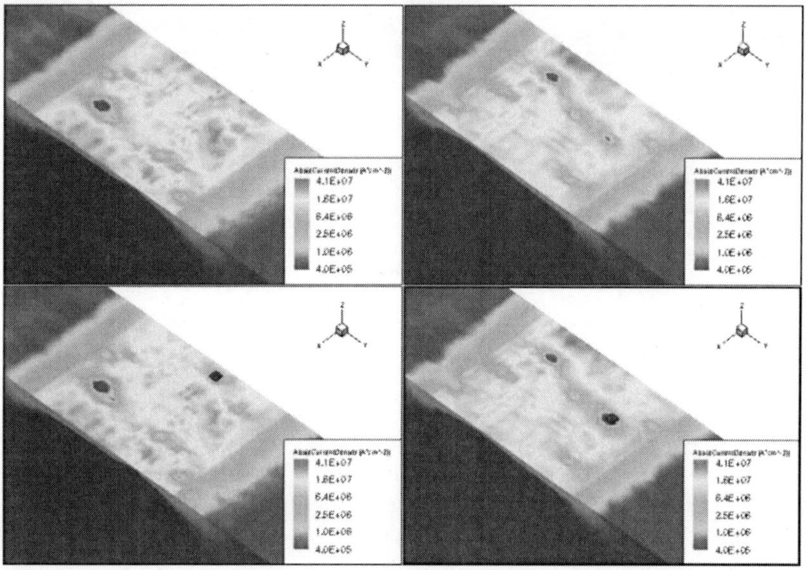

Figure 5. Current density profiles in the channel of decananometer Flash memories with and without a charged RTN trap. Left-hand side = high V_T shift case, right-hand side = low V_T shift case. Upper figures represent the empty trap case, lower figures the charged trap ones. The trap position is marked by the black squares in the lower figures.

Monte Carlo results for the statistical distribution of the V_T shift due to a single RTN trap, ΔV_T^1, are shown in Figure 6: an exponential distribution was found, which can be easily characterized by its slope λ, usually expressed in mV/decade. A thorough parametric analysis of λ was carried out in [113,121], resulting in the following dependence:

$$\lambda = \frac{K}{\alpha_G} \frac{t_{ox}^{\alpha} \sqrt{N_A}}{W\sqrt{L}}, \tag{3}$$

where t_{ox} is the tunnel oxide thickness, N_A the average substrate doping, α_G the control-gate coupling coefficient and K is a constant dependent on device design, doping profile and so on. This relation has proven itself to reproduce quite successfully the scaling dependence of the RTN slope over several different technologies.

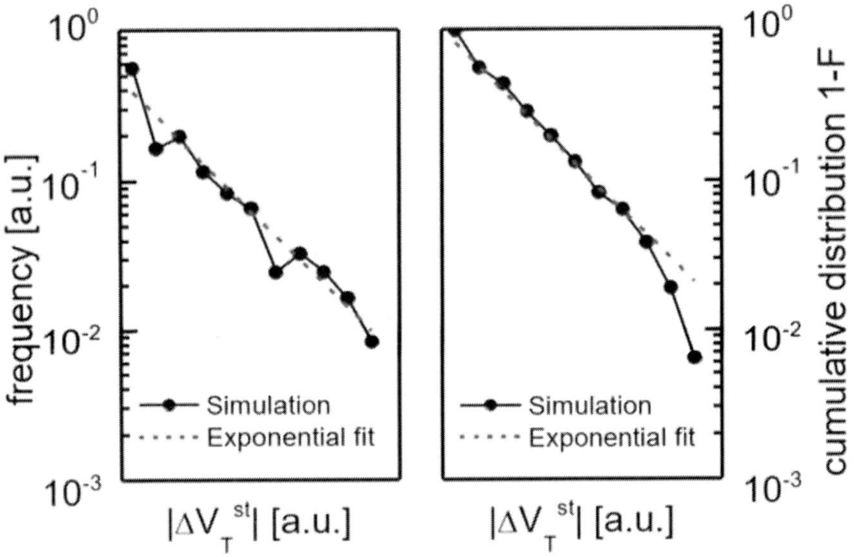

Figure 6. Probability density function (**left**) and cumulative distribution (**right**) for the threshold voltage shift determined by the charging of a single interfacial trap, ΔV_T^1, in a decananometer NAND Flash cell [112], © 2008 IEEE.

3.2.2. Main Experimental Data

The first experimental data on the statistical distribution of RTN amplitude in Flash memory arrays were published in 2006 [95]. For each cell in the array, the authors applied a staircase control-gate bias, recording the V_T value at which the cell switched from on to off state. Because of RTN, different V_T s are recorded every time the experiment is repeated, and their maximum and minimum values will give the amplitude of the RTN fluctuation (two staircases are actually employed, with increasing and decreasing control-gate bias, but the working principle does not change). Results for the fluctuation amplitude on a 90 nm cell are reported in Figure 7: note that the majority of cells exhibit negligible V_T fluctuations, but a small number of them can reach V_T shifts of a few hundred mV. Moreover, the number of cells exhibiting this behavior increases with P/E cycling.

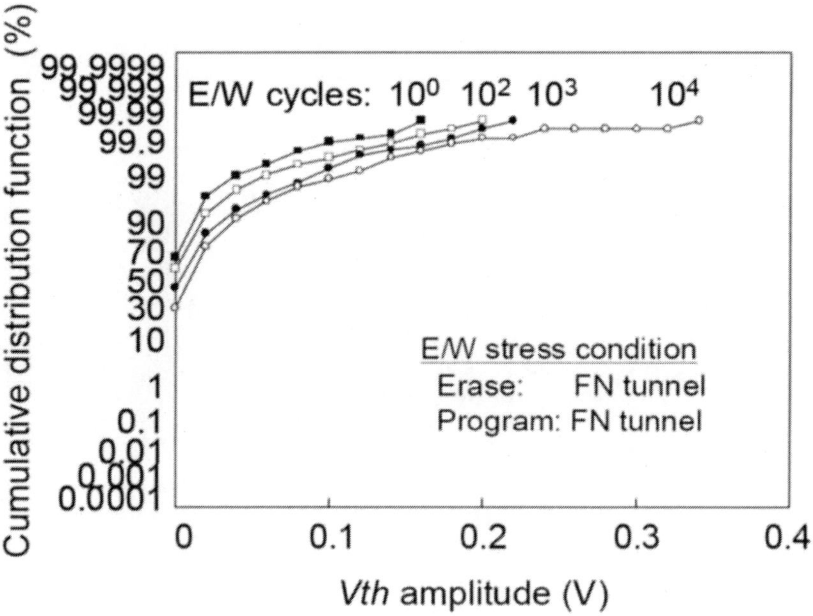

Figure 7. Experimental cumulative distribution function for the V_T shift induced by RTN in a Flash array [95], © 2007 IEEE.

A simpler characterization technique for RTN is reported in Figure 8 [96], where the V_T values at two different times are recorded for each cell. The RTN amplitude is now defined as the V_T difference between the two values. Note that this technique does not measure the *maximum* RTN fluctuation amplitude, as accomplished by [90,95,97,99], because a noisy cell can give either positive, negative or zero fluctuation, depending upon the sampling time (see the different cases Figure 8). However, this technique will return the actual RTN amplitude distribution that matters under real operating conditions, when a cell is read after being programmed.

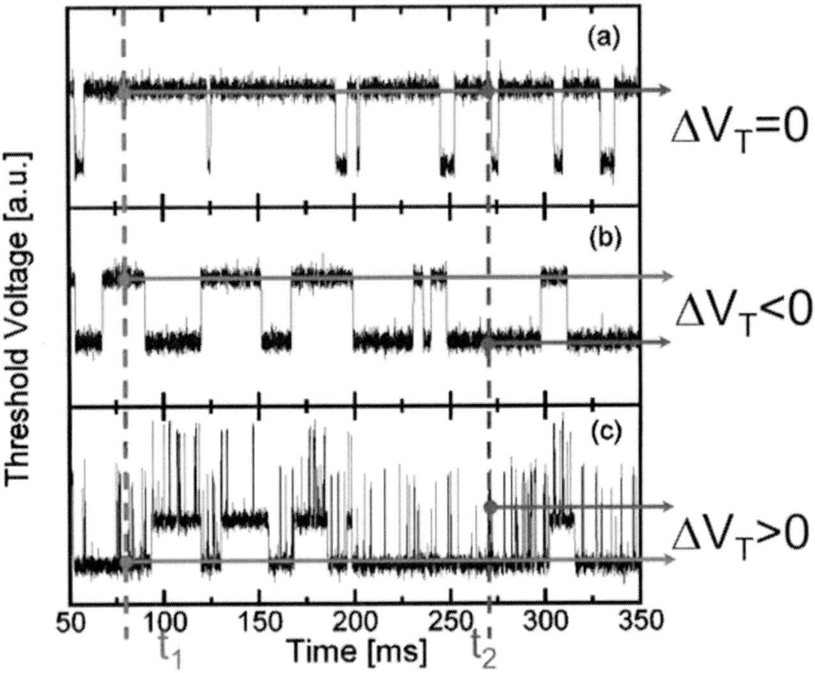

Figure 8. Time dependence of V_T for three Flash cells affected by RTN. V_T is sampled at times t_1 and t_2 and the RTN amplitude is defined as the V_T difference. Note that (**a**) null, (**b**) negative and (**c**) positive ΔV_T can result.

Typical results obtained with this technique are shown in Figure 9 for NOR (left) and NAND (right) Flash arrays [96,122,123]. Figures show the cumulative density function (cdf) and its complement, highlighting clear exponential tails down to very low probability values [124] for both high and low ΔV_T values. The region around $\Delta V_T = 0$ is affected by Gaussian noise, and the exponential tails depart from such region at different probabilities. Note also that when ΔV_T is evaluated between the nth and the first read operation, the tails slightly increase with n. Differences between NOR and NAND RTN levels and amplitude are determined by cell design parameters (see Equation (3)) as well as by the cycling conditions.

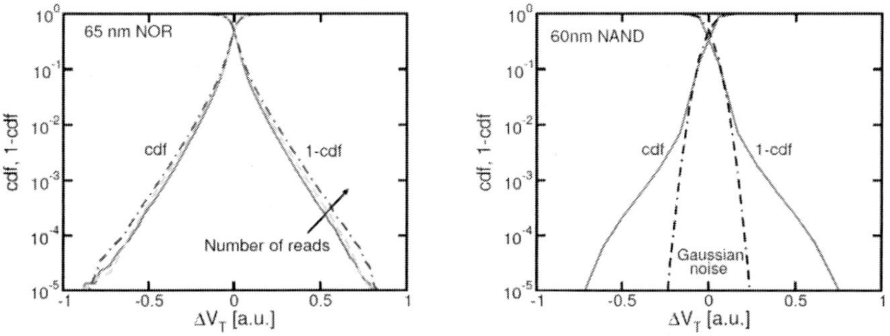

Figure 9. Experimental results for the ΔV_T statistical distribution on a 65 nm NOR (**left**) and a 60 nm NAND (**right**) Flash array, obtained following the procedure highlighted in Figure 8. Both the cumulative distribution function (cdf) and its complement are displayed, to highlight the exponential tails. Results for different numbers of reads are also shown for the NOR case.

Figure 7 shows that RTN depends on P/E cycling. An increase in the tail height was also reported in [98,123,125], where different technologies were compared. Figure 10 (left) shows experimental data for RTN in a NAND array for different cycling conditions: note that the exponential tails grow while their slope is hardly affected. This is reported in the right-hand side of Figure 10 where the slope λ of the exponential tails and the probability at a fixed $|\Delta V_T|$ value are shown as a function of NC, for the same technology. Note the negligible dependence of λ on NC, confirming that cycling results in a parallel shift of the RTN tails. The increase in tail height follows a power-law dependence with exponent 0.29 for this technology [123]. Different values were found in NOR arrays, owing once more to differences in cell design and cycling conditions. Such dependences prompted several works and studies on the optimization of the cell process/architecture in order to reduce the extent of the RTN [124,126,127,128,129,130,131,132,133,134]. In the NAND array, moreover, the RTN amplitude was shown to depend on the cell position and state along the string [135,136] (because of the different transconductances, depending on the source and drain series resistances) and on the state of cells on adjacent BLs [137] (because of the modification in the electron density profile induced by electrostatic interference).

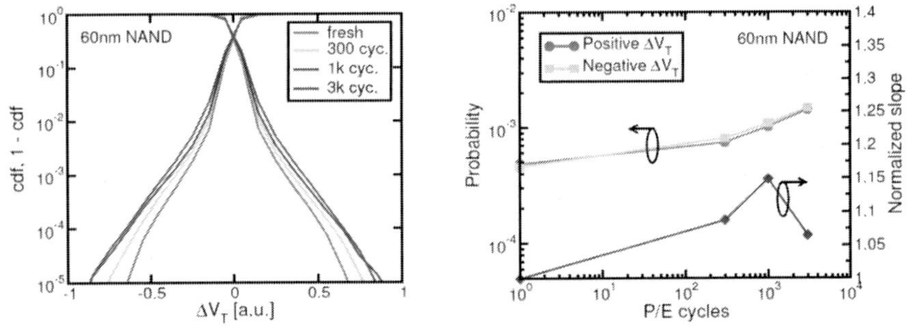

Figure 10. Experimental data for the RTN distribution in NAND arrays as a function of NC (**left**) and RTN tail height at a fixed $|\Delta V_T|$ value and slope as a function of NCfor a 60 nm NAND (**right**) Flash array [123], © 2008 IEEE.

The cycling dependence was also investigated in [138], in connection with the temperature dependence of RTN. Two cycling schemes were adopted, involving different temperatures and cycling times in accordance with the activation energy of damage recovery which will be discussed in Section 3.4. Then, RTN was read at room temperature. Figure 11 (left) shows that no significant difference can be detected in the RTN cdfs, meaning that the high-temperature phase applied during the 500 h scheme is not annealing any defect responsible for RTN. The right-hand side of Figure 11 shows instead the dependence upon the read temperature. Again, no dependence appears when the temperature is changed from room temperature to 55∘C. This result may seem surprising, as it is well known that trap capture and emission times are temperature-dependent [136,139,140], but comes from the distribution of such times among the cells, as discussed in the next section.

3.2.3. Models

The experimental data just reported call for a consistent model able to explain the different features of the phenomenon. While data shown in Figure 7 can be reproduced by a simple exponential fit [98], the explicit time dependence captured by Figure 8 and Figure 9 requires a time-dependent analysis, that was first put forward in [96,122]. The starting point is the exponential distribution of V_T shift extracted from simulation results described in Section 3.2.1. Such a distribution describes a positive V_T shift between time t_1 and t_2 if an empty trap at t_1 is filled at t_2 and a negative shift if the opposite is true. If the trap state does

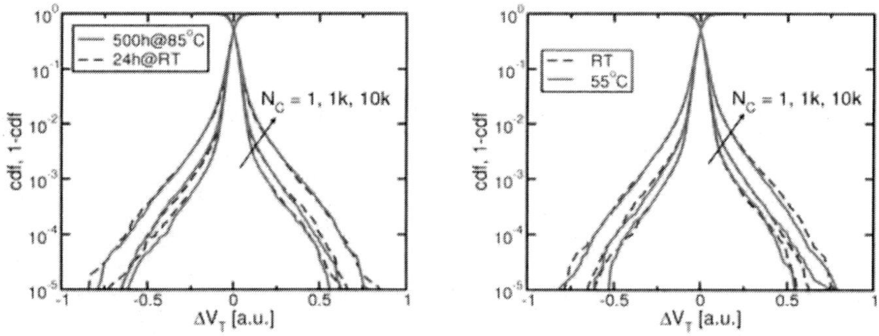

Figure 11. RTN cdfs for two cycling schemes and room-temperature read (**left**) and for one cycling scheme and different read temperatures (**right**). No significant differences appear in all cases.

not change, a zero V_T shift holds. The final time-dependent distribution is then the one shown in Figure 12 (left), where the positive and negative exponentials and the zero-shift delta function must be weighted by their occurrence probabilities, representing their areas. This is accomplished describing the RTN process as a two-state Markov process [141,142], as depicted in Figure 12 (right). The two states correspond to the empty (low-V_T, state 0) and filled (high-V_T, state 1) trap. Considering constant capture and emission times τ_c and τ_e, the transition probabilities between state i and j, P_{ij}, over a time interval T become:

$$P_{11}(\tau_c, \tau_e) = \frac{1}{\tau_c + \tau_e}\left(\tau_c\, e^{-T/\tau_{eq}} + \tau_e\right) \tag{4a}$$

$$P_{10}(\tau_c, \tau_e) = 1 - P_{11} = \frac{\tau_c}{\tau_c + \tau_e}\left(1 - e^{-T/\tau_{eq}}\right) \tag{4b}$$

$$P_{00}(\tau_c, \tau_e) = \frac{1}{\tau_c + \tau_e}\left(\tau_e\, e^{-T/\tau_{eq}} + \tau_c\right) \tag{4c}$$

$$P_{01}(\tau_c, \tau_e) = 1 - P_{00} = \frac{\tau_e}{\tau_c + \tau_e}\left(1 - e^{-T/\tau_{eq}}\right) \tag{4d}$$

where $\tau_{eq}^{-1} = \tau_c^{-1} + \tau_e^{-1}$. Note that these expressions describe the probabilities to find the trap in a certain state after a time interval T, given the initial condition, irrespective of the number of transitions taking place in-between.

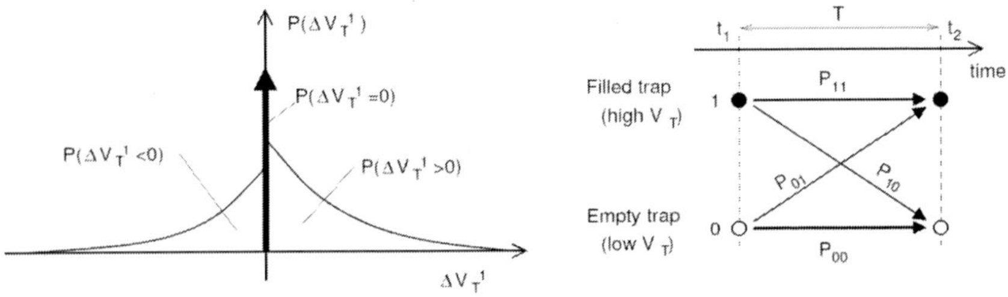

Figure 12. PDF (Probability Density Function) of the single-trap threshold voltage shift $P(\Delta V_T^1)$ (**left**) and Markov process view of the single-trap RTN waveform between times t_1 and t_2=t_1+T (**right**).

From (Section 3.2.3), the probability for a single trap to provide a V_T shift, ΔV_T^1, can be easily calculated:

$$P(\Delta V_T^1 > 0|\tau_c, \tau_e) = P_0^{t_1} P_{01}(\tau_c, \tau_e) \quad (5a)$$

$$P(\Delta V_T^1 < 0|\tau_c, \tau_e) = P_1^{t_1} P_{10}(\tau_c, \tau_e) \quad (5b)$$

$$P(\Delta V_T^1 = 0|\tau_c, \tau_e) = P_0^{t_1} P_{00}(\tau_c, \tau_e) + P_1^{t1} P_{11}(\tau_c, \tau_e), \quad (5c)$$

where $P_i^{t_1}$ is the probability for the trap to be in state i=0,1 at time t_1.

Once the analysis for a given set of τ_c, τ_e is assessed, we must account for their statistical distribution among the traps. The original approach in [122] relied on the classical theory of direct tunneling into oxide defects [80,81], explicitly linking τ_c and τ_e to the space and energy position of the trap inside the oxide. Later results have, however, shown that capture/emission times in MOSFETs are not compatible with such a description [143,144], due to device variability effects on the time constants [145,146,147] and structural relaxation of defects [140,148], generating a large spread in time constants even for traps located close to the silicon/oxide interface and disrupting the classical correlation between τ_c and τ_e. The study of the microscopic properties of RTN traps is still an active research topic [149,150,151,152,153,154,155,156], but for our purposes we will adopt a pragmatic approach, assuming given distributions of capture/emission time constants that may be consistent with observations, without linking them to any particular trap location. A constant PDF on a log-time scale was successfully assumed in [138], such as the one that will be used to describe detrapping after cycling in Section 3.4.2. It is worth noting that the constant trap density adopted in [122] implicitly provides a similar time constant distribution, due to the exponential nature of the tunneling probability, the only difference being the lack of correlation between and and between the time constant and the spatial location of traps. Within this framework, the expression for the single-trap V_T shift pdf, $P(\Delta V_T^1)$, becomes

$$P(\Delta V_T^1) = \int\int P(\Delta V_T^1|\tau_c, \tau_e)p(\tau_c, \tau_e)d\tau_c d\tau_e, \tag{6}$$

Where $p(\tau_c, \tau_e)$ is the probability for a trap to possess such capture/emission time constants.

Once $P(\Delta V_T^1)$ is computed, multiple trap effects can be easily accounted for assuming statistical independence of their V_T shifts. This means that the overall V_T shift is the sum of the individual contributions, which translates into a convolution operation for pdfs. In particular, for N_T independent traps, each with the same $P(\Delta V_T^1)$ pdf, we have

$$P(\Delta V_T^{N_t}) = \bigotimes_1^{N_t} P(\Delta V_T^1) \tag{7}$$

and, since the number of traps per cell can be assumed to be statistically distributed according to a Poisson statistics $p(N_T)$, we finally get the RTN pdf:

$$P(\Delta V_T) = \sum_{N_t=1}^{\infty} P(\Delta V_T^{N_t})p(N_t). \tag{8}$$

Typical comparisons between model results and experimental data are reported in Figure 13 for NAND [138] and NOR [122] arrays. The left-hand side shows that a nice agreement can be achieved for different technologies and cycling conditions. In particular, because of the exponential nature of $P(\Delta V_T^1)$, its slope ⍰ matches what is obtained from array data and resulting from the convolution operation in (13). The right-hand size is instead related to data reported in Figure 9 (left), where RTN distributions for increasing read operations (i.e., time) are reported, and shows the probabilities at fixed |⍰V_T| values as a function of the number of read operations. The increase in the RTN figure is well reproduced by the model and can now be explained: the increase in the elapsed time between the read operations (i.e., T in Equation (Section 3.2.3)) allows for more and more traps to be activated, as their becomes comparable with T. This increases the randomness of , enlarging the RTN distribution. This effect was nicely shown in [96,122] as an extension with time of the oxide region whose traps are sampled by RTN.

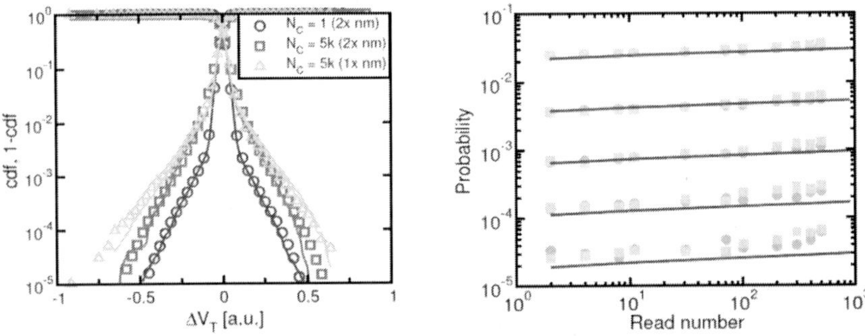

Figure 13. (**Left**) Comparison between experimental data and model for the RTN distribution in and NAND arrays and different ; (**Right**) same comparison for the tail probability at a fixed as a function of the number of reads for a 65 nm NOR Flash array.

Finally, the temperature-independent RTN behavior observed in [138] was explained in terms of a lack in correlation between the time constants and their activation energy values; therefore, temperature will not affect the overall time constant distribution. Of course, different defects will contribute to the RTN fluctuations, as some are accelerated and will end up being filtered by the read time while slow ones will come into play as a result of thermal activation.

3.2.4. Effect on Programmed V_T Distribution

Figure 14 shows pictorially what the effect of RTN on the V_T levels is and how it affects the array reliability, for the case of an MLC array. Exponential tails due to RTN extend from the ideal Gaussian distribution, their magnitude increasing with N_c and time. This results in a reduced separation between adjacent V_T levels (the so-called *window*), compromising the accuracy of the read operation and leading eventually to a read failure: the dashed lines in Figure 14 indicate the reference V_T levels used during read, to which the cell V_T is compared. When a distribution crosses a reference level, a read error is bound to occur.

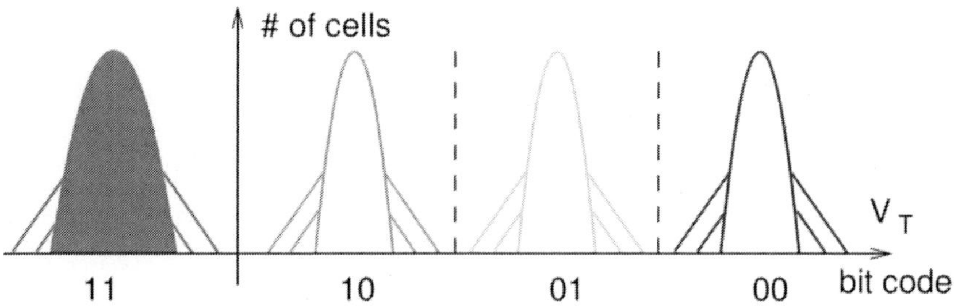

Figure 14. V_T distribution of an MLC array in the presence of RTN. Dashed lines represent the reference levels used during read.

To investigate this issue in more detail, Figure 15 shows the effects of RTN on the programmed V_T distribution [157] following ISPP and successive read. The left-hand side shows the V_T pdf right at the end of the ISPP algorithm, for two values of the ISPP step: V_{S1} and $V_{s2} > V_{s1}$. In the ideal case, the algorithm is expected to result in a V_T distribution exceeding the program-verify level V_T by an amount at most equal to the ISPP step (dashed lines in Figure 15). In reality, the distributions are widened by the program noise (not discussed here, see [158,159]) and by RTN and display the characteristic exponential tails. Note that no cell can have values below the program-verify level, as the algorithm would apply an additional step to them. However, a negative RTN tail also arises as soon as the cells are read (right-hand side of Figure 15). This figure also shows that the RTN tail height is smaller for larger values of the ISPP step; however, increasing its value will lead to an overall larger V_T distribution. On the other hand, reducing the ISPP step is effective—apart from the increase in program time—only until its value becomes comparable with the RTN distribution width.

3.3. Retention after Cycling and SILC

Data retention specifies the maximum period of time that allows a correct retrieval of stored data, and is determined by the faint but nonzero leakage of electrons from the floating gate. Early experimenters ascribed such leakage to electron tunneling through the dielectric and realized that temperature [160] or bias [161] could be used to accelerate the phenomenon, while devising appropriate design criteria [162,163,164]. In particular, it was found that a ten-year retention requirement could be satisfied with a minimum tunnel oxide thickness

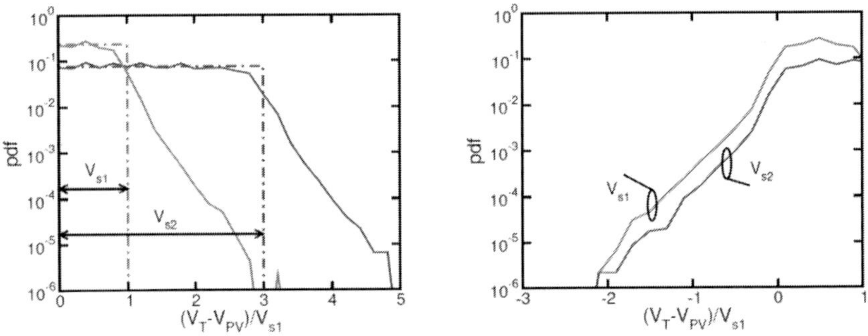

Figure 15. pdf at the end of the ISPP operation (**left**) and after a successive read operation (**right**). Results are shown for two values of the ISPP step V_{S1} and $V_{s2} > V_{s1}$.

around – nm [56,165], much thinner than the 7–8 nm usually adopted. However, the oxide thickness has barely scaled over the many Flash technology generations, remaining far away from its theoretical limit, because it was soon realized [70,166,167] that repeated P/E cycles led to degradation of the tunnel oxide characteristics, enhanced leakage and worse retention. The enhanced oxide leakage at low fields after electrical stress is known as *stress-induced leakage current*, or SILC [168,169]. SILC has been largely studied on capacitors and described in terms of a trap-assisted tunneling (TAT) process through oxide traps [170,171,172,173,174,175,176,177,178,179] (see Figure 17a for a pictorial view of the process), in agreement with experimental evidence [180,181]. In the realm of non-volatile memories, SILC can lead to charge loss/gain from/to the floating gate after stress and to retention errors. To keep this contribution negligible, the oxide thickness of Flash cells has barely scaled with the technology generations, going from about 10 nm in the first prototype [3] to about 7–8 nm in the latest nodes [182]. However, experimental data [183,184,185,186,187,188] have demonstrated that, even if the SILC is reduced, there is a small number of cells that can exhibit a high leakage after stress, in analogy with the statistical behavior of RTN. An example of this effect is reported in Figure 16, where thin-oxide (nm) NOR Flash arrays were cycled heavily (10^4 P/E cycles) before being subjected to positive (left) or negative (right) oxide field to induce a V_T shift [189]: note the distribution tails made up of cells featuring an enhanced leakage with respect to the main part of the distribution, whose shift is due to the intrinsic FN tunnelig current.

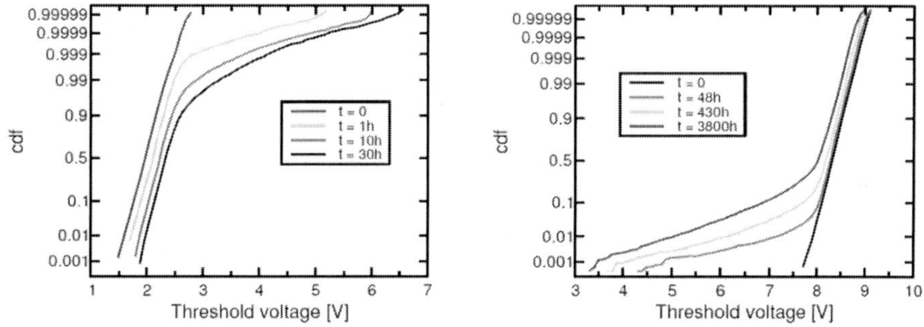

Figure 16. V_T cdf over time for nm NOR Flash cells arrays after cycling. Positive control-gate bias (5.5 V) was applied in the left-hand side figure, while the right-hand side shows results of a retention experiment [189], © 2001 IEEE.

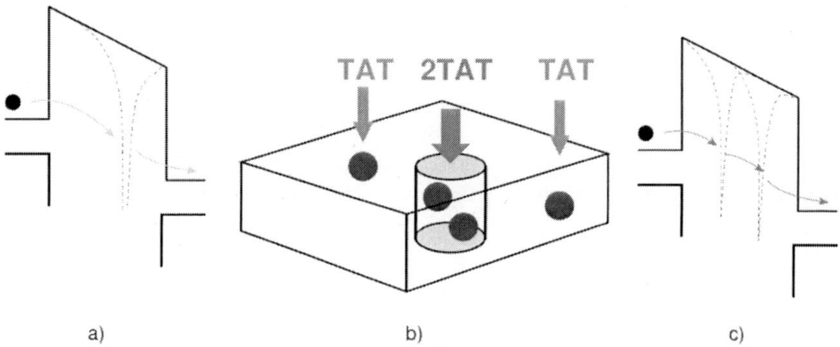

Figure 17. Pictorial view of the SILC distribution in Flash cells. The center picture (**b**) shows a trap distribution in the tunnel oxide region of the cell. One- and two-trap-assisted tunneling mechanisms (respectively TAT and 2TAT) are depicted as (**a**,**c**), respectively.

This enhanced leakage, called *anomalous SILC*, has been also identified as a trap-related phenomenon, because of its low activation energy [188,189] and fluctuations resulting from capture/emission processes [185,190]. However, a two-trap assisted tunneling process (2TAT) has now been invoked to explain the experimental data of individual cells [191,192], as depicted in Figure 17c. From this view, a picture of the SILC distribution in a Flash array can be built, assuming a statistical distribution of stress-generated oxide defects (Figure 17b): those cells that happen to have two traps close enough to allow significant tunneling will be affected by anomalous SILC [193,194], while the majority will only experience SILC or the FN tunneling current. Figure 18 shows an example of gate leakage current of two cells affected by anomalous SILC compared with the 2TAT model predictions (left), and (right) the statistical distribution of SILC at a given control-gate bias for different oxide thicknesses, again compared with results of the statistical model. From there, reliability extrapolations can be performed to estimate the number of failures. Other extrapolations procedures have been proposed, based on the same model or on empirical approximations [195,196,197,198,199,200,201,202,203], while addressing the optimization of the P/E waveforms and cell design [204,205,206,207].

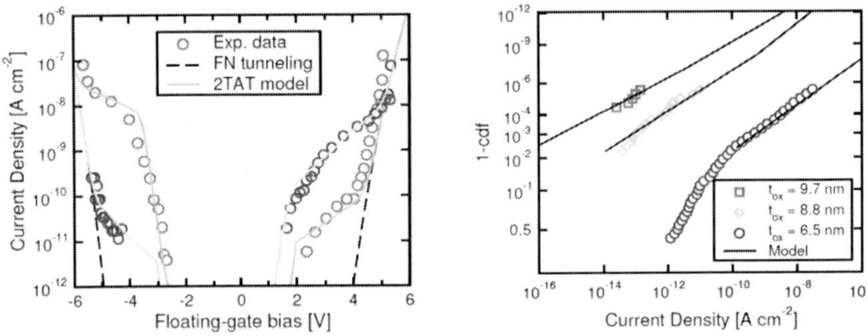

Figure 18. (**Left**) average gate current density vs. floating-gate bias for two cells affected by anomalous SILC, together with model predictions; (**Right**) cdf of the array leakage currents at fixed bias for different gate oxide thicknesses and model results. In all cases, symbols = experimental data, lines = model results.

Like all reliability issues, SILC becomes more of a concern in scaled MLCs or TLCs, because of the reduced separation between V_T levels and smaller floating-gate capacitance, meaning that fewer electrons are stored into the floating gate and that even microscopic leakage currents can cause significant damage. In today's technologies, SILC remains dominant in tail cells [208], that are usually covered by ECC, while its importance as a reliability issue is somewhat decreased with respect to phenomena affecting the entire distribution, such as the next one [209].

3.4. Charge Detrapping

Early investigators of SILC in stressed oxides observed a transient component in the leakage current, that was attributed to charging and discharging of oxide traps [210,211,212,213,214] and was exploited to extract information on the oxide quality [215,216]. The detrapping current followed a $1/t$ dependence down to very short times [217], in agreement with tunneling models [218]. The impact of this phenomenon on Flash reliability was first discussed in [219,220], pointing out that the log-time dependence of the V_T shift could lead to retention errors even in the absence of SILC and could dominate the charge loss with respect to SILC itself [63]. Moreover, as charge is trapped into the oxide during P/E cycling and detrapped in-between, a dependence on the cycling pattern arises, which complicates the development of comprehensive models. Such task began in [221,222] and developed into a full model accounting for the major experimental evidence in [223,224,225].

3.4.1. Main Experimental Data

Figure 19 shows experimental data from a retention experiment at room temperature on a 16 nm NAND vehicle after 10^4 P/E cycles [226]. Note the difference with respect to Figure 16 (right): no specific tail is detected here, but rather the entire distribution is affected, although different charge loss rates can be detected between the main part and the lower extreme of the distribution (see right-hand side of Figure 19). Note also the logarithmic dependence on time of ⍰V_T, which can be fully characterized by its slope ⍰.

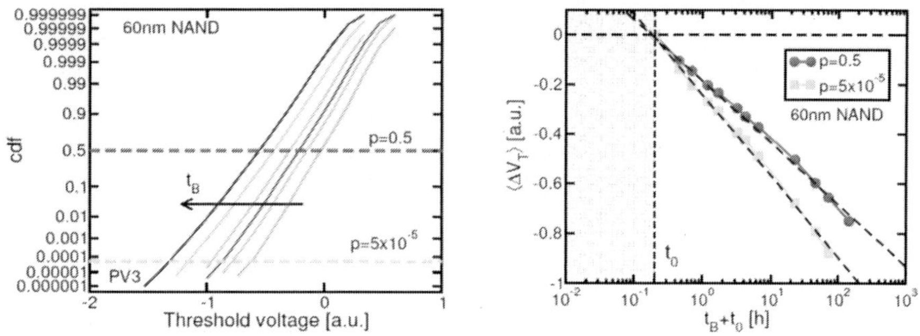

Figure 19. (Left) V_T cdf over time for 60 nm NAND Flash cell arrays after cycling; **(Right)** V_T shift vs. time extracted from the cdfs at fixed probability levels [226], © 2011 IEEE.

The dependence of charge detrapping on the cycling conditions is reported in Figure 20: its left-hand side shows that ⍰V_T increases with N_c, as more electrons are trapped during P/E operations and emitted afterwards. The right-hand side reports the dependence of ⍰ on V_T: a square-root law was found as soon as the cycling-induced damage overcomes the native detrapping, in agreement with observations on RTN-related defects [138]. This dependence is also independent of the programmed level and temperature (note that the curves for different programmed levels have been vertically shifted to highlight the square-root law and avoid excessive overlap). Figure 21 shows instead the dependence on the bake conditions: note that the detrapping dynamics depend on the time t_0 elapsing between the last program and the first read operation (left-hand side), resulting in a shift of the detrapping curve along the log-time axis by a quantity exactly equal to t_0. Also,⍰V_T depends upon the bake temperature T_B (right-hand side of Figure 21), demonstrating that detrapping is thermally activated: the ⍰V_T curves at different T_B are shifted according to an Arrhenius law with activation energy $E_A \approx$ 1.1eV [221,222,228,229], a single detrapping curve can be obtained for an equivalent T_B. This value of activation energy had been also observed in earlier retention experiments [51]. Careful account of additional effects which may compromise a reliable extraction of E_A, such as interface-state and mobility recovery [230,231] or disturbs [232], is required.

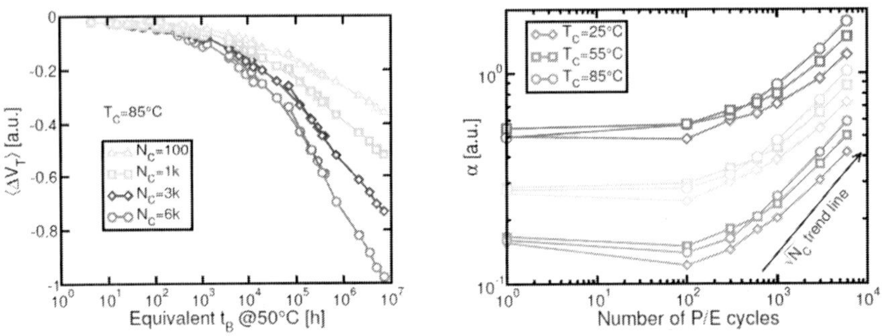

Figure 20. (Left) ⍰V_T average transients measured on 16 nm NAND Flash arrays for increasing number of P/E cycles [227], © 2016 IEEE; (**Right**) dependence of the slope ⍰ on N_C for different programmed levels (colors) and temperatures (symbols). Curves for different programmed levels have been vertically shifted for better clarity.

The last data to be discussed relate to the cycling conditions dependence. Figure 22 shows ⍰V_T for different values of the cycling time t_C (left) and temperature T_C (right). Note that a smaller ⍰V_T is detected if cycling is distributed over a longer period of time or performed at higher temperatures. This is consistent with a picture that assumes electron detrapping between one cycle and the next one, reducing the number of trapped electrons at the end of the last cycle that are then lost during the bake phase. Coherently with this interpretation, the activation energy for cycling was found to be the same as for bake. These findings have important implications for the device reliability, that will be discussed after reviewing the quantitative understanding of the phenomenon together with other experimental data.

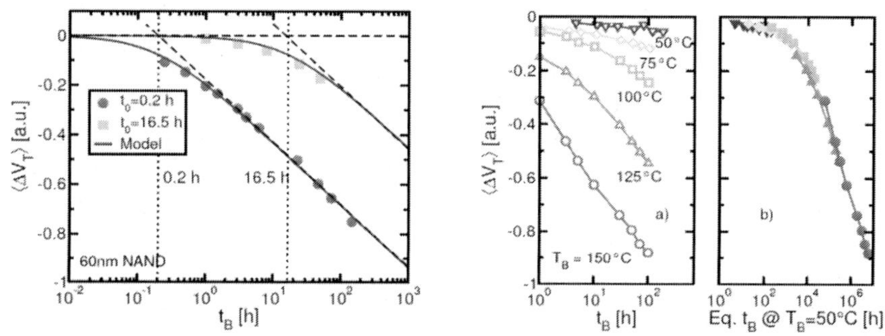

Figure 21. Dependence of ▯V_T on bake conditions, namely time t_0 [226], © 2011 IEEE (**left**) and bake temperature T_B (**right**). Data are referred to 60 and 16 nm NAND Flash vehicles. Data on the right-hand side show a) experimental data for different T_B and b) the same data when t_B is shifted on log scale according to E_A.

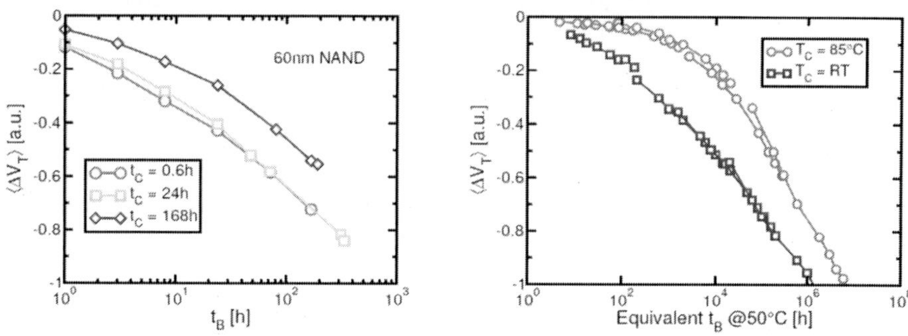

Figure 22. Dependence of ▯V_T on cycling conditions, namely total cycling time t_C(**left**) and cycling temperature T_C [227], © 2016 IEEE (**right**).

3.4.2. Models

We start this Section with the description of a compact model [221,222,226] that captures the main features of the data and can be used for some first-order extrapolations, while moving to more refined descriptions of the underlying physics later on. To this aim, we assume a log-time dependence of the shift that, considering that the first read operation is performed at time t_0 after the end of the cycling phase, leads to

$$\Delta V_T(t_B) = V_T(t_0 + t_B) - V_T(t_0) = -\alpha \ln(t_0 + t_B) + \alpha \ln(t_0) = -\alpha \ln\left(1 + \frac{t_B}{t_0}\right), \tag{9}$$

which is what we observe. The dependences of ⍰V_T on N_C and probability level (see Figure 19) can then be simply inserted into the expression of ⍰. To account for the thermal activation of detrapping during bake, which results in a rigid shift of the detrapping curve in a semilog plot (see right-hand side of Figure 21), we can now modify the expression of the term t_0, which is the zero-loss intercept of the log-time curve (see Figure 21, left). This results in:

$$t_0^* = t_0\, e^{E_A(1/k_B T_B - 1/k_B T_{ref})}; \quad \Delta V_T(t_B) = -\alpha \ln\left(1 + \frac{t_B}{t_0^*}\right), \tag{10}$$

Where t_0 is now the initial wait time at the reference temperature T_{ref} and k_B is the Boltzmann constant. Following this line of thought, even data in Figure 22 can be represented by a dependence of the intercept t_0^* on cycling conditions. To do this, it is assumed that damage creation during P/E cycles and recovery between them can be described as a fast-cycling experiment (i.e., with negligible delay between cycles) generating the whole oxide damage, followed by a single recovery phase of a certain duration. Combined with the thermal activation term, this results in the final expression

$$t_B^* = t_0^* + A t_c\, e^{E_A(1/k_B T_B - 1/k_B T_C)}; \quad \Delta V_T(t_B) = -\alpha \ln\left(1 + \frac{t_B}{t_B^*}\right), \tag{11}$$

which allows to completely describe the ⍰V_T dynamics as a function of the various experimental parameters once the fitting parameter A is selected. This means that the term $\ln(1 + t_B/t_B^*)$, called the *universal damage metric* or UDM, should now account for all ⍰V_T transients measured on a specific technology, irrespective of cycling and bake conditions. This is demonstrated in Figure 23(left). Figure 23 (right) shows instead the value of t_B^* for different cycling conditions.

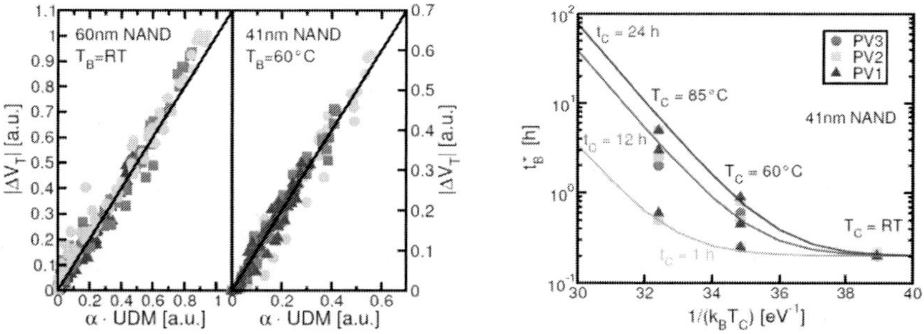

Figure 23. (**Left**) ⍰V_T transients as a function of the UDM term (symbols represent the different programmed levels); (**Right**) Values of the parameter t_B^* extracted from experimental data (symbols) and Equation (11) (lines) [226], © 2011 IEEE.

The experimental data presented so far and the physical understanding leading to Equation (11) led to the recognition that traditional testing procedures based on fast cycling and long retention times were magnifying the V_T loss with respect to real usage conditions, when cycling takes place over the functional life of the device (see the effect of increasing t_C in Figure 22, left). The concept of *distributed cycling* conditions was then proposed in [221] as a way to better emulate the real array behavior by performing either a uniform cycling over a longer time or several groups of fast cycles preceded by bake times, usually at high temperature to accelerate the charge loss. However, the previous model shows its limitations when dealing with highly non-uniform cycling patterns, calling for a more comprehensive interpretation of the trapping/detrapping physics, that was pursued in [223,224,225,227,233]. Such a model was built upon a few phenomenological assumptions: a Poisson distribution for the number of trapped electrons and a uniform distribution over a log-time axis of their detrapping time constant ⍰$_d$. This led to an expression for ⍰ [223,224]

$$\alpha \propto \langle \Delta V_T^1 \rangle \langle n_t \rangle, \tag{12}$$

which directly links the slope of the ⍰V_T transient to the product of the average (indicated by the operator) $\langle \cdot \rangle$) ΔV_T^1 and n_t. Via a clever analytical manipulation, an expression for the entire statistical distribution of ⍰V_T can be obtained:

$$p(\Delta V_T(t)) = \mathscr{F}^{-1}\left\{ e^{\langle n_d(t) \rangle} \left(\mathscr{F}\{p(\Delta V_T^1)\} - 1 \right) \right\}, \tag{13}$$

Where $\mathscr{F}$ and $\mathscr{F}^{-1}$ are the Fourier transform and its inverse and the average number of detrapped electrons, $\langle n_d(t)\rangle$, is given by

$$\langle n_d(t)\rangle = \int \langle n_t(\log_{10}\tau_d)\rangle \left(1 - e^{-t/\tau_d}\right) d\log_{10}\tau_d, \quad (14)$$

an activation energy E_A = 1.1 eV being embedded into ⍰$_d$. Charge trapping is also inserted in a phenomenological manner, in agreement with the observed square-root dependence on N_C [221,222], resulting in the following expression for the average number of electrons being trapped at cycle N_C:

$$\Delta n_t(\tau_d) = \frac{\eta^2}{2\langle n_t(\tau_d, N_C - 1)\rangle}. \quad (15)$$

In Equation (15), ⍰ ≈ 0.001 is a fitting parameter, and the expression is modified for small values of to avoid the divergence in ⍰n_t (see [224]). Typical results for the n_t distribution as a function of time (fast cycling conditions, where P/E cycles are conducted before the retention test) are shown in Figure 24, left: at a given time t, because of the detrapping term 1-$e^{-t/⍰d}$ in Equation (14), only electrons having $\tau_d \ll t$ are detrapped, while those with $\tau_d \gg t$ are unaffected. Nice agreement with experimental data (Figure 24, right) confirms the validity of this approach.

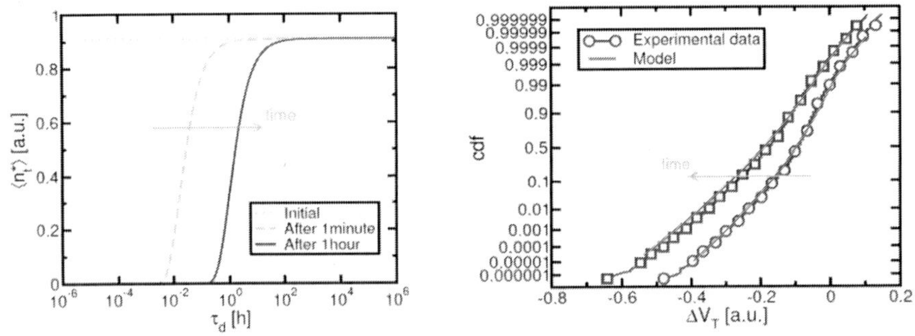

Figure 24. (Left) distribution of n_t(⍰$_d$) for different bake times after fast cycling; **(Right)** comparison between experimental data and model results for ⍰V_T.

The power of this model can be best appreciated when dealing with so-called *distributed cycling* schemes, where P/E cycles are spread over longer times and separated by bake phases at different temperatures, to emulate the real behavior of a device. Such schemes cannot be managed by the compact expression given by Equation (11), but are easily dealt with by the model synthesized by Equations (12)–(15). As a reference, we take three schemes that are deemed almost equivalent by the model adopted in the JEDEC standard [235] and are depicted in Figure 25: all involve fast-cycling phases at room temperature (marked as blue) separated by high temperature bakes (marked as red). In all cases, the array is subjected to $N_C = 10k$ cycles, but their distribution over time is different. Results for at the end of each process are reported in Figure 26 (left): different ?V_T transients are obtained with the three schemes, featuring nonetheless the same asymptotic slope. The model nicely captures the differences and explains them in terms of the different distributions of n_t(?$_d$), as a consequence of the trapping/detrapping phases. Note also that the asymptotic value of n_t is the same for all schemes, as it is due to the total number of cycles, explaining the uniqueness in the asymptotic slope (see Equation (12)).

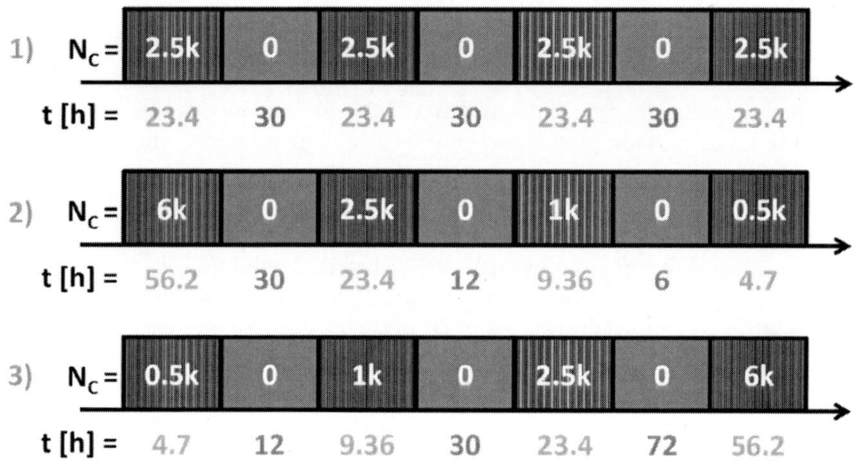

Figure 25. View of the considered distributed-cycling schemes that are almost equivalent from the viewpoint of the JEDEC standard [236].

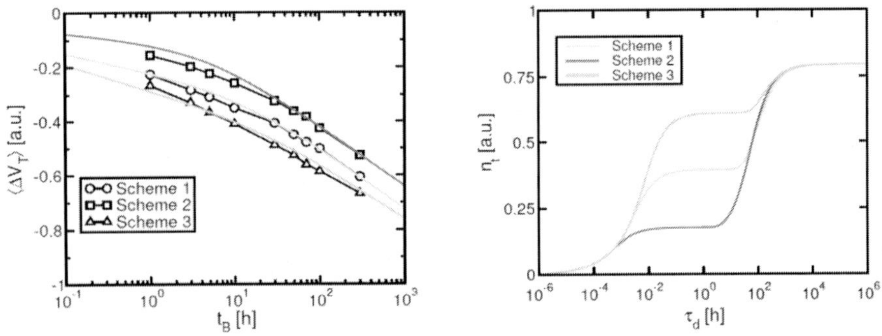

Figure 26. (**Left**) comparison between experimental data (symbols) and model results (lines) for ΔV_T after the distributed cycling schemes in Figure 25 [234], © 2014 IEEE; (**Right**) distribution of $n_t(\tau_d)$ at the end of the cycling schemes.

Although the above-mentioned model is successful in describing distributed-cycling procedures and provides a useful support for developing appropriate qualification schemes, it still misses some features, namely the programmed-level dependence of charge detrapping and ⍰. In particular, it was noted that charge detrapping is enhanced with the oxide electric field and that little or no detrapping takes place during bake phases performed with cells in the erased state [227,237], suggesting that holes may play a non-negligible role in what is usually described as an electron-only detrapping process [238]. These developments led to a recent refinement of the above-mentioned model [225,233] that goes beyond the classical vision of pure charge detrapping, considering multiple-state defects related to structural relaxation of the oxide network. A pictorial view of the proposed defect states can be seen in Figure 27, where the lower states $D2$ and $N2$ represent the usual trapping/detrapping phenomena, for which the model discussed so far still holds (besides a minor modification to account for trapping). Now, however, the reservoir for charged defects that can exchange electrons with the substrate (state $D2$) is dictated by a structural relaxation process from an initial state $D1$ generated by P/E cycling, allowing a higher degree of flexibility in the description of the ΔV_T transients. By assigning a log-time distribution to the structural relaxation time constants while retaining nearly-constant values for the trapping and detrapping time constants (whose ratio is assumed to be affected by the oxide field), a greatly improved fitting was achieved, even over fine details of the experimental data. However, the assumptions of the model still need further investigations to be confirmed, and charge detrapping is likely to remain an active area of research in Flash reliability for quite some time to come.

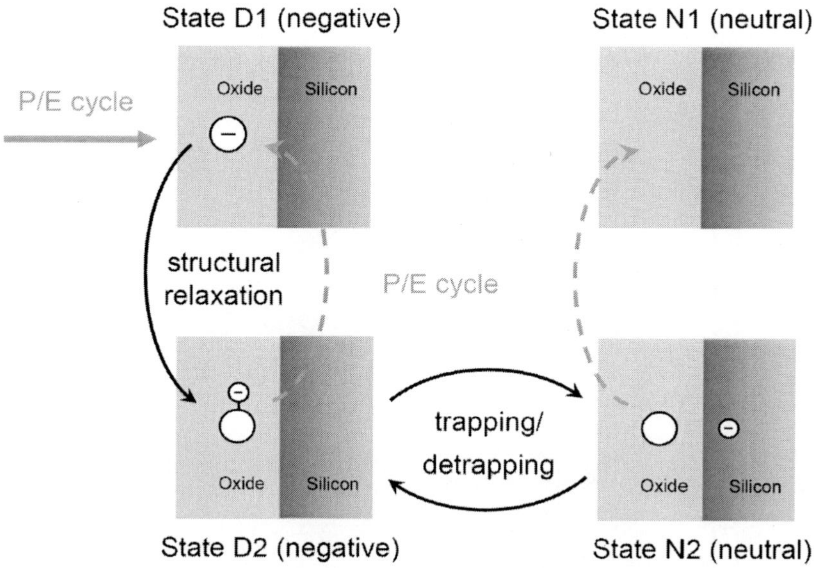

Figure 27. Pictorial view of the multiple-state defects responsible for charge trapping/detrapping according to [225,227,233].

3.4.3. Effect on V_T Distribution after Cycling

From the previous results, it is clear that charge detrapping leads to both an average charge loss and an enlargement of the V_T distribution, due to the randomness of the detrapping process. A comparative analysis of the impact of charge detrapping on the width of the V_T distribution after cycling and bake was reported in [239] and shown in Figure 28, normalized to the ISPP step V_S. The two bottom-most contributions are related to the time-0 placement of the V_T distribution—namely, the ISPP step V_S and the fluctuation in the number of injected electrons per step [158,159,240]—and are not discussed here. The remaining part is made up of a native RTN contribution plus the additional terms induced by cycling. Note that RTN is not dependent on the V_T level while detrapping increases with such level, as previously discussed, becoming a non-negligigble error source for the higher V_T levels. A similar breakdown of the distribution width was presented in [241], without accounting for detrapping. Other studies also identified charge detrapping as the dominant error source in highly-cycled NAND samples [242].

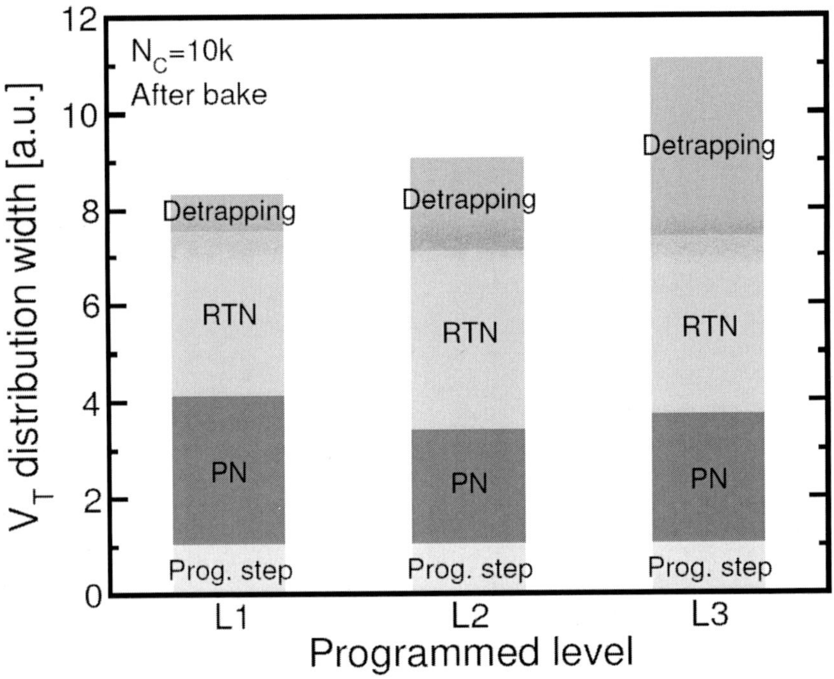

Figure 28. Contributions to the width of the V_T distribution after cycling and bake for different programmed levels [239], © 2015 IEEE.

To overcome this limitation, several solutions have been proposed, including a dummy read [243] or a moving reference [244] scheme and a *self-healing* approach [245], where a built-in heater in the proximity of the tunnel oxide is used to accelerate damage recovery and prolong the endurance. The insertion of specific block healing cycles during run time was shown to increase lifetime and performance of NAND devices [246], suggesting that distributed healing schemes could be devised to maximize their efficacy while reducing the energy dissipation of such cycles [247].

3.5. Interface State Recovery

In analogy with oxide defects, interface states are created during P/E cycles and recover in-between. Besides their effect on the endurance figure of NAND devices, discussed in Section 3.1, they also influence the V_T shift after cycling. In fact, it has been proposed that interface state recovery represents the major source of V_T instability in 90 nm NAND

Flash [63,209,248]. Interface state annealing manifests itself via a reduction of the activation energy, down from the eV associated to detrapping to about eV [230,231] or even eV [249], and are also characterized by a non-uniform distribution over the cell area, usually peaked at the STI edges [74,250]. However, it has been suggested that their effect is dominant for high values of the read current only [230,231], while playing a minor role with respect to charge detrapping as the read current is reduced.

3.6. Disturbs

The arrangements of Flash cells into compact arrays implies that many cells other than the selected ones are biased during read and program operations. Although such biases should not affect the cell state, degradation in the tunnel oxide characteristics following P/E cycling can reduce the cell immunity to such disturbs. The NAND layout generates three types of disturbs, namely related to the read, program and pass operation, that are discussed hereafter.

3.6.1. Read Disturb

The bias condition of a NAND block during read has been discussed in Section 2.2 and is shown in Figure 29: a suitable read voltage is applied to the selected WLs to sense whether the cells in the page are in the on or off condition, while all other cells in the string are biased to a pass voltage higher than their maximum V_T. The pass bias is hence analogous to a program operation, albeit performed at a lower gate bias, and as such it can induce a V_T shift in the cells. This *read disturb* (RD) cumulates with the number of read operations and could eventually lead to a failure. Note that this is a recoverable error, that is nulled after the block is erased and reprogrammed.

The oxide thickness and bias conditions are designed to assure that no significant V_T shift is expected after RD. However, this might no longer be the case after heavy P/E cycling, when the gate oxide quality is degraded and SILC and charge detrapping arise, enhancing the conduction at low fields (see the *I–V* characteristics in Figure 18, left). RD is hence exacerbated in cycled cells [251,252,253] and in read-intensive applications such as database access, explaining why a lower endurance is sometimes specified in such cases.

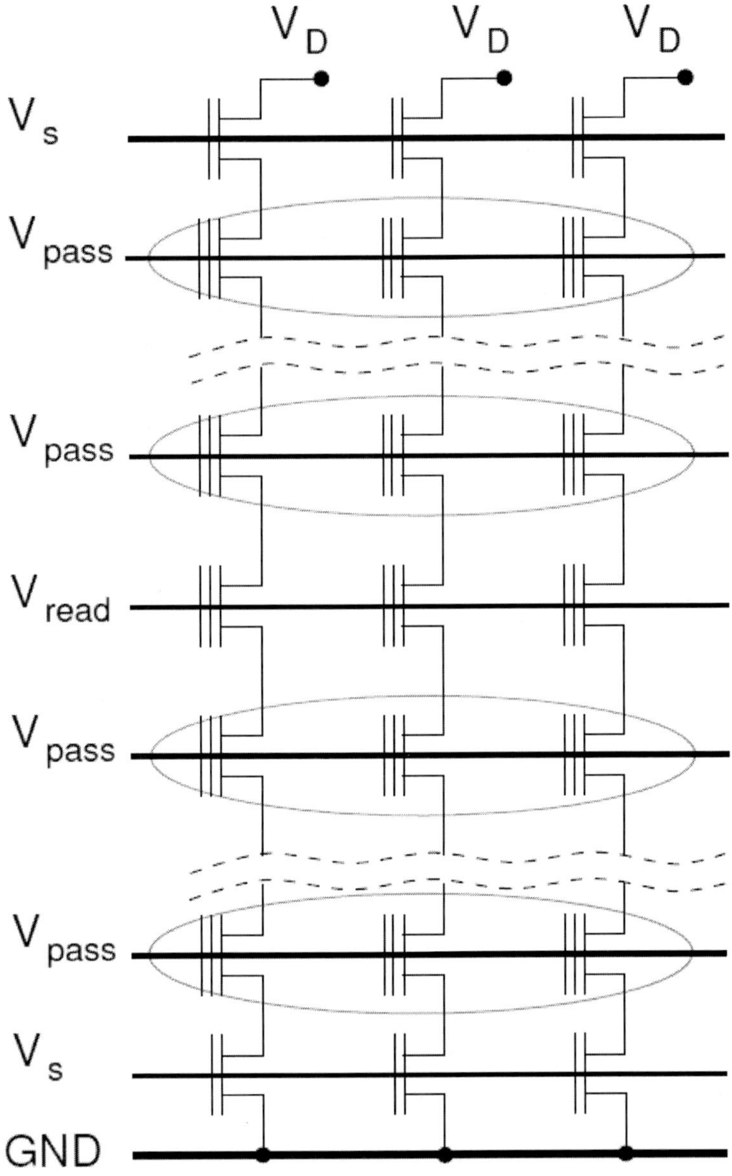

Figure 29. NAND array bias condition during read. Cells subjected to RD are circled.

Experimental data for the memory lifetime report a power-law dependence on N_c, with exponent ranging from -0.3 to -0.4 [251,254]. Data for the V_T shift as a function of the number of reads feature different dependences [255,256], while raw bit-error rate (RBER) data vs. the number of reads reveal a linear dependence whose slope increases with cycles [60,255,257], with a higher BER observed on SLCs than on MLCs. As with SILC, several process optimization [258,259], voltage tuning [255,260] or system-level correction [261] proposals have been put forward to alleviate the issue.

Typical RD characteristics are reported in Figure 30 [255]: the V_T shift is larger for the erased state and decreases for higher programmed V_T levels, in agreement with the reduction on the tunnel oxide field. As a consequence, more errors are expected in the logical pages corresponding to those states. It has also been found [262] that repeated RD on partially-programmed blocks leads to an increased BER when the remaining pages are programmed, suggesting that extensive RD can lead to a degradation of the program characteristic of the cell. In any case, it has been claimed [61] that RD errors are not the prevalent ones in SSDs used in servers.

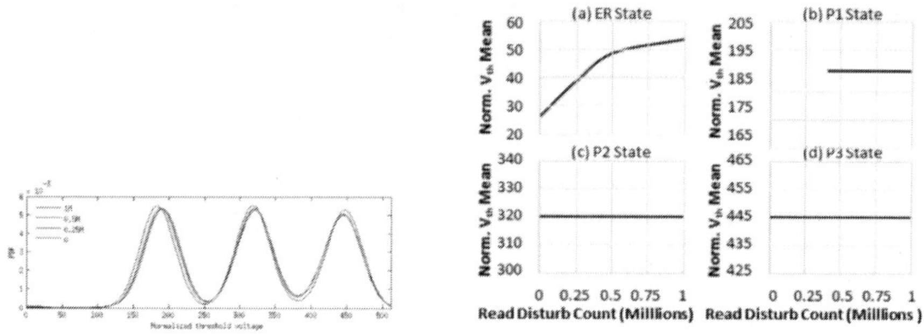

Figure 30. (**Left**) V_T distribution for increasing RD time; (**Right**) mean V_T shift of the different levels as a function of the number of reads [255], © 2015 IEEE.

3.6.2. Program and Pass Disturbs

When programming a page, we must ensure that only the desired BLs are actually programmed, while other cells on the same page that must remain in their original state are not affected by such operation. Since electrons are injected into the floating gate via FN tunneling, which is a process strongly dependent on the oxide field, we can inhibit programming by reducing such a field, i.e., the gate-to-channel voltage drop. The simpler solution is to disconnect (i.e., ground) the SSL (see Figure 1) and raise the inhibited BLs to a high voltage [263,264], while biasing the unselected WLs too, to avoid unintentional erase. However, this solution was soon abandoned because of power and time issues related to the charging of highly-capacitive BLs and because of high-voltage limitations in scaled technologies. In replacement, a *self boosting*technique was proposed [265], illustrated in Figure 31: BLs to be programmed are grounded, which turns on their DSL and maintains the string channel to ground (green line). Note that a pass voltage (different from the one used during read) is applied to the unselected WLs. Inhibited BLs, instead, are biased at V_{cc}, which precharges their channel. When the high passing voltage is applied to the WLs, however, the channel potential (red line) is lifted by capacitive coupling, turning off the DSL and reaching a value high enough to suppress electron injection into the floating gate.

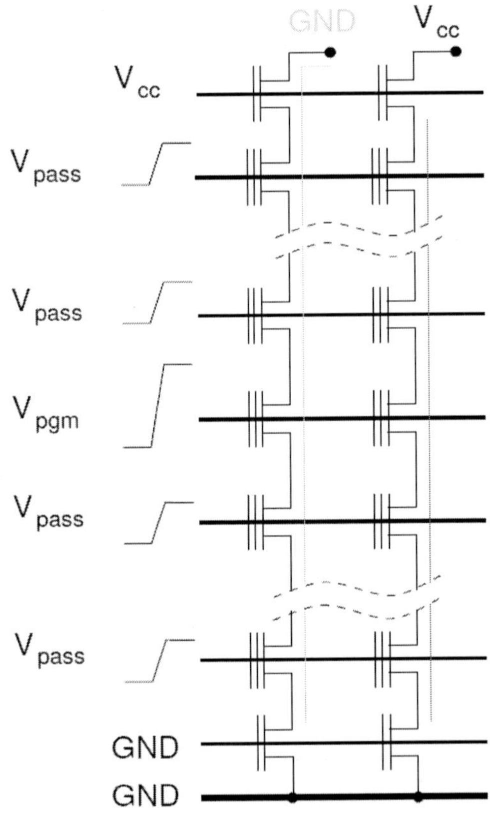

Figure 31. NAND array bias condition during program, highlighting the self-boosting effect.

This solution generates two kinds of disturbs, highlighted in Figure 32: a program disturb on the cells on the programmed WL (encircled in green), related to the potential difference between program and boosted bias, and a pass disturb on the cells located on the same (grounded) BL as the programmed ones (red circles), related to the pass voltage. The nature of such disturbs is just the same as the RD, namely a soft-programming at low voltage, with the difference that now the applied field is higher with respect to RD, while the number of cycles is significantly lower. Note also that an increase in the pass voltage will lead to an increase in the channel potential, reducing the program disturb but raising the pass one. This typical behavior is shown in Figure 33 (left). While a logarithmic dependence of V_T on the number of cycles has been reported [266], the dependence of BER is more complex, as shown on the right of Figure 33 [257]. To limit the pass disturb, improved boosting schemes have been developed, including local boosting [36,267], ramped WL voltage [268], dynamic pass voltage biasing [269] or negative WL pre-biasing [270]. Interestingly, a self-boosting scheme during read was also proposed, to alleviate read-disturb issues [271].

The boosting operation in scaled NAND devices is also responsible for additional reliability concerns. First, the string select transistors must withstand the boosted potential with minimum leakage current, as such current will discharge the channel and increase program disturb (for this reason, such devices are hard to scale). However, under program inhibit conditions, the SSL transistor is subjected to gate-induced drain leakage (GIDL) due to band-to-band tunneling, injecting hot electrons toward the first WL (see Figure 32) and into its floating gate, when erased [266]. This effect leads to an enhanced program disturb at the first WL and is usually curbed by increasing the SSL-to-WL spacing [266] or employing one [272] or two [273] dummy edge WLs. In fact, because of short-channel effects in scaled technologies, the actual electron injection point into the floating gate can move farther into the string, affecting even inner cells [274]. Moreover, such hot carriers may be trapped into the oxide and be subsequently released, giving rise to detrapping-related retention degradation [275]. The full picture is further complicated by electrostatic coupling between adjacent strings, via the STI isolation: when a string is programmed, its channel is kept at zero voltage and it will behave like the gate of a parasitic transistor whose gate oxide is the STI isolation. Adjacent inhibited cells will therefore experience large vertical GIDL current, between the (boosted) surface of the channel and its bulk [276]. Air-gap isolation as well as optimization of channel implant and STI depth uniformity [277] have been suggested as remedies. A related effect has been noted in [278], where a negative V_T shift contribution was found and explained in terms of hot-hole injection from the STI edge to the control-gate [279,280]. All this suggests that the detailed leakage mechanisms affecting the boosted channel potential must be studied with extreme care [281].

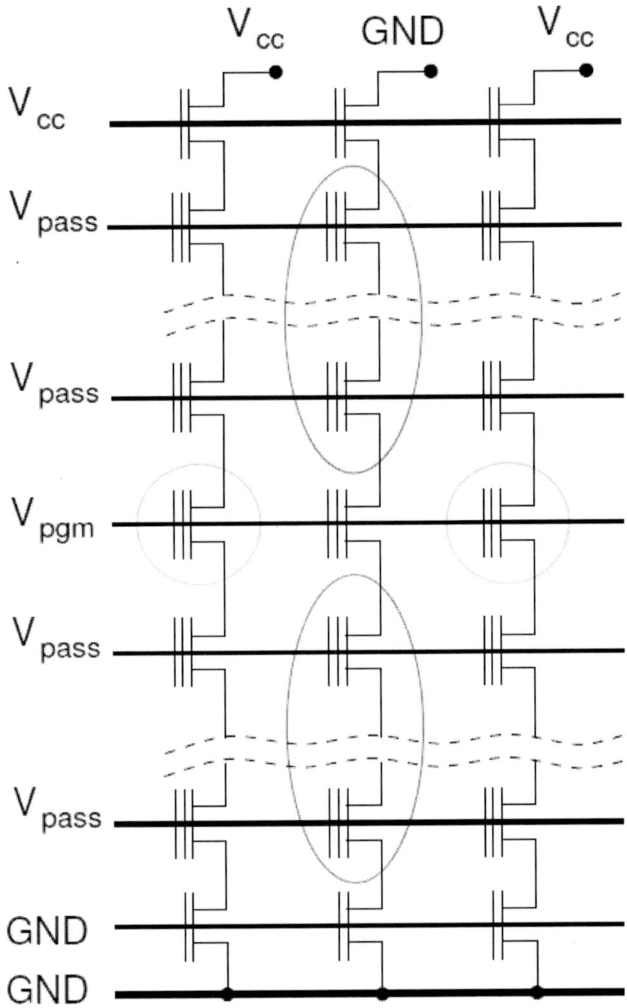

Figure 32. NAND array bias condition during program. Cells subjected to program (green) and pass (red) disturbs are circled.

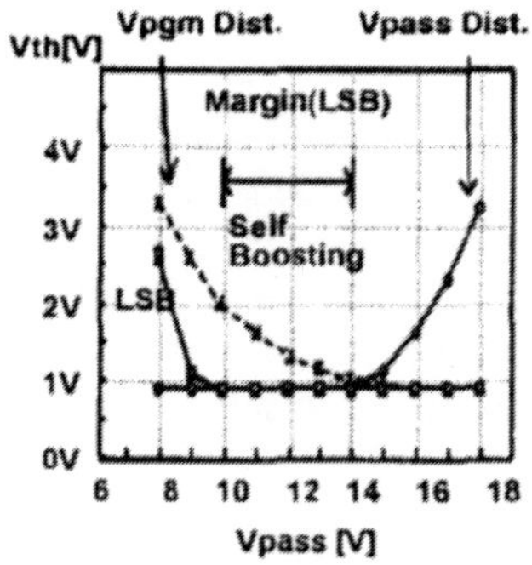

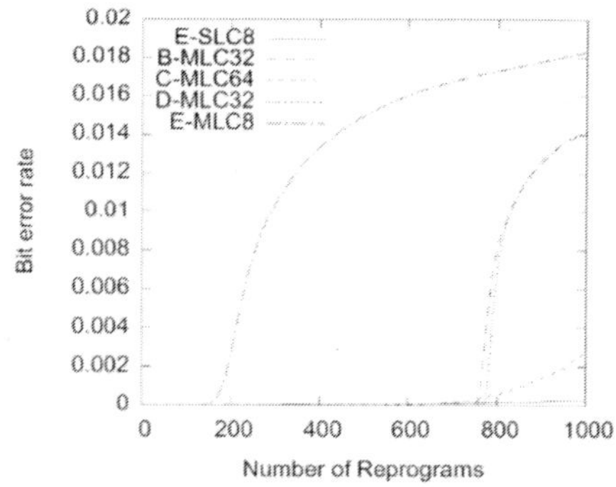

Figure 33. (**Left**) V_T shift due to program and pass disturbs as a function of the pass voltage [36], © 1996 IEEE; (**Right**) BER due to program disturb as a function of the number of P/E cycles [257], © 2009 ACM.

In local self-boosting schemes, where the channel of the programmed WL is isolated from the rest of the string by grounding the neighboring WLs and is boosted by the programmed Wl, other issues arise, such as hot-electron injection via drain-induced barrier lowering (DIBL) [282,283], that paved the way for proposals of hot-carrier based programming of NAND cells [284,285].

In conclusion, this overview demonstrates that many reliability issues have threatened the scaling of the planar NAND technology. In turn, its successful scaling down to the 15 nm node is a tribute to the effectiveness of the many different solutions adopted to keep the failure rate under control. NAND-based SSDs are today the main candidates for mass storage, having reached reliability performance comparable to, or in some aspects even better than, HDDs [62]. In the following, we will extend this discussion to the current leading-edge technology for NAND: 3D.

4. 3D NAND RELIABILITY

In this Section, we will mainly focus our attention on the most promising 3D NAND architecture, i.e., the one based on vertical strings of cells and parallel planes connecting the control gates [23,24,25]. A schematic view of the memory array organization is shown in Figure 34 (left), but we will leave the details about the array operation to the specific literature [28]. The individual cell is a vertical-channel transistor with polysilicon body, usually shaped as a hollow cylinder filled with dielectric [286], whose cross-section is depicted on the right of Figure 34. Such a device was shown to exhibit superior performance with respect to a conventional vertical nanowire transistor, because of the high defectivity in the central region that plagued the performance of the latter structure. The gate stack of today's 3D NAND can be based on either a floating gate [23,287,288,289,290,291], similar to planar NAND devices, or a charge-trap stack similar to an oxide/nitride/oxide (ONO) layer, where the charge is stored in traps within the nitride layer [10,11,22,292].

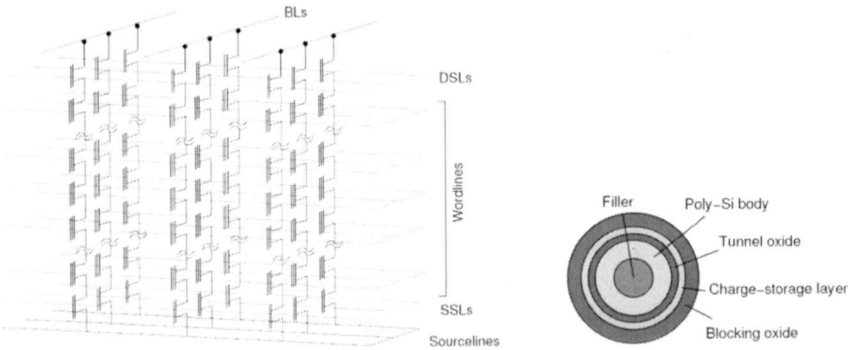

Figure 34. **(Left)** schematic view of the array organization in a 3D NAND; **(Right)** schematic cross-section of the typical NAND cell used in 3D arrays.

As far as reliability is concerned, all mechanisms effective in planar NAND cells are also present in 3D devices, though their impact is affected by the new cell geometry and design and array operation. Unfortunately, the very limited amount of data publicly available makes it hard to conduct an exhaustive assessment of the role played by each of them. On a general standpoint, the 3D cylindrical cell footprint and channel length can be made larger than their planar counterparts [22,291], as the 3D architecture relieves the burden of scaling the cell area by placing it onto the increase in the number of stacked layers. As a consequence, a better reliability is achieved in 3D NAND cells, which resulted in the development of QLCs. However, the vertical polysilicon channel and the layer stacking give rise to new and unexpected issues, providing ever newer ground for research. In the following, we will review the reliability issues in 3D NAND within the same framework adopted so far: we will only discuss issues that become relevant with P/E cycling or repetitive data read, neglecting those associated, for example, with placement, string resistance and time-0 variability.

4.1. Endurance

Figure 35 shows the endurance performance of a 3D NAND cell as a function of [286] and the distribution for a NAND array after heavy cycling [22]. Because of the larger cell size, 3D NAND are reported to have better endurance properties with respect to their planar counterpart, as also claimed in [293,294,295,296], where an endurance of 5-7 k P/E cycles is reported for a TLC 3D NAND. All these cells feature an ONO stack. Retention data on a floating-gate based 3D NAND [290] show comparable results, with higher shift but larger voltage window.

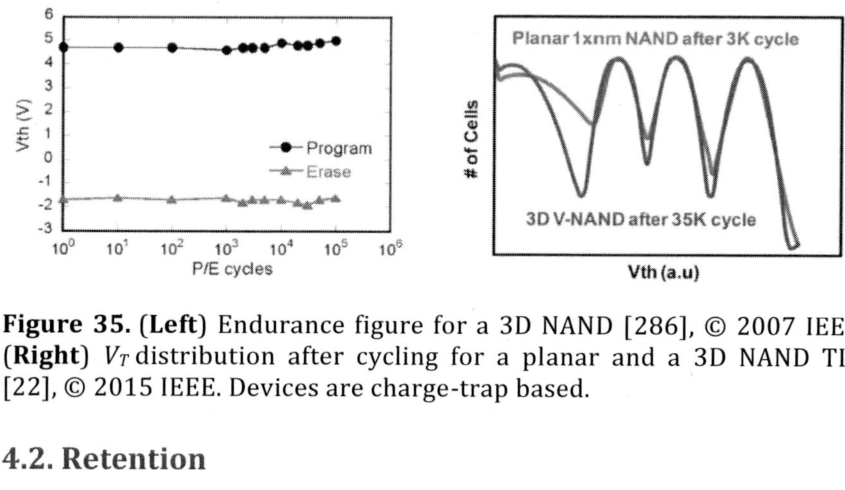

Figure 35. (**Left**) Endurance figure for a 3D NAND [286], © 2007 IEEE; (**Right**) V_T distribution after cycling for a planar and a 3D NAND TLC [22], © 2015 IEEE. Devices are charge-trap based.

4.2. Retention

Floating-gate based 3D NAND suffer from anomalous SILC after heavy P/E cycling, though this effect is not expected to be the dominant cause of retention loss. Charge-trap based memories are claimed to be insensitive to such a loss [297], due to the localized nature of storage, although it has been shown [298] that defect build-up in the tunnel oxide eventually leads to TAT mechanism enabling charge loss from the nitride layer. In any case, to our knowledge, there are no published reports of anomalous SILC in 3D NAND, so it seems safe to extend to this technology the conclusion drawn for planar arrays, where other effects are dominating retention.

A comparison between 2D and floating-gate based 3D NAND with respect to detrapping is reported in Figure 36 (left), showing a smaller spread for the 3D array, even if subjected to a higher number of P/E cycles [299]. This is once again the result of the larger cell in 3D technology, i.e., larger gate capacitance, minimizing the effect of single-charge detrapping onto V_T. Published single-cell data (Figure 36) suggest that a good intrinsic retention can be achieved, as expected from a floating-gate design [289].

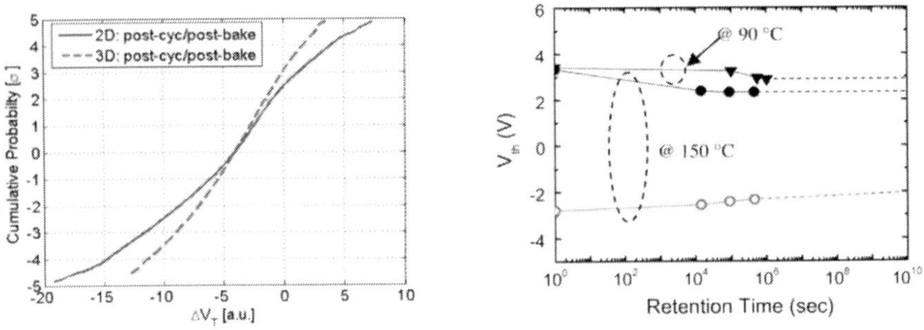

Figure 36. **(Left)** comparison between 2D and 3D cycled NAND arrays for V_T shift induced by charge detrapping [299], © 2015 IEEE; **(Right)** retention for different temperatures [289], © 2013 IEEE. All devices are floating-gate based.

The retention properties of charge-trap based planar devices have been extensively investigated as a function of temperature and bias [300], pointing out the roles of vertical charge loss through the tunnel oxide and lateral charge migration in the nitride [301,302,303,304], which also reduces the stored charge located above the channel region, that determines V_T. In 3D devices, the importance of such effects is shown in Figure 37, where lateral diffusion along the nitride is modulated by the state of adjacent cells [305]: having all cells programmed decreases the concentration gradient, slowing down diffusion and reducing the V_T shift. Similar results obtained via simulations were also presented in [306,307,308] and, changing the device size, in [309]. Note also the thermal activation, accelerating both leakage mechanisms and making pattern-dependent charge spreading effects more relevant at high temperatures. The resulting multiple activation energies are discussed in [310]. An analogous charge-gain phenomenon for the low- V_T state was reported in [311]. Such a feature had been already discussed in planar nitride memories and floating-gate devices, being interpreted in terms of recovery of negative bias-temperature instability [312,313].

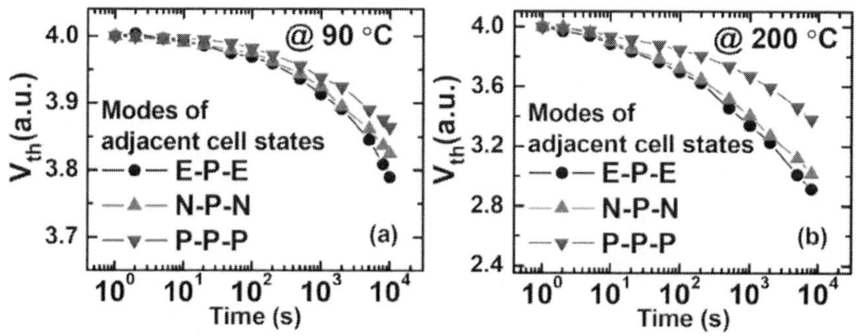

Figure 37. Charge retention characteristics for a charge-trap based 3D NAND as a function of the state of the adjacent cells and for different temperatures [305], © 2015 IEEE.

A related effect that was first observed in planar devices [314] is a fast charge loss taking place on a s timescale [315]. Currently, there is no consensus on the root cause of this phenomenon, ascribed to either electron detrapping from shallow traps in the nitride [314] or in the tunnel oxide [316], or to lateral electron migration enhanced by holes injected during erase that are not fully compensated by electrons [315]. The effect is however dependent on the nature of the gate stack, being smaller in conventional ONO stacks with respect to band-engineered ones, and its magnitude increases with cycling.

4.3. Cell Current Fluctuations

Cell current fluctuations in 3D cells, related to capture and emission of single carriers, present some distinct features that differentiate them from what is usually seen in planar devices. These can be ascribed to the trap location, either at one of the oxide interfaces (tunnel oxide and filler) or within the polysilicon itself, because of the polycrystalline nature of the conductive channel, and to 3D percolation as well as the conduction mechanisms of electrons in the polysilicon. These factors are discussed hereafter.

4.3.1. Random Telegraph Noise

RTN in silicon nanowire structures has received attention in the past and single-trap characterization in vertical structures for 3D NAND has been also presented [317,318,319]. Statistical data for RTN are shown in [320,321,322] and are presented in Figure 38 (left) for different temperatures. Note that the usual exponential distribution of amplitude appears, whose slope is greatly reduced with respect to planar technologies [299,323,324]. This is due not only to the larger cell size, but also to the different doping of the channel and nature of the electron conduction process in the subthreshold region: a uniform conduction over the channel area is expected in 3D cells [325], as opposed to a surface conduction in planar devices. The effect of the improved RTN figure on the V_T distribution width is shown on the right of Figure 38, making clear that a tighter V_T distribution can be achieved with 3D NAND. This is confirmed by data in Figure 39, where the V_T distribution width in 2D and 3D NAND is compared: note the improvement in both RTN and detrapping contributions.

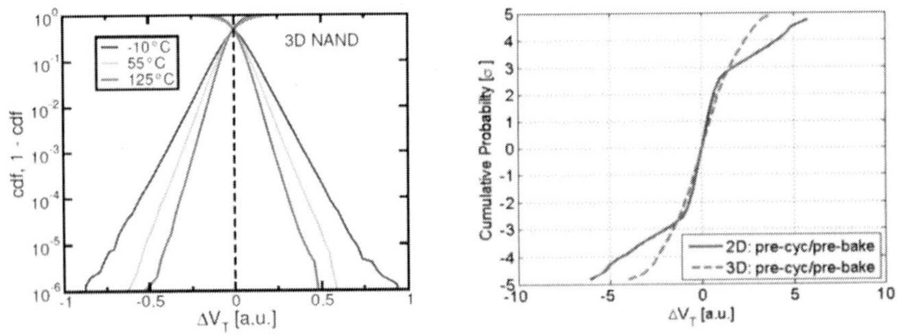

Figure 38. (Left) RTN cdf after cycling for a planar and a 3D NAND as a function of temperature [324]; **(Right)** ΔV_T cdf for a planar and a 3D NAND [299], © 2015 IEEE. Devices are floating-gate based.

4.3.2. Polysilicon Grains

The temperature dependence of RTN in 3D NAND that can be seen in Figure 38 (left) is quite different with respect to what is reported for planar devices (see Figure 11): while RTN in planar NAND is temperature-independent, its slope decreases at higher temperatures in 3D NAND (similar dependences were also reported by [326,327]) and affect all cells in the array (see the unique exponential distribution in 3D NAND compared to the exponential tails departing from a central Gaussian distribution in the 2D case). The explanation for the exponential distribution of the RTN ΔV_T that was put forward in Section 3.2.3 ascribes such fluctuations to current percolation around the

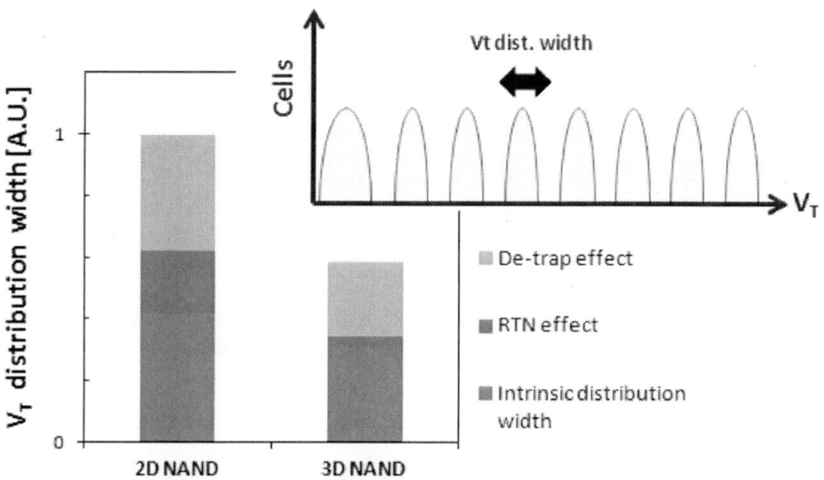

Figure 39. V_T Contributions to the distribution width in planar and 3D NAND [299], © 2015 IEEE.

charged traps, meaning that different conduction mechanisms and/or trap locations should now be considered and reminding us that RTN remains a powerful tool for probing electron conduction in microscopic devices. In 3D NAND devices, in particular, the vertical string is usually made of polycrystalline silicon, because of the fabrication process. Such a material is composed of many single-crystal "grains" having different orientations, separated by grain boundaries that are highly-defective regions [328]. As a matter of fact, since its early use in analog resistors or thin-film transistors, it was recognized that polysilicon characteristics such as low-frequency noise are dominated by capture/emission at the grain boundary interfaces [329,330]. More recently, the effect of polysilicon grain traps charging and discharging on the transient BL current has been addressed [331,332].

A characterization of electron conduction in vertical poly-Si channels was presented in [333], comparing different technological processes and concluding that charge capture/emission from traps within the poly-Si is the main source of current fluctuations, as opposed to traps at the silicon/oxide interfaces. RTN improves for larger cells and smaller average grain size, as shown in Figure 40 [326,334] (where it was measured above the threshold). Other works also reported poor reliability and higher variability in 3D NAND for large poly-Si grain size [317,335]. This is because the larger number of grains associated with their smaller size results in a self-averaging effect, as discussed in [336] for the case of a polysilicon gate in a planar transistor.

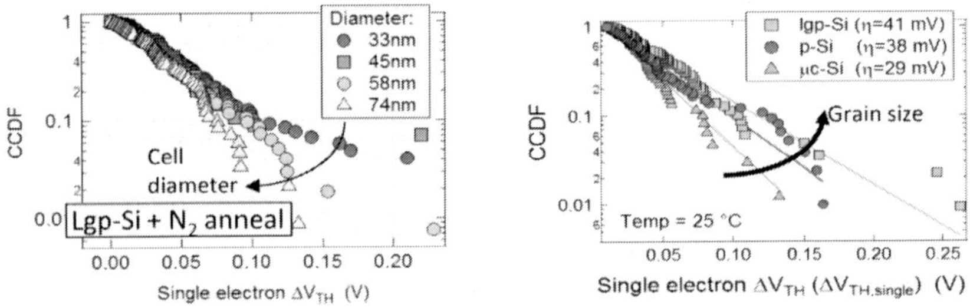

Figure 40. RTN complementary cdf for vertical poly-Si channel transistors having different channel diameter (**left**) and poly-Si grain size (**right**) [334], © 2012 IEEE.

These features call for a thorough analysis of the conduction mechanism in poly-Si channels, accounting for grain boundaries and related traps. The current understanding of the phenomenon is that defects at grain boundaries can trap electrons and act as percolation centers, while causing a degradation in the carrier mobility with respect to its bulk value [337,338]. Charge transport is actually limited by the energy barriers at the grain boundaries (due to the trapped charge) rather than by transport within the grains, except for very large grain size [339]. Several works have exploited numerical simulations to investigate the effect of grain boundaries on variability in nanowires [340,341,342,343,344,345] and 3D NAND devices [346,347]. However, several important features of such models still have to be assessed, such as the grain size [348,349], the density and energy distribution of grain boundary traps [350,351], and the mobility degradation and conduction process at the grain boundaries [352,353].

4.4. Disturbs

In the 3D NAND architecture of Figure 34, strings are vertically aligned while control gates run over parallel planes [28], meaning that read and program/pass disturbs are much more complex with respect to a planar layout, as different pages now share the same WL. Moreover, the onset of disturbs is strongly related to the actual array architecture and internal subdivision as well as operation mode; therefore, extreme care should be taken before generalizing the following results. A schematic of the different types of program disturb is reported in Figure 41 [294,354]: an *x* mode for cells on the same selector line but different BLs, analogous to conventional 2D arrays, a *y* mode for cells on the same BL, and a mixed mode for all other cells sharing the programmed WL. The modes differ for the BL bias (selected cell and *y* mode disturb = 0; *x* andmodes =

high) and the drain selector bias (selected cell and *x* mode disturb = high; *y* and modes = 0), meaning that different boosting levels are achieved. As a matter of fact, the channel is first precharged to a high voltage in *x*-mode disturb before being boosted, while it is boosted from ground level during *y* mode, resulting in a lower potential (i.e., increased disturbs) in the latter case, as shown in Figure 42 (left) [355].

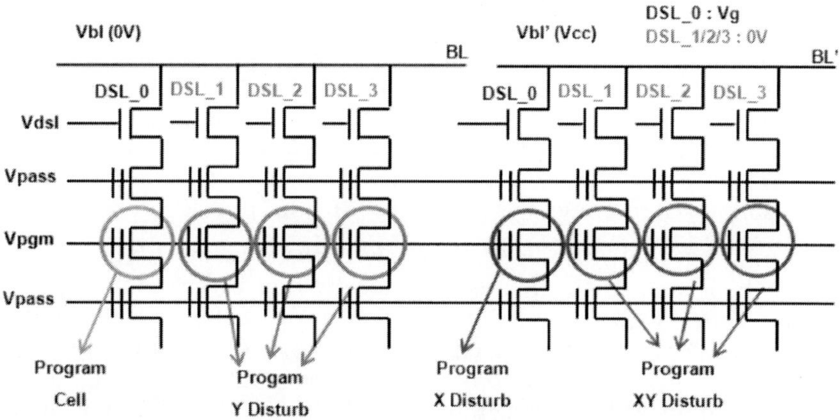

Figure 41. Schematic view of 3D NAND program disturb modes [354], © 2012 IEEE.

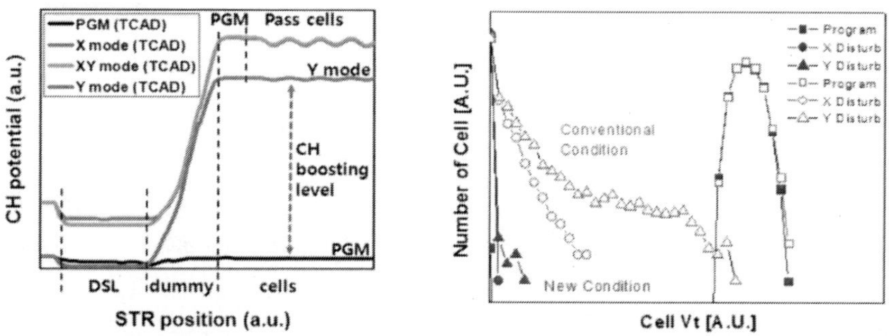

Figure 42. (**Left**) Channel potential for different cell modes during boosting in a 3D NAND [355], © 2013 IEEE; (**Right**) V_T pdf for programmed and disturbed cells in a 3D NAND [354], © 2012 IEEE. Devices are charge-trap based.

The boosting efficiency of 3D NAND is affected by different factors, such as the coupling coefficient and the leakage current. A high coupling coefficient is expected in 3D NAND because of the floating channel lacking a silicon substrate [356]. On the other hand, the worse subthreshold slope of 3D polysilicon devices with respect to planar devices increases the leakage current that discharges the boosted channel potential [294,354]. Clever device optimization, including selector design and insertion of dummy cells to relieve edge WL disturbs, eventually leads to much alleviated program disturbs, as shown in Figure 42 (right) [354]. Other optimization schemes such as an initial channel precharge phase [357] and selector doping [358] have been also proposed. In analogy with planar devices, suitable management schemes for 3D NAND have been also proposed [359,360]. Similar considerations apply to vertical-gate structures, in which channel arrays are horizontally stacked with vertical WLs [361,362].

A study of read disturb in 3D NAND was presented in [363], highlighting an enhanced disturb on unselected cells due to channel boosting. A related effect was discussed in [364], where it is shown that the trailing edge of the WL bias at the end of a read phase can draw the channel potential negative, as it becomes floated when high- cells along the string are turned off. This can become an issue if the read operation is followed by a program one, as in program-verify steps: the channel of program-inhibited cells will now reach a lower boosted potential, resulting in increased program disturbs. Suitable program schemes must therefore be adopted to avoid this issue.

5. CONCLUSIONS

NAND reliability limitations arise from the high-field operation needed to store and remove electrons to/from the floating gate and its undesirable effect on the tunnel oxide, whose insulating properties degrade with usage. This results in a series of issues affecting endurance, retention and read accuracy which were thoroughly reviewed in this paper, with an eye on their physical origin and dependence on cell/array parameters. Physical understanding of those issues led to clever solutions in terms of array organization and operation, boosted by process/technology advancements, that have pushed the planar technology to the 15 nm node, after which the conventional scaling path could no longer be followed.

The revolutionary move to 3D NAND resulted in a larger cell size, relieving the impact of traditional reliability issues, but gave rise to new and subtle reliability concerns, owed to the floating body and polysilicon channel of 3D NAND cells. These reliability issues are reviewed in this paper, comparing their relevance with what is obtained on planar devices. It is clear that 3D NAND allowed the cell architecture to be engineered and allowed the array operation to increase the overall reliability of the cell, resulting in improved performance in terms of storage density, as testified by the impressive development of QLCs, and read/write throughput. A key understanding of the physical mechanisms at the basis of the cell operation and degradation with usage remains of utmost importance to support the development of ever-improving solid-state storage solutions.

ACKNOWLEDGMENTS

The authors would like to acknowledge Angelo Visconti, Paolo Tessariol, Emilio Camerlenghi and Akira Goda from Micron Technology Inc. for the fruitful and long-lasting collaboration, and Davide Resnati and Gianluca Nicosia from Politenico di Milano for their support. Useful comments on ECCs from an anonymous Reviewer are also gratefully acknowledged.

REFERENCES

1. Harari, E. Electrically Erasable Non-Volatile Semiconductor Memory. U.S. Patent 4,115,914, 26 September 1978.
2. Masuoka, F.; Asano, M.; Iwahashi, H.; Komuro, T.; Tanaka, S. A new Flash E^2PROM cell using triple polysilicon technology. In Proceedings of the 1984 International Electron Devices Meeting, San Francisco, CA, USA, 9–12 December 1984; pp. 464–467.
3. Masuoka, F.; Momodomi, M.; Iwata, Y.; Shirota, R. New ultra high density EPROM and Flash EEPROM with NAND structure cell. In Proceedings of the 1987 International Electron Devices Meeting, New York, NY, USA, 6–9 December 1987; pp. 552–555.
4. Micheloni, R.; Marelli, A.; Eshghi, K. (Eds.) *Inside Solid State Drives (SSDs)*; Springer: Heidelberg/Berlin, Germany, 2013.
5. Fontana, R.E.; Decad, G.M.; Hetzler, S.R. Volumetric density trends (TB/in.^{3}) for storage components: Tape, hard disk drives, NAND, and blu-ray. *J. Appl. Phys.* **2015**, *117*, 17E301-1–17E301-4.
6. Tokutomi, T.; Doi, M.; Hachiya, S.; Kobayashi, A.; Tanakamaru, S.; Takeuchi, K. Enterprise-grade 6x fast read and 5x highly reliable SSD with TLC NAND-Flash memory for Big-Data storage. In Proceedings of the 2015 IEEE International Solid-State Circuits Conference (ISSCC), San Francisco, CA, USA, 22–26 February 2015; pp. 140–142.
7. Kim, H.J.; Lim, J.D.; Lee, J.W.; Na, D.H.; Shin, J.H.; Kim, C.H.; Yu, S.W.; Shin, J.Y.; Lee, S.K.; Rajagopal, D.; et al. 1 GB/s 2 Tb NAND Flash multi-chip package with frequency-boosting interface chip. In Proceedings of the 2015 IEEE International Solid-State Circuits Conference (ISSCC), San Francisco, CA, USA, 22–26 February 2015; pp. 138–140.

8. Choi, S.; Kim, D.; Choi, S.; Kim, B.; Jung, S.; Chun, K.; Kim, N.; Lee, W.; Shin, T.; Jin, H.; et al. A 93.4 mm^2 64 Gb MLC NAND-Flash memory with 16 nm CMOS technology. In Proceedings of the 2014 IEEE International Solid-State Circuits Conference (ISSCC), San Francisco, CA, USA, 9–13 February 2014; pp. 328–329.
9. Helm, M.; Park, J.K.; Ghalam, A.; Guo, J.; wan Ha, C.; Hu, C.; Kim, H.; Kavalipurapu, K.; Lee, E.; Mohammadzadeh, A.; et al. A 128 Gb MLC NAND-Flash device using 16 nm planar cell. In Proceedings of the 2014 IEEE International Solid-State Circuits Conference (ISSCC), San Francisco, CA, USA, 9–13 February 2014; pp. 326–327.
10. Lee, S.; Lee, J.-Y.; Park, I.-H.; Park, J.; Yun, S.-W.; Kim, M.-S.; Lee, J.-H.; Kim, M.; Lee, K.; Kim, T.; et al. A 128 Gb 2b/cell NAND Flash memory in 14nm technology with t_{PROG} = 640 μs and 800 MB/s I/O rate. In Proceedings of the 2014 IEEE International Solid-State Circuits Conference (ISSCC), San Francisco, CA, USA, 31 January–4 February 2016; pp. 138–139.
11. Sako, M.; Watanabe, Y.; Nakajima, T.; Sato, J.; Muraoka, K.; Fujiu, M.; Kouno, F.; Nakagawa, M.; Masuda, M.; Kato, K.; et al. A low-power 64 Gb MLC NAND-Flash memory in 15 nm CMOS technology. In Proceedings of the 2015 IEEE International Solid-State Circuits Conference (ISSCC), San Francisco, CA, USA, 22–26 February 2015; pp. 128–129.
12. Prall, K.; Parat, K. 25 nm 64 Gb MLC NAND technology and scaling challenges. In Proceedings of the 2010 IEEE International Electron Devices Meeting (IEDM), San Francisco, CA, USA, 6–8 December 2010; pp. 102–105.
13. Goda, A.; Parat, K. Scaling directions for 2D and 3D NAND cells. In Proceedings of the 2012 IEEE International Electron Devices Meeting (IEDM), San Francisco, CA, USA, 10–13 December 2012; pp. 2.1.1–2.1.4.
14. Lee, S. Scaling challenges in NAND Flash device toward 10 nm technology. In Proceedings of the 2012 4th IEEE International Memory Workshop (IMW), Milan, Italy, 20–23 May 2012; pp. 1–4.
15. Monzio Compagnoni, C.; Goda, A.; Spinelli, A.S.; Feeley, P.; Lacaita, A.L.; Visconti, A. Reviewing the evolution of the NAND Flash technology. *Proc. IEEE* **2017**.
16. Endoh, T.; Kinoshita, K.; Tanigami, T.; Wada, Y.; Sato, K.; Yamada, K.; Yokoyama, T.; Takeuchi, N.; Tanaka, K.; Awaya, N.; et al. Novel ultra high density Flash memory with a stacked-surrounding gate transistor (S-SGT) structured cell. In Proceedings of the 2001 International Electron Devices Meeting (IEDM'01), Washington, DC, USA, 2–5 December 2001; pp. 33–36.
17. Jung, S.M.; Jang, J.; Cho, W.; Cho, H.; Jeong, J.; Chang, Y.; Kim, J.; Rah, Y.; Son, Y.; Park, J.; et al. Three dimensionally stacked NAND Flash memory technology using stacking single crystal Si layers on ILD and TANOS structure for beyond 30 nm node. In Proceedings of the 2006 International Electron Devices Meeting (IEDM), San Francisco, CA, USA, 11–13 December 2006; pp. 37–40.
18. Lai, E.K.; Lue, H.T.; Hsiao, Y.H.; Hsieh, J.Y.; Lu, C.P.; Wang, S.Y.; Yang, L.W.; Yang, T.; Chen, K.C.; Gong, J.; et al. A multi-layer stackable thin-film transistor (TFT) NAND-type Flash memory. In Proceedings of the 2006 International Electron Devices Meeting (IEDM), San Francisco, CA, USA, 11–13 December 2006; pp. 41–44.
19. Park, K.T.; Kim, D.; Hwang, S.; Kang, M.; Cho, H.; Jeong, Y.; Seo, Y.I.; Jang, J.; Kim, H.S.; Jung, S.M.; et al. A 45 nm 4Gb 3-dimensional double-stacked multi-Level NAND Flash memory with shared bitline structure. In Proceedings of the 2008 IEEE International Solid-State Circuits Conference—Digest of Technical Papers, San Francisco, CA, USA, 3–7 February 2008; pp. 510–511.
20. Tanaka, H.; Kido, M.; Yahashi, K.; Oomura, M.; Katsumata, R.; Kito, M.; Fukuzumi, Y.; Sato, M.; Nagata, Y.; Matsuoka, Y.; et al. Bit cost scalable technology with punch and plug process for ultra high density Flash memory. In Proceedings of the 2007 IEEE Symposium on VLSI Technology, Kyoto, Japan, 12–14 June 2007; pp. 14–15.
21. Park, K.T.; Han, J.M.; Kim, D.; Nam, S.; Choi, K.; Kim, M.-S.; Kwak, P.; Lee, D.; Choi, Y.H.; Kang, K.M.; et al. Three-dimensional 128Gb MLC vertical NAND Flash-

memory with 24-WL stacked layers and 50MB/s high-speed programming. In Proceedings of the 2014 IEEE International Solid-State Circuits Conference (ISSCC), San Francisco, CA, USA, 9–13 February 2014; pp. 334–335.
22. Park, K.T.; Nam, S.; Kim, D.; Kwak, P.; Lee, D.; Choi, Y.H.; Choi, M.H.; Kwak, D.H.; Kim, D.H.; Kim, M.S.; et al. Three-dimensional 128 Gb MLC vertical NAND Flash memory with 24-WL stacked layers and 50 MB/s high-speed programming. *IEEE J. Solid-State Circuits* **2015**, *50*, 204–213.
23. Tanaka, T.; Helm, M.; Vali, T.; Ghodsi, R.; Kawai, K.; Park, J.K.; Yamada, S.; Pan, F.; Einaga, Y.; Ghalam, A.; et al. A 768Gb 3b/cell 3D-floating-gate NAND Flash memory. In Proceedings of the 2014 IEEE International Solid-State Circuits Conference (ISSCC), San Francisco, CA, USA, 31 January–4 February 2016; pp. 142–143.
24. Yamashita, R.; Magia, S.; Higuchi, T.; Yoneya, K.; Yamamura, T.; Mizukoshi, H.; Zaitsu, S.; Yamashita, M.; Toyama, S.; Kamae, N.; et al. A 512Gb 3b/cell Flash memory on 64-word-line-layer BiCS technology. In Proceedings of the 2017 IEEE International Solid- State Circuits Conference (ISSCC), San Francisco, CA, USA, 5–9 February 2017; pp. 196–198.
25. Kim, C.; Cho, J.H.; Jeong, W.; Park, I.H.; Park, H.W.; Kim, D.H.; Kang, D.; Lee, S.; Lee, J.S.; Kim, W.; et al. A 512 Gb 3b/cell 64-stacked WL 3D V-NAND Flash memory. In Proceedings of the 2017 IEEE International Solid- State Circuits Conference (ISSCC), San Francisco, CA, USA, 5–9 February 2017; pp. 202–204.
26. Shirota, R. Developments in 3D-NAND Flash technology. In *Advances in Non-Volatile Memory and Storage Technology*; Nishi, Y., Ed.; Elsevier: Amsterdam, The Netherlands, 2014; Chapter 2; pp. 27–74.
27. Aritome, S. *NAND Flash Memory Technologies*; John Wiley & Sons: Hoboken, NJ, USA, 2015.
28. Micheloni, R. (Ed.) *3D Flash Memories*; Springer: Heidelberg/Berlin, Germany, 2016.
29. Lee, J.; Jang, J.; Lim, J.; Shin, Y.G.; Lee, K.; Jung, E. A new ruler on the market: 3D-NAND flash for high-density memory and its technology evolution and challenges on the future. In Proceedings of the 2016 International Electron Devices Meeting (IEDM), San Francisco, CA, USA, 3–7 December 2016; pp. 11.2.1–11.2.4.
30. *Nonvolatile Memory Technology with Emphasis on Flash*; Brewer, J.E.; Gill, M. (Eds.) Wiley-IEEE Press: Hoboken, NJ, USA, 2008.
31. *Inside NAND Flash Memories*; Micheloni, R.; Crippa, L.; Marelli, A. (Eds.) Springer: Heidelberg/Berlin, Germany, 2010.
32. Pavan, P.; Bez, R.; Olivo, P.; Zanoni, E. Flash memory cells—An overview. *Proc. IEEE* **1997**, *85*, 1248–1271.
33. Aritome, S.; Satoh, S.; Maruyama, T.; Watanabe, H.; Shuto, S.; Hemink, G.J.; Shirota, R.; Watanabe, S.; Masuoka, F. A 0.67 μm^2 self-aligned shallow trench isolation cell (SA-STI cell) for 3 V-only 256 Mbit NAND EEPROMs. In Proceedings of the 1994 International Electron Devices Meeting (IEDM), San Francisco, CA, USA, 11–14 December 1994; pp. 61–64.
34. Naso, G.; Botticchio, L.; Castelli, M.; Cerafogli, C.; Cichocki, M.; Conenna, P.; D'Alessandro, A.; Santis, L.D.; Di Cicco, D.; Di Francesco, W.; et al. A 128 Gb 3 b/cell NAND Flash design using 20 nm planar-cell technology. In Proceedings of the 2013 IEEE International Solid-State Circuits Conference (ISSCC), San Francisco, CA, USA, 17–21 February 2013; pp. 218–219.
35. Bauer, M.; Alexis, R.; Atwood, G.; Baltar, B.; Fazio, A.; Frary, K.; Hensel, M.; Ishac, M.; Javanifard, J.; Landgraf, M.; et al. A Multilevel-cell 32 Mb Flash memory. In Proceedings of the 1995 IEEE International Solid-State Circuits Conference, San Francisco, CA, USA, 15–17 February 1995; pp. 132–133.
36. Jung, T.S.; Choi, Y.J.; Suh, K.D.; Suh, B.H.; Kim, J.K.; Lim, Y.H.; Koh, Y.N.; Park, J.W.; Lee, K.J.; Park, J.H.; et al. A 117-mm^{2}3.3-V only 128-Mb multilevel NAND Flash

memory for mass storage applications. *IEEE J. Solid-State Circuits* **1996**, *31*, 1575–1583.
37. Shibata, N.; Maejima, H.; Isobe, K.; Iwasa, K.; Nakagawa, M.; Fujiu, M.; Shimizu, T.; Honma, M.; Hoshi, S.; Kawaai, T.; et al. A 70 nm 16 Gb 16-level-cell NAND flash memory. *IEEE J. Solid-State Circuits* **2008**, *43*, 929–937.
38. Micheloni, R.; Crippa, L. Multi-bit NAND Flash memories for ultra-high density storage devices. In *Advances in Non-Volatile Memory and Storage Technology*; Nishi, Y., Ed.; Woodhead Publishing: Amsterdam, The Netherlands, 2014; Chapter 3; pp. 75–119.
39. Aritome, S.; Shirota, R.; Kirisawa, R.; Endoh, T.; Nakayama, R.; Sakui, K.; Masuoka, F. A reliable bi-polarity write/erase technology in Flash EEPROMs. In Proceedings of the 1990 International Electron Devices Meeting (IEDM), San Francisco, CA, USA, 9–12 December 1990; pp. 111–114.
40. Hemink, G.J.; Tanaka, T.; Endoh, T.; Aritome, S.; Shirota, R. Fast and accurate programming method for multi-level NAND EEPROMs. In Proceedings of the 1995 Symposium on VLSI Technology—Digest of Technical Papers, Kyoto, Japan, 6–8 June 1995; pp. 129–130.
41. Micheloni, R.; Marelli, A.; Ravasio, R. *Error-Correction Codes for Non-Volatile Memories*; Springer: Heidelberg/Berlin, Germany, 2008.
42. Zhao, K.; Zhao, W.; Sun, H.; Zhang, T.; Zhang, X.; Zheng, N. LDPC-in-SSD: Making advanced error correction codes work effectively in solid state drives. In Proceedings of the 11th USENIX Conference on File and Storage Technologies, San Jose, CA, USA, 12–15 February 2013; pp. 243–256.
43. Wang, J.; Vakilinia, K.; Chen, T.Y.; Courtade, T.; Dong, G.; Zhang, T.; Shankar, H.; Wesel, R. Enhanced precision through multiple reads for LDPC decoding in Flash memories. *IEEE J. Sel. Areas Commun.* **2014**, *32*, 880–891.
44. Lin, W.; Yen, S.W.; Hsu, Y.C.; Lin, Y.H.; Liang, L.C.; Wang, T.C.; Shih, P.Y.; Lai, K.H.; Cheng, K.Y.; Chang, C.Y. A low power and ultra high reliability LDPC error correction engine with digital signal processing for embedded NAND Flash controller in 40nm CMOS. In Proceedings of the 2014 Symposium on VLSI Technology (VLSI-Technology): Digest of Technical Papers, Honolulu, HI, USA, 9–12 June 2014; pp. 101–102.
45. Haymaker, K.; Kelley, C.A. Structured bit-interleaved LDPC codes for MLC Flash memory. *IEEE J. Sel. Areas Commun.* **2014**, *32*, 870–879.
46. Cho, S.-G.; Kim, D.; Choi, J.; Ha, J. Block-wise concatenated BCH codes for NAND Flash memories. *IEEE Trans. Commun.* **2014**, *62*, 1164–1177.
47. Kim, D.; Ha, J. Quasi-primitive block-wise concatenated BCH codes for NAND Flash memories. In Proceedings of the 2014 IEEE Information Theory Workshop (ITW), Hobart, Australia, 2–5 November 2014; pp. 611–615.
48. Spinner, J.; Rajab, M.; Freudenberger, J. Construction of high-rate generalized concatenated codes for applications in non-volatile Flash memories. In Proceedings of the 2016 IEEE Information Theory Workshop (ITW), Cambridge, UK, 11–14 September 2016.
49. Spinner, J.; Freudenberger, J.; Shavgulidze, S. A soft input decoding algorithm for generalized concatenated codes. *IEEE Trans. Commun.* **2016**, *64*, 3585–3595.
50. Garvin, P.K.; Stanard, H.D. Method and System for Managing Bad Areas in Flash Memories. U.S. Patent 6,260,156 B1, 10 July 2001.
51. Verma, G.; Mielke, N. Reliability performance of ETOX based Flash memories. In Proceedings of the 26th Annual Proceedings, International Reliability Physics Symposium, Monterey, CA, USA, 12–14 April 1988; pp. 158–166.
52. Aritome, S.; Shirota, R.; Hemink, G.; Endoh, T.; Masuoka, F. Reliability issues of Flash memory cells. *Proc. IEEE* **1993**, *81*, 776–788.
53. Cappelletti, P.; Modelli, A. Flash memory reliability. In *Flash Memories*; Cappelletti, P., Golla, C., Olivo, P., Zanoni, E., Eds.; Kluwer: Dordrecht, The Netherlands, 1999.

54. Tao, G.; Valk, H.; Scarpa, A.; van Marwijk, L.; Kuper, F.G.; Dormans, G.J.M.; Verhaar, R.D.J. Reliability aspects of advanced embedded floating-gate non-volatile memories with uniform channel FN tunneling for both program and erase. In Proceedings of the Non-Volatile Semiconductor Memory Workshop, Monterey, CA, USA, 12–15 August 2001; pp. 130–132.
55. Cappelletti, P.; Bez, R.; Modelli, A.; Visconti, A. What we have learned on Flash memory reliability in the last ten years. In Proceedings of the 2004 International Electron Devices Meeting (IEDM), San Francisco, CA, USA, 13–15 December 2004; pp. 489–492.
56. Ielmini, D.; Spinelli, A.S.; Lacaita, A.L. Recent developments on Flash memory reliability. *Microelectron. Eng.* **2005**, *80C*, 321–328.
57. Monzio Compagnoni, C.; Spinelli, A.S.; Lacaita, A.L.; Ghetti, A.; Visconti, A. Emerging constraints on NAND Flash memory reliability. In *Nonvolatile Memories: Materials, Devices and Applications*; American Scientific Publishers: Valencia, CA, USA, 2012; pp. 267–293.
58. Zambelli, C.; Olivo, P. SSD reliability. In *Inside Solid-State Drives (SSD)s*; Micheloni, R., Marelli, A., Eshghi, K., Eds.; Springer: Heidelberg/Berlin, Germany, 2013; Chapter 8; pp. 203–231.
59. Park, Y.; Lee, J.; Cho, S.S.; Jin, G.; Jung, E. Scaling and reliability of NAND Flash devices. In Proceedings of the 2014 IEEE International Reliability Physics Symposium (IRPS), Waikoloa, HI, USA, 1–5 June 2014; pp. 2E.1.1–2E.1.4.
60. Mielke, N.; Marquart, T.; Wu, N.; Kessenich, J.; Belgal, H.; Schares, E.; Trivedi, F.; Goodness, E.; Nevill, L.R. Bit error rate in NAND Flash memories. In Proceedings of the IEEE International Reliability Physics Symposium, Phoenix, AZ, USA, 27 April–1 May 1 2008; pp. 9–19.
61. Meza, J.; Wu, Q.; Kumar, S.; Mutlu, O. A large-scale study of Flash memory failures in the field. In Proceedings of the ACM SIGMETRICS—International Conference on Measurement and Modeling of Computer Systems, Portland, OR, USA, 15–19 June 2015; pp. 177–190.
62. Schroeder, B.; Lagisetty, R.; Merchant, A. Flash reliability in production: the expected and the unexpected. In Proceedings of the 14th USENIX Conference on File and Storage Technologies (FAST'16), Santa Clara, CA, USA, 22–25 February 2016; pp. 67–80.
63. Lee, J.D.; Choi, J.H.; Park, D.; Kim, K. Effects of interface trap generation and annihilation on the data retention characteristics of Flash memory cells. *IEEE Trans. Device Mater. Reliab.* **2004**, *4*, 110–117.
64. Itsumi, M. Positive and negative charging of thermally-grown SiO_2 induced by Fowler-Nordheim emission. *J. Appl. Phys.* **1981**, *52*, 3491–3497.
65. Maserjian, J.; Zamani, N. Behavior of the Si/SiO_2 interface observed by Fowler-Nordheim tunneling. *J. Appl. Phys.* **1982**, *53*, 559–567.
66. Maserjian, J.; Zamani, N. Observation of positively charged state generation near the Si/SiO_2 interface during Fowler-Nordheim tunneling. *J. Vac. Sci. Technol.* **1982**, *20*, 743–746.
67. Lai, S.K. Interface trap generation in silicon dioxide when electrons are captured by trapped holes. *J. Appl. Phys.* **1983**, *54*, 2540–2546.
68. Olivo, P.; Riccò, B.; Sangiorgi, E. Electron trapping/detrapping within thin SiO_2 films in the high field tunneling regime. *J. Appl. Phys.* **1983**, *54*, 5267–5276.
69. Chen, I.C.; Holland, S.; Hu, C. Electron-trap generation by recombination of electrons and holes in SiO_2. *J. Appl. Phys.* **1987**, *61*, 4544–4548.
70. Euzent, B.; Jeng, C.; Lee, J.; Boruta, N. Reliability aspects of a floating gate E^2PROM. In Proceedings of the IEEE International Reliability Physics Symposium (IRPS), Orlando, FL, USA, 7–9 April 1981; pp. 11–16.
71. Witters, J.S.; Groeseneken, G.; Maes, H.E. Degradation of tunnel-oxide floating-gate EEPROM devices and the correlation with high field-current-induced degradation of thin gate oxides. *IEEE Trans. Electron Devices* **1989**, *36*, 1663–1682.

72. Park, Y.B.; Schroder, D.K. Degradation of thin tunnel gate oxide under constant Fowler-Nordheim current stress for a Flash EEPROM. *IEEE Trans. Electron Devices* **1998**, *45*, 1361–1368.
73. Oyama, K.; Shirai, H.; Kodama, N.; Kanamori, K.; Saitoh, K.; Hisamune, Y.S.; Okazawa, T. A novel erasing technology for 3.3 V Flash memory with 64 Mb capacity and beyond. In Proceedings of the 1992 International Electron Devices Meeting (IEDM), San Francisco, CA, USA, 13–16 December 1992; pp. 607–610.
74. Fayrushin, A.; Seol, K.; Na, J.; Hur, S.; Choi, J.; Kim, K. The new program/erase cycling degradation mechanism of NAND Flash memory devices. In Proceedings of the 2009 International Electron Devices Meeting (IEDM), Baltimore, MD, USA, 7–9 December 2009; pp. 823–826.
75. Fayrushin, A.; Lee, C.H.; Park, Y.; Choi, J.H.; Chung, C. Unified endurance degradation model of floating gate NAND Flash memory. *IEEE Trans. Electron Devices* **2013**, *60*, 2031–2037.
76. Shirota, R.; Yang, B.J.; Chiu, Y.Y.; Chen, H.T.; Ng, S.F.; Wang, P.Y.; Chang, J.H.; Kurachi, I. New method to analyze the shift of floating gate charge and generated tunnel oxide trapped charge profile in NAND Flash memory by program/erase endurance. *IEEE Trans. Electron Devices* **2015**, *62*, 114–120.
77. Chang, Y.H.; Chang, L.P. Efficient wear leveling in NAND Flash memory. In *Inside Solid State Drives (SSDs)*; Micheloni, R., Marelli, A., Eshghi, K., Eds.; Springer: Heidelberg/Berlin, Germany, 2013.
78. Ralls, K.S.; Skocpol, W.J.; Jackel, L.D.; Howard, R.E.; Fetter, L.A.; Epworth, R.W.; Tennant, D.M. Discrete resistance switching in submicrometer silicon inversion layers: individual interface traps and low-frequency (1/f?) noise. *Phys. Rev. Lett.* **1984**, *52*, 228–231.
79. Kirton, M.J.; Uren, M.J. Noise in solid-state microstructures: a new perspective on individual defects, interface states and low-frequency (1/f) noise. *Adv. Phys.* **1989**, *38*, 367–468.
80. Hung, K.K.; Ko, P.K.; Hu, C.; Cheng, Y.C. Flicker noise characterization of advanced MOS technologies. In Proceedings of the 1988 International Electron Devices Meeting (IEDM), San Francisco, CA, USA, 11–14 December 1988; pp. 34–37.
81. Hung, K.K.; Ko, P.K.; Hu, C.; Cheng, Y.C. Random telegraph noise of deep-submicrometer MOSFET's. *IEEE Electron Device Lett.* **1990**, *11*, 90–92.
82. Ohata, A.; Toriumi, A.; Iwase, M.; Natori, K. Observation of random telegraph signals: anomalous nature of defects at the Si/SiO_2 interface. *J. Appl. Phys.* **1990**, *68*, 200–204.
83. Shi, Z.M.; Mieville, J.P.; Barrier, J.; Dutoit, M. Low frequency noise and quantum transport in 0.1 μm n-MOSFET's. In Proceedings of the 1991 International Electron Devices Meeting (IEDM), Washington, DC, USA, 8–11 December 1991; pp. 363–366.
84. Fang, P.; Hung, K.K.; Ko, P.K.; Hu, C. Hot-electron-induced traps studied through the random telegraph noise. *IEEE Electron Device Lett.* **1991**, *12*, 273–275.
85. Simoen, E.; Diericks, B.; Claeys, C.L.; Declerck, G.J. Explaining the amplitude of RTS noise in submicrometer MOSFET's. *IEEE Trans. Electron Devices* **1992**, *39*, 422–429.
86. Tsai, M.H.; Ma, T.P.; Hook, T.B. Channel length dependence of random telegraph signal in sub-micron MOSFET's. *IEEE Electron Device Lett.* **1994**, *15*, 504–506.
87. Uren, M.J.; Kirton, M.J.; Collins, S. Anomalous telegraph noise in small-area silicon metal-oxide- semiconductor field-effect transistors. *Phys. Rev. B* **1988**, *37*, 8346–8350.
88. Kirton, M.J.; Uren, M.J.; Collins, S.; Schultz, M.; Karmann, A.; Scheffer, K. Individual defects at the $Si{:}SiO_2$ interface. *Semicond. Sci. Technol.* **1989**, *4*, 1116–1126.
89. Nakamura, H.; Yasuda, N.; Taniguchi, K.; Hamaguchi, C.; Toriumi, A. Existence of double-charged oxide traps in submicron MOSFET's. *Jpn. J. Appl. Phys.* **1989**, *28*, L2057–L2060.

90. Tega, N.; Miki, H.; Osabe, T.; Kotabe, A.; Otsuga, K.; Kurata, H.; Kamohara, S.; Tokami, K.; Ikeda, Y.; Yamada, R. Anomalously large threshold voltage fluctuation by complex random telegraph signal in floating gate Flash memory. In Proceedings of the 2006 International Electron Devices Meeting (IEDM), San Francisco, CA, USA, 11–13 December 2006; pp. 491–494.
91. Fantini, P.; Ghetti, A.; Marinoni, A.; Ghidini, G.; Visconti, A.; Marmiroli, A. Giant random telegraph signals in nanoscale floating-gate devices. *IEEE Electron Device Lett.* **2007**, *28*, 1114–1116.
92. Tega, N.; Miki, H.; Yamaoka, M.; Kume, H.; Mine, T.; Ishida, T.; Mori, Y.; Yamada, R.; Torii, K. Impact of threshold voltage fluctuation due to random telegraph noise on scaled-down SRAM. In Proceedings of the IEEE International Reliability Physics Symposium (IRPS), Phoenix, AZ, USA, 27 April–1 May 1 2008; pp. 541–546.
93. Toh, S.O.; Tsukamoto, Y.; Guo, Z.; Jones, L.; King Liu, T.J.; Nikolic, B. Impact of random telegraph signals on V_{min} in 45 nm SRAM. In Proceedings of the 2009 International Electron Devices Meeting (IEDM), Baltimore, MD, USA, 7–9 December 2009; pp. 767–770.
94. Tanizawa, M.; Ohbayashi, S.; Okagaki, T.; Sonoda, K.; Eikyu, K.; Hirano, Y.; Ishikawa, K.; Tsuchiya, O.; Inoue, Y. Application of a statistical compact model for random telegraph noise to scaled-SRAM Vmin analysis. In Proceedings of the 2010 Symposium on VLSI Technology (VLSI-Technology), Honolulu, HI, USA, 15–17 June 2010; pp. 95–96.
95. Kurata, H.; Otsuga, K.; Kotabe, A.; Kajiyama, S.; Osabe, T.; Sasago, Y.; Narumi, S.; Tokami, K.; Kamohara, S.; Tsuchiya, O. The impact of random telegraph signals on the scaling of multilevel Flash memories. In Proceedings of the 2006 Symposium on VLSI Technology (VLSI-Technology), Honolulu, HI, USA, 13–15 June 2006; pp. 112–113.
96. Gusmeroli, R.; Monzio Compagnoni, C.; Riva, A.; Spinelli, A.S.; Lacaita, A.L.; Bonanomi, M.; Visconti, A. Defects spectroscopy in SiO_2 by statistical random telegraph noise analysis. In Proceedings of the 2006 International Electron Devices Meeting (IEDM), San Francisco, CA, USA, 11–13 December 2006; pp. 483–486.
97. Kurata, H.; Otsuga, K.; Kotabe, A.; Kajiyama, S.; Osabe, T.; Sasago, Y.; Narumi, S.; Tokami, K.; Kamohara, S.; Tsuchiya, O. Random telegraph signal in Flash memory: Its impact on scaling of multilevel Flash memory beyond the 90-nm node. *IEEE J. Solid-State Circuits* **2007**, *42*, 1362–1369.
98. Fukuda, K.; Shimizu, Y.; Amemiya, K.; Kamoshida, M.; Hu, C. Random telegraph noise in Flash memories—Model and technology scaling. In Proceedings of the 2007 International Electron Devices Meeting (IEDM), Washington, DC, USA, 10–12 December 2007; pp. 169–172.
99. Miki, H.; Osabe, T.; Tega, N.; Kotabe, A.; Kurata, H.; Tokami, K.; Ikeda, Y.; Kamohara, S.; Yamada, R. Quantitative analysis of random telegraph signals as fluctuations of threshold voltages in scaled Flash memory cells. In Proceedings of the IEEE International Reliability Physics Symposium (IRPS), Phoenix, AZ, USA, 15–19 April 2007; pp. 29–35.
100. Kandiah, K.; Deighton, M.O.; Whiting, F.B. A physical model for random telegraph signal currents in semiconductor devices. *J. Appl. Phys.* **1989**, *62*, 937–948.
101. Mueller, H.H.; Schulz, M. Conductance modulation of submicrometer metal-oxide-semiconductor field effect transistors by single-electron trapping. *J. Appl. Phys.* **1996**, *79*, 4178–4180.
102. Lust, L.M.; Kakalios, J. Dynamic percolation model of conductance fluctuations in hydrogenated amorphous silicon. *Phys. Rev. Lett.* **1995**, *75*, 2192–2195.
103. Brews, J.R. Theory of the carrier-density fluctuations in an IGFET near threshold. *J. Appl. Phys.* **1975**, *46*, 2181–2192.
104. Arnold, E. Conduction mechanisms in bandtails at the Si-SiO_2 interface. *Surf. Sci.* **1976**, *58*, 60–70.

105. Wong, H.S.; Taur, Y. Three-dimensional "atomistic" simulation of discrete random dopant distribution effects in sub-0.1 μm MOSFET's. In Proceedings of the 1993 International Electron Devices Meeting (IEDM), Washington, DC, USA, 5–8 December 1993; pp. 705–708.
106. Asenov, A. Random dopant induced threshold voltage lowering and fluctuations in sub-0.1μm MOSFET's: A 3-D "atomistic" simulation study. *IEEE Trans. Electron Devices* **1998**, *45*, 2505–2513.
107. Asenov, A.; Brown, A.R.; Davies, J.H.; Saini, S. Hierarchical approach to "atomistic" 3-D MOSFET simulation. *IEEE Trans. Comput.-Aided Des.* **1999**, *18*, 1558–1565.
108. Asenov, A.; Saini, S. Suppression of random dopant-induced threshold voltage fluctuations in sub-0.1μm MOSFET's with epitaxial and δ-doped channels. *IEEE Trans. Electron Devices* **1999**, *46*, 1718–1724.
109. Vandamme, L.K.J.; Sodini, D.; Gingl, Z. On the anomalous behavior of the relative amplitude of RTS noise. *Solid-State Electron.* **1998**, *42*, 901–905.
110. Asenov, A.; Balasubramaniam, R.; Brown, A.R.; Davies, J.H.; Saini, S. Random telegraph signal amplitudes in sub 100 nm (decanano) MOSFETs: A 3D "atomistic" simulation study. In Proceedings of the 2000 International Electron Devices Meeting (IEDM), San Francisco, CA, USA, 10–13 December 2000; pp. 279–282.
111. Asenov, A.; Balasubramaniam, R.; Brown, A.R.; Davies, J.H. RTS amplitudes in decananometer MOSFETs: 3-D simulation study. *IEEE Trans. Electron Devices* **2003**, *50*, 839–845.
112. Ghetti, A.; Bonanomi, M.; Monzio Compagnoni, C.; Spinelli, A.S.; Lacaita, A.L.; Visconti, A. Physical modeling of single-trap RTS statistical distribution in Flash memories. In Proceedings of the IEEE International Reliability Physics Symposium (IRPS), Phoenix, AZ, USA, 27 April–1 May 1 2008; pp. 610–615.
113. Ghetti, A.; Monzio Compagnoni, C.; Spinelli, A.S.; Visconti, A. Comprehensive analysis of random telegraph noise instability and its scaling in deca-nanometer Flash memories. *IEEE Trans. Electron Devices* **2009**, *56*, 1746–1752.
114. Alexander, C.L.; Brown, A.R.; Watling, J.R.; Asenov, A. Impact of single charge trapping nano- MOSFETs—Electrostatics versus transport effects. *IEEE Trans. Nanotechnol.* **2005**, *4*, 339–344.
115. Alexander, C.; Brown, A.R.; Watling, J.R.; Asenov, A. Impact of scattering in "atomistic" device simulations. *Solid-State Electron.* **2005**, *49*, 733–739.
116. Kovac, U.; Alexander, C.; Roy, G.; Riddet, C.; Cheng, B.; Asenov, A. Hierarchical simulation of statistical variability: from 3-D MC with "ab initio" ionized impurity scattering to statistical compact models. *IEEE Trans. Electron Devices***2010**, *57*, 2418–2426.
117. Kovac, U.; Alexander, C.; Asenov, A. Statistical estimation of electrostatic and transport contributions to device parameter variation. In Proceedings of the International Workshop on Computational Electronics (IWCE), Pisa, Italy, 27–29 October 2010; pp. 235–238.
118. Ancona, M.G.; Tiersten, H.F. Macroscopic physics of the silicon inversion layer. *Phys. Rev. B* **1987**, *35*, 7959–7965.
119. Ancona, M.G.; Yu, Z.; Dutton, R.W.; Voorde, P.J.V.; Cao, M.; Vook, D. Density-gradient analysis of MOS tunneling. *IEEE Trans. Electron Devices* **2000**, *47*, 2310–2319.
120. Asenov, A.; Slavcheva, G.; Brown, A.R.; Davies, J.H.; Saini, S. Increase in the random dopant induced threshold fluctuations and lowering in sub-100 nm MOSFETs due to quantum effects: A 3-D density-gradient simulation study. *IEEE Trans. Electron Devices* **2001**, *48*, 722–729.
121. Ghetti, A.; Monzio Compagnoni, C.; Biancardi, F.; Lacaita, A.L.; Beltrami, S.; Chiavarone, L.; Spinelli, A.S.; Visconti, A. Scaling trends for random telegraph noise in deca-nanometer Flash memories. In Proceedings of the 2008

International Electron Devices Meeting (IEDM), San Francisco, CA, USA, 15–17 December 2008; pp. 835–838.

122. Monzio Compagnoni, C.; Gusmeroli, R.; Spinelli, A.S.; Lacaita, A.L.; Bonanomi, M.; Visconti, A. Statistical model for random telegraph noise in Flash memories. *IEEE Trans. Electron Devices* **2008**, *55*, 388–395.

123. Monzio Compagnoni, C.; Spinelli, A.S.; Beltrami, S.; Bonanomi, M.; Visconti, A. Cycling effect on the random telegraph noise instabilities of NOR and NAND Flash arrays. *IEEE Electron Device Lett.* **2008**, *29*, 941–943.

124. Spinelli, A.S.; Monzio Compagnoni, C.; Gusmeroli, R.; Ghidotti, M.; Visconti, A. Investigation of the random telegraph noise instability in scaled Flash memory arrays. *Jpn. J. Appl. Phys.* **2008**, *47*, 2598–2601.

125. Seidel, K.; Hoffmann, R.; Löhr, D.A.; Melde, T.; Czernohorsky, M.; Paul, J.; Beug, M.F.; Beyer, V. Analysis of trap mechanisms responsible for random telegraph noise and erratic programming on sub-50 nm floating gate flash memories. In Proceedings of the 2009 10th Annual Non-Volatile Memory Technology Symposium (NVMTS), Portland, OR, USA, 25–28 October 2009; pp. 67–71.

126. Kwon, W.; Song, Y.H.; Cai, Y.; Ryu, W.; Jang, Y.; Shin, S.; Jun, J.; Lee, S.A.; Park, C.K.; Lee, W.S. 3-D channel structure Flash having short channel effect immunity and low random telegraph noise. In Proceedings of the 2008 10th Annual Non-Volatile Memory Technology Symposium (NVMTS), Pacific Grove, CA, USA, 11–14 November 2008; pp. 20–21.

127. Wang, R.V.; Lee, Y.H.; Lu, Y.L.R.; McMahon, W.; Hu, S.; Ghetti, A. Shallow trench isolation edge effect on random telegraph signal noise and implications for Flash memory. *IEEE Trans. Electron Devices* **2009**, *56*, 2107–2113.

128. Lee, C.H.; Yang, I.C.; Lee, C.; Cheng, C.H.; Chong, L.H.; Chen, K.F.; Huang, J.S.; Ku, S.H.; Zous, N.K.; Huang, I.J.; et al. Junction optimization for Reliability issues in floating gate NAND Flash cells. In Proceedings of the IEEE International Reliability Physics Symposium (IRPS), Monterey, CA, USA, 10–14 April 2011; pp. 6B.3.1–6B.3.5.

129. Kim, T.; He, D.; Morinville, K.; Sarpatwari, K.; Millemon, B.; Goda, A.; Kessenich, J. Tunnel oxide nitridation effect on the evolution of V_t instabilities (RTS/QED) and defect characterization for sub-40 nm Flash memory. *IEEE Electron Device Lett.* **2011**, *32*, 999–1001.

130. Kim, T.; Franklin, N.; Srinivasan, C.; Kalavade, P.; Goda, A. Extreme short-channel effect on RTS and inverse scaling behavior: source–drain implantation effect in 25-nm NAND Flash memory. *IEEE Electron Dev. Lett.* **2011**, *32*, 1185–1187.

131. Ghetti, A.; Amoroso, S.M.; Mauri, A.; Monzio Compagnoni, C. Impact of nonuniform doping on random telegraph noise in Flash memory devices. *IEEE Trans. Electron Devices* **2012**, *59*, 309–315.

132. Amoroso, S.M.; Ghetti, A.; Brown, A.R.; Mauri, A.; Monzio Compagnoni, C.; Asenov, A. Impact of cell shape on random telegraph noise in decananometer Flash memories. *IEEE Trans. Electron Devices* **2012**, *59*, 2774–2779.

133. Kim, S.; Lee, M.; Choi, G.B.; Lee, J.; Lee, Y.; Cho, M.; Ahn, K.O.; Kim, J. RTS noise reduction of 1Y-nm floating gate NAND Flash memory using process optimization. In Proceedings of the IEEE International Reliability Physics Symposium (IRPS), Monterey, CA, USA, 19–23 April 2015; pp. MY.8.1–MY.8.4.

134. Tomita, T.; Miyaji, K. Channel doping concentration and cell program state dependence on random telegraph noise spatial amd statistical distribution in 30 nm NAND flash memory. *Jap. J. Appl. Phys.* **2015**, *54*, 04DD02-1–04DD02-6.

135. Joe, S.M.; Yi, J.H.; Park, S.K.; Kwon, H.I.; Lee, J.H. Position-dependent threshold-voltage variation by random telegraph noise in NAND Flash memory strings. *IEEE Electron Device Lett.* **2010**, *31*, 635–637.

136. Joe, S.M.; Yi, J.H.; Park, S.K.; Shin, H.; Park, B.G.; Park, Y.J.; Lee, J.H. Threshold voltage fluctuation by random telegraph noise in floatin gate NAND Flash memory string. *IEEE Trans. Electron Devices* **2011**, *58*, 67–73.

137. Joe, S.M.; Jeong, M.K.; Jo, B.S.; Han, K.R.; Park, S.K.; Lee, J.H. The effect of adjacent bit-line cell interference on random telegraph noise in NAND Flash memory cell strings. *IEEE Trans. Electron Devices* **2012**, *59*, 3568–3573.
138. Miccoli, C.; Paolucci, G.M.; Monzio Compagnoni, C.; Spinelli, A.S.; Goda, A. Cycling pattern and read/bake conditions dependence of random telegraph noise in decananometer NAND Flash arrays. In Proceedings of the IEEE International Reliability Physics Symposium (IRPS), Monterey, CA, USA, 19–23 April 2015; pp. MY.9.1–MY.9.6.
139. Miki, H.; Tega, N.; Yamaoka, M.; Frank, D.J.; Bansal, A.; Kobayashi, M.; Cheng, K.; D'Emic, C.P.; Ren, Z.; Wu, S.; et al. Statistical measurement of random telegraph noise and its impact in scaled-down high-k/metal-gate MOSFETs. In Proceedings of the 2012 IEEE International Electron Devices Meeting (IEDM), San Francisco, CA, USA, 10–13 December 2012; pp. 450–453.
140. Grasser, T. Stochastic charge trapping in oxides: From random telegraph noise to bias temperature instabilities. *Microelectron. Reliab.* **2012**, *52*, 39–70.
141. Machlup, S. Noise in semiconductors: spectrum of a two-parameter random signal. *J. Appl. Phys.* **1954**, *25*, 341–343.
142. Yuzhelevski, Y.; Yuzhelevski, M.; Jung, G. Random telegraph noise analysis in time domain. *Rev. Sci. Instrum.* **2000**, *71*, 1681–1688.
143. Campbell, J.P.; Qin, J.; Cheung, K.P.; Yu, L.C.; Suehle, J.S.; Oates, A.; Sheng, K. Random telegraph noise in highly scaled nMOSFETs. In Proceedings of the IEEE International Reliability Physics Symposium (IRPS), Montreal, QC, Canada, 26–30 April 2009; pp. 382–388.
144. Campbell, J.P.; Yu, L.C.; Cheung, K.P.; Qin, J.; Suehle, J.S.; Oates, A.; Sheng, K. Large random telegraph noise in sub-threshold operation of nano-scale nMOSFETs. In Proceedings of the 2009 IEEE International Conference on IC Design & Technology (ICICDT), Austin, TX, USA, 18–20 May 2009; pp. 17–20.
145. Castellani, N.; Monzio Compagnoni, C.; Mauri, A.; Spinelli, A.S.; Lacaita, A.L. Three-dimensional electrostatics- and atomistic doping-induced variability of RTN time constants in nanoscale MOS devices—Part I: Physical investigation. *IEEE Trans. Electron Devices* **2012**, *59*, 2488–2494.
146. Monzio Compagnoni, C.; Castellani, N.; Mauri, A.; Spinelli, A.S.; Lacaita, A.L. Three-dimensional electrostatics- and atomistic doping-induced variability of RTN time constants in nanoscale MOS devices—Part II: Spectroscopic implications. *IEEE Trans. Electron Devices* **2012**, *59*, 2495–2500.
147. Adamu-Lema, F.; Monzio Compagnoni, C.; Amoroso, S.M.; Castellani, N.; Gerrer, L.; Markov, S.; Spinelli, A.S.; Lacaita, A.L.; Asenov, A. Accuracy and issues of the spectroscopic analysis of RTN traps in nanoscale MOSFETs. *IEEE Trans. Electron Devices* **2013**, *60*, 833–839.
148. Grasser, T.; Kaczer, B.; Goes, W.; Reisinger, H.; Aichinger, T.; Hehenberger, P.; Wagner, P.J.; Schanovsky, F.; Franco, J.; Luque, M.T.; et al. The paradigm shift in understanding the bias temperature instability: From reaction-diffusion to switching oxide traps. *IEEE Trans. Electron Devices* **2011**, *58*, 3652–3666.
149. Nagumo, T.; Takeuchi, K.; Hase, T.; Hayashi, Y. Statistical characterization of trap position, energy, amplitude and time constants by RTN measurement of multiple individual traps. In Proceedings of the 2010 IEEE International Electron Devices Meeting (IEDM), San Francisco, CA, USA, 6–8 December 2010; pp. 628–631.
150. Abe, K.; Teramoto, A.; Sugawa, S.; Ohmi, T. Understanding of traps causing random telegraph noise based on experimentally extracted time constants and amplitude. In Proceedings of the IEEE International Reliability Physics Symposium (IRPS), Monterey, CA, USA, 10–14 April 2011; pp. 381–386.
151. Yonezawa, A.; Teramoto, A.; Obara, T.; Kuroda, R.; Sugawa, S.; Ohmi, T. The study of time constant analysis in random telegraph noise at the subthreshold voltage region. In Proceedings of the IEEE International Reliability Physics Symposium (IRPS), Monterey, CA, USA, 14–18 April 2013; pp. XT.11.1–XT.11.6.

152. Guo, S.; Ren, P.; Wang, R.; Yu, Z.; Luo, M.; Zhang, X.; Huang, R. A new efficient method for characterizing time constants of switching oxide traps. In Proceedings of the 2014 IEEE International Reliability Physics Symposium (IRPS), Waikoloa, HI, USA, 1–5 June 2014; pp. XT.14.1–XT.14.4.
153. Grasser, T.; Rott, K.; Reisinger, H.; Waltl, M.; Franco, J.; Kaczer, B. A unified perspective of RTN and BTI. In Proceedings of the 2014 IEEE International Reliability Physics Symposium (IRPS), Waikoloa, HI, USA, 1–5 June 2014; pp. 4A.5.1–4A.5.7.
154. Chen, J.; Higashi, Y.; Kato, K.; Mitani, Y. Further understandings on random telegraph signal noise through comprehensive studies on large time constant variation and its strong correlations to thermal activation energies. In Proceedings of the 2014 Symposium on VLSI Technology (VLSI-Technology): Digest of Technical Papers, Honolulu, HI, USA, 9–12 June 2014; pp. 164–165.
155. Chen, M.J.; Tu, K.C.; Wang, H.H.; Chen, C.L.; Lai, S.Y.; Liu, Y.S. A statistical model for the headed and tail distributions of random telegraph signal magnitudes in nanoscale MOSFETs. *IEEE Trans. Electron Devices* **2014**, *61*, 2495–2502.
156. Chen, M.J.; Tu, K.C.; Chuang, L.Y.; Wang, H.H. Graphically transforming Mueller-Schulz percolation criteria to random telegraph signal magnitudes in scaled FETs. *IEEE Electron Dev. Lett.* **2015**, *36*, 217–219.
157. Monzio Compagnoni, C.; Ghidotti, M.; Lacaita, A.L.; Spinelli, A.S.; Visconti, A. Random telegraph noise effect on the programmed threshold-voltage distribution of Flash memories. *IEEE Electron Dev. Lett.* **2009**, *30*, 984–986.
158. Monzio Compagnoni, C.; Spinelli, A.S.; Gusmeroli, R.; Beltrami, S.; Ghetti, A.; Visconti, A. Ultimate accuracy for the NAND Flash program algorithm due to the electron injection statistics. *IEEE Trans. Electron Devices* **2008**, *55*, 2695–2702.
159. Monzio Compagnoni, C.; Gusmeroli, R.; Spinelli, A.S.; Visconti, A. Analytical model for the electron-injection statistics during programming of nanoscale NAND Flash memories. *IEEE Trans. Electron Devices* **2008**, *55*, 3192–3199.
160. Alexander, R.M. Accelerated testing in FAMOS deviecs—8K EPROM. In Proceedings of the 16th Annual IEEE International Reliability Physics Symposium, San Diego, CA, USA, 18–20 April 1978; pp. 229–232.
161. Horiuchi, M.; Katto, H. FCAT—A low-voltage high-speed alterable *n*-channel nonvolatile memory device. *IEEE Trans. Electron Devices* **1979**, *26*, 914–918.
162. Wang, S.T. Charge retention of floating-gate transistors under applied bias conditions. *IEEE Trans. Electron Devices***1980**, *27*, 297–299.
163. Bhattacharyya, A. Modelling of write/erase and charge retention characteristics of floating gate EEPROM devices. *Solid-State Electron.* **1984**, *26*, 899–906.
164. Papadas, C.; Pananakakis, G.; Ghibaudo, G.; Riva, C.; Pio, F.; Ghezzi, P. Modeling of the intrinc retention characteristics of FLOTOX EEPROM cells under elevated temperature conditions. *IEEE Trans. Electron Devices* **1995**, *42*, 678–681.
165. Watanabe, H.; Ishihara, T.; Matsunaga, Y.; Matsuzawa, K.; Matsushita, D.; Muraoka, K. Numerical study of data retention due to direct tunneling for nonvolatile memory cell. *IEEE Trans. Electron Devices* **2005**, *52*, 955–961.
166. Dockerty, R.C. Degradation mechanisms in rewritable n-channel FAMOS devices. In Proceedings of the 13th Annual IEEE International Reliability Physics Symposium, Downey, CA, USA, 1–3 April 1975; pp. 6–9.
167. Baglee, D.A.; Smayling, M.C. The effects of write/erase cycling on data loss in EEPROMs. In Proceedings of the 1985 IEEE International Electron Devices Meeting (IEDM), New York, NY, USA, 1–4 December 1985; pp. 624–626.
168. Olivo, P.; Nguyen, T.N.; Riccò, B. High-field-induced degradation in ultra-thin SiO_2 films. *IEEE Trans. Electron Devices* **1988**, *35*, 2259–2267.
169. Naruke, K.; Taguchi, S.; Wada, M. Stress induced leakage current limiting to scale down EEPROM tunnel oxide thickness. In Proceedings of the 1988 International Electron Devices Meeting (IEDM), San Francisco, CA, USA, 11–14 December 1988; pp. 424–427.

170. Moazzami, R.; Hu, C. Stress-induced current in thin silicon dioxide films. In Proceedings of the 1992 International Electron Devices Meeting (IEDM), San Francisco, CA, USA, 13–16 December 1992; pp. 139–142.
171. Ghetti, A.; Alam, M.A.; Bude, J.; Venturi, F. Assessment of quantum yield experiments via full band Monte Carlo simulations. In Proceedings of the 1997 International Electron Devices Meeting (IEDM), Washington, DC, USA, 7–10 December 1997; pp. 873–876.
172. Sakakibara, K.; Ajika, N.; Eikyu, K.; Ishikawa, K.; Miyoshi, H. A quantitative analysis of time-decay reproducible stress-induced leakage current in SiO_2 films. *IEEE Trans. Electron Devices* **1997**, *44*, 1002–1007.
173. Chou, A.I.; Lai, K.; Kumar, K.; Chowdhury, P.; Lee, J.C. Modeling of stress-induced leakage current in ultrathin oxides with the trap-assisted tunneling mechanism. *Appl. Phys. Lett.* **1997**, *70*, 3407–3409.
174. Kamohara, S.; Park, D.; Hu, C. Deep-trap SILC (Stress-Induced Leakage Current) model for nominal and weak oxides. In Proceedings of the IEEE International Reliability Physics Symposium, Reno, NV, USA, 31 March–2 April 1998; pp. 57–61.
175. Riccò, B.; Torelli, G.; Lanzoni, M.; Manstretta, A.; Maes, H.E.; Montanari, D.; Modelli, A. Nonvolatile multilevel memories for digital applications. *Proc. IEEE* **1998**, *86*, 2399–2421.
176. Larcher, L.; Paccagnella, A.; Scarpa, A.; Ghidini, G. A new model of tunneling current and SILC in ultra-thin oxides. In Proceedings of the 1988 International Electron Devices Meeting (IEDM), San Francisco, CA, USA, 11–14 December 1988; pp. 901–904.
177. Takagi, S.; Yasuda, N.; Toriumi, A. A new *I–V* model for stress-induced leakage current including inelastic tunneling. *IEEE Trans. Electron Devices* **1999**, *46*, 348–354.
178. Ielmini, D.; Spinelli, A.S.; Lacaita, A.L.; Ghidini, G. Impact ionization and stress-induced leakage current in thin gate oxides. In Proceedings of the First European Workshop on Ultimate Integration of Silicon, Grenoble, France, 20–21 January 2000; pp. 85–88.
179. Ielmini, D.; Spinelli, A.S.; Lacaita, A.L. Experimental evidence for recombination-assisted leakage in thin oxides. *Appl. Phys. Lett.* **2000**, *76*, 1719–1721.
180. Satake, H.; Toriumi, A. Common origin for stress-induced leakage current and electron trap generation in SiO_2. *Appl. Phys. Lett.* **1995**, *67*, 3489–3490.
181. Alers, G.B.; Weir, B.E.; Alam, M.A.; Timp, G.L.; Sorch, T. Trap assisted tunneling as a mechanism of degradation and noise. In Proceedings of the IEEE International Reliability Physics Symposium, Reno, NV, USA, 31 March–2 April 1998; pp. 76–79.
182. James, D. Recent advances in memory technology. In Proceedings of the 2013 24th Annual SEMI Advanced Semiconductor Manufacturing Conference (ASMC), Saratoga Springs, NY, USA, 14–16 May 2013; pp. 386–395.
183. Yamada, S.; Amemiya, K.; Yamane, T.; Hazama, H.; Hashimoto, K. Non-uniform current flow through thin oxide after Fowler-Nordheim current stress. In Proceedings of the IEEE International Reliability Physics Symposium, Dalas, TX, USA, 30 April 30–2 May 1996; pp. 108–112.
184. Arai, F.; Maruyama, T.; Shirota, R. Extended data retention process technology for highly reliable Flash EEPROMs of 10^6 to 10^7 W/E cycles. In Proceedings of the IEEE International Reliability Physics Symposium, Reno, NV, USA, 31 March–2 April 1998; pp. 378–382.
185. Sakura, T.; Utsunomiya, H.; Kamakura, Y.; Taniguchi, K. A detailed study of soft- and pre-soft-breakdowns in small geometry MOS structures. In Proceedings of the 1988 International Electron Devices Meeting (IEDM), San Francisco, CA, USA, 11–14 December 1988; pp. 183–186.
186. Tsuji, N.; Sakakibara, K.; Ajika, N.; Miyoshi, H. Microscopic and statistical approach to SILC characteristics. In Proceedings of the 1998 Symposium on VLSI

Technology (VLSI-Technology): Digest of Technical Papers, Honolulu, HI, USA, 9–11 June 1998; pp. 196–197.
187. Kameyama, H.; Okuyama, Y.; Kamohara, S.; Kubota, K.; Kume, H.; Okuyama, K.; Manabe, Y.; Nozoe, A.; Uchida, H.; Hidaka, M.; et al. A new data retention mechanism after endurance stress on Flash memory. In Proceedings of the IEEE International Reliability Physics Symposium, San Jose, CA, USA, 10–13 April 2000; pp. 194–199.
188. Kumagai, Y.; Teramoto, A.; Inatsuka, T.; Kuroda, R.; Suwa, T.; Sugawa, S.; Ohmi, T. Evaluation for anomalous stress-induced leakage current of gate SiO_2 fimls using array test pattern. *IEEE Trans. Electron Devices* **2011**, *58*, 3307–3313.
189. Modelli, A.; Gilardoni, F.; Ielmini, D.; Spinelli, A.S. A new conduction mechanism for the anomalous bits in thin oxide Flash EEPROMs. In Proceedings of the IEEE International Reliability Physics Symposium, Orlando, FL, USA, 30 April–3 May 2001; pp. 61–66.
190. Inatsuka, T.; Kumagai, Y.; Kuroda, R.; Teramoto, A.; Sugawa, S.; Ohmi, T. Recovery characteristics of anomalous stress induced leakage current of 5.6nm oxide films. *Jpn. J. Appl. Phys.* **2012**, *51*, 04DC02.
191. Ielmini, D.; Spinelli, A.S.; Lacaita, A.L.; Modelli, A. A new two-trap tunneling model for the anomalous SILC in flash memories. *Microelectron. Eng.* **2001**, *59*, 189–195.
192. Ielmini, D.; Spinelli, A.S.; Lacaita, A.L.; Modelli, A. Modeling of anomalous SILC in flash memories based on tunneling at multiple defects. *Solid-State Electron.* **2002**, *46*, 1749–1756.
193. Ielmini, D.; Spinelli, A.S.; Lacaita, A.L.; Modelli, A. Statistical modeling of reliability and scaling projections for Flash memories. In Proceedings of the 2001 International Electron Devices Meeting (IEDM'01), Washington, DC, USA, 2–5 December 2001; pp. 703–706.
194. Ielmini, D.; Spinelli, A.S.; Lacaita, A.L.; Modelli, A. A statistical model for SILC in Flash memories. *IEEE Trans. Electron Devices* **2002**, *49*, 1955–1961.
195. Kuhn, P.J.; Hoefler, A.; Harp, T.; Hornung, B.; Paulsen, R.; Burnett, D.; Higman, J.M. A reliability methodology for low temperature data retention in floating gate non-volatile memories. In Proceedings of the IEEE International Reliability Physics Symposium, Orlando, FL, USA, 30 April–3 May 2001; pp. 266–270.
196. Hoefler, A.; Higman, J.M.; Harp, T.; Kuhn, P.J. Statistical modeling of the program/erase cycling acceleration of low temperature data retention in floating gate nonvolatile memories. In Proceedings of the IEEE International Reliability Physics Symposium, Dallas, TX, USA, 7–11 April 2002; pp. 21–25.
197. Schuler, F.; Degraeve, R.; Hendrickx, P.; Wellekens, D. Physical description of anomalous charge loss in floating gate based NVM's and identification of its dominant parameter. In Proceedings of the IEEE International Reliability Physics Symposium, Dallas, TX, USA, 7–11 April 2002; pp. 26–33.
198. Degraeve, R.; Schuler, F.; Lorenzini, M.; Wellekens, D.; Hendrickx, P.; Houdt, J.V.; Haspeslagh, L.; Groeseneken, G.; Tempel, G. Analytical model for failure rate prediction due to anomalous charge loss of flash memories. In Proceedings of the 2001 International Electron Devices Meeting (IEDM'01), Washington, DC, USA, 2–5 December 2001; pp. 699–702.
199. Belgal, H.P.; Righos, N.; Kalastirsky, I.; Peterson, J.J.; Shiner, R.; Mielke, N. A new reliability model for post-cycling charge retention of flash memories. In Proceedings of the IEEE International Reliability Physics Symposium, Dallas, TX, USA, 7–11 April 2002; pp. 7–20.
200. Larcher, L.; Pavan, P. Statistical simulations to inspect and predict data retention and program disturbs in Flash memories. In Proceedings of the 2003 International Electron Devices Meeting (IEDM), Washington, DC, USA, 8–10 December 2003; pp. 165–168.
201. Shim, B.S.; Jin, S.; Park, Y.J.; Min, H.S. A new statistical model for SILC distribution of Flash memory and the effect of spatial trap distribution. In Proceedings of the

International Conference on Simulation of Semiconductor Processes and Devices (SISPAD), Monterery, CA, USA, 6–8 September 2006; pp. 326–329.

202. Serov, A.; Shin, D.; Kim, D.S.; Kim, T.; Lee, K.H.; Park, Y.K.; Yoo, M.H.; Kim, T.; Sung, S.K.; Lee, C.H. Statistical retention modeling in floating-gate cell: ONO scaling. In Proceedings of the IEEE International Reliability Physics Symposium (IRPS), Montreal, QC, Canada, 26–30 April 2009; pp. 887–890.

203. Amoroso, S.; Gerrer, L.; Adamu-Lema, F.; Markov, S.; Asenov, A. Impact of statistical variability and 3D electrostatics on post-cycling anomalous charge loss in nanoscale Flash memories. In Proceedings of the IEEE International Reliability Physics Symposium (IRPS), Monterey, CA, USA, 14–18 April 2013; pp. 3B.4.1–3B.4.6.

204. Kubota, T.; Ando, K.; Muramatsu, S. The effect of the floating gate / tunnel SiO_2 interface on FLASH memory data retention reliability. In Proceedings of the IEEE International Reliability Physics Symposium, Dalas, TX, USA, 30 April 30–2 May 1996; pp. 12–16.

205. Park, M.; Suh, K.; Kim, K.; Hur, S.H.; Kim, K.; Lee, W.S. The effect of trapped charge distributions on data retention characteristics of NAND Flash memory cells. *IEEE Electron Device Lett.* **2007**, *28*, 750–752.

206. Park, M.; Ahn, E.; Cho, E.; Kim, K.; Lee, W.S. The effect of negative V_{TH} of NAND Flash memory cells on data retention characteristics. *IEEE Electron Device Lett.* **2009**, *30*, 155–157.

207. Lee, M.Y.; Hsiao, W.H.; Chen, R.P.; Kuo, L.K.; Dai, S.Q.; Ting, R.C.; Chen, C.L.; Chu, D.G.; Lu, C.Y. Anomalous single bit retention induced by asymmetric STI-corner-thinning for floating gate Flash memories. In Proceedings of the 19th IEEE International Symposium on the Physical and Failure Analysis of Integrated Circuits, Singapore, 2–6 July 2012; pp. 263–267.

208. Kang, D.; Lee, K.; Seo, S.; Kim, S.; Lee, J.S.; Bae, D.S.; Li, D.H.; Hwang, Y.; Shin, H. Generation dependence of retention characteristics in extremely scaled NAND Flash memory. *IEEE Trans. Electron Devices* **2013**, *34*, 1139–1141.

209. Yang, H.; Kim, H.; Park, S.-I.; Kim, J.; Lee, S.H.; Choi, J.K.; Hwang, D.; Kim, C.; Park, M.; Lee, K.H.; et al. Reliability issues and models of sub-90nm NAND Flash memory cells. In Proceedings of the 8th International Conference on Solid-State and Integrated Circuit Technology Proceedings, Shanghai, China, 23–26 October 2006; pp. 760–762.

210. Dumin, D.J.; Maddux, J.R. Correlation of stress-induced leakage current in thin oxides with trap generation inside the oxides. *IEEE Trans. Electron Devices* **1993**, *40*, 986–993.

211. Scott, R.S.; Dumin, D.J. The transient nature of excess low-level leakage currents in thin oxides. *J. Electrochem. Soc.* **1995**, *142*, 586–590.

212. Scott, R.S.; Dumin, D.J. The superposition of transient low-level leakage currents in stressed silicon oxides. *Solid-State Electron.* **1995**, *38*, 1325–1328.

213. Scott, R.S.; Dumin, N.A.; Hughes, T.W.; Dumin, D.J.; Moore, B.T. Properties of high-voltage stress generated traps in thin silicon oxide. *IEEE Trans. Electron Devices* **1996**, *43*, 1133–1142.

214. Scott, R.S.; Dumin, D.J. The charging and discharging of high-voltage stress-generated traps in thin silicon oxide. *IEEE Trans. Electron Devices* **1996**, *43*, 130–136.

215. Spinelli, A.S.; Lacaita, A.L.; Rigamonti, M.; Ghidini, G. Experimental method for the determination of the energy distribution of stress-induced oxide traps. *IEEE Electron Device Lett.* **1999**, *20*, 106–108.

216. Spinelli, A.S.; Lacaita, A.L.; Rigamonti, M.A.; Ielmini, D.; Ghidini, G. Separation of electron and hole traps by transient current analysis. *Microelectron. Eng.* **1999**, *48*, 151–154.

217. Monzio Compagnoni, C.; Spinelli, A.S.; Bianchini, A.; Lacaita, A.L.; Spiga, S.; Fanciulli, M. Characterization of transient currents in HfO_2 capacitors in the short timescale. *Microelectron. Eng.* **2006**, *83*, 1927–1930.

218. Manzini, S.; Modelli, A. Tunneling discharge of trapped holes in silicon dioxide. In *Insulating Films on Semiconductors*; Elsevier Science: Amsterdam, The Netherlands, 1983; pp. 112–115.
219. Yamada, R.; Mori, Y.; Okuyama, Y.; Yugami, J.; Nishimoto, T.; Kume, H. Analysis of detrap current due to oxide traps to improve Flash memory retention. In Proceedings of the IEEE International Reliability Physics Symposium, San Jose, CA, USA, 10–13 April 2000; pp. 200–204.
220. Yamada, R.; King, T. Variable stress-induced leakage current and analysis of anomalous charge loss for flash memory application. In Proceedings of the IEEE International Reliability Physics Symposium, Dallas, TX, USA, 30 March–4 April 2003; pp. 491–496.
221. Mielke, N.; Belgal, H.; Kalastirsky, I.; Kalavade, P.; Kurtz, A.; Meng, Q.; Righos, N.; Wu, J. Flash EEPROM threshold instabilities due to charge trapping during program/erase cycling. *IEEE Trans. Device Mater. Reliab.* **2004**, *4*, 335–344.
222. Mielke, N.; Belgal, H.; Fazio, A.; Meng, Q.; Righos, N. Recovery effects in the distributed cycling of Flash memories. In Proceedings of the IEEE International Reliability Physics Symposium, San Jose, CA, USA, 26–30 March 2006; pp. 29–35.
223. Paolucci, G.M.; Monzio Compagnoni, C.; Miccoli, C.; Spinelli, A.S.; Lacaita, A.L.; Visconti, A. Revisiting charge trapping/detrapping in Flash memories from a discrete and statistical standpoint—Part I: V_T instabilities. *IEEE Trans. Electron Devices* **2014**, *61*, 2802–2810.
224. Paolucci, G.M.; Monzio Compagnoni, C.; Miccoli, C.; Spinelli, A.S.; Lacaita, A.L.; Visconti, A. Revisiting charge trapping/detrapping in Flash memories from a discrete and statistical standpoint—Part II: On-field operation and distributed-cycling effects. *IEEE Trans. Electron Devices* **2014**, *61*, 2811–2819.
225. Resnati, D.; Monzio Compagnoni, C.; Paolucci, G.M.; Miccoli, C.; Barber, J.; Bertuccio, M.; Beltrami, S.; Lacaita, A.L.; Spinelli, A.S.; Visconti, A. A Step ahead toward a new microscopic Picture for Charge Trapping/Detrapping in Flash memories. In Proceedings of the IEEE International Reliability Physics Symposium, Pasadena, CA, USA, 17–21 April 2016; pp. 6C-3-1–6C-3-7.
226. Miccoli, C.; Monzio Compagnoni, C.; Beltrami, S.; Spinelli, A.S.; Visconti, A. Threshold-voltage instability due to damage recovery in nanoscale NAND Flash memories. *IEEE Trans. Electron Devices* **2011**, *58*, 2406–2414.
227. Resnati, D.; Nicosia, G.; Paolucci, G.M.; Visconti, A.; Monzio Compagnoni, C. Cycling-induced charge trapping/detrapping in Flash memories—Part I: Experimental evidence. *IEEE Trans. Electron Devices* **2016**, *63*, 4753–4760.
228. Ghidini, G.; Sebastiani, A.; Brazzelli, D. Stress induced leakage current and bulk oxide trapping: Temperature evolution. In Proceedings of the IEEE International Reliability Physics Symposium, Dallas, TX, USA, 7–11 April 2002; pp. 415–416.
229. Lee, K.; Kang, M.; Seo, S.; Kang, D.; Kim, S.; Li, D.H.; Shin, H. Activation energies (E_a) of failure mechanisms in advanced NAND Flash cells for different generations and cycling. *IEEE Trans. Electron Devices* **2013**, *60*, 1099–1107.
230. Paolucci, G.M.; Miccoli, C.; Monzio Compagnoni, C.; Crespi, L.; Spinelli, A.S.; Lacaita, A.L. Investigation of cycling-induced V_T instabilities in NAND Flash cells via compact modeling. In Proceedings of the 2012 4th IEEE International Memory Workshop (IMW), Milan, Italy, 20–23 May 2012; pp. 194–197.
231. Paolucci, G.M.; Miccoli, C.; Monzio Compagnoni, C.; Spinelli, A.S.; Lacaita, A.L. String current in decananometer NAND Flash arrays: A compact-modeling investigation. *IEEE Trans. Electron Devices* **2012**, *59*, 2331–2337.
232. Miccoli, C.; Monzio Compagnoni, C.; Chiavarone, L.; Beltrami, S.; Lacaita, A.L.; Spinelli, A.S.; Visconti, A. Reliability characterization issues for nanoscale Flash memories: A case study on 45-nm NOR devices. *IEEE Trans. Device Mater. Reliab.* **2013**, *13*, 362–369.
233. Resnati, D.; Nicosia, G.; Paolucci, G.M.; Visconti, A.; Monzio Compagnoni, C. Cycling-induced charge trapping/detrapping in Flash memories—Part II: Modeling. *IEEE Trans. Electron Devices* **2016**, *63*, 4761–4768.

234. Paolucci, G.M; Monzio Compagnoni, C.; Miccoli, C.; Bertuccio, M.; Beltrami, S.; Barber, J.; Kessenich, J.; Lacaita, A.L.; Spinelli, A.S.; Visconti, A. A new spectral method to modeling charge trapping/detrapping in NAND Flash memories. In Proceedings of the IEEE International Reliability Physics Symposium, Waikoloa, HI, USA, 1–5 June 2014; pp. 2E-2-1–2E-2-6.

235. JEDEC Standard JESD22-A117C. *Electrically Erasable Programmable ROM (EEPROM) Program/Erase Endurance and Data Retention Stress Test*; Technical Report; JEDEC Solid State Technology Association: Arlington, VA, USA, 2011.

236. JEDEC Standard JESD22-A117A. *Electrically Erasable Programmable ROM (EEPROM) Program/Erase Endurance and Data Retention Stress Test*; Technical Report; JEDEC Solid State Technology Association: Arlington, VA, USA, 2006.

237. Lee, W.H.; Hur, C.H.; Lee, H.M.; Yoo, H.; Lee, S.E.; Lee, B.Y.; Park, C.; Kim, K. Post-cycling data retention failure in multilevel NOR Flash memory with nitrided tunnel-oxide. In Proceedings of the IEEE International Reliability Physics Symposium (IRPS), Montreal, QC, Canada, 26–30 April 2009; pp. 907–908.

238. Lakshmanna, V.; Vengurlekar, A.S. Logarithmic detrapping response for holes injected into SiO_2 and the influence of thermal activation and electric fields. *J. Appl. Phys.* **1988**, *63*, 4548–4554.

239. Paolucci, G.M.; Monzio Compagnoni, C.; Spinelli, A.S.; Lacaita, A.L.; Goda, A. Fitting cells into a narrow V_T interval: Physical constraints along the lifetime of an extremely scaled NAND Flash memory array. *IEEE Trans. Electron Devices* **2015**, *62*, 1491–1497.

240. Monzio Compagnoni, C.; Chiavarone, L.; Calabrese, M.; Gusmeroli, R.; Ghidotti, M.; Lacaita, A.L.; Spinelli, A.S.; Visconti, A. Granular electron injection and random telegraph noise impact on the programming accuracy of NOR Flash memories. In Proceedings of the IEEE International Reliability Physics Symposium (IRPS), Montreal, QC, Canada, 26–30 April 2009; pp. 274–279.

241. Aritome, S.; Kikkawa, T. Scaling challenge of self-aligned STI cell (SA-STI cell) for NAND Flash memories. *Solid-State Electron.* **2013**, *82*, 54–62.

242. Lee, K.; Kang, D.; Shin, H.; Kwon, S.; Kim, S.; Hwang, Y. Analysis of failure mechanisms in erased state of sub 20-nm NAND Flash memory. In Proceedings of the 2014 44th European Solid State Device Research Conference (ESSDERC), Venezia, Italy, 22–26 September 2014; pp. 58–61.

243. Wu, D.Y.; Chen, S.F.; Lin, C.J.; King, Y.C. Dummy read scheme for lifetime improvement of MLC NAND Flash memories. *IEEE Trans. Device Mater. Reliab.* **2016**, *16*, 583–587.

244. Lee, C.; Lee, S.K.; Ahn, S.; Lee, J.; Park, W.; Cho, Y.; Jang, C.; Yang, C.; Chung, S.; Yun, I.S.; et al. A 32-Gb MLC NAND Flash memory with Vth endurance enhancing schemes in 32 nm CMOS. *IEEE J. Solid-State Circuits* **2011**, *46*, 97–106.

245. Lue, H.T.; Du, P.Y.; Chen, C.P.; Chen, W.C.; Hsieh, C.C.; Hsiao, Y.H.; Shih, Y.H.; Lu, C.Y. Radically extending the cycling endurance of Flash memory (to >100M cycles) by using built-in thermal annealing to self-heal the stress-induced damage. In Proceedings of the 2012 IEEE International Electron Devices Meeting (IEDM), San Francisco, CA, USA, 10–13 December 2012; pp. 199–202.

246. Chang, Y.M.; Chang, Y.H.; Chen, J.J.; Kuo, T.W.; Li, H.P.; Lue, H.T. On trading wear-leveling with heal-leveling. In Proceedings of the 2014 51st ACM/EDAC/IEEE Design Automation Conference (DAC), San Francisco, CA, USA, 1–5 June 2014; pp. 478–483.

247. Chen, R.; Wang, Y.; Liu, D.; Shao, Z.; Jiang, S. Heating dispersal for self-healing NAND Flash memory. *IEEE Trans. Comput.* **2017**, *66*, 361–367.

248. Lee, J.D.; Choi, J.H.; Park, D.; Kim, K. Degradation of tunnel oxide by FN current stress and its effects on data retention characteristics of 90 nm NAND Flash memory cells. In Proceedings of the IEEE International Reliability Physics Symposium (IRPS), Dallas, TX, USA, 30 March–4 April 2003; pp. 497–501.

249. Lee, K.; Kang, M.; Seo, S.; Li, D.H.; Kim, J.; Shin, H. Analysis of failure mechanisms and extraction of activation energies (E_a) in 21-nm NAND Flash cells. *IEEE Electron Device Lett.* **2013**, *34*, 48–50.
250. Yan, C.R.; Chen, J.F.; Lee, Y.J.; Liao, Y.J.; Lin, C.Y.; Chen, C.Y.; Lin, Y.C.; Chen, H.H. Extraction and analysis of interface states in 50-nm NAND Flash devices. *IEEE Trans. Electron Devices* **2013**, *60*, 992–997.
251. Dunn, C.; Kaya, C.; Lewis, T.; Strauss, T.; Schreck, J.; Hefley, P.; Middennorf, M.; San, T. Flash EEPROM disturb mechanism. In Proceedings of the IEEE International Reliability Physics Symposium (IRPS), San Jose, CA, USA, 11–14 April 1994; pp. 299–308.
252. Kato, M.; Miyamoto, N.; Kume, H.; Satoh, A.; Adachi, T.; Ushiyama, M.; Kimura, K. Read-disturb degradation mechanism due to electron trapping in the tunnel oxide for low-voltage flash memories. In Proceedings of the 1994 International Electron Devices Meeting (IEDM), San Francisco, CA, USA, 11–14 December 1994; pp. 45–48.
253. Endoh, T.; Shimizu, K.; Iizuka, H.; Masuoka, F. A new write/erase method to improve the read disturb characteristics based on the decay phenomena of stress leakage current for Flash memories. *IEEE Trans. Electron Devices* **1998**, *45*, 98–104.
254. Wada, T. Acceleration method for gate-disturb degradation on embedded flash EEPROM. *Microelectron. Reliab.* **2000**, *40*, 1279–1283.
255. Cai, Y.; Luo, Y.; Ghose, S.; Mutlu, O. Read disturb errors in MLC NAND Flash memory: characterization, mitigation, and recovery. In Proceedings of the 28th International Conference on Defects in Semiconductors, Matsue City, Japan, 27–31 July 2015; pp. 438–449.
256. Kang, D.; Lee, K.; Kwon, S.; Kim, S.; Hwang, Y.; Shin, H. Analysis of read disturbance mechanism in retention of sub-20 nm NAND flash memory. *Jpn. J. Appl. Phys.* **2015**, *54*, 04DD03-1–04DD03-4.
257. Grupp, L.M.; Caulfield, A.M.; Coburn, J.; Swanson, S.; Yaakobi, E.; Wolf, P.H.S.J.K. Characterizing Flash memory: Anomalies, observations, and applications. In Proceedings of the 2009 42nd Annual IEEE/ACM International Symposium on Microarchitecture (MICRO), New York, NY, USA, 12–16 December 2009; pp. 24–33.
258. Goda, A.; Moriyama, W.; Hazama, H.; Iizuka, H.; Shimizu, K.; Aritome, S.; Shirota, R. A novel surface-oxidized barrier-SiN cell technology to improve endurance and read-disturb characteristics for gigabit NAND Flash memories. In Proceedings of the 2000 International Electron Devices Meeting (IEDM), San Francisco, CA, USA, 10–13 December 2000; pp. 771–774.
259. Lee, Y.H.; McMahon, W.; Lu, Y.L.R.; Tewg, J.Y.J.; Ma, S.T. On the scaling of Flash cell spacer for gate disturb and charge retention otimization. *IEEE Trans. Electron Devices* **2009**, *56*, 1959–1965.
260. Joe, S.M.; Jeong, M.K.; Kang, M.; Han, K.R.; Park, S.K.; Lee, J.H. New read schemes using boosted channel potential of adjacent bit-line Strings in NAND Flash memory. *IEEE Electron Device Lett.* **2012**, *33*, 1198–1200.
261. Deguchi, Y.; Tokutomi, T.; Takeuchi, K. System-level error correction by read-disturb error model of 1Xnm TLC NAND Flash memory for read-intensive enterprise solid-state drives (SSDs). In Proceedings of the IEEE International Reliability Physics Symposium, Pasadena, CA, USA, 17–21 April 2016; pp. MY-6-1–MY-6-4.
262. Papandreou, N.; Parnell, T.; Mittelholzer, T.; Pozidis, H.; Griffin, T.; Tressler, G.; Fisher, T.; Camp, C. Effect of read disturb on incomplete blocks in MLC NAND Flash arrays. In Proceedings of the IEEE International Memory Workshop (IMW), Paris, France, 15–18 May 2016; pp. 117–120.
263. Tanaka, T.; Momodomi, M.; Iwata, Y.; Tanaka, Y.; Oodaira, H.; Itoh, Y.; Shirota, R.; Ohuchi, K.; Masuoka, F. A 4-Mbit NAND-EEPROM with tight programmed V_T distribution. In Proceedings of the 1990 Symposium on VLSI

Technology (VLSI-Technology): Digest of Technical Papers, Honolulu, HI, USA, 4–7 June 1990; pp. 105–106.

264. Tanaka, T.; Tanaka, Y.; Nakamura, H.; Oodaira, H.; Aritome, S.; Shirota, R.; Masuoka, F. A quick intelligent program architecture for 3 V-only NAND-EEPROMs. In Proceedings of the 1992 Symposium on VLSI Technology (VLSI-Technology): Digest of Technical Papers, Seattle, WA, USA, 2–4 June 1992; pp. 20–21.

265. Suh, K.D.; Suh, B.H.; Lim, Y.H.; Kim, J.K.; Choi, Y.J.; Koh, Y.N.; Lee, S.S.; Kwon, S.C.; Choi, B.S.; Yum, J.S.; et al. A 3.3 V 32 Mb NAND Flash memory with incremental step pulse programming scheme. *IEEE J. Solid-State Circuits* **1995**, *30*, 1149–1156.

266. Lee, J.D.; Lee, C.K.; Lee, M.W.; Kim, H.S.; Park, K.C.; Lee, W.S. A new programming disturbance phenomenon in NAND Flash memory by source/drain hot-electrons generated by GIDL current. In Proceedings of the 21st IEEE Non-Volatile Semiconductor Memory Workshop, Monterey, CA, USA, 12–16 February 2006; pp. 31–33.

267. Oh, D.; Lee, C.; Lee, S.; Kim, T.K.; Song, J.; Choi, J. A new self-boosting phenomenon by source/drain depletion cut-off in NAND Flash memory. In Proceedings of the 22nd IEEE Non-Volatile Semiconductor Memory Workshop, Monterey, CA, USA, 26–30 August 2007; pp. 39–41.

268. Cho, T.; Lee, Y.T.; Kim, E.; Lee, J.; Choi, S.; Lee, S.; Kim, D.H.; Han, W.K.; Lim, Y.H.; Lee, J.D.; et al. A 3.3 V 1 Gb multi-level NAND Flash memory with non-uniform threshold voltage distribution. In Proceedings of the 1995 IEEE International Solid-State Circuits Conference, San Francisco, CA, USA, 5–7 February 2001; pp. 1–2.

269. Park, K.T.; Song, Y.; Kang, M.; Lee, S.; Lim, Y.H.; Suh, K.D.; Chung, C. Dynamic Vpass controlled program scheme and optimized erase Vth control for high program inhibition in MLC NAND Flash memories. *IEEE J. Solid-State Circuits* **2010**, *45*, 2165–2172.

270. Shirota, R.; Huang, C.H.; Nagai, S.; Sakamoto, Y.; Li, F.H.; Mitiukhina, N.; Arakawa, H. A new programming scheme for the improvement of program disturb characteristics in scaled NAND Flash memory. *IEEE Trans. Electron Devices***2012**, *59*, 2767–2773.

271. Kang, M.; Park, K.T.; Song, Y.; Hwang, S.; Choi, B.Y.; Song, Y.; Lee, Y.T.; Kim, C. Improving read disturb characteristics by self-boosting read scheme for multilevel NAND Flash memories. *Jpn. J. Appl. Phys.* **2009**, *48*, 04C062-1–04C062-6.

272. Park, K.T.; Lee, S.; Sel, J.S.; Choi, J.; Kim, K. Scalable wordline shielding using dummy cell beyond 40 nm NAND Flash memory for eliminating abnormal disturb of edge memory cell. *Jpn. J. Appl. Phys.* **2007**, *46*, 2188–2192.

273. Kanda, K.; Koyanagi, M.; Yamamura, T.; Hosono, K.; Yoshihara, M.; Miwa, T.; Kato, Y.; Mak, A.; Chan, S.L.; Tsai, F.; et al. A 120 mm^2 16 Gb 4-MLC NAND Flash memory with 43 nm CMOS technology. In Proceedings of the 2008 IEEE International Solid-State Circuits Conference (ISSCC), San Francisco, CA, USA, 3–7 February 2008; pp. 430–431.

274. Park, I.H.; Hahn, W.G.; Song, K.W.; Choi, K.H.; Choi, H.K.; Lee, S.B.; Lee, C.S.; Song, J.H.; Han, J.M.; Kyoung, K.H.; et al. A new GIDL phenomenon by field effect of neighboring cell transistors and its control solutions in sub-30 nm NAND Flash devices. In Proceedings of the 2012 Symposium on VLSI Technology (VLSI-Technology), Honolulu, HI, USA, 12–14 June 2012; pp. 23–24.

275. Park, M.; Park, S.; Cho, S.; Lee, D.K.; Jeong, Y.; Hong, C.; Lee, H.S.; Cho, M.K.; Ahn, K.O.; Koh, Y. NAND Flash reliability degradation induced by HCI in boosted channel potential. In Proceedings of the IEEE International Reliability Physics Symposium (IRPS), Garden Grove, CA, USA, 2–6 May 2010; pp. 975–976.

276. Lee, C.; Hwang, J.; Fayrushin, A.; Kim, H.; Son, B.; Park, Y.; Jin, G.; Jung, E. Channel coupling phenomenon as scaling barrier of NAND Flash memory beyond 20nm

node. In Proceedings of the 2013 5th IEEE International Memory Workshop (IMW), Monterey, CA, USA, 26–29 May 2013; pp. 72–75.

277. Chandrasekaran, S.; Venkatesan, S.; Eagle, O.H.; Iyengar, V.V.; Reyes, A.B.; Gowda, S.S. The effect of shallow trench isolation improvement on program disturb response in 20 nm NAND Flash technology. In Proceedings of the 2014 IEEE Workshop On Microelectronics And Electron Devices (WMED), Boise, ID, USA, 18 April 2014; pp. 1–4.

278. Seo, S.; Kim, H.; Park, S.; Lee, S.; Aritome, S.; Hong, S. Novel negative Vt shift program disturb phenomena in 2X 3X NAND Flash memory cells. In Proceedings of the IEEE International Reliability Physics Symposium (IRPS), Monterey, CA, USA, 10–14 April 2011; pp. 641–644.

279. Aritome, S.; Seo, S.; Kim, H.S.; Park, S.K.; Lee, S.K.; Hong, S. Novel negative V_T shift phenomenon of program-inhibit cell in 2*X*–3*X*-nm self-aligned STI NAND Flash memory. *IEEE Trans. Electron Devices* **2012**, *59*, 2950–2955.

280. Zambelli, C.; Andrian, F.; Aritome, S.; Olivo, P. Compact modeling of negative V_T shift disturb in NAND Flash memories. *IEEE Trans. Electron Devices* **2016**, *63*, 1516–1523.

281. Torsi, A.; Zhao, Y.; Liu, H.; Tanzawa, T.; Goda, A.; Kalavade, P.; Parat, K. A program disturb model and channel leakage current study for sub-20 nm NAND Flash cells. *IEEE Trans. Electron Devices* **2011**, *58*, 11–16.

282. Oh, D.; Lee, S.; Lee, C.; Song, J.; Lee, W.; Choi, J. Program disturb phenomenon by DIBL in MLC NAND Flash device. In Proceedings of the 2008 Joint Non-Volatile Semiconductor Memory Workshop (NVSMW) & International Conference on Memory Technology and Design (ICMTD), Opio, France, 18–21 May 2008; pp. 5–7.

283. Kang, M.; Hahn, W.; Park, I.H.; Park, J.; Song, Y.; Lee, H.; Eun, C.; Ju, S.; Choi, K.; Lim, Y.; et al. DIBL-induced program disturb characteristics in 32-nm NAND Flash memory array. *IEEE Trans. Electron Devices* **2011**, *58*, 3626–3629.

284. Tsai, W.J.; Huang, J.S.; Tsai, P.H.; Yan, S.G.; Cheng, C.H.; Cheng, C.C.; Chen, Y.J.; Lee, C.H.; Han, T.T.; Lu, T.C.; et al. A novel low-voltage low-power programming method for NAND Flash cell by utilizing self-boosting channel potential for carrier heating. *IEEE Trans. Electron Devices* **2011**, *58*, 1620–1627.

285. Ahn, S.T.; Mun, K.; Lee, K.W.; Cho, G.; Park, S.K.; Lee, S.; Hong, S. Advanced hot-carrier injection programming scheme for sub 20nm NAND Flash cell and beyond. In Proceedings of the 2012 4th IEEE International Memory Workshop (IMW), Milan, Italy, 20–23 May 2012; pp. 29–32.

286. Fukuzumi, Y.; Katsumata, R.; Kito, M.; Kido, M.; Sato, M.; Tanaka, H.; Nagata, Y.; Matsuoka, Y.; Iwata, Y.; Aochi, H.; et al. Optimal integration and characteristics of vertical array devices for ultra-high density, bit-cost scalable Flash memory. In Proceedings of the 2007 International Electron Devices Meeting (IEDM), Washington, DC, USA, 10–12 December 2007; pp. 449–452.

287. Whang, S.; Lee, K.; Shin, D.; Kim, B.; Kim, M.; Bin, J.; Han, J.; Kim, S.; Lee, B.; Jung, Y.; et al. Novel 3-Dimensional dual control-gate with surrounding floating-gate (DC-SF) NAND Flash cell for 1Tb file storage application. In Proceedings of the 2010 IEEE International Electron Devices Meeting (IEDM), San Francisco, CA, USA, 6–8 December 2010; pp. 668–671.

288. Noh, Y.; Ahn, Y.; Yoo, H.; Han, B.; Chung, S.; Shim, K.; Lee, K.; Kwak, S.; Shin, S.; Choi, I.; et al. A new metal control gate last process (MCGL process) for high performance DC-SF (dual control gate with surrounding floating gate) 3D NAND Flash memory. In Proceedings of the 2012 Symposium on VLSI Technology (VLSI-Technology), Honolulu, HI, USA, 12–14 June 2012; pp. 19–20.

289. Aritome, S.; Whang, S.; Lee, K.; Shin, D.; Kim, B.; Kim, M.; Bin, J.; Han, J.; Kim, S.; Lee, B.; et al. A novel three-dimensional dual control-gate with surrounding floating-gate (DC-SF) NAND flash cell. *Solid-State Electron.* **2013**, *79*, 166–171.

290. Aritome, S.; Noh, Y.; Yoo, H.; Choi, E.S.; Joo, H.S.; Ahn, Y.; Han, B.; Chung, S.; Shim, K.; Lee, K.; et al. Advanced DC-SF cell technology for 3-D NAND Flash. *IEEE Trans. Electron Devices* **2013**, *60*, 1327–1333.
291. Parat, K.; Dennison, C. A floating gate based 3-D NAND technology with CMOS under array. In Proceedings of the 2015 International Electron Devices Meeting (IEDM), Washington, DC, USA, 7–9 December 2015; pp. 48–51.
292. Kang, D.; Jeong, W.; Kim, C.; Kim, D.H.; Cho, Y.S.; Kang, K.T.; Ryu, J.; Kang, K.M.; Lee, S.; Kim, W.; et al. 256 Gb 3 b/cell V-NAND Flash memory with 48 stacked WL layers. In Proceedings of the 2014 IEEE International Solid-State Circuits Conference (ISSCC), San Francisco, CA, USA, 31 January–4 February 2016; pp. 130–131.
293. Choi, J.; Seol, K.S. 3D approaches for non-volatile memory. In Proceedings of the 2011 Symposium on VLSI Technology (VLSI-Technology), Kyoto, Japan, 14–16 June 2011; pp. 178–179.
294. Choi, E.S.; Park, S.K. Device considerations for high density and highly reliable 3D NAND Flash cell in near future. In Proceedings of the 2012 IEEE International Electron Devices Meeting (IEDM), San Francisco, CA, USA, 10–13 December 2012; pp. 211–214.
295. Im, J.W.; Jeong, W.P.; Kim, D.H.; Nam, S.W.; Shim, D.K.; Choi, M.H.; Yoon, H.J.; Kim, D.H.; Kim, Y.S.; Park, H.W.; et al. A 128Gb 3b/cell V-NAND Flash memory with 1Gb/s I/O rate. In Proceedings of the 2015 IEEE International Solid-State Circuits Conference (ISSCC), San Francisco, CA, USA, 22–26 February 2015; pp. 130–132.
296. Jeong, W.; Im, J.W.; Kim, D.H.; Nam, S.W.; Shim, D.K.; Choi, M.H.; Yoon, H.J.; Kim, D.H.; Kim, Y.S.; Park, H.W.; et al. A 128 Gb 3 b/cell V-NAND Flash memory with 1 Gb/s I/O rate. *IEEE J. Solid-State Circuits* **2016**, *51*, 204–212.
297. Janai, M.; Eitan, B. The kinetics of degradation of data retention of post-cycled NROM non-volatile memory products. In Proceedings of the IEEE International Reliability Physics Symposium (IRPS), San Jose, CA, USA, 17–21 April 2003; pp. 175–180.
298. Lee, M.C.; Wong, H.Y. Investigation on the origin of the anomalous tail bits on nitrided charge trap flash memory. *Microelectron. Reliab.* **2015**, *55*, 337–341.
299. Goda, A.; Miccoli, C.; Monzio Compagnoni, C. Time dependent threshold-voltage fluctuations in NAND Flash memories: From basic physics to impact on array operation. In Proceedings of the 2015 International Electron Devices Meeting (IEDM), Washington, DC, USA, 7–9 December 2015; pp. 374–377.
300. Monzio Compagnoni, C.; Spinelli, A.S.; Lacaita, A.L. Experimental study of data retention in nitride memories by temperature and field acceleration. *IEEE Electron Device Lett.* **2007**, *28*, 628–630.
301. Shappir, A.; Levy, D.; Shacham-Diamand, Y.; Lusky, E.; Bloom, I.; Eitan, B. Spatial characterization of localized charge trapping and charge redistribution in the NROM device. *Solid-State Electron.* **2004**, *48*, 1489–1495.
302. Kang, C.; Choi, J.; Sim, J.; Lee, C.; Shin, Y.; Park, J.; Sel, J.; Jeon, S.; Park, Y.; Kim, K. Effects of lateral charge spreading on the reliability of TANOS (TaN/AlO/SiN/Oxide/Si) NAND Flash memory. In Proceedings of the IEEE International Reliability Physics Symposium, Phoenix, AZ, USA, 15–19 April 2007; pp. 167–170.
303. Maconi, A.; Arreghini, A.; Monzio Compagnoni, C.; Van den bosch, G.; Spinelli, A.S.; Van Houdt, J.; Lacaita, A.L. Comprehensive investigation of the impact of lateral charge migration on retention performance of planar and 3D SONOS devices. *Solid-State Electron.* **2012**, *74*, 64–70.
304. Liu, Y.H.; Jiang, C.M.; Chen, W.C.; Wang, T.; Tsai, W.J.; Lu, T.C.; Chen, K.C.; Lu, C.Y. Electric field induced nitride trapped charge lateral migration in a SONOS Flash memory. *IEEE Electron Device Lett.* **2017**, *38*, 48–51.
305. Kang, H.J.; Choi, N.; Joe, S.M.; Seo, J.H.; Choi, E.; Park, S.K.; Park, B.G.; Lee, J.H. Comprehensive analysis of retention characteristics in 3-D NAND Flash memory

cells with tube-type poly-Si channel structure. In Proceedings of the 2015 Symposium on VLSI Technology (VLSI-Technology), Kyoto, Japan, 15–18 June 2015; pp. 182–183.

306. Lun, Z.; Liu, S.; He, Y.; Hou, Y.; Zhao, K.; Du, G.; Liu, X.; Wang, Y. Investigation of retention behavior for 3D charge trapping NAND Flash memory by 2D self-consistent simulation. In Proceedings of the 2014 International Conference on Simulation of Semiconductor Processes and Devices (SISPAD), Yokohama, Japan, 9–11 September 2014; pp. 141–144.

307. Li, X.; Huo, Z.; Jin, L.; Wang, Y.; Liu, J.; Jiang, D.; Yang, X.; Liu, M. Investigation of charge loss mechanisms in 3D TANOS cylindrical junction-less charge trapping memory. In Proceedings of the IEEE 12th International Conference on Solid - State and Integrated Circuit Technology (ICSICT), Guilin, China, 28–31 October 2014; pp. 720–722.

308. Oh, D.; Lee, B.; Kwon, E.; Kim, S.; Cho, G.; Park, S.; Lee, S.; Hong, S. TCAD simulation of data retention characteristics of charge trap device for 3-D NAND Flash memory. In Proceedings of the IEEE International Memory Workshop (IMW), Monterey, CA, USA, 17–20 May 2015; pp. 113–116.

309. Kim, S.H.; Yang, S.D.; Kim, J.S.; Jeong, J.K.; Lee, H.D.; Lee, G.W. Charge spreading effect of stored charge on retention characteristics in SONOS NAND Flash memory devices. *Trans. Electr. Electron. Mater.* **2015**, *16*, 183–186.

310. Park, S.; Choi, S.; Jun, K.S.; Kim, H.; Rhee, S.; Park, Y.J. Investigation on multiple activation energy of retention in charge trapping memory using self-consistent simulation. In Proceedings of the 2014 44th European Solid State Device Research Conference (ESSDERC), Venezia, Italy, 22–26 September 2014; pp. 50–53.

311. Jeong, M.K.; Kang, H.J.; Joe, S.M.; Park, S.K.; Park, B.G.; Lee, J.H. Trap profiling in nitride storage layer in 3-D NAND Flash memory using retention characteristics and AC-g_m method. *IEEE Electron Device Lett.* **2015**, *36*, 561–563.

312. Janai, M.; Bloom, I. Charge gain, NBTI, and random telegraph noise in EEPROM Flash memory devices. *IEEE Electron Device Lett.* **2010**, *31*, 1038–1040.

313. Janai, M.; Bloom, I.; Shur, Y. Charge gain, NBTI recovery and random telegraph noise in localized-trapping NVM devices. In Proceedings of the IEEE International Reliability Physics Symposium (IRPS), Monterey, CA, USA, 10–14 April 2011; pp. 819–823.

314. Chen, C.P.; Lue, H.T.; Hsieh, C.C.; Chang, K.P.; Hsieh, K.Y.; Lu, C.Y. Study of fast initial charge loss and its impact on the programmed states Vt distribution of charge-trapping NAND Flash. In Proceedings of the 2010 IEEE International Electron Devices Meeting (IEDM), San Francisco, CA, USA, 6–8 December 2010; pp. 118–121.

315. Choi, B.; Jang, S.H.; Yoon, J.; Lee, J.; Jeon, M.; Lee, Y.; Han, J.; Lee, J.; Kim, D.M.; Kim, D.H.; et al. Comprehensive evaluation of early retention (fast charge loss within a few seconds) characteristics in tube-type 3-D NAND Flash memory. In Proceedings of the 2016 Symposium on VLSI Technology (VLSI-Technology), Honolulu, HI, USA, 14–16 June 2016; pp. 61–62.

316. Subirats, A.; Arreghini, A.; Degraeve, R.; Linten, D.; den bosch, G.V.; Furnemont, A. In depth analysis of post-program V_T instability after electrical stress in 3D SONOS memories. In Proceedings of the IEEE International Memory Workshop (IMW), Paris, France, 15–18 May 2016; pp. 84–87.

317. Jeong, M.K.; Joe, S.M.; Seo, C.S.; Han, K.R.; Choi, E.; Park, S.K.; Lee, J.H. Analysis of random telegraph noise and low frequency noise properties in 3-D stacked NAND Flash memory with tube-type poly-Si channel structure. In Proceedings of the 2012 Symposium on VLSI Technology (VLSI-Technology), Honolulu, HI, USA, 12–14 June 2012; pp. 55–56.

318. Jeong, M.K.; Joe, S.M.; Jo, B.S.; Kang, H.J.; Bae, J.H.; Han, K.R.; Choi, E.; Cho, G.; Park, S.K.; Park, B.G.; et al. Characterization of traps in 3-D stacked NAND Flash memory devices with tube-type poly-Si channel structure. In Proceedings of the

2012 IEEE International Electron Devices Meeting (IEDM), San Francisco, CA, USA, 10–13 December 2012; pp. 207–210.
319. Kang, D.; Lee, C.; Hur, S.; Song, D.; Choi, J.H. A new approach for trap analysis of vertical NAND Flash cell using RTN characteristics. In Proceedings of the 2014 IEEE International Electron Devices Meeting (IEDM), San Francisco, CA, USA, 15–17 December 2014; pp. 367–370.
320. Nowak, E.; Kim, J.H.; Kwon, H.; Kim, Y.G.; Sim, J.S.; Lim, S.H.; Kim, D.S.; Lee, K.H.; Park, Y.K.; Choi, J.H.; et al. Intrinsic fluctuations in vertical NAND Flash memories. In Proceedings of the 2012 Symposium on VLSI Technology (VLSI-Technology), Honolulu, HI, USA, 12–14 June 2012; pp. 21–22.
321. Chou, Y.L.; Wang, T.; Lin, M.; Chang, Y.W.; Liu, L.; Huang, S.W.; Tsai, W.J.; Lu, T.C.; Chen, K.C.; Lu, C.Y. Poly-silicon trap position and pass voltage effects on RTN amplitude in a vertical NAND Flash cell string. *IEEE Electron Device Lett.* **2016**, *37*, 998–1001.
322. Hsieh, C.C.; Lue, H.T.; Hsu, T.H.; Du, P.Y.; Chiang, K.H.; Lu, C.Y. A Monte Carlo simulation method to predict large-density NAND product memory window from small-array test element group (TEG) verified on a 3D NAND Flash test chip. In Proceedings of the 2016 Symposium on VLSI Technology (VLSI-Technology), Honolulu, HI, USA, 14–16 June 2016; pp. 63–64.
323. Cappelletti, P. Non volatile memory evolution and revolution. In Proceedings of the 2015 International Electron Devices Meeting (IEDM), Washington, DC, USA, 7–9 December 2015; pp. 241–244.
324. Resnati, D.; Goda, A.; Spinelli, A.S.; Monzio Compagnoni, C. Thermal effects in NAND Flash memories: A comparison between 2-D and 3-D arrays. *IEEE Electron Device Lett.* **2017**. submitted.
325. Paolucci, G.M.; Spinelli, A.S.; Monzio Compagnoni, C.; Tessariol, P. A semi-analytical model for macaroni MOSFETs with application to vertical Flash memories. *IEEE Trans. Electron Devices* **2016**, *63*, 1871–1876.
326. Toledano-Luque, M.; Degraeve, R.; Roussel, P.J.; Luong, V.; Tang, B.; Lisoni, J.G.; Tan, C.L.; Arreghini, A.; Van den bosch, G.; Groeseneken, G.; et al. Statistical spectroscopy of switching traps in deeply scaled vertical poly-Si channel for 3D memories. In Proceedings of the 2013 International Electron Devices Meeting (IEDM), Washington, DC, USA, 9–11 December 2013; pp. 562–565.
327. Lee, C.M.; Tsui, B.Y. Random telegraph signal noise arising from grain boundary traps in nano-scale poly-Si nanowire thin-film transistors. In Proceedings of the 2013 Symposium on VLSI Technology (VLSI-Technology), Kyoto, Japan, 11–13 June 2013; pp. 47–48.
328. Seto, J.Y.W. The electrical properties of polycrystalline silicon films. *J. Appl. Phys.* **1975**, *46*, 5247–5254.
329. Hashimoto, T.; Aoki, M.; Yamanaka, T.; Kamigaki, Y.; Nishida, T. $1/f$ noise in poly-Si MOSFET and strong correlation with oxide traps and grain boundaries. In Proceedings of the 1994 Symposium on VLSI Technology (VLSI-Technology): Digest of Technical Papers, Honolulu, HI, USA, 7–9 June 1994; pp. 87–88.
330. Brederlow, R.; Weber, W.; Dahl, C.; Schmitt-Landsiedel, D.; Thewes, R. A physically based model for low-frequency noise of poly-silicon resistors. In Proceedings of the 1988 International Electron Devices Meeting (IEDM), San Francisco, CA, USA, 11–14 December 1988; pp. 89–92.
331. Kang, H.J.; Jeong, M.K.; Joe, S.M.; Seo, J.H.; Park, S.K.; Jin, S.H.; Park, B.G.; Lee, J.H. Effect of traps on transient bit-line current behavior in word-line stacked NAND Flash memory with poly-Si body. In Proceedings of the 2014 Symposium on VLSI Technology (VLSI-Technology): Digest of Technical Papers, Honolulu, HI, USA, 9–12 June 2014; pp. 24–25.
332. Tsai, W.J.; Lin, W.L.; Cheng, C.C.; Ku, S.H.; Chou, Y.L.; Liu, L.; Hwang, S.W.; Lu, T.C.; Chen, K.C.; Wang, T.; et al. Polycrystalline-silicon channel trap induced transient read instability in a 3D NAND Flash cell string. In Proceedings of the 2016

International Electron Devices Meeting (IEDM), San Francisco, CA, USA, 3–7 December 2016; pp. 11.3.1–11.3.4.

333. Degraeve, R.; Toledano-Luque, M.; Suhane, A.; den bosch, G.V.; Arreghini, A.; Tang, B.; Kaczer, B.; Roussel, P.; Kar, G.S.; Houdt, J.V.; Groeseneken, G. Statistical characterization of current paths in narrow poly-Si channels. In Proceedings of the 2011 International Electron Devices Meeting (IEDM), Washington, DC, USA, 5–7 December 2011; pp. 287–290.

334. Toledano-Luque, M.; Degraeve, R.; Kaczer, B.; Tang, B.; Roussel, P.J.; Weckx, P.; Franco, J.; Arreghini, A.; Suhane, A.; Kar, G.S.; et al. Quantitative and predictive model of reading current variability in deeply scaled vertical poly-Si channel for 3D memories. In Proceedings of the 2012 IEEE International Electron Devices Meeting (IEDM), San Francisco, CA, USA, 10–13 December 2012; pp. 203–206.

335. Kim, S.Y.; Park, J.K.; Hwang, W.S.; Lee, S.J.; Lee, K.H.; Pyi, S.H.; Cho, B.J. Dependence of grain size on the performance of a polysilicon channel TFT for 3D NAND Flash memory. *J. Nanosci. Nanotechnol.* **2016**, *16*, 5044–5048.

336. Brown, A.R.; Roy, G.; Asenov, A. Poly-Si-gate-related variability in decananometer MOSFETs with conventional architecture. *IEEE Trans. Electron Devices* **2007**, *54*, 3056–3063.

337. Baccarani, G.; Riccó, B.; Spadini, G. Transport properties of polycrystalline silicon films. *J. Appl. Phys.* **1978**, *49*, 5565–5570.

338. Lu, N.C.C.; Gerzberg, L.; Lu, C.Y.; Meindl, J.D. Modeling and optimization of monolithic polycrystalline silicon resistors. *IEEE Trans. Electron Devices* **1981**, *28*, 818–830.

339. Farmakis, F.V.; Brini, J.; Kamarinos, G.; Angelis, C.T.; Dimitriadis, C.A.; Miyasaka, M.; Ouisse, T. Grain and grain-boundary control of the transfer characteristics of large-grain polycrystalline silicon thin-film transistors. *Solid-State Electron.* **2000**, *44*, 913–916.

340. Hsiao, Y.H.; Lue, H.T.; Chen, W.C.; Chen, C.P.; Chang, K.P.; Shih, Y.H.; Tsui, B.Y.; Lu, C.Y. Modeling the variability caused by random grain boundary and trap-location induced asymmetrical read behavior for a tight-pitch vertical gate 3D NAND Flash memory using double-gate thin-film transistor (TFT) device. In Proceedings of the 2012 IEEE International Electron Devices Meeting (IEDM), San Francisco, CA, USA, 10–13 December 2012; pp. 609–612.

341. Hsiao, Y.H.; Lue, H.T.; Chen, W.C.; Chang, K.P.; Shih, Y.H.; Tsui, B.Y.; Hsieh, K.Y.; Lu, C.Y. Modeling the impact of random grain boundary traps on the electrical behavior of vertical gate 3-D NAND Flash memory devices. *IEEE Trans. Electron Devices* **2014**, *61*, 2064–2070.

342. Yang, C.W.; Su, P. Simulation and investigation of random grain-boundary-induced variabilities for stackable NAND Flash using 3-D Voronoi grain patterns. *IEEE Trans. Electron Devices* **2014**, *61*, 1211–1214.

343. Kim, J.; Lee, J.; Oh, H.; Rim, T.; Baek, C.K.; Meyyappan, M.; Lee, J.S. The variability due to random discrete dopant and grain boundary in 3D NAND unit cell. In Proceedings of the 12th International Naval Engineering Conference and Exhibition, Amsterdam, The Netherlands, 20–22 May 2014; pp. 66–68.

344. Wang, P.Y.; Tsui, B.Y. A novel approach using discrete grain-boundary traps to study the variability of 3-D vertical-gate NAND Flash memory cells. *IEEE Trans. Electron Devices* **2015**, *62*, 2488–2493.

345. Degraeve, R.; Clima, S.; Putcha, V.; Kaczer, B.; Roussel, P.; Linten, D.; Groeseneken, G.; Arreghini, A.; Karner, M.; Kernstock, C.; et al. Statistical poly-Si grain boundary model with discrete charging defects and its 2D and 3D implementation for vertical 3D NAND channels. In Proceedings of the 2015 International Electron Devices Meeting (IEDM), Washington, DC, USA, 7–9 December 2015; pp. 121–124.

346. Lun, Z.; Shen, L.; Cong, Y.; Du, G.; Liu, X.; Wang, Y. Investigation of the impact of grain boundary on threshold voltage of 3-D MLC NAND Flash memory. In

Proceedings of the 2015 Silicon Nanoelectronics Workshop (SNW), Kyoto, Japan, 14–15 June 2015; pp. 35–37.

347. Oh, H.; Kim, J.; Lee, J.; Rim, T.; Baek, C.K.; Lee, J.S. Effects of single grain boundary and random interface traps on electrical variations of sub-30 nm polysilicon nanowire structures. *Microelectron. Eng.* **2016**, *149*, 113–116.

348. Degraeve, R.; Toledano-Luque, M.; Arreghini, A.; Tang, B.; Capogreco, E.; Lisoni, J.; Roussel, P.; Kaczer, B.; Van den bosch, G.; Groeseneken, G.; et al. Characterizing grain size and defect energy distribution in vertical SONOS poly-Si channels by means of a resistive network model. In Proceedings of the 2013 International Electron Devices Meeting (IEDM), Washington, DC, USA, 9–11 December 2013; pp. 558–561.

349. Yoo, W.S.; Ishigaki, T.; Ueda, T.; Kang, K.; Kwak, N.Y.; Sheen, D.S.; Kim, S.S.; Ko, M.S.; Shin, W.S.; Lee, B.S.; et al. Grain size monitoring of 3D Flash memory channel poly-Si using multiwavelength Raman spectroscopy. In Proceedings of the 14th Annual Non-Volatile Memory Technology Symposium (NVMTS), Jeju, Korea, 27–29 October 2014; pp. 44–47.

350. Amit, I.; Englander, D.; Horvitz, D.; Sasson, Y.; Rosenwaks, Y. Density and energy distribution of interface states in the grain boundaries of polysilicon nanowire. *Nano Lett.* **2014**, *14*, 6190–6194.

351. Shamir, A.; Amit, I.; Englander, D.; Horvitz, D.; Rosenwaks, Y. Potential barrier height at the grain boundaries of a poly-silicon nanowire. *Nanotechnology* **2015**, *26*, 355201-1–355201-6.

352. He, D.; Okada, N.; Fortmann, C.M.; Shimizu, I. Carrier transport in polycrystalline silicon films deposited by a layer-by-layer technique. *J. Appl. Phys.* **1994**, *76*, 4728–4733.

353. Walker, A.J.; Herner, S.B.; Kumar, T.; Chen, E.H. On the conduction mechanism in polycrystalline silicon thin-film transistors. *IEEE Trans. Electron Devices* **2004**, *51*, 1856–1866.

354. Shim, K.S.; Choi, E.S.; Jung, S.W.; Kim, S.H.; Yoo, H.S.; Jeon, K.S.; Joo, H.S.; Oh, J.S.; Jang, Y.S.; Park, K.J.; et al. Inherent issues and challenges of program disturbance of 3D NAND Flash cell. In Proceedings of the 2012 4th IEEE International Memory Workshop (IMW), Milan, Italy, 20–23 May 2012; pp. 95–98.

355. Yoo, H.; Choi, E.; Oh, J.; Park, K.; Jung, S.; Kim, S.; Shim, K.; Joo, H.; Jeon, K.; Seo, M.; et al. Modeling and optimization of the chip level program disturbance of 3D NAND Flash memory. In Proceedings of the 2013 5th IEEE International Memory Workshop (IMW), Monterey, CA, USA, 26–29 May 2013; pp. 147–150.

356. Komori, Y.; Kido, M.; Kito, M.; Katsumata, R.; Fukuzumi, Y.; Tanaka, H.; Nagata, Y.; Ishiduki, M.; Aochi, H.; Nitayama, A. Disturbless Flash memory due to high boost efficiency on BiCS structure and optimal memory film stack for ultra high density storage device. In Proceedings of the 2008 International Electron Devices Meeting (IEDM), San Francisco, CA, USA, 15–17 December 2008; pp. 847–850.

357. Seo, J.Y.; Kim, Y.; Park, B.G. New program inhibition scheme for high boosting efficiency in three-dimensional NAND array. *Jpn. J. Appl. Phys.* **2014**, *53*, 070304-1–070304-3.

358. Choe, B.I.; Park, B.G.; Lee, J.H. Body doping profile of select device to minimize program disturbance in three-dimensional stack NAND Flash. *Jpn. J. Appl. Phys.* **2013**, *52*, 06GE02-1–06GE02-6.

359. Chang, Y.M.; Chang, Y.H.; Kuo, T.W.; Li, H.P.; Li, Y.C. A disturb-alleviation scheme for 3D Flash memory. In Proceedings of the 2013 IEEE/ACM International Conference on Computer-Aided Design (ICCAD), San Jose, CA, USA, 18–21 November 2013; pp. 421–428.

360. Chang, Y.M.; Chang, Y.H.; Kuo, T.W.; Li, Y.C.; Li, H.P. Disturbance relaxation for 3D Flash memory. *IEEE Trans. Comput.* **2016**, *65*, 1467–1483.

361. Chang, K.P.; Lue, H.T.; Chen, C.P.; Chen, C.F.; Chen, Y.R.; Hsiao, Y.H.; Hsieh, C.C.; Shih, Y.H.; Yang, T.; Chen, K.C.; et al. Memory architecture of 3D vertical gate (3DVG) NAND Flash using plural island-gate SSL decoding method and study of

its program inhibit characteristics. In Proceedings of the 2012 4th IEEE International Memory Workshop (IMW), Milan, Italy, 20–23 May 2012; pp. 25–28.

362. Yeh, T.H.; Chen, W.C.; Hsu, T.H.; Du, P.Y.; Hsieh, C.C.; Lue, H.T.; Shih, Y.H.; King, Y.C.; Lu, C.Y. Z-interference and Z-disturbance in vertical gate-type 3-D NAND. *IEEE Trans. Electron Devices* **2016**, *63*, 1047–1053.

363. Choe, B.I.; Lee, J.K.; Park, B.G.; Lee, J.H. Suppression of read disturb fail caused by boosting hot carrier injection effect for 3-D stack NAND Flash memories. *IEEE Electron Device Lett.* **2014**, *35*, 42–44.

364. Kim, Y.; Kang, M. Down-coupling phenomenon of floating channel in 3D NAND Flash memory. *IEEE Electron Device Lett.* **2016**, *37*, 1566–1569.

Index

A

B

C

D

E

F